Geoheritage and Geotourism Resources: Education, Recreation, Sustainability

Geoheritage and Geotourism Resources: Education, Recreation, Sustainability

Editors

Hara Drinia
Panagiotis Voudouris
Assimina Antonarakou

MDPI • Basel • Beijing • Wuhan • Barcelona • Belgrade • Manchester • Tokyo • Cluj • Tianjin

Editors
Hara Drinia
National and Kapodistrian University of Athens
Greece

Panagiotis Voudouris
National and Kapodistrian
University of Athens
Greece

Assimina Antonarakou
National and Kapodistrian
University of Athens
Greece

Editorial Office
MDPI
St. Alban-Anlage 66
4052 Basel, Switzerland

This is a reprint of articles from the Special Issue published online in the open access journal *Geosciences* (ISSN 2076-3263) (available at: https://www.mdpi.com/journal/geosciences/special_issues/geoheritage_geotourism_resources).

For citation purposes, cite each article independently as indicated on the article page online and as indicated below:

LastName, A.A.; LastName, B.B.; LastName, C.C. Article Title. *Journal Name* **Year**, *Volume Number*, Page Range.

ISBN 978-3-0365-4849-4 (Hbk)
ISBN 978-3-0365-4850-0 (PDF)

Cover image courtesy of Panagiotis Voudouris.

© 2022 by the authors. Articles in this book are Open Access and distributed under the Creative Commons Attribution (CC BY) license, which allows users to download, copy and build upon published articles, as long as the author and publisher are properly credited, which ensures maximum dissemination and a wider impact of our publications.

The book as a whole is distributed by MDPI under the terms and conditions of the Creative Commons license CC BY-NC-ND.

Contents

About the Editors vii

Hara Drinia, Panagiotis Voudouris and Assimina Antonarakou
Editorial of Special Issue—"Geoheritage and Geotourism Resources: Education, Recreation, Sustainability"
Reprinted from: *Geosciences* **2022**, *12*, 251, doi:10.3390/geosciences12060251 1

George Zafeiropoulos, Hara Drinia, Assimina Antonarakou and Nikolaos Zouros
From Geoheritage to Geoeducation, Geoethics and Geotourism: A Critical Evaluation of the Greek Region
Reprinted from: *Geosciences* **2021**, *11*, 381, doi:10.3390/geosciences11090381 5

Edyta Pijet-Migoń and Piotr Migoń
Geoheritage and Cultural Heritage—A Review of Recurrent and Interlinked Themes
Reprinted from: *Geosciences* **2022**, *12*, 98, doi:10.3390/geosciences12020098 25

Anna V. Mikhailenko, Vladimir A. Ermolaev and Dmitry A. Ruban
Bridges as Geoheritage Viewpoints in the Western Caucasus
Reprinted from: *Geosciences* **2021**, *11*, 377, doi:10.3390/geosciences11090377 53

Niki Evelpidou, Anna Karkani, Maria Tzouxanioti, Evangelos Spyrou, Alexandros Petropoulos and Lida Lakidi
Inventory and Assessment of the Geomorphosites in Central Cyclades, Greece: The Case of Paros and Naxos Islands
Reprinted from: *Geosciences* **2021**, *11*, 512, doi:10.3390/geosciences11120512 65

Fabio L. Bonali, Elena Russo, Fabio Vitello, Varvara Antoniou, Fabio Marchese, Luca Fallati, Valentina Bracchi, Noemi Corti, Alessandra Savini, Malcolm Whitworth, Kyriaki Drymoni, Federico Pasquaré Mariotto, Paraskevi Nomikou, Eva Sciacca, Sofia Bressan, Susanna Falsaperla, Danilo Reitano, Benjamin van Wyk de Vries, Mel Krokos, Giuliana Panieri, Mathew Alexander Stiller-Reeve, Giuseppe Vizzari, Ugo Becciani and Alessandro Tibaldi
How Academics and the Public Experienced Immersive Virtual Reality for Geo-Education
Reprinted from: *Geosciences* **2022**, *12*, 9, doi:10.3390/geosciences12010009 93

Evangelos Spyrou, Maria V. Triantaphyllou, Theodora Tsourou, Emmanuel Vassilakis, Christos Asimakopoulos, Aliki Konsolaki, Dimitris Markakis, Dimitra Marketou-Galari and Athanasios Skentos
Assessment of Geological Heritage Sites and Their Significance for Geotouristic Exploitation: The Case of Lefkas, Meganisi, Kefalonia and Ithaki Islands, Ionian Sea, Greece
Reprinted from: *Geosciences* **2022**, *12*, 55, doi:10.3390/geosciences12020055 115

Vasilis Golfinopoulos, Penelope Papadopoulou, Eleni Koumoutsou, Nickolas Zouros, Charalampos Fassoulas, Avraam Zelilidis and George Iliopoulos
Quantitative Assessment of the Geosites of Chelmos-Vouraikos UNESCO Global Geopark (Greece)
Reprinted from: *Geosciences* **2022**, *12*, 63, doi:10.3390/geosciences12020063 147

Rosendo Mendoza, Javier Rey, Julián Martínez and M. Carmen Hidalgo
Geological and Mining Heritage as a Driver of Development: The NE Sector of the Linares-La Carolina District (Southeastern Spain)
Reprinted from: *Geosciences* **2022**, *12*, 76, doi:10.3390/geosciences12020076 175

Charalampos Fassoulas, Emmanouel Nikolakakis and Spiridon Staridas
Digital Tools to Serve Geotourism and Sustainable Development at Psiloritis UNESCO Global Geopark in COVID Times and Beyond
Reprinted from: *Geosciences* **2022**, *12*, 78, doi:10.3390/geosciences12020078 . . . . **189**

George Zafeiropoulos and Hara Drinia
Comparative Analysis of Two Assessment Methods for the Geoeducational Values of Geosites: A Case Study from the Volcanic Island of Nisyros, SE Aegean Sea, Greece
Reprinted from: *Geosciences* **2022**, *12*, 82, doi:10.3390/geosciences12020082 . . . . **211**

Francesca Filocamo, Carmen Maria Rosskopf, Vincenzo Amato and Massimo Cesarano
A Step towards a Sustainable Tourism in Apennine Mountain Areas: A Proposal of Geoitinerary across the Matese Mountains (Central-Southern Italy)
Reprinted from: *Geosciences* **2022**, *12*, 100, doi:geosciences12020100 . . . . **233**

Nikolaos Vlachopoulos and Panagiotis Voudouris
Preservation of the Geoheritage and Mining Heritage of Serifos Island, Greece: Geotourism Perspectives in a Potential New Global Unesco Geopark
Reprinted from: *Geosciences* **2022**, *12*, 127, doi:10.3390/geosciences12030127 . . . . **261**

Katia Hueso-Kortekaas and Emilio Iranzo-García
Salinas and "Saltscape" as a Geological Heritage with a Strong Potential for Tourism and Geoeducation
Reprinted from: *Geosciences* **2022**, *12*, 141, doi:10.3390/geosciences12030141 . . . . **289**

Efthymios Georgousis, Maria Savelidi, Socrates Savelides, Spyros Mosios, Maximos-Vasileios Holokolos and Hara Drinia
How Greek Students Perceive Concepts Related to Geoenvironment: A Semiotics Content Analysis
Reprinted from: *Geosciences* **2022**, *12*, 172, doi:10.3390/geosciences12040172 . . . . **311**

About the Editors

Hara Drinia

Hara Drinia is a Professor of Palaeoecology and Sedimentology at the Department of Geology and Geoenvironment at the National and Kapodistrian University of Athens. She teaches a wide range of undergraduate and postgraduate courses including Sedimentology, Palaeoecology, Stratigraphy, Marine Ecosystems, Environmental Education, Geoscience Teaching, Geological Heritage and Geoconservation. Her strong academic performance, combined with her scientific research, is strong evidence of her ongoing national and international research activity and contributions to the science of geology. She has been actively involved in many international scientific conferences and in regional, national, and international meetings. She has a strong record of participating in research programs and initiatives and has been the Principal Investigator of several research projects. Regarding her wider contribution to the geological community, it is worth noting that Prof. Drinia is an active member of eight geosciences committees and has been a member of the organizing committee of eight international and national geological conferences. She is also actively involved in the Socrates / Erasmus educational exchange programs in Switzerland, Germany and Italy.

Panagiotis Voudouris

Panagiotis Voudouris is a Professor of Mineralogy and Petrology at the Department of Geology and Geoenvironment, National and Kapodistrian University of Athens (Greece). He graduated in this university in 1986 and received his Doctorate (1993) in Mineralogy and Economic Geology from the University of Hamburg (Germany), where he also served as a Visiting Professor. His main research interests are in the fields of ore mineralogy, genesis of magmatic–hydrothermal ore deposits, the mineralogy and genesis of gemstones, and promotion of mineralogical and mining heritage in Greece. He is a member of the Editorial Board of the journal Minerals and Director of the Mineralogy and Petrology Museum, National and Kapodistrian University of Athens. He has published 99 scientific papers in international refereed journals, is the co-author of four IMA-approved new mineral species, and has co-authored five books. The International Mineralogical Association adopted a new mineral name in his honor (Voudourisite).

Assimina Antonarakou

Assimina Antonarakou is a Professor of Marine Geology–Micropaleontology–Didactics on Geosciences, and Vice President of the Faculty of Geology and Geoenvironment at the University of Athens. Her PhD thesis dealt with Miocene cyclic sedimentary successions of the eastern Mediterranean in terms of orbital periodicities and paleoclimatic variations based on planktonic foraminiferal assemblages. Her main research topics are summarized as follows: planktonic foraminiferal eco-biostratigraphy, geobiology and paleoceanography; astronomical frequencies in paleoclimates; extreme geological events; marine environmental monitoring; ocean dynamics and sea-level changes; natural and human environmental stressors; and foraminiferal trace metals and stable isotopes. She has participated in several national and international projects focused on multiproxy ecosystem responses to past and present environmental events, and she is the co-author of more than 50 peer-reviewed publications in international journals.

Editorial

Editorial of Special Issue—"Geoheritage and Geotourism Resources: Education, Recreation, Sustainability"

Hara Drinia *, Panagiotis Voudouris and Assimina Antonarakou

Department of Geology and Geoenvironment, National and Kapodistrian University of Athens, 157 72 Athens, Greece; voudouris@geol.uoa.gr (P.V.); aantonar@geol.uoa.gr (A.A.)
* Correspondence: cntrinia@geol.uoa.gr

In recent years, the interest of society in the geoenvironment is constantly increasing. Concepts such as "geosites", "geoparks", and "geodiversity" are linked to new local economic and cultural growth of the areas. Significant geosites are recognized worldwide through the activity of geoparks and benefit from the exchange of information, skills, experience and staff between geoparks. Geotourism is a form of tourism that allows the discovery of the geological peculiarities of the visited territories, combined with other natural and human resources. Geo-education is at the core of the interest and operation of geoparks, which are considered as ideal destinations for educational activities.

Zafeiropoulos et al. [1] highlight the importance of geoenvironmental education in promoting and preserving geological heritage and geoethical values, and they present the current situation in Greece. Greece is known for its exceptional and rare natural beauty, abundance of natural resources, and remarkable geological features. As a result, this country has already established six global geoparks. The significance of establishing a legal framework for geotope protection is highlighted by the fact that the promotion and rational management of geological heritage create opportunities for sustainable development as well as quality tourism (geotourism) through nature protection and education. Such initiatives can not only improve geological heritage protection, but also play an important role in its sustainable development.

Pijet-Migoń and Migoń [2] identify the primary and secondary themes at the geoheritage—cultural heritage interface and offer examples of specific topics and approaches. Intangible cultural heritage is also examined in the context of geoheritage. In the final section of the paper, various classifications of geoheritage—cultural heritage linkages are proposed, but it is concluded that themes and fields of inquiry overlap and interlink, making a single classification system impractical. Instead, a mind map is provided to demonstrate these various connections. The article states recommendations for future research based on the findings of the review and the identification of research gaps and under-researched areas.

Mikhailenko et al. [3] introduce the concept of bridge-based geoheritage viewpoints in the geologically rich Western Caucasus region (southwestern Russia). Eleven bridges were evaluated semi-quantitatively using a new method. The findings indicated that bridges have varying but moderate utility as geoheritage perspectives. Bridges differ not only in terms of the quality of the views they provide, but also in terms of their accessibility. In some cases, mandatory permissions and entrance fees reduce this property. Bridge-based geoheritage viewpoints are important for geotourism development because they help to establish optimal and comfortable routes.

Evelpidou et al. [4] inventory the main geomorphosites of the islands of Paros and Naxos in the central Aegean Sea, evaluating their scientific and added values using qualitative and quantitative criteria. The findings revealed that, in addition to being of high scientific interest, most geomorphosites have high ecological value and could potentially lead to a significant increase in island tourism. The outcomes of this work aim to raise

Citation: Drinia, H.; Voudouris, P.; Antonarakou, A. Editorial of Special Issue—"Geoheritage and Geotourism Resources: Education, Recreation, Sustainability". *Geosciences* **2022**, *12*, 251. https://doi.org/10.3390/geosciences12060251

Received: 8 June 2022
Accepted: 13 June 2022
Published: 16 June 2022

Publisher's Note: MDPI stays neutral with regard to jurisdictional claims in published maps and institutional affiliations.

Copyright: © 2022 by the authors. Licensee MDPI, Basel, Switzerland. This article is an open access article distributed under the terms and conditions of the Creative Commons Attribution (CC BY) license (https://creativecommons.org/licenses/by/4.0/).

awareness about the geomorphological heritage of the central Cyclades and provide a framework for its promotion, protection, and management.

According to Bonali et al. [5], immersive virtual reality has the potential to introduce students, academics, and others to interesting geological sites that they may not have had the opportunity to visit previously. These authors demonstrate the importance of immersive VR as a tool for: popularizing Earth Sciences teaching and research by making geological key areas available to the public in the form of 3D models and scientific explanations of geological processes; including people with motor disabilities who would not otherwise have access to dangerous/remote areas (e.g., tectonically or volcanically active). As a result, immersive VR can be viewed as a game-changing tool for improving democratic access to information and experience, as well as for promoting inclusivity and accessibility in geoeducation while reducing travel saving time, and carbon footprints.

Spyrou et al. [6] investigate and evaluate the scientific, environmental, cultural, economic, and aesthetic value of several geosites on the Greek islands of Lefkas, Meganisi, Kefalonia, and Ithaki. The most representative geological sites (e.g., geomorphology, tectonics, stratigraphy, and paleontology) have been chosen, mapped, and assessed, and indicative georoutes have been proposed, which could aid the island's geotouristic promotion to future geologist and non-geologist visitors.

Golfinopoulos et al. [7] evaluated the geosites of the Chelmos-Vouraikos UNESCO Global Geopark (UGGp) using a well-established methodology for evaluating geopark geosites. The assessment of the geopark's 40 geosites revealed geosites with high educational and touristic value, as well as geosites with high protection-need value. As a result, the assessment results will be used for the planning of the effective management of the geosites based on strengths and weaknesses, promoting the geopark and contributing to the sustainable development of local communities. These authors concluded that the use of such evaluation methodologies should be regarded as critical for the development, protection, and promotion of geoparks.

The study of Mendoza et al. [8] aimed to assess the geological features of the Linares-La Carolina mining district's northeastern sector and relate them with the mining activities of the district's main veins. This old mining region's educational and tourism potential is highlighted. Finally, points of special interest are noted in ways to construct a guided tour for the visitor.

Fassoulas et al. [9] present the benefits of new digital applications designed by Psiloritis UGGp in their study. These were created as part of the RURITAGE project, which puts an emphasis on rural revitalization through natural and cultural resources. The authors investigated the impact of this technology on geopark promotion and visibility, knowledge communication, local economic and tourism support, and its commitment to local sustainable development and growth in COVID-19 and post-pandemic times.

The main goal of Zafeiropoulos and Drinia [10] article is to evaluate geosites using two quantitative assessment methodologies that approach a geosite's geoeducational value in different ways. The first method is a general-purpose method (G-P method) that is designed to assess any type of geosite while considering a wide range of criteria. It is one of the most widely used inventory methods. The second method, the M-GAM (Modified Geosite Assessment Model), incorporates the perspectives of both experts and visitors and is being used for the first time in Greece. The ultimate purpose is to analyze the results of the two methodologies and determine which method is best for determining a geosite's educational value. Nisyros Island was chosen as a case study.

The study of Filocamo et al. [11] is concerned with the enhancement of the geoheritage of the Matese Mountains, one of southern Italy's most suggestive and integral mountain areas. This mountain area shares many of the hardships and limitations that characterize other mountain regions and, more broadly, inner areas, such as land abandonment, population decline, marginality, mobility limitations, and inaccessibility. The authors propose a geoitinerary that could be used to promote geotourism in the Matese area. The authors propose a geoitinerary that could be used to promote geotourism in the Matese area. This geoitinerary can help to develop sustainable geotourism and related associate activities within the nascent Matese National Park, assisting in the creation of a tourist offering capable of attracting visitors interested in geology and other natural or cultural resources. Future developments of this study will aim to connect the geoitinerary with visits to other sites of natural/cultural interest, as well as to create a network with other trails, in order to encourage tourists to stay for several days and favor overnight stays.

The paper of Vlachopoulos and Voudouris [12] focuses on Serifos' (Aegean Sea, Greece) geological and mining heritage, with the goal of integrating the island into the international geoparks environment in the near future. Six geotrails were created during this study to connect cultural and ecological sites with the geological heritage. The geodiversity is explained along the routes, as well as its relationship with the surrounding biodiversity and the region's historical and cultural aspects. Historical conditions determine the dialectic relationship between humans and nature in the proposed geocultural routes.

Hueso-Kortekaas and Iranzo-Garcia [13] state that saline and saline landscapes are geo-heritage sites that have significant socio-economic implications beyond salt production, particularly in tourism and education. They have implications for the identity of their communities as cultural landscapes.

Finally, Georgousis et al. [14] investigate students' perceptions of geodiversity, geoheritage, geoethics, and geotourism to design a geoeducation program within the constraints of an experimental school. They used the educational technique of creating cognitive conflicts in order to promote scientific perceptions of these concepts when designing this geoeducation program. Thus, research questions were identified, leading to the research assessing the current latent state of students' perceptions regarding thematic areas of concepts and identifying concepts whose perceptions can be used in the educational process to achieve effective cognitive conflicts to promote scientific perceptions of them. The qualitative research strategy approach, specifically the hybrid technique of semiotics content analysis in conjunction with thematic analysis, was chosen.

Funding: This research received no external funding.

Conflicts of Interest: The authors declare no conflict of interest.

References

1. Zafeiropoulos, G.; Drinia, H.; Antonarakou, A.; Zouros, N. From Geoheritage to Geoeducation, Geoethics and Geotourism: A Critical Evaluation of the Greek Region. *Geosciences* **2021**, *11*, 381. [CrossRef]
2. Pijet-Migoń, E.; Migoń, P. Geoheritage and Cultural Heritage—A Review of Recurrent and Interlinked Themes. *Geosciences* **2022**, *12*, 98. [CrossRef]
3. Mikhailenko, A.; Ermolaev, V.; Ruban, D. Bridges as Geoheritage Viewpoints in the Western Caucasus. *Geosciences* **2021**, *11*, 377. [CrossRef]
4. Evelpidou, N.; Karkani, A.; Tzouxanioti, M.; Spyrou, E.; Petropoulos, A.; Lakidi, L. Inventory and Assessment of the Geomorphosites in Central Cyclades, Greece: The Case of Paros and Naxos Islands. *Geosciences* **2021**, *11*, 512. [CrossRef]
5. Bonali, F.; Russo, E.; Vitello, F.; Antoniou, V.; Marchese, F.; Fallati, L.; Bracchi, V.; Corti, N.; Savini, A.; Whitworth, M.; et al. How Academics and the Public Experienced Immersive Virtual Reality for Geo-Education. *Geosciences* **2022**, *12*, 9. [CrossRef]
6. Spyrou, E.; Triantaphyllou, M.; Tsourou, T.; Vassilakis, E.; Asimakopoulos, C.; Konsolaki, A.; Markakis, D.; Marketou-Galari, D.; Skentos, A. Assessment of Geological Heritage Sites and Their Significance for Geotouristic Exploitation: The Case of Lefkas, Meganisi, Kefalonia and Ithaki Islands, Ionian Sea, Greece. *Geosciences* **2022**, *12*, 55. [CrossRef]
7. Golfinopoulos, V.; Papadopoulou, P.; Koumoutsou, E.; Zouros, N.; Fassoulas, C.; Zelilidis, A.; Iliopoulos, G. Quantitative Assessment of the Geosites of Chelmos-Vouraikos UNESCO Global Geopark (Greece). *Geosciences* **2022**, *12*, 63. [CrossRef]
8. Mendoza, R.; Rey, J.; Martínez, J.; Hidalgo, M. Geological and Mining Heritage as a Driver of Development: The NE Sector of the Linares-La Carolina District (Southeastern Spain). *Geosciences* **2022**, *12*, 76. [CrossRef]
9. Fassoulas, C.; Nikolakakis, E.; Staridas, S. Digital Tools to Serve Geotourism and Sustainable Development at Psiloritis UNESCO Global Geopark in COVID Times and Beyond. *Geosciences* **2022**, *12*, 78. [CrossRef]
10. Zafeiropoulos, G.; Drinia, H. Comparative Analysis of Two Assessment Methods for the Geoeducational Values of Geosites: A Case Study from the Volcanic Island of Nisyros, SE Aegean Sea, Greece. *Geosciences* **2022**, *12*, 82. [CrossRef]
11. Filocamo, F.; Rosskopf, C.; Amato, V.; Cesarano, M. A Step towards a Sustainable Tourism in Apennine Mountain Areas: A Proposal of Geoitinerary across the Matese Mountains (Central-Southern Italy). *Geosciences* **2022**, *12*, 100. [CrossRef]

12. Vlachopoulos, N.; Voudouris, P. Preservation of the Geoheritage and Mining Heritage of Serifos Island, Greece: Geotourism Perspectives in a Potential New Global Unesco Geopark. *Geosciences* **2022**, *12*, 127. [CrossRef]
13. Hueso-Kortekaas, K.; Iranzo-García, E. Salinas and "Saltscape" as a Geological Heritage with a Strong Potential for Tourism and Geoeducation. *Geosciences* **2022**, *12*, 141. [CrossRef]
14. Georgousis, E.; Savelidi, M.; Savelides, S.; Mosios, S.; Holokolos, M.; Drinia, H. How Greek Students Perceive Concepts Related to Geoenvironment: A Semiotics Content Analysis. *Geosciences* **2022**, *12*, 172. [CrossRef]

Review

From Geoheritage to Geoeducation, Geoethics and Geotourism: A Critical Evaluation of the Greek Region

George Zafeiropoulos [1,*], Hara Drinia [1,*], Assimina Antonarakou [1] and Nikolaos Zouros [2]

1. Department of Geology and Geoenvironment, National and Kapodistrian University of Athens, 15784 Athens, Greece; aantonar@geol.uoa.gr
2. Department of Geography, University of Aegean, 81100 Mytilini, Greece; nzour@aegean.gr
* Correspondence: georzafeir@geol.uoa.gr (G.Z.); cntrinia@geol.uoa.gr (H.D.)

Citation: Zafeiropoulos, G.; Drinia, H.; Antonarakou, A.; Zouros, N. From Geoheritage to Geoeducation, Geoethics and Geotourism: A Critical Evaluation of the Greek Region. *Geosciences* **2021**, *11*, 381. https://doi.org/10.3390/geosciences11090381

Academic Editors: Karoly Nemeth and Jesus Martinez-Frias

Received: 9 August 2021
Accepted: 6 September 2021
Published: 9 September 2021

Publisher's Note: MDPI stays neutral with regard to jurisdictional claims in published maps and institutional affiliations.

Copyright: © 2021 by the authors. Licensee MDPI, Basel, Switzerland. This article is an open access article distributed under the terms and conditions of the Creative Commons Attribution (CC BY) license (https://creativecommons.org/licenses/by/4.0/).

Abstract: The purpose of this review is, initially, to emphasize the importance of geoenvironmental education for the promotion and preservation of geological heritage and geoethical values, and based on these, to present the current situation in Greece. Geoeducation is a broader component of environmental education which aims to promote the geological heritage of a place and its geoconservation. It is a key integral tool for tackling environmental issues and therefore further assisting in sustainable development. Greece is known for its exceptional and rare natural beauty, as well as for the abundance of natural resources and its remarkable geological features. For this reason, six global geoparks have already been established in this country. However, its nature protection is mainly considered as the protection of biodiversity, while the term "geodiversity" is almost absent in Greek law. The importance of establishing a legal framework for the protection of geotopes is underlined by the fact that their promotion and rational management create opportunities for sustainable development, as well as to become quality tourist destinations (geotourism) through nature protection and education. Geodiversity can gain public attention and have a positive impact on geotopes protection. Such initiatives can not only improve the protection of geological sites, but also play an important role in their sustainable development.

Keywords: geoheritage; geoconservation; geoeducation; geotourism; sustainable development; Greece

1. Introduction

In the last few decades and under the UNESCO (United Nations Educational Scientific and Cultural Organization) initiative, there has been an increasing attempt to establish environmental education and its framework in various ways. More precisely, a meeting in Belgrade (Serbia) in 1976, under the auspices of UNESCO, formalized a set of documents known as Charter of Statutes of Belgrade [1], explicitly mentions the need for a significant change in human attitudes by taking more effective action for the global environmental issue through universal effort at school units. As a result of this effort, a strategy has been established that will allow some organizations and individuals to present more systematic content on environmental challenges and how to address them. In this way, a more harmonious balance between environment and human activity could be reached. This has resulted in several environmental programs in schools, with the goal of rationalizing future citizens to assist them in adopting a more positive attitude toward the needs of our planet and our society, after having received appropriate and necessary knowledge of such matters in advance.

It is worth mentioning that UNESCO never gave up on the idea and initiative of the environmental education; in fact, in 1977 [2], in the context of the International Conference in Tbilisi (Georgia), the participants agreed that environmental education should be treated as a distinct scientific field of paramount importance that should be integrated into educational programs rather than being treated as an afterthought.

Along with environmental education, an international effort was launched to establish and protect the geological heritage. In 1972, the Convention on the Protection of the World Cultural and Natural Heritage took place in Paris (France), and some years later, in 1991, the International Declaration on the Rights of the Memory of the Earth took place in Digne (France) [3]. Through these conferences and their declarations, a European initiative for the protection of geoheritage and geoconservation was launched, with the goal of protecting exceptional geological areas that reflect the evolution of biotic and abiotic factors. In 2000, the European Geoparks Network (EGN) was founded to promote a more systematic and collective process of development and, of course, to ensure geodiversity [4]. Following this, in 2004, the Global Geoparks Network was established, which, along with the EGN, aims to promote the concept of geological heritage in the scientific community and the public, as well as to promote sustainable development in areas that host the geoparks [5].

Despite these initiatives to promote and protect the geological heritage, environmental education does not deepen directly into issues related to geoethics, geodiversity, and geoheritage. There is a reasonable need to promote geoeducation, which will deal exclusively with the above concepts and will be the primary tool for the first transmission of knowledge and highlighting the importance of places of intense geological interest, which, in turn, will offer the opportunity of geotourism services [6]. Specifically, geotourism encourages various forms of geoeducation in order to organize geosites to be open for the public and offering educational and recreational activities [7,8].

Finally, the need for a more rational assessment of the geoenvironmental status of our planet, as well as the need for more effective management of issues connected to geoenvironmental conservation, led to the development of a new field of geosciences: Geoethics. The first function of geoethics is to improve the social profile and role of geoscience. Moreover, it contributes to forwarding the sustainable use of natural sources with harmonious operation between human activity and environment. Consequently, this field can accelerate strategies and methods that will respect geoheritage and its prospective [9].

The purpose of this review is, first, to highlight the concepts of geological heritage and geoconservation and the importance of geoeducation for the proper promotion and utilization of geological heritage for the benefit of the common good. The terms "geotourism" and "geoethics", as well as their interconnectedness, emerge from this review. Particular emphasis is placed on the situation prevailing in Greece regarding the protection and use of geological heritage, geoeducation, and most importantly, the current legal framework.

2. Literature Overview

2.1. Geoheritage and Geoconservation

"Geoheritage" is a new term that assumes complete perception of Man for nature and the environment [10]. As stated by Carcavilla et al. [11] "Geological-geomorphological heritage is the collection of geotopes, deposits, forms, and processes that comprise the geological history of each region, and the concept of preserving geological-geomorphological heritage is a cultural concept".

Geoheritage aims to highlight the diversity of our planet to illustrate the importance of the biotic and abiotic factors, which document the historical evolution of the Earth [12]. Furthermore, geoheritage focuses on the important geological elements, such as rocks, minerals, and fossils that interpret the effects of past and present actions, which have shaped landforms and other geomorphological structures [13]. The value of geological heritage is further underlined in report from UNESCO [14], according to which geological heritage is characterized as the whole of the most interesting geological sites (geotopes, geoparks, and geological natural monuments) that deserve to be preserved for scientific, didactic, historical, aesthetic, and cultural reasons. There is also a reference in the European Manifesto on Earth Heritage and Geodiversity [15] that argues that the heritage of Earth interconnects the Earth, its people, and their culture; that is, it forms the cornerstone and foundation of our society.

The link between geological heritage and geodiversity, on the other hand, is quite complicated and encompasses all the elements that contribute to the creation and development of the Earth [16] (Figure 1). Geodiversity is a crucial component of the Earth system and is described as the variability of abiotic nature or the abiotic diversity of the surface of the Earth. Geodiversity, along with biodiversity, constitutes the natural diversity of our planet.

Geoconservation is a relatively new scientific field that has emerged in recent decades due to the growing importance of conservation and sustainable use of environmental resources [17].

The concept of "geoconservation" can be defined as an activity or group of actions that contribute to the conservation, rational management, and protection of geological structures that present geodiversity and hence have scientific and educational value [18]. The term "geoconservation" first appeared in the 1990s e.g., [18]: more specifically, Semeniuk [19] and Semeniuk and Semeniuk [20] reported that geoconservation is concerned with the conservation and preservation of the features of Earth for educational, scientific, and hereditary purposes. Etymologically, this term combines conservation specifically with geological features and parameters. The goal of geoconservation is to identify, protect, and manage valuable parts of geodiversity. According to the international literature, geoconservation is a broad field that deals with concerns such as environmental management, geological hazards (geohazards), and sustainable development [21]. Thus, it becomes clear that geoconservation is initially part of geodiversity along with biodiversity, which together constitute the two major environmental components. There is also the preservation of the geoheritage that highlights the geological history of the Earth.

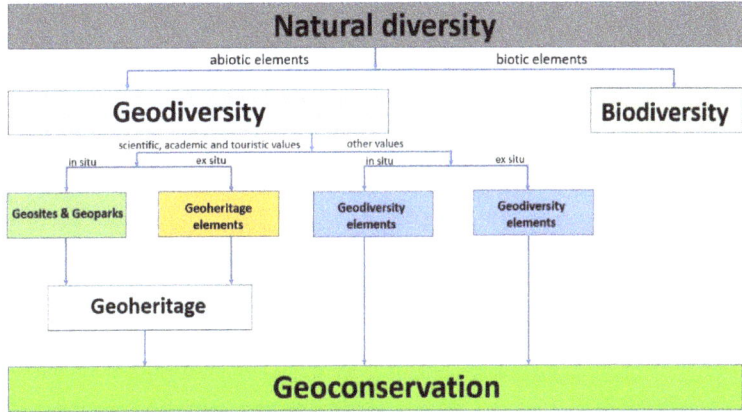

Figure 1. Conceptual framework of geodiversity, geological heritage, and geoconservation. Modified from [22].

The first steps for the implementation and dissemination of geoheritage and geoconservation are drawn from historical data: a typical example took place at the beginning of the 19th century, when over-exploitation in the quarry in the Salisbury area of Edinburgh, Scotland, had a huge negative impact on the geomorphology of the adjacent city, resulting in a decision in 1819 with legal coverage to protect the characteristic rock and to avoid further deterioration of the area [23]. Later, Germany established, in 1836, the first geological natural monument in the world in the Siebengebirge area, located southeast of the city of Bonn: there were several quarries in this area where the mineral wealth exploitation over a long period of time ended unpleasantly. The initiative was taken by the then Prussian government to protect the region. In 1872 in USA, the Yellowstone National Park was founded due to its scenic beauty and the many geological formations and phenomena observed in it [24]. As early as the 1870s in Switzerland, Fritz Muhlberg started a campaign mainly for the protection of erratic stones (massive irregular masses) that were systematically exploited [25]. In fact, at the same time, a committee is set up in Scotland to propose measures for the protection and preservation of these stones [26]. Finally, it is worth mentioning that, according to the data at hand, Tanzania is considered the leader in the battle for land conservation [18], where an intense effort is in place to protect geological areas such as Ngorongoro lengai in Tanzania, where an active volcano is also located, and to protect fauna and wildlife.

2.2. The Need for More Systematic Use of Geoeducation and Its Awareness

The traditional educational system cannot highlight the importance and components of geological heritage. As a result, it is vital to make geoeducation more widely available, as well as to integrate it into special curricula programs at various school levels. In this way, there will be a major opportunity for future citizens to be informed about issues that raise geological and cultural interest. It is worth noting here that the European and Global Geopark networks allow the full development of geoeducation, as there is an on-site opportunity for both the public and the scientific community to be informed through educational and cultural activities [27,28]. In addition, the diverse geotopes of geoparks, geosites, and geotrails are valuable tools that may be used by professional geoguides to educate visitors about their importance and impact on the ecosystem.

According to the planning and the agenda of UNESCO [29], education and sustainable development are key objectives for the World Geoparks. Rational information and geoeducation regarding sustainable development can deposit a wide range of uses through geoparks, where the geological and cultural heritage is accentuated. As a result, the immediate goal of a geopark is to assist students who visit them in gaining a better understanding of sustainability and its positive prospects, with the goal of achieving better life conditions for future generations.

Thus, it becomes even clearer that geoeducation constitutes the main tool for transmitting knowledge and, at the same time, emphasizing the importance of geoheritage and geoconservation. Specifically, geoeducation can address the following points: knowledge and awareness of the value of geological monuments; direct experience and understanding of the historical evolution of the planet, and thus the importance of geoheritage reflected in the rocks; and the establishment of natural history museums for the promotion and more systematic identification of areas of intense geological interest and awareness and perception of the geoethical dimension of important geological sites. Furthermore, the aforementioned points, along with the presence of geotopes and geoparks, are appropriate elements for in situ geoeducation both locally and regionally or even internationally. The harmonious coexistence of people with their environment presupposes a thorough understanding of the fundamental of geological processes active in the formation of the planet. This knowledge ensures an attitude for the protection of the environment and strengthens the view of citizens on issues of protection of natural and geological monuments.

It should also be mentioned that the natural history museums that are part of a geopark can widely contribute, with educational activities, special learning programs, outdoor exercises, seminars for teachers and students, organization of conferences and lectures, elaboration and support of research or school programs, cooperation with environmental education centers, creation of interactive educational material, cooperation with global environmental management institutes, and finally, cooperation with universities. In addition, the dissemination of geoeducation can be done in various ways, such as with a series of guided geotrails, knowledge transfer through educational activities organized by qualified teaching staff of each geopark and addressed to schools, and departments of universities and research institutes. In this way, geoeducation can be promoted.

Following that, the transmission and preservation of geoheritage and related concepts can be combined with strategic applications and means that could contribute to the development of the local community.

2.3. The Positive Impact of Geotourism

Geotourism is a relatively new and ever-changing phenomenon. As a result, it is natural that different approaches exist, owing primarily to the geological peculiarities of the areas involved. This means, among other things, that no universally accepted definition of geotourism exists. This could be considered a type of alternative tourism that combines tourism and geology. On the one hand, tourism is a recreational activity based on subjective and aesthetic criteria. Geology, on the other hand, is a science that employs objective criteria. As a result, tourism and geology are two very different disciplines that can coexist and form geotourism, a new emerging type of "environmentally innovative" tourism.

Geotourism is a relatively new form of alternative tourism with significant European and global development potential. It first appeared at the beginning of the 21st century, especially with the appearance and institutionalization of geoparks, which are areas with important geological heritage and rich natural and cultural environments, which, through nature protection and education, contribute to the development of responsible tourism, strengthening the local economy and sustainable development [30].

To date, many interpretations of the concept of geotourism have been provided. Thus, geotourism is a subset of ecotourism that occurs in areas with significant geological monuments [31,32] and "prioritizes the interactive experience through contact with the geoenvironment and the cultural elements that form the unique identity of each place". In other words, it focuses on the characteristics of the environment of an area with emphasis on landscape and geoenvironment, which includes not only geological elements but also all other elements of cultural and natural heritage, which are closely linked and interdependent with the respective geological environment of a place [33,34].

A key component for the development of geotourism is the understanding of the identity or character of an area. To achieve this, geotourism is based on the idea that the environment consists of abiotic, biotic, and cultural elements (Figure 2). This "ABC" approach of Dowling [35] includes the abiotic elements of geology and climate, the biotic elements of animals (fauna) and plants (flora), and the cultural or human components of the past and present. Geotourism argues that, to fully understand and appreciate the environment, we must first know the abiotic elements of geology and climate, as these determine the biotic elements of the animals and plants that live there.

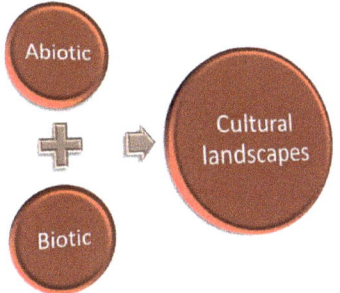

Figure 2. The ABC Approach of Dowling [35].

Consequently, the combination of the abiotic and biotic components of the environment determines the cultural landscape and how people lived in the area in the past, as well as how they live there today.

Furthermore, modern geotourism is distinguished by three major components known as the 3Gs: geointerpretation, geodiversity, and geohistory [8].

For the development of geotourism, there must be systematic steps towards the geoconservation of important landscapes and the rational utilization of geotopes and geoparks, so that they together become important destinations of tourist interest. It is worth mentioning that geotourism constitutes a new field of applied geology called tourism geology, under the broader field of geoconservation. In other terms, geotourism deals with the application of geological knowledge to promote ecotourism activities through a systematic research and geological characterization of new and existing tourist destinations. The tourism packages that result from these efforts are called geotourism activities [36]. Moreover, Hose [8] defines geotourism as the provision of opportunities that allow tourists to acquire specific knowledge and understanding of the geology, geotopes, and geodiversity of an area, beyond the level of simple aesthetic appreciation [37]. In this formulation, it should be emphasized that the concept of geotourism is based on the principles of sustainability, the protection of tourism services offered, the development of ecological awareness of visitors, the understanding and respect of local cultural elements, and finally, aims at a quality offer of knowledge and recreational activities and not in reckless quantitative issues. For the above reasons, geotourism is one alternative form of tourism with constructive outcomes.

In conclusion, geotourism and its consequences can promote local sustainable development, but must be in line with certain principles that respect the environment and geoheritage. For this reason, there is a holistic approach by the scientific community so that the activities developed with geotourism have some specific characteristics. Firstly, it must provide an integrated management approach that considers the constraints of the carrying capacity of the geological, natural, social, and cultural reserve of the environment to entertain visitors. Further, there must also be a strategy of rational management of natural resources to reduce any kind of waste from such activities. In addition, geotourism should create new development opportunities suitable for the environment and the local character of the area, to create new jobs positions. In this way can geotourism contribute to the local economy by promoting local employment, using local products or skills, and creating new added value. Moreover, geotourism products or facilities should aim to provide education, especially for the benefit of young people and students, to encourage people to understand and learn more about geology and the environment in general using modern means. Thereafter, recreational activities can also be developed using electronic applications to better understand geological concepts, as well as predetermined walking routes or climbing, in such a way as to live unique experiences. To summarize, geotourism promotion actions will enhance respect for traditions and customs of the region and emphasize authentic values in raising awareness of geodiversity and environmental protection to visitors [38].

2.4. Geoethics

Geoethics is an emerging field that examines many aspects of the interactions of geoscientists with society and the environment. It addresses the moral, social, and cultural implications of geoscience research and practice in collaboration with Sociology and Philosophy, providing an opportunity for geologists to recognize their social role and responsibilities in the course of their work. According to Peppoloni and Di Capua [39] "Geological culture and geoethics can strengthen the bonds between people and their land, between their places of origin and their own memories" by recognizing the value of the geological heritage of a region. Education can also convey messages to people about environmental issues and the sustainable use of natural resources, such as the consequences of geological heritage destruction [40]. Geoethics is thus a tool for raising public awareness of issues concerning geopolitical resources and the geoenvironment. An ethical approach must emphasize the importance of nature as a sensual, contemplative, spiritual, religious, and aesthetic experience that is passed down to future generations, rather than just the economic viability of natural resources [41–43].

The claim that nature has an intrinsic value that should be protected is often based on spiritual or metaphysical beliefs, but it also stems from human moral considerations and responsibilities to the natural world, as well as the preservation of natural diversity and cultural heritage [44–46].

In the expanding field of geoethics, geotourism plays a cultural role. According to Peppoloni and Di Capua [39], geoethics promotes geoeducation through the development of tourism and UNESCO World Geoparks, with the goal of raising awareness, values, and responsibility for geological heritage, particularly among young people.

Consequently, it is understood that this new scientific field constitutes the forerunner for the most effective sustainability and its components (environment, economy, society). Therefore, due to the necessity created by the massive and systematic use of planet Earth, this new field focuses on the need for a more specialized knowledge of sustainability with the ultimate goal of disseminating knowledge through academia to society (Figure 3) [9].

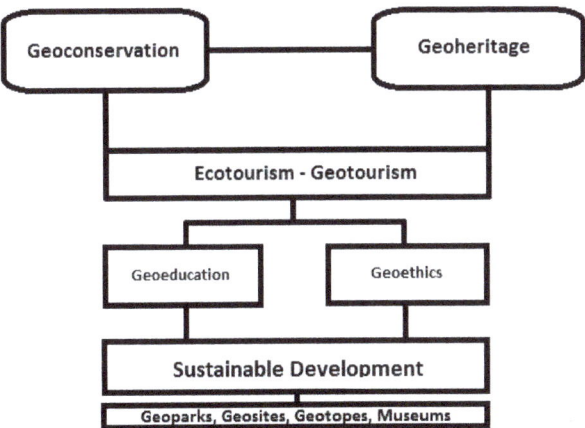

Figure 3. Interconnection and components of rational use between geoconservation and geoheritage.

Geoethics is primarily concerned with the most critical and pressing environmental issues, such as the greenhouse effect, climate change, pollution, and waste management problems. It also aims to promote critical thinking about the use of the natural resources of the earth, the development of environmentally friendly technological methods, and the dissemination of knowledge and information about natural hazards [39]. The incorporation of geoethics into geotourism activities can raise the necessary awareness of sustainability, so that people in a local community perceive the rational use of resources and not take advantage of them. Accordingly, it is fully understood that human society must, in every action, respect the concepts of geodiversity and biodiversity and operate with respect, without negatively affecting any form of mapping of locations and formations that testify to the geohistorical evolution of the Earth and the monuments of geoheritage. Thus, the local community will be able to continue an activity, which will be based on the principles of geoethics and hence future citizens will be able to reap the benefits of the above actions [46] without negative influences or results.

3. The Geological Heritage of GREECE and Its Peculiarities

3.1. The Geological Setting of Greece

The Late Cenozoic evolution of Greece has been controlled by the northward subduction of the African plate beneath the Aegean lithosphere [47–50] (Figure 4). As the African plate moves northwards, strike slip displacements along the Dead Sea fault zone causes compression between the Arabian and the Eurasian plates [51]. This process results in a crustal thickening of East Turkey. As a result, the Aegean and Anatolia plates are being pushed westwards, causing the extension of the Aegean region towards the eastern Mediterranean (gravitational spreading, Figure 5).

The gravitational spreading of the Aegean region towards the eastern Mediterranean is evidenced by the presence of a dense pattern of normal faults (Upper Miocene) and horst and graben structures [48,50,52,53]. Due to the extensional faulting, the crustal thickness of the Aegean plate is reduced, reflecting crustal attenuation [54,55]. For instance, the thickness of the Aegean crust in the Cretan Sea is not more than 20 km, which means that the crustal stretching must have been considerable there. According to Makris [53,54], the crustal thickness of the Aegean region is not constant, expressing the variability of the amount of stretching of the Aegean crust. The same author believes that this fact is due to spatial variations in the mode of tectonics in the Aegean plate. Indeed, apart from normal faulting, the Aegean region was also affected by rotational deformation since the Middle Tertiary. Angeller et al. [56] described a 28° rotation of the Aegean area with respect to Eurasia, around a pole situated in the southern Adriatic Sea. This rotation caused the progressive extension of an inner landmass and the continuous readjustment to this extension of the Hellenic Trench [57]. Since the Mio–Pliocene boundary (5 Ma), another clockwise rotation of 26° occurred in the western and northwestern Greece [58,59]. Palaeomagnetic studies show that this rotation did not affect the southern and southeastern part of the Hellenic Arc (Crete and Rhodes), so that a structural discontinuity between the western and the eastern part of the Hellenic Arc must be assumed [60]. Because of this discontinuity, the western segment of the Hellenic Arc was marked by a compressional phase since the Mio–Pliocene boundary, which may be related to the continental collision of the Aegean domain with the Apulian continental margin (found west of the Ionian Sea) [61,62], whereas the southern segment remained under extensional conditions, except for short intervals of compression [63].

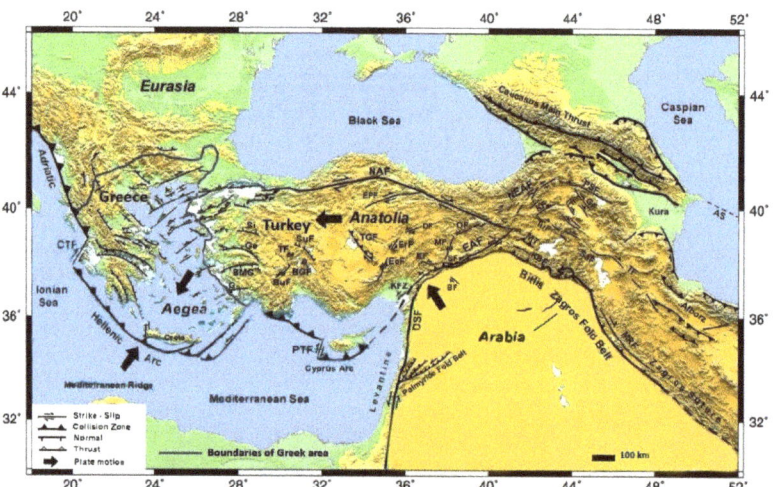

Figure 4. Sketch map depicting the geodynamic setting of Greece ([64], modified by us).

Although the Aegean region is characterized by extensional tectonics, resulting in a steady subsidence, the Hellenic Arc has an elevated position relative to the Cretan Sea in the north (Figure 5). Angelier and Le Pichon [64], Angelier [65], Le Pichon and Angelier [66], and Angelier et al. [56] attributed this uplift of the Hellenic Arc to a mechanism of crustal underplating, at least since the Middle Miocene. This would mean that sediments were removed from the subducting plate (African) to form the new basement of the Hellenic Arc [67].

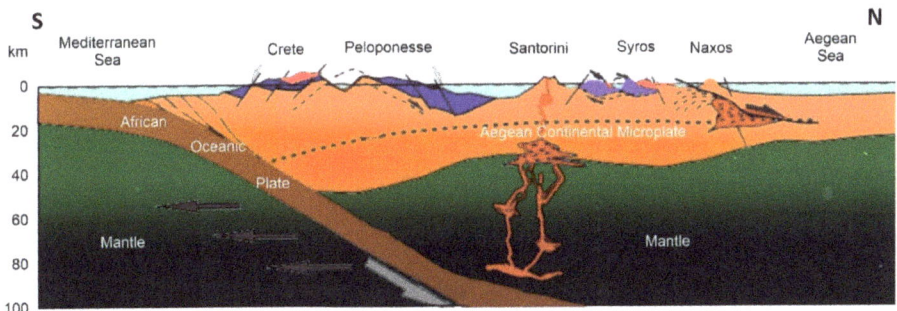

Figure 5. Schematic reconstruction of the northward dipping Hellenic Subduction system (after [68], modified).

3.2. The Geological Heritage of Greece

For the geological scientific community, Greece is a geological museum and a natural laboratory, where the course of the subduction of the African plate beneath the Eurasian can be studied [46]. Furthermore, the intense geotectonic processes in Greece, together with its archaeological and cultural wealth, make it a location deserving of preservation and promotion.

Greece has a range of geoforms and formations worthy of protection for world science and research, (e.g., [46]). This is owing to its location within the convergence zone between two tectonic plates (African and Aegean microplates). As a result, Greece is well-known among geologists around the world as a "natural geological laboratory" that provides valuable information about global geodynamic processes due to its intense earthquake activity, volcanoes, diverse sediment processes, and littoral dynamics, among other things. At the same time, Greece is a great geomuseum, housing "moments" of the dynamic evolution of Earth, from the Proterozoic to the present. These "moments" are represented by geosites and geomorphosites, which are "portions of the geosphere that exhibit a particular importance for the understanding of Earth history", [69,70]. These geosites or geotopes are scientifically, aesthetically, culturally, and ecologically significant (Figure 6). However, ignorance frequently destroys these geosites in an irreversible manner. The Meteora, the Olympus Mountain, the Samaria Gorge in Crete, the ancient Lavrion mines in the Sounion National Park, the Petrified Forest of Lesvos, the Vicos and Aoos Gorges in Epirus, the Diros Caves in Peloponnesus, the Santorini Volcanic Caldera, the Prespes Lakes in West Macedonia, and the Falakron Mountain—Aggitis Karstic System in East Macedonia are some of the most significant geosites, which constitute well-known, legally protected and developed tourist attractions with thousands of visitors each year [5,46]. Moreover, Greece hosts six of the global geoparks of UNESCO: the Petrified Forest on the island of Lesvos, the Vikos–Aoos National Park in Epirus, the Chelmos-Vouraikos National Park in the Peloponnese, the Psiloritis National Park on Crete, the Sitia Nature Park on Crete, and the Grevena–Kozani Geopark.

Figure 6. Selection of some representative geotopes of Greece. (**a**) Meteora geomorphes; (**b,c**) Limnos island, lavas in Fakos area; (**d**) Limnos island, weathered Oligocene sandstones; (**e**) Milos island, pumiceous pyroclastics; (**f**) spherical granites in Tinos island; (**g**) Naxos island, Migmatinte and Kinidatos marbles (personal photo archive P. Voudouris, H. Drinia).

The UNESCO Global Geopark of Lesvos Island (since 2004), formerly known as the Lesvos Petrified Forest Geopark, is a founding member of the Geoparks Network. It hosts an ancient forest that was preserved by a massive volcanic eruption that occurred 20 million years ago. Furthermore, in the Lesvos Island UNESCO Global Geopark area, there are discoveries of the oldest known land mammal (*Prodeinotherium bavaricum*) 19 million years ago in Greece; impressive fossils of animals that lived on Lesvos 2 million years ago; numerous volcanic sites and thermal springs, witnesses of intense volcanic activity (21.5–16.2 million years ago); and faults and landscapes created by tectonic activity [71].

The Psiloritis UNESCO Global Geopark (since 2004), is located on the Greek island of Crete, covering an area of approximately 1200 km^2. It encompasses the entire central region of the island, including the entire area of Mountain Idi (Psiloritis), Crete's highest point, reaching a height of 2456 m. The Geopark is distinguished by its exceptional geodiversity. This is reflected in a wide range of volcanic, sedimentary, and metamorphic rocks dating from the Permian to the Pleistocene (300 to 1 million years ago), as well as spectacular folds and faults, fascinating caves, and deep gorges with a diverse biodiversity. These are exposed in a number of excellent outcrops and cross-sections that provide insight into the mountain-building processes of Earth [72].

The Vikos–Aoos UNESCO Global Geopark is located in Ioannina, Epirus, in northwestern Greece. It is located in the northwest corner of the Pindus Mountain Range and is distinguished by high rugged relief and an impressive landscape. It includes Mt. Smolikas (2637 m asl), Greece's second highest peak, Mt. Tymfi (2497 m asl), and the two spectacular gorges of Vikos and Aoos. The Vikos–Aoos UNESCO Global Geopark is composed of deep sea sedimentary rocks that were folded and faulted by the powerful compressive movements that prevailed in the Greek area 20 million years ago as a result of the collision of African and Eurasian plates. An ophiolitic complex is part of the UNESCO Global Geopark [73].

The Chelmos–Vouraikos UNESCO Global Geopark (since 2009), is located in North Peloponnese, Greece, approximately 200 km from Athens. Chelmos Mountain evolved many distinct forms over millions of years due to the action of water and other natural factors, such as the impressive Vouraikos gorge, the beautiful Cave of the Lakes, the cold springs of Aroanios river, the mythical waters of Styx, and the Tsivlos and Doxa lakes. Lakes can be found not only on the surface, but also underground, and the Cave of the Lakes can be discovered by visiting three of its thirteen underground lakes [74].

The Sitia UNESCO Global Geopark is located on the easternmost edge of Crete and is characterized by numerous Pleistocene mammal fossil sites, the discovery of three *Deinotherium giganteum* fossils, the extensive cave systems, and the paleo-shorelines. The abundant karstic structures on the limestone environment constitute the most profound geological feature. To date, more than 170 caves and numerous gorges have been discovered in the surrounding area [75].

The Grevena–Kozani UNESCO Global Geopark (since 2021) is one of the global geoheritage sites related to the birth of plate tectonic theory, as well as the expression of this tectonic legacy as the source of exceptional landforms and unique ecological systems. The region contains the oldest rocks discovered in Greece to date, as well as sites that reveal the geologic history and rifting processes surrounding the "birth" of the Tethyan Ocean and Europe as an independent continental mass [76].

3.3. Institutional Framework–Legislation

Greece has been identifying natural areas and placing them under special protection since 1937. Natural areas are designated as protected areas via existing national legislation, international conventions, or international or European efforts. Furthermore, the sites of the Natura 2000 network are important locations for the conservation of natural habitats and wildlife and plants. In many cases, the same area is frequently listed at the national, European, and international levels. In terms of national legislation, the declaration of protected areas in various categories of protection was based primarily on Forest Law up to 1986. Law No. 996/1971 establishes National Woodland Parks, Aesthetic Forests, Natural Monuments and Landmarks. Law No. 177/75, as revised by Law N. 2637/1998, establishes wildlife refuges, controlled hunting areas, and game breeding stations. The Environmental Protection Law was afterwards enacted. Following the IUCN standards, Law N. 1650/86 established five distinct categories of protected areas:

- Absolute Nature Reserve Area.
- Nature reserve Area.
- National Park.
- Protected significant natural formation and protected landscape.
- Ecodevelopment Area

Until 1986, some geotopes were protected by forest and archaeological legislation. The geological heritage of Greece is not officially protected, despite the fact that it is well-known and documented. Greece has adopted all relevant international conventions, and the institutional system in place ensures that even individual geotopes surrounded by incompatible practices of the gentle and sustainable development model can be preserved and enhanced (i.e., urban environment, industrial park, etc.). The new Biodiversity Law 3937/2011 has significantly strengthened the protective system established by Law 1650/86 for the category "natural formations-landscapes- components of landscapes". This law expressly protects functional portions of nature or human creations that are scientifically, ecologically, geologically, geomorphologically, or aesthetically significant, and so contributes to the conservation of natural processes and the protection of natural resources.

Despite the efforts of the scientific community, no national geosite inventory exists. The first systematic registration of geological monuments was carried out in 1982 by the Institute of Geology and Mineral Exploration (IGME) of Greece on behalf of the Ministry of Culture. This, however, focused on the monumental character of the selected geomorphs. A broader effort was made as part of the GEOSITES program of UNESCO, which aimed to create a global list of geosites. The Greek Geosites project was developed by the Institute of Geological and Mineral Research (IGME), which participated in the ProGEO Executive Secretariat as coordinator of the South East Europe Working Group [70]. In 1995, IGME attempted to develop a working group for the protection of the geological heritage of Greece, which was later expanded with the participation of scientists from universities and other institutions. Over time, IGME has made a significant contribution to the protection of national Geo heritage (Geotopes–Geopaths–Geoparks) having prepared comprehensive management plans for its systematic registration and promotion. The initiative of IGME in 2005 to include a project, on one hand, for the systematic registration of geotopes based on geoscientific, educational, or tourist value, and on the other hand, for the selection of sites as potential geoparks, in the Third Community Support Framework, deserves special mention. Since then, the issue has been included in various sub-projects of the NSRF, resulting in the completion of the systematic inventory project and the development of an interactive GIS system, which describes the geotopes registered so far, the geopaths, and the institutionalized geoparks of Greece. The GIS presents both the geospatial data and the metadata and information about the European geoparks as well as the Greek geopaths that have been investigated by IGME.

The Commission for the Enhancement of Geological and Geomorphological Heritage was established by the Geological Society of Greece in 2004 to coordinate scientific activities in geoconservation.

The first attempt at an open discussion on the issue of preservation of the geological heritage in Greece took place in 1996 in Ermoupolis, Syros. Four years later, in 2000, the mentioned terminology was consolidated by the Academy of Athens. In this way, it was realized that there was a more systematic way of recording potential geotopes, with the goal of designating a part of them as geoparks.

In addition, it is worth mentioning that the large presence of strong geological monuments in Greece resulted in the recognition of six geological parks (Figure 7). Initially, in 2000, the area of the Petrified Forest of Lesvos was recognized as the first geological park in Greece and was a founding member of the European Geoparks Network. Further, in 2012, the entire island of Lesvos was designated as "Lesvos Island UNESCO World Geopark". Additionally, in 2001 followed the recognition of the Psiloritis geopark (Psiloritis Natural Park) in Crete. It should be emphasized that these first two geoparks contributed to the establishment and characterization of the other geoparks both in Greece and at European level. Later in 2009, the Chelmos–Vouraikos Geopark in the Peloponnese was recognized; in 2010, the Vikos–Aoos Geopark in Epirus; in 2015, the Sitia Geopark in Crete; and finally, in 2021, the Grevena–Kozani Geopark was included.

Figure 7. Satellite photo of Greece, indicating the location of Geoparks: GP1—Lesvos, GP2—Psiloritis, GP3—Vikos-Aoos, GP4— Chelmos-Vouraikos, GP5—Sitia and GP6—Grevena-Kozani.

3.4. Geoeducation in Greece

Geological education seeks to promote geological thinking through geological knowledge, with the goal of positively influencing issues of public concern. This knowledge is imparted in the Greek educational system through a course in primary education. However, this course is taught by unskilled staff as part of the Geography course and even at certain times during the thematic units [77]. Moreover, in secondary education, particularly in the first and second grade of high school, the subjects of Geology and Geography are only taught for one and two hours, respectively, per week. These two curricula do not place enough emphasis on geoheritage, the palaeontological significance of specific geological sites, and fossilized areas that represent the evolution of our planet [78]. As a result, the education of Greek students in geological and geoenvironmental issues is limited to inadequate [79,80]. Because of this, when students complete elementary school, they lack a fundamental understanding of geosciences, which are crucial in the daily lives of citizens [79,80]. According to the statistical study of Georgousis et al. [81], it appears that most Greek school children and students lack adequate knowledge and understanding of geoheritage and its significance. The outcome of this research indicates how important it is to introduce and implement geoenvironmental education rather than simply environmental education. It is also important to have well-trained staff in the Greek education system, able to transfer the necessary knowledge and the importance of geoenvironmental concepts to students. According to Georgousis et al. [81], in which 612 pupils and students participated, a lack of participation of pupils, primarily in geoenvironmental education programs, was initially apparent. However, the findings of the study revealed that 43 percent of students participate in school-based environmental education programs. Then, with a percentage of 30%, it appeared that students participate in cultural educational programs, and finally, only 27% of students participate in educational programs that take place outside school activities (Figure 8) [81].

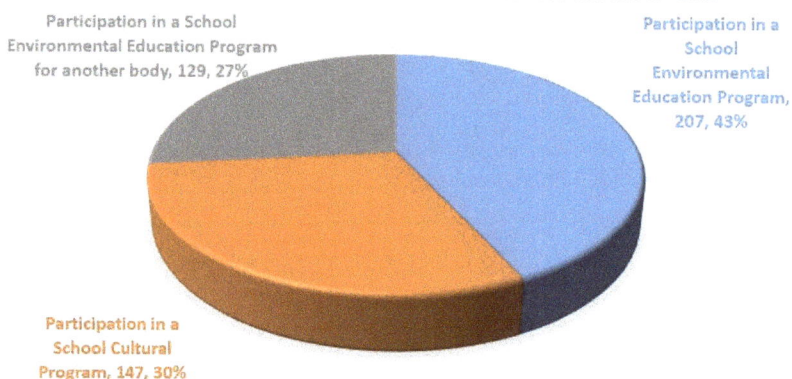

Figure 8. The results of student participation in various school educational programs, based on Georgousis et al. [81].

Furthermore, the findings of this study are critical in terms of the most important aspects that emerge from the implementation of geoeducation. Twenty-one percent of participants stated that geoeducation is related to environmental heritage and geoheritage (Figure 9). According to 19% of participants, geoeducation is related to the proper use of geological sources. Seventeen percent of participants believe that geoeducation is associated with hazard awareness. A percentage of 12 percent relates the concepts of geosciences with everyday life and the ability to mitigate risk. Finally, 11% of participants stated that geoeducation is related to broader geoscientific knowledge, while 8% thinks that geoeducation is related to capacity in sustainable approaches [81].

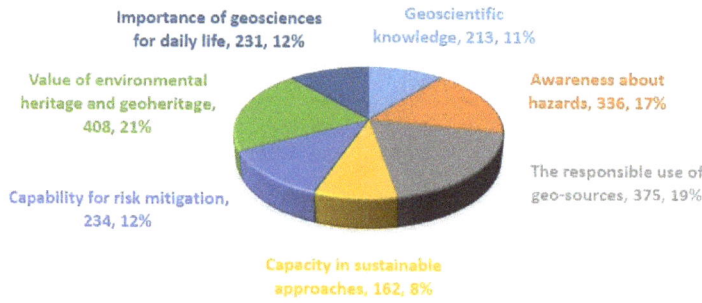

Figure 9. The most important aspects of geoeducation as reported by 612 pupils and students [81].

To summarize, there is a clear lack of geoenvironmental education in the Greek educational system, which means that the next generation does not understand the value and importance of protecting geological heritage. Understanding the value of Greek geological heritage will lead to the emergence of other aspects of Greek society, such as cultural and archaeological heritage.

3.5. State of Development of Geotourism in Greece

Tourism is one of the most important pillars of growth and revenue for the Greek economy. The contribution of tourism to GDP was estimated at 16.2% for 2009, which corresponds to approximately 35 billion EUR [82]. Greece has established itself among the top tourist destinations internationally. However, the future of the Greek tourism product should not be considered guaranteed. Maintaining high growth rates of tourist arrivals and receipts, despite increasing competition, is vital to the national economy.

Geotourism is an alternative form of tourism based on the sustainable development and preservation of geological monuments, the natural environment, the cultural heritage, and the landscape. With its rich natural and cultural heritage, Greece has the potential to develop this specialized form of tourism. Geotourism has its own dynamics and is vying for a market share. Geotopes, geoparks, and natural history museums are the most common locations for geotourism activities, and as a result, geotourism can be a valuable opportunity for local community development with numerous benefits.

Crete is the region with the highest geotourism activity in the country [83]. The South Aegean region also presents high geotourism activity. The regions of Ionian Islands and Attica present medium to high geotourist activity. The regions of Epirus, Peloponnese, and Thessaly present average geotourist activity.

Therefore, it is understood that the geotopes of Greece can play an important role in strengthening its tourism economy. It is a local low-cost tourism product that facilitates regional development [83]. Geotourism is a "new" challenge for Greece, not only because it can redistribute the tourist product of the country to areas that were previously not tourist destinations, but also because it can stimulate a new quality tourist flow in the country. The usual forms of geotourism in Greece are cave tourism (only in tourist caves) and hot spring and spa tourism [84,85]. Greek geotopes, however, can support other activities, targeting a wide range of citizens. In recent years, there has been adequate mobility in Greece for the promotion of geological heritage and geotope management, primarily through the IGME project. This project aims to systematically register geotopes based on criteria such as geoscientific, educational, or tourist value, to identify georoutes, to promote and include Geoparks in the global and European network, to select areas as potential geoparks, etc.

Geotourism, in general, can be one of the main pillars of rural tourism in Greece. In most cases, the "geotourism product" is part of the overall rural tourism product [86] (agrotourism, cultural tourism, adventure tourism), from which it cannot and therefore should not be separated. There are also instances where the "geotourism product" can be incorporated into the main tourism product of the country, to the extent that the dominant and most popular form of tourism (sun and sea) is differentiated and enriched in the direction of sustainable tourism development.

4. Discussion

It is well documented in the scientific community that Greece, as a tectonically active country, constitutes a Natural Geological Laboratory and Museum, which must and can be used for information and education and consequently public awareness.

The Greek natural environment, particularly the living flora and fauna, has been documented, studied, and evaluated. The geoenvironment, on the other hand, despite being the foundation of fauna and flora as well as the framework of spatial planning and development policies, has not yet been fully studied and researched, and its protection is insufficient. This is evident from the lack of geoenvironmental education within the curriculum of Greek schools and universities [81]. The question is whether sustainable development and interdisciplinary knowledge and interpretation of the environment can be understood in the absence of knowledge and integrated geoenvironmental management.

According to the literature review and our personal opinion, the following are some of the reasons for Greece's lack of geoenvironmental promotion, protection, and integrated management:

- Since the establishment of the modern Greek state, Greek geoscientists have played a minor or non-existent role in the effort to locate raw materials, such as energy and quarry reserves, as well as in the study and construction of large development projects;
- The rapid passage of Greek society from the phase of survival in the phase of consumerism–eudemonism;
- The environment is viewed as a consumer good;
- The misconception that the geoenvironment accepts and offers endlessly;
- The time lag with which our country implemented international conventions and European Union legislation concerning the natural environment;
- Inadequate understanding of the importance of geological knowledge and heritage;
- The failure of the Greek state and society to recognize the importance of the role of geoscientists, geological knowledge, and its applications in the country's development process;
- The absence of a geoenvironmental pedagogical, educational, and recreational dimension;
- Greece's inability to function as an open natural geological museum;
- The exploitative and piecemeal approach to the geoenvironment;
- Inability to adapt the activities and services provided by each geoproduct and resource category (spa, caves, gorges, waterfalls, thermal springs, etc.) to changes in Greek society;
- The inability of the Greek society to identify with new data on multiculturalism, mobility, the global market, and the economy.

It is common for society to disregard the values contained in the geoenvironment, which is characterized by a wide range of phenomena and processes, and is a witness to human civilization and geological phenomena. All this makes it a destination for scientific study and observation of phenomena, as well as recreational space and human contact with nature. The geoenvironment, with its constraints and opportunities, serves as the foundation for development planning. For this reason, the involvement of geoscientists in the problems of the Greek development model of Greece, as well as the increased social interest in the environment and its protection, especially in recent years, require the review of attitudes and policies related to the geoenvironment.

Based on the above findings, the goals for the near future should be:

- The recording and investigation of the existing institutional framework of the geoenvironment;
- The recording of the existing reality of the utilized or not, but visitable natural habitats—the caves, the waterfalls, the gorges, and the thermo-mineral waters, as well as the man-made habitats—the museums of natural history, canals, etc.;
- Exploring the development dimension of geotopes as attractions at local, regional, and national scale;
- The promotion of geological heritage and the determination of methods for its protection;
- The emergence of geotourism;
- The use of geological monuments for educational purposes in primary and secondary schools, particularly in environmental education;
- The possibility of using the geoenvironment as an open geological laboratory-museum for higher education;
- Assessing the need for new specialties such as land conservators, tour guides, etc.;
- The diffusion of geological knowledge, science, and their applications in everyday life.
- Exploring the relationships and interactions between the geoenvironment and folklore, literature, mythology, religion, photography, and philately;

- The investigation of the relationship of the geoenvironment with the activities of man in his free time.

The designation of a location as a geosite does not ensure its preservation and protection. Geoconservation cannot be successful unless the public and the state are made aware of its importance. Through educational programs, particularly environmental education courses, young people can learn about their own local geological heritage. Education, particularly environmental education, can be a driving force behind successful land conservation.

The preservation of geological and geomorphological heritage through environmental education can be combined with strategies aimed at the socio-economic development of each region (e.g., development of alternative forms of tourism, such as cultural tourism, ecotourism, etc.) contributing to promoting the harmonious human–nature relationship.

It should be emphasized once more that the current system that governs education has degraded the position of geology in the Greek educational system. It is also requested that scientists with critical and creative thinking be trained to contribute to the resolution of energy issues or issues of management of Greece's abundant natural resources, in order to provide Greece with the necessary economic impetus for long-term development and sustainability.

5. Conclusions

The aim of this review has been based on the current system in force for geological heritage, geodiversity, geoethics, and geoeducation, to demonstrate the current situation regarding the promotion and utilization of the geological heritage of Greece. The country is characterized by a complicated geological context and evolution, as well as a wide range of geomorphological processes that have resulted in a high level of geodiversity. As a result, it has many spectacular landscapes and outstanding or unique geosites, which, due to the lack of the appropriate legislation framework, are poorly managed and protected.

Greece occupies a unique position on the European geological map, with extensive areas of important geotopes containing, among other things, paleontological remains, rare geomorphological structures, and thousands of caves. Furthermore, this is the primary reason why the country has been a member of the UNESCO European Network of Geoparks since November 2015, with six geoparks (the fossil forest of Lesvos, Psiloritis of Crete, the area of Sitia, the National Forest of Vikos–Aaros, Vouraikos Peloponnese, and last but not least, the Grevena–Kozani geopark) among 81 geoparks in 26 countries in Europe.

Although the provisions of Law 3937/11 resulted in the definition, characterization, and process of characterization of geotopes, there is still an institutional and legal gap in geotope protection. The importance of establishing a legal framework for geotope protection is highlighted by the fact that their promotion and rational management creates opportunities for sustainable development, particularly in the rural sector, as well as becoming quality tourist destinations (geotourism) through nature protection and education. However, at national level, the importance of geotopes has, so far, been underestimated. The institutional framework that exists for the protection and preservation of the geological heritage is based on the necessity of issuing administrative acts and, to date, this has not been activated.

Finally, the need for geoenvironmental education at all levels of the Greek education system is emphasized which will lead to geoenvironmentally responsible citizens with developed geoethical values.

Author Contributions: Conceptualization, G.Z. and H.D.; methodology, G.Z.; investigation, G.Z.; resources, H.D. and N.Z.; writing—original draft preparation, G.Z.; writing—review and editing, G.Z, H.D., A.A. and N.Z.; supervision, H.D. All authors have read and agreed to the published version of the manuscript.

Funding: This research received no external funding.

Data Availability Statement: The data presented in this study are available on request from the corresponding author.

Acknowledgments: The authors gratefully thank the journal editor and the three reviewers for their thorough consideration of this paper.

Conflicts of Interest: The authors declare no conflict of interest.

References

1. UNESCO—U.N. Environment program. In Proceedings of the International Workshop on Environmental Education at Belgrade, Belgrade, Yugoslavia, 13–22 October 1975; pp. 1–4.
2. UNESCO. In *Proceedings of the Intergovernmental Conference Education, Tbilisi, Georgia, 14–26 October 1977*; UNESCO: Paris, France, 1978; pp. 1–96.
3. Martini, G. (Ed.) Actes du Premier Symposium International sur la Protection au Patrimonie Geologique. In *Proceedings of the First Symposium on Earth Heritage Conservation, Digne, France, 11–16 June 1991*; Memoires de la Societe geologique de France; Société géologique de France: Paris, France, 1993; Numero Special 165; p. 276.
4. Zouros, N. The European Geoparks Network. *Episodes* **2004**, *27*, 165–171. [CrossRef]
5. Zouros, N.; Martini, G. Introduction to the European Geoparks Network. In Proceedings of the 2nd European Geoparks Network Meeting: Lesvos, Natural History Museum of the Lesvos Petrified Forest, Lesvos, Greece, 3–7 October 2003; Zouros, N., Martini, G., Frey, M.-L., Eds.; pp. 17–21.
6. Farsani, N.D.; Coelho, C.; Costa, C. Geotourism and Geoparks as novel strategies for socio-economic development in rural areas. *Int. J. Tour. Res.* **2011**, *13*, 68–81. [CrossRef]
7. Dowling, R.; Newsome, D. (Eds.) *Geotourism*; Elsevier/Heineman: Oxford, UK, 2006.
8. Hose, T. A3G's for Modern. Geotourism. *Geoheritage* **2012**, *4*, 7–24. [CrossRef]
9. Vascanelos, C.; Moutinho, S.; Torres, J. Geothics in the context of Sustainability and its Teaching across the Curriculum. In Proceedings of the 8th International Conference of Education, Research and Innovation, Seville, Spain, 16–18 November 2015.
10. Santangelo, N.; Valente, E. Geoheritage and Geotourism. *Resources* **2020**, *9*, 80. [CrossRef]
11. Carcavilla, L.; Durán Valsero, J.; García-Cortés, Á.; López-Martínez, J. Geological Heritage and Geoconservation in Spain: Past, Present, and Future. *Geoheritage* **2009**, *1*, 75–91. [CrossRef]
12. Gordon, J.E. Geoheritage, Geotourism and the Cultural Landscape: Enhancing the Visitor Experience and Promoting Geoconservation. *Geosciences* **2018**, *8*, 136. [CrossRef]
13. McBriar, M. Foreword. In *A Report Prepared for the Australian Heritage Commission by the Standing Committee for the Geological Heritage of the Geological Society of Australia Inc.*; Joyce, E.B., Ed.; Standing Committee for Geological Heritage of the Geological Society of Australia Inc.: Sydney, Australia, 1995.
14. UNESCO. *UNESCO Geoparks Programme—A New Initiative to Promote a Global Network of Geoparks Safeguarding and Developing Selected Areas Having Significant Geological Features*; Document 156 EX/11 Rev., Executive Board, 156th session; UNESCO: Paris, France, 1999; p. 4.
15. European Manifesto on Earth Heritage and Geodiversity. 2004. Available online: http://www.progeo.pt/pdfs/Manifesto_EH.pdf (accessed on 9 May 2021).
16. Gray, M. *Geodiversity: Valuing and Conserving Abiotic Nature*; John Wiley: Chichester, UK, 2004.
17. Sharples, C.; Concepts and Principles of Geoconservation. Tasmanian Parks and Wildlife Service. 2002. Available online: http://www.dpipwe.tas.gov.au/Documents/geoconservation.pdf (accessed on 3 July 2021).
18. Brocx, M.; Semeniuk, V. Geoheritgae and geoconservation—History, definition, scope and scale. *J. R. Soc. West. Aust.* **2007**, *90*, 53–87.
19. Semeniuk, V. The linkage between biodiversity and geodiversity. In *Pattern and Process—Towards a Regional Approach to National Estate Assessment of Geodiversity*; Canberra ACT: Environment Australia Technical Series No 2; Eberhard, R., Ed.; Australian Heritage Commission: Canberra, Australia, 1996; pp. 51–58.
20. Semeniuk, V.; Semeniuk, C.A. Human impacts on globally to regionally significant geoheritage features of the Swan Coastal Plain and adjoining coastal zone, southwestern Australia. In *Gondwana to Greenhouse: Australian Environmental Geoscience*; Gostin, V.A., Ed.; Geological Society of Australia: Sydney, Australia, 2001; pp. 181–199.
21. Prosser, C.D.; Bridgland, D.R.; Brown, E.J.; Larwood, J.G. Geoconservation for science and society: Challenges and opportunities. *Proc. Geol. Assoc.* **2011**, *122*, 337–342. [CrossRef]
22. Lazzari, M. Geosites, cultural tourism and sustainability in Gargano National park (southern Italy): The case of the Salata (Vieste) geoarchaeological site. *Rend. Online Soc. Geol. It.* **2013**, *28*, 97–101.
23. McMillan, A.A.; Gillanders, R.J.; Fairhurst, J.A. *The Building Stones of Edinburgh*, 2nd ed.; Edinburgh Geological Society: Edinburgh, UK, 1999.
24. Schullery, P. *Searching for Yellowstone: Ecology and Wonder in the Last Wilderness*; Mariner Books: Boston, MA, USA, 1999.
25. Jackli, H. Opening address. In *Moraines & Varves*; Schlüchter, C., Ed.; Balkema: Rotterdam, The Netherlands, 1979; pp. 5–7.
26. Gordon, J.E. Conservation of geomorphology and Quaternary sites in Great Britain: An overview of site assessment. In *Conserving our Landscape*; Stevens, C., Gordon, J.E., Macklin, M.G., Eds.; English Nature: Peterborough, UK, 1994; pp. 11–21.
27. Mc Keever, P.; Zouros, N. Geoparks: Celebrating earth heritage, sustaining local communities. *Episodes* **2005**, *28*, 274–278. [CrossRef] [PubMed]

28. EGN. European Geoparks Magazine No 7. 2005, p. 36. Available online: http://www.europeangeoparks.org (accessed on 9 May 2021).
29. UNESCO. Transforming our World: The 2030 Agenda for Sustainable Development. 2005. Available online: https://sustainabledevelopment.un.org/content/documents/21252030%20Agenda%20for%20Sustainable%20Development%20web.pdf (accessed on 9 May 2021).
30. Herrera-Franco, G.; Montalván-Burbano, N.; Carrión-Mero, P.; Apolo-Masache, B.; Jaya-Montalvo, M. Research trends in geotourism: A bibliometric analysis using the scopus database. *Geosciences* **2020**, *10*, 379. [CrossRef]
31. Robinson, A.M. Geotourism: Who is a geotourist. In Proceedings of the Inaugural National Conference on Green Travel, Climate Change and Ecotourism, Adelaide, Australia, 17–20 November 2008; pp. 909–923.
32. Farsani, N.T.; Coelho, C.; Costa, C. Rural geotourism: A new tourism product. *Acta Geoturistica* **2013**, *4*, 1–10.
33. Joyce, E.B. Geological heritage of Australia: Selecting the best for Geosites and World Heritage and telling the story for geotourism and Geoparks. *ASEG Ext. Abstr.* **2006**, *2006*, 1–4. [CrossRef]
34. Kanellopoulou, G. The challenge of geotourism in Greece: Geocultural routes in Zagori region (NW Greece). In Proceedings of the IMIC, 1st International Conference on Experiential Tourism, Santorini, Greece, 9–11 October 2015; pp. 9–11.
35. Dowling, R.K. Global geotourism—An emerging form of sustainable tourism. *Czech J. Tour.* **2013**, *2*, 59–79. [CrossRef]
36. Boo, E. Planning for Ecotourism. *Parks* **1991**, *2*, 4–9.
37. Imbrahim, K.; Hanzah, M. Geotourism: An Effective Approach towards Conservation of Geological Heritage. In *Symposium of Indonesia–Malaysia Culture, Badung, Malaysia*; Unpublished paper; 1993.
38. Inskeep, E. *Tourism Planning: An Integrated and Sustainable Development Approach*; Van Nostrand Reinhold: New York, NY, USA, 1997.
39. Peppoloni, S.; Di Capua, G. The meaning of geoethics. In *Ethical Challenges and Case Studies in Earth Sciences*; Wyss, M., Peppoloni, S., Eds.; Elsevier: Amsterdam, The Netherlands, 2015; pp. 3–14.
40. Brocx, M.; Semeniuk, V. The '8Gs'—A blueprint for Geoheritage, Geoconservation, Geo-education and Geotourism. *Aust. J. Earth Sci.* **2019**, *66*, 803–821. [CrossRef]
41. Chan, K.M.; Balvanera, P.; Benessaiah, K.; Chapman, M.; Díaz, S.; Gómez-Baggethun, E. Luck GW Opinion: Why protect nature? Rethinking values and the environment. *Proc. Natl. Acad. Sci. USA* **2016**, *113*, 1462–1465. [CrossRef] [PubMed]
42. Slaymaker, O.; Spencer, T.; Embleton-Hamann, C. *Geomorphology and Global Environmental Change*; Cambridge University Press: Cambridge, UK, 2015. [CrossRef]
43. Winter, C. The intrinsic, instrumental and spiritual values of natural area visitors and the general public: A comparative study. *J. Sustain. Tour.* **2007**, *15*, 599–614. [CrossRef]
44. Papayannis, T.; Howard, P. Editorial: Nature as Heritage. *Int. J. Herit. Stud.* **2007**, *13*, 298–307. [CrossRef]
45. Vucetich, J.; Bruskotter, J.; Nelson, M. Evaluating whether nature's intrinsic value is an axiom of or anathema to conservation. *Conserv. Biol.* **2015**, *29*, 321–332. [CrossRef] [PubMed]
46. Drinia, H.; Tsipra, T.; Panagiaris, G.; Patsoules, M.; Papantoniou, C.; Magganas, A. Geological heritage of syros island, cyclades complex, Greece: An assessment and geotourism perspectives. *Geoscience* **2021**, *11*, 138. [CrossRef]
47. Makris, J. Geophysical investigation of the Hellenides. *Hambg. Geophys. Einzelschr.* **1977**, *34*, 124.
48. McKenzie, D. Some remarks on the development of sedimentary basins. *Earth Planet. Sci. Lett.* **1978**, *40*, 25–32. [CrossRef]
49. Papazachos, B.C.; Comninakis, P.E. Geotectonic significance of the deep seismic zones in the Aegean Area. In *Proceedings Symposium Thera and the Aegean World I*; Cambridge University Press: Cambridge, UK, 1978; pp. 121–129.
50. Pichon, X.L.; Angelier, J. The Hellenic Arc and Trench System: A key to the neotectonic evolution of the Eastern Mediterranean. *Tectonophysics* **1979**, *60*, 1–42. [CrossRef]
51. Molnar, P.; Tapponier, P. Cenozoic tectonics of Asia: Effects of a continental collision. *Science* **1975**, *189*, 419. [CrossRef]
52. McKenzier, D.P. Active tectonics of the Mediterranean region. *Geophys. J. Int.* **1972**, *30*, 109–185. [CrossRef]
53. Angelier, J.; Dumont, J.F.; Karamanderesi, H.; Poisson, A.; Simsek, S.; Uysal, S. Analyses of fault mechanisms and expansion of southwestern Anatolia since the Late Miocene. *Tectonophysics* **1981**, *75*, T1–T9. [CrossRef]
54. Makris, J. The crust and upper mantle of the Aegean region from deep seismic soundings. *Tectonophysics* **1978**, *46*, 269–284. [CrossRef]
55. Makris, J. Geophysical studies and geodynamic implications for the evolution of the Hellenides. In *Geological Evolution of the Mediterranean Basin*; Stanley, D.J., Wezel, F.C., Eds.; Springer: New York, NY, USA, 1985.
56. Angeller, J.; Lyberis, N.; Pichon, X.L.; Barrier, E.; Huchon, P. The tectonic development of the Hellenic Arc and the Sea of Crete: A synthesis. *Tectonophysics* **1982**, *86*, 159–196. [CrossRef]
57. Jarnet, M. *Paleoomagnetisme et Neotectonique de l' ile de Corfou*; Theses Univ. Paris X; Paris Nanterre University: Paris, France, 1982; p. 146.
58. La, J.C.; Jamet, M.; Sorel, D.; Valente, J.P. First paleomagnetic results from Mio-Pliocene series of the Hellenic sedimentary arc. *Tectonophysics* **1982**, *86*, 45–67.
59. Valente, J.P.; Laj, C.; Sorel, D.; Roy, S.; Valet, J.P. Paleomagnelic results from Mio-Pliocene marine sedimentary series in Crete. *Earth Plan. Sei. Lett.* **1982**, *57*, 159–172. [CrossRef]
60. Mercier, J.L.; Carey, E.; Philip, H.; Sorel, D. La neotectonique plio-quaternaire de l' arc egeen externe et de la mer Egee et ses relations avec la seismicite. *Bull. Soc. Geol. Fr.* **1976**, *7*, 355–372. [CrossRef]

61. Sorel, D. *Etude Neotectonique dans l' arc Egeen Externe Occidental: Les iles Ioniennes de Kephallinia et Zakynthos et l Elide Occidental*; These Univ. Paris XI (Orsay); Paris-Sud University: Paris, France, 1976; p. 196.
62. Paquin, C.; Bloyet, J.; Angelidis, C. Tectonic stresses on the boundary of the Aegean domain: "in situ" measurements by overcoring. *Tectonophysics* **1984**, *110*, 145–150. [CrossRef]
63. Taymaz, T.; Yilmaz, Y.; Dilek, Y. The Geodynamics of the Aegean and Anatolia. *Geol. Soc. Lond. Spec. Publ.* **2007**, *291*, 1–16. [CrossRef]
64. Angelier, J.; Pichon, X.L. Neotectonique horizontale et verticale de l'Egée: Subduction et expansion. Aubouin, J. Debelmas and M. Latreille (Coordinators), Géologie des Chaînes Alpines Issues de la Téthys. In Proceedings of the 26th International Geological Congress, Mém BRGM, Paris, France, 7–17 July 1980; Volume 115, pp. 249–260.
65. Angelier, J. Analyse quantitative des relations entre deformation horizontale et movements vertlcaux: 1' extension Egeen, la subsidence de la mer de Crete et la surrection de 1' arc hellenique. *Ann. Göophys.* **1981**, *37*, 327–345.
66. Huchon, P.; Lybéris, N.; Angelier, J.; Le Pichon, X.; Renard, V. Tectonics of the hellenic trench: A synthesis of sea-beam and submersible observations. *Tectonophysics* **1982**, *86*, 69–112. [CrossRef]
67. Barton, M.; Salters, V.; Huijsmans, J. Sr-isotope and Trace Element Evidence for the Role of Continental Crust in Calc-Alkaline Volcanism on Santorini and Milos, Aegean Sea, Greece. *Earth Planet. Sci. L* **1983**, *63*, 273–291. [CrossRef]
68. Trotet, F.; Vidal, O.; Jolivet, L. Exhumation of Syros and Sifnos metamorphic rocks (Cyclades, Greece). New constraints on the P-653 T paths. *Eur. J. Mineral.* **2001**, *13*, 901–920. [CrossRef]
69. Goudie, A.G. *Encyclopedia of Geomorphology*; Routledge: London, UK, 2004.
70. Theodosiou, E.; Fermeli, G.; Koutsouveli, A. *Our Geological Heritage*; Kaleidoskopion: Athens, Greece, 2006.
71. Zouros, N. Lesvos petrified forest geopark, Greece: Geoconservation, geotourism, and local development. In *The George Wright Forum*; George Wright Society: Hancock, MI, USA, 2010; pp. 19–28.
72. Fassoulas, C.; Zouros, N. Evaluating the influence of Greek geoparks to the local communities. *Bull. Geol. Soc. Greece* **2010**, *43*, 896–906. [CrossRef]
73. Stergiou, L.; Chatzipetros, A.; Telbisz, T.; Mindszenty, A. The Gamila Peak Spherical Concretions and the Oxia Hanging Valley at the Vikos-Aoos Geopark. In Proceedings of the 15th International Congress of the Geological Society of Greece, Athens, Greece, 22–24 May 2019; Volume 7, pp. 722–723, Special Publication.
74. Liapi, E. Study of plant macro-remains from the Upper Miocene–Lower Pliocene lignite deposits of Kalavrita basin. Master's Thesis, Applied Ecology-Ecosystems and Biological Resources Management' of G.S.P Program of Biology Department, University of Patras, Patras, Greece, 2020.
75. Fassoulas, C.; Staridas, S.; Perakis, V.; Mavrokosta, C. Revealing the geoheritage of Eastern Crete, through the development of Sitia Geopark, Crete, Greece. *Bull. Geol. Soc. Greece* **2013**, *47*, 1004–1016. [CrossRef]
76. Rassios, A.; Grieco, G.; Batsi, A.; Myhill, R.; Ghikas, D. Preserving the non-preservable geoheritage of the aliakmon river: A case study in geoeducation leading to cutting-edge science. *Bull. Geol. Soc. Greece* **2016**, *50*, 255–264. [CrossRef]
77. Rokka, A. Geology in primary education; Potential and perspectives. *Bull. Geol. Soc. Greece* **2018**, *34*, 819–823. [CrossRef]
78. Meléndez, G.; Fermeli, G.; Koutsouveli, A. Analyzing Geology textbooks for secondary school curricula in Greece and Spain: Educational use of geological heritage. *Bull. Geol. Soc. Greece* **2007**, *40*, 1819–1832. [CrossRef]
79. Trikolas, K.; Ladas, I. The necessity of teaching earth sciences in secondary education. In Proceedings of the 3rd International GEOschools Conference, Teaching Geosciences in Europe from Primary to Secondary School, Athens, Greece, 28–29 September 2013; pp. 73–76.
80. Fermeli, G.; Meléndez, G.; Calonge, A.; Dermitzakis, M.; Steininger, F.; Koutsouveli, A.; Neto de Carvalho, C.; Rodrigues, J.; D'Arpa, C.; Di Patti, C. GEOschools: Innovative teaching of geosciences in secondary schools and raising awareness on geoheritage in the society. In *Avances y Retos en la Conservación del Patrimonio Geológico en España. Actas de la IX Reunión Nacional de la Comisión de Patrimonio Geológico (Sociedad Geológica de España)*; Fernández-Martínez, E., Castaño de Luis, R., Eds.; Universidad de León: León, Spain, 2011; pp. 120–124. ISBN 978-84-9773-578-0.
81. Georgousis, E.; Savelides, S.; Mosios, S.; Holokolos, M.-V.; Drinia, H. The Need for Geoethical Awareness: The Importance of Geoenvironmental Education in Geoheritage Understanding in the Case of Meteora Geomorphes, Greece. *Sustainability* **2021**, *13*, 6626. [CrossRef]
82. Skoumpi, M.; Tsartas, P.; Sarantakou, E.; Pagoni, M. Wellness Tourism Resorts: A Case Study of an Emerging Segment of Tourism Sector in Greece. In *Culture and Tourism in a Smart, Globalized, and Sustainable World*; Springer Proceedings in Business and Economics; Katsoni, V., van Zyl, C., Eds.; Springer: Cham, Switzerland, 2021; pp. 477–495. [CrossRef]
83. Skentos, A. Geotopes of Greece. Master's Thesis, University of Athens, Athens, Greece, 2012.
84. Georgakopoulou, S.; Delitheou, V. Alternative forms of sustainable development: The case of thermal tourism. *Int. J. Environ. Sustain. Dev.* **2020**, *19*, 367–377. [CrossRef]
85. Karamani, P.; Drinia, H.; Panagiaris, G. Geotourism as a tool for the protection and promotion of the Cave of Galaxidi. In Proceedings of the 15th International Congress of the Geological Society of Greece Athens, Athens, Greece, 22–24 May 2019; pp. 22–24.
86. Newsome, D.; Dowling, R.K. *Geotourism: The Tourism of Geology and Landscape*; Goodfellow Publishers: Oxford, UK, 2010.

Review

Geoheritage and Cultural Heritage—A Review of Recurrent and Interlinked Themes

Edyta Pijet-Migoń [1] and Piotr Migoń [2,*]

[1] WSB University of Wrocław, ul. Fabryczna 29-31, 53-609 Wrocław, Poland; edyta.migon@wsb.wroclaw.pl
[2] Institute of Geography and Regional Development, University of Wrocław, pl. Uniwersytecki 1, 50-137 Wrocław, Poland
* Correspondence: piotr.migon@uwr.edu.pl

Abstract: Relationships between geoheritage and cultural heritage are being increasingly explored and have become one of the mainstreams within studies of geoheritage and geodiversity. In this review paper, we identify the main and secondary themes at the geoheritage—cultural heritage interface and provide examples of specific topics and approaches. These themes include added cultural value to geoheritage sites, geoheritage in urban spaces, cultural landscapes, and the contribution of geoheritage to their identity, mining and quarrying heritage, linkages with natural disasters, history of science, and art. Intangible cultural heritage is also reviewed in the geoheritage context. In the closing part of the paper, various classifications of geoheritage—cultural heritage linkages are proposed, although it is concluded that themes and fields of inquiry are overlapping and interlinked, rendering one classification system not very feasible. Instead, a mind map to show these diverse connections is offered. The paper closes with recommendations for future studies, arising from this review and the identification of research gaps and under-researched areas.

Keywords: urban geoheritage; building stone; cultural landscapes; mining; quarrying; geotourism; UNESCO World Heritage; natural disasters

1. Introduction

Even though the term "geoheritage" tends to emphasize the rock, fossil and landform record created by natural processes during protracted intervals of geological evolution, it was never disassociated from cultural heritage and some definition proposals include cultural significance [1]. Relationships between geoheritage and cultural heritage are multiple: spatial, conceptual, causal, and thematic [2,3]. Likewise, part of geoconservation is located at the interface with cultural heritage, as the introduction of certain conservation measures may necessitate a good understanding of the local cultural context, including the intangible heritage of indigenous societies [4], whereas on the other hand, preservation of built-up cultural heritage requires adequate consideration of building-stone heritage. Finally, geoparks—as means of transferring knowledge of geosciences to the society and increasing awareness of geoheritage—strongly emphasize the ABC concept that highlights linkages between abiotic, biotic and cultural components [5]. In this review, we outline the multiple linkages between geoheritage and cultural heritage, identifying the main themes that explore this interface and provide selected examples, mainly taken from publications published in the last decade or so.

This review does not pretend to be exhaustive in the coverage of recent literature, which is growing at an unprecedented rate and hence, difficult to follow. Likewise, we do not employ a bibliometric and text-mining approach, even though we acknowledge it can be an interesting step towards recognition of thematic preferences among researchers involved in geoheritage issues. In our view, the main contribution of this paper is the identification of the diversity of topics and approaches. In doing so, we build upon the work of Reynard and Giusti [3], who listed three main questions pertinent to the cultural value of geoheritage: (1) how geological processes and geoheritage affect culture, (2) how culture affects geoheritage perception and management, and (3) how culture and geological heritage are integrated. Although the themes presented in this paper can be subsumed under these three lines of inquiry, they are sufficiently broad and multifaceted to deserve more systematic presentation.

2. Geoheritage Sites of Additional Cultural Significance

It has long been recognized that localities appreciated by geoscientists for their geoheritage values may also carry significant cultural values, with or without causal connection between the two (Figure 1). Reasons for cultural significance are varied, both intangible and tangible, and may include spiritual (religious) importance, connection with local legends, famous personalities and historical events such as battles or meetings, occurrence of architectural heritage, typically defensive structures, shrines and temples, tombs, or even entire settlements, the presence of rock art, and former mining grounds.

Figure 1. Geomorphosites with additional cultural values. (**a**) cinder cone of Zebín in the Bohemian Paradise UNESCO Global Geopark is an important geosite, with the medieval church at the foot (right) and a baroque chapel on the top, (**b**) conglomerate rock city of Belogradchik, Bulgaria, with remnants of the Kaleto fortress incorporated into it, (**c**) a large erratic boulder in central Poland (Budziejewko village), associated with various local legends, (**d**) Solfatara in Pozzuoli, Italy, is a world-renowned volcanic geosite, visited and used for therapeutic reasons since antiquity (all photographs by the authors).

Landforms, rather than geological outcrops, represent most of geoheritage sites with these associations, which is understandable given their often prominent position within a regional landscape. Thus, these geoheritage sites may be also considered as geomorphosites [6]. Typical geomorphosites of this kind are distinctive terrain elevations (volcanoes and volcanic necks, inselbergs, dome-shaped hills, mesas, solitary pinnacles), crags and tors, cliff lines, both coastal and inland, canyons, gorges, and necks of entrenched meanders. Relevant smaller landforms include erratic boulders and their clusters, curiously shaped weathering features, rock shelters, gullies, and dolines. Caves may also belong to geoheritage of mixed significance [7], if they contain archaeological remains, examples of rock art (paintings, petroglyphs), or were used as hideouts or hermitage sites.

The dual, globally outstanding value, involving both geoheritage and cultural heritage, is demonstrated at UNESCO World Heritage mixed properties. Their total number is 39, as of early 2022 (www.unesco.org; accessed on 22 January 2022), although not all of them are recognized for geoheritage. Among sites recently presented are the travertine depositional features associated with an ancient town of Hierapolis in Pamukkale, Turkey [8], and conglomerate towers crowned by Orthodox Christian monasteries at Meteora in Greece [9,10], whereas the great sandstone sceneries of Tassill n'Ajjer (Algeria) and Ennedi (Chad) that include impressive rock art of outstanding value are still waiting for proper presentation of their geoheritage.

Some of the recently published studies emphasised spiritual [11] and military use of caves [12], their importance for archaeology and human history [13], the use of natural configuration of rock landforms to insert defensive structures such as castles and fortresses [14], and incorporation of modern architecture into natural rock-cut scenery [15]. Large erratic boulders in the formerly glaciated European lowlands areas have often acquired special cultural significance, being associated with pre-Christian cults, beliefs in supernatural powers, legendary or historical events [16,17]. These associations are also reflected in their names, containing references to evil spirits or saints. Huge granite boulders in granite denudational landscapes, especially if associated with intriguing microrelief (circular weathering pits, karren) may also bear similar associations, for instance in the Waldviertel area of Austria [18]. Multiple connections between volcanic phenomena and related landforms and diverse human activities over centuries were presented for the Campi Flegrei area in southern Italy [19], whereas in the Swabian Alb, southern Germany, various karstic sites (caves, tufa cascades, springs) have strong associations with local cultural heritage [20,21].

3. Urban Geoheritage and Heritage Stones

Exploration of geoheritage in an urban context is among the most frequently addressed themes at the geoheritage/cultural heritage interface. It is close to urban geomorphology on the one hand, which examines how landform patterns influence location and growth of towns and cities [22], but crosscuts the heritage stone issues, ranging from conservation challenges at historical monuments to the identity of a city due to preferential use of a certain kind of building stone. Thus, specific subjects also vary in the spatial scale of inquiry, from a singular building to the whole-city layout and appearance (Figure 2).

Some towns and cities are presented as examples of how urban planning, often going back to medieval times or even beyond, was adjusted to natural topography and spatial domains of ongoing geomorphic process, thereby enhancing local geoheritage rather than erasing it [23–25]. In Rio de Janeiro, this symbiosis between superb granite geomorphology, with its numerous steep-sided domes [26,27], and specific urban design and culture was even emphasised in the justification of outstanding universal value, decisive for the city's inscription on the UNESCO World Heritage list as an urban cultural landscape (Figure 2a).

However, accelerated urban growth coupled with increasing engineering abilities has led to the situation that natural landform features no longer control pathways of urban development, but may have been eliminated if that was considered necessary. The respective engineering solutions include valley filling, channel relocation, slope grading and hilltop trimming, with the common denominator being the ultimate disappearance of a landform. In specific places, these disappeared landforms may have been of geoheritage value. Recently reviewed examples include Rome [28] and Genoa in Italy [29], Lausanne in Switzerland [30], Mexico City [31], Perugia [32], Toruń in Poland [33], Brno in the Czech Republic [34], and Auckland in New Zealand [35].

Figure 2. Examples of urban geoheritage of different kinds and scale. (**a**) the city of Rio de Janeiro developed amidst large granite-gneiss domes, (**b**) Tiberina Island in the centre of Rome is a much altered natural gravel bar of the river Tiber, (**c**) a romanesque cathedral in Modena, Italy, built of various building stones, (**d**) Parc des Buttes-Chaumont—an old gypsum quarry in the centre of Paris, converted into a municipal park (all photographs by the authors).

Despite the inevitable loss of geodiversity and geoheritage due to urbanization, or perhaps in response to it, a voluminous literature concerns recognition of geoheritage sites within urban space. Some publications deal with classic localities, appreciated for more than a century, such as the giant glacial potholes in Luzern, Switzerland [36], but there is an increasing trend to present less evident cities and towns to an international audience. For some of these, local language literature may have been available earlier, but broadly accessible sources were missing. Places of recent interest include São Paulo [37], Lisbon [38,39], Poznań [40], Segovia [41], Oslo [42], Khorramabad in Iran [43], Brno [34,44], Ljubljana in Slovenia [45], and Zagreb in Croatia [46]. Apart from large cities, small towns are also occasionally evaluated in this context, such as Pruszków in Poland [47]. However, in many of these cases the assessment of urban geoheritage is limited to sites of geological interest, mainly natural or artificial bedrock outcrops, whereas geomorphological heritage and viewing points allowing for examination of geomorphological scenery are considered less often. This imbalance may be partly related to the intensity of urban development but may also reflect specific interests of particular authors. Thus, the diversity of urban geoheritage was presented by Pica et al. [28] for the heart of Rome (ancient Rome), where geomorphological, hydrological, and stratigraphical geosites were identified (Figure 2b), whereas in Mexico City, attention was directed to the remnants of the lake, the use of ornamental stones, and old lava flows, the latter encroaching onto a pre-Hispanic ceremonial centre at Cuicuilco, with its partly buried round pyramid [31]. In Segovia, as many as 94 potential geosites belonging to 16 thematic categories were documented [41], whereas in Brno, 89 localities were analysed in terms of scientific (geoheritage, geodiversity) and added values, the latter including cultural associations [48].

A theme in its own right is heritage stone, usually analysed in an urban context (Figure 2c), even though its use in vernacular architecture is also addressed [38,49–60]. De Wever et al. [50] reviewed the significance of geosites and heritage stones in a more general way, providing numerous examples from European cities of how background geological knowledge may be used to design rock-focused geotrails, leading visitors to famous historical buildings erected from characteristic stones and to the very sites of stone extraction, that is ancient quarries now located within city limits or just outside. These examples include the towns of Salamanca in Spain, Bath in England and Paris, where an old gypsum quarry at Buttes-Chaumont was converted into a large park (Figure 2d). Some publications are deliberately focused on one or two rock types, most used in a region. In the Segovia province in Spain, rock colour determines the visual appearance of villages, named as 'red hamlets' (*pueblos negros*) and 'black hamlets' (*pueblos rojos*). These differences reflect the use of traditional building materials, Miocene red gossan breccias and Ordovician-Silurian black slates, respectively [61]. Further examples include the use of Arrábida Breccia in manueline-style buildings in central Portugal [62], Red Ereño limestone in the Basque Country, Spain [63], and green phyllite used for roofing in Lugo in Galicia, Spain [64]. This geoheritage-oriented approach to building heritage is not limited to Europe. Numerous recent papers have documented heritage stone use in India, including basalt [65], marble [66,67], sandstone in Rajasthan province [68], limestone [69], slate [70] and quartzite [71]. Other studies have been focused on particular regions rather than specific building stones, as exemplified by a recent thorough study of historic stones used in construction of churches and chapels in West Sussex, England [72], examination of geological foundations of Japanese castles [73], or inclusion of more than 30 historical structures into a comprehensive geoheritage and geodiversity assessment in the Sudetic Foreland Geopark, Poland [74]. It is also observed that building stones, as well as pavement tiles, may contain fossils and hence, are suitable for palaeontological education for the general public [75–78]. Valentino et al. [79] described special mobile applications designed to help recognizing rock types used in architecture. The range of heritage stone themes was recently expanded by consideration of lighthouses [80] and cemeteries, where various rock types are used as tombstones and in sepulchral art. A notion of 'cemeterial geotourism' was recently proposed [81,82].

Geosites other than rock outcrops, landforms and buildings are seldom addressed. However, the significance of hydrological sites, mainly springs and ancient wells, has been realized in Rome [28], Perugia [83], and Lisbon [84].

4. Cultural Landscapes

The relevance of geoheritage to cultural landscapes and various relationships between the two have been recently comprehensively reviewed by Gordon [2]. The contribution of abiotic components is very diverse, ranging from crucial underpinning of the landscape [85,86], evident in its rock- and/or landform-controlled appearance, to incorporation of minor and not necessarily very significant elements into a landscape that is dominated by anthropic elements, with all possible intermediate situations. Consequently, the spatial extent and specific characteristics of cultural landscapes twinned with geoheritage vary too (Figure 3). Although a complete typology is difficult to propose, notable examples include: (a) rugged countryside adjusted to become an agricultural area through an introduction of terracing and stone walls, (b) romantic parks and gardens built around distinctive natural landforms, (c) rural settlements constructed with the pervasive use of distinctive stone types, (d) certain mineral extraction areas, where technology required modification of natural relief (e.g., salt extraction ponds—salinas), (e) adjustments of natural topography to better serve defensive purposes, (f) rugged rock landscape, as settings of monastic complexes and other sites of spiritual significance, and (g) cave dwellings incorporated into natural rock cliffs. Landscapes dominated by the legacy of mining may be also considered as cultural landscapes, even though alteration of the natural environment was usually quite substantial. This topic, closely linked with mining heritage issues, is presented more thoroughly in Section 5. Less evident and rarely explored in geoheritage context are land-use histories related to specific soils, giving rise to 'soilscapes' [87].

Cultural landscapes associated with agricultural production are perhaps best exemplified by rice fields and vineyards, the latter often located on steep hillsides of favourable soil and topoclimatic characteristics, significantly modified through terracing. As of 2021, 14 wine-producing regions in several European countries (Portugal, Spain, France, Switzerland, Italy, Germany) were listed as UNESCO World Heritage properties, recognized as cultural landscapes of outstanding universal value [88,89]. The analysis of documents demonstrated that the underpinning role of geology and landforms is increasingly recognized and recent nomination files for World Heritage give due attention to the abiotic factors [89]. An excellent example of very tight links between geology, landforms and land use is provided by the vineyards of Lavaux at Lake Geneva in Switzerland, inscribed in 2007 [90]. Here stepped hillsides are natural landforms and reflect repeating appearance of stronger conglomerates and less resistant sandstones and mudstones in the vertical geological profile. The former support mid-slope cliffs, whereas the latter underlie slope sections of lower inclination, developed as vineyard plots (Figure 3a). The terraced slopes of the famous Alto Douro wine district in Portugal is another recently presented case of close relationships between abiotic factors and wine culture [91]. Winescapes involving geoheritage value are not limited to World Heritage properties and famous wine regions but occur in other regions too [92–95].

Figure 3. Cultural landscapes illustrating various connections with abiotic nature. (**a**) terraced slopes of Lavaux Vineyards in Switzerland, with the glacial trough of Lake Geneva in the background, (**b**) stone walls built of local basalt and pieces of corals are part of endangered cultural heritage of Penghu Islands, Taiwan, (**c**) granite inselberg of Monsanto, Portugal, with the village built almost exclusively from local granite (left) and castle ruins on the hilltop, (**d**) salt pans near Trapani, Sicily, Italy (all photographs by the authors).

The geoheritage context of stone-walled countryside sceneries is represented by the use of local stone to build the walls (Figure 3b), where availability of stone of specific shapes and dimensions as well as their other properties dictated construction technologies applied in specific localities [96–98]. Consequent to the realization that stone walls are part of combined geo-cultural heritage are studies of their degradation after abandonment, which is considered as a loss of value [99,100].

Incorporation of natural geomorphic and hydrological features into park and garden layouts was part of the 19th century romantic concept of 'tamed wilderness', implemented in Central and Western Europe. Using a specific example from an intramontane basin in the Sudetes it was shown how denudational granite landforms such as domes, tors, boulder piles, open and roofed clefts, overhangs, and minor weathering features became integral parts of landscape parks designed around royal and aristocratic residences [101].

Stone type used for construction of houses, street paving, plot-bounding walls and small architecture was long recognized as a factor defining the identity of rural settlements. This connection between abiotic and cultural spheres is often emphasized in European Geoparks, exemplified by heritage stone villages such as Monsanto in Naturtejo Geopark in Portugal, almost perfectly blended with the granite outcrops on the slopes of this impressive inselberg (Figure 3c), slate villages in the region of Valdeorras, Galicia, in Spain [102], or heritage villages in the Courel Mountains UNESCO Global Geopark, also in Spain [103].

Adaptation of existing caves, rock shelters and overhangs to serve as dwelling places, storage rooms and shrines has a long tradition in various parts of the world, especially where rock is simultaneously soft enough to allow for excavations, but sufficiently strong not to suffer from immediate wall or roof collapse. Tuffs and certain variants of sandstones fulfil these requirements and host extensive underground spaces used by people either in the past or, less commonly, until now. The archetypal example is the rock landscape of central Cappadocia, with its underground cities, rock-hewn churches, houses and pigeonholes [104], whereas multiple uses of natural and excavated hollows in sandstones were presented in the pictorial atlas of sandstone landforms of Czechia [105]. In limestone, in turn, natural caves were ready-to-use places for habitation, as documented from many countries worldwide, particularly from the Mediterranean realm. Further examples of cave dwellings were recently provided from the Basilicata region and Amendolara in southern Italy [106,107], Matmata in Tunisia [108], Vardzia and Uplistsikhe in Georgia [109], and from the Mekong Delta in south-east Asia [11].

A small-scale example of adaptation of natural landforms to serve economic purposes is offered by rocky shore platforms, modified into patterns of interconnected shallow pans to allow for salt harvesting [110–112]. Locally, salt pans were also developed in inland sites (Figure 3d) [113].

Defensive cultural landscapes may take different shapes and exploit various natural features. In the coastal context, Schembri and Spiteri [114] examined the pattern of drowned valleys (rias) on the north-eastern coast of Malta and how it underpinned the long-term strategy to fortify the city of Valletta and adjacent settlements, then the headquarters of the St. John's Order. Another great example is provided by Hadrian's Wall in northern England, where the 2nd century AD fortifications from the times of the Roman Empire were erected on the top of a long dolerite ridge, which is a distinctive rock-controlled landform and regional landmark [115]. In a similar vein, the association of the walled city of Luxembourg with the natural scenery of sandstone cliffs and canyons was examined [116]. All these three localities are UNESCO World Heritage cultural properties. Even though the geological underpinnings themselves are not considered of outstanding universal value, they were crucial for the appearance and significance of these defensive facilities.

Human perception of some rugged rock landscapes as locations of spiritual significance, present in various cultures and religions across the world, resulted in visitations, erection of temples and hermitages, and eventually transformation of the entire physical landscape, which however retained the main scenery values. The most impressive examples have the status of UNESCO World Heritage properties, such as Meteora in Greece, where monasteries were built atop rock pillars built of conglomerates [10]. Many impressive rock landscapes in China, valued for their geology and geomorphology, are closely associated with Taoism and Buddhism, as reflected in the presence of temples, shrines, tombs, sacred wells etc., connected by an intricate network of paths and staircases, with accompanying minor monuments and statues [117]. The theme of sacred meaning of natural landforms, present in different cultural environments of the world, and how they are used by people nowadays, was comprehensively reviewed by Kiernan [118].

5. Mining and Quarrying Heritage

The use of natural mineral resources is another frequently explored theme at the interface of geoheritage and cultural heritage. Two interlinked aspects may be distinguished within these inquiries, related to on-site exploitation and testified by mines, quarries and other sites of extraction, as well as to the fate of rock blocks and other materials retrieved from sites of exploitation, including means of their transportation to destination sites. The latter bridges the gap between mining heritage and heritage stone, covered in Section 3.

Within this broad theme, on the one hand, it is the rock itself and its various properties that are of interest. An important point is that quarries and mines often offer excellent opportunities to examine rock complexes in three dimensions, over areas usually larger than natural outcrops [119–121]. Hence, geologists have a long-standing interest in monitoring quarry operations, to update three-dimensional views of rock masses and all inherent structures. It is also not surprising that abandoned quarries (and occasionally working ones as well) are listed as geosites within regional or national inventories [122,123]. On the other hand, more relevant to this paper, quarrying and mining are parts of cultural heritage, with tangible evidence not limited to the quarry faces/mine adits themselves, but extended over processing plants, mine buildings, transportation routes, waste heaps, collapse hollows above ancient exploitation chambers, etc. [124–126]. Quarry layouts and mine galleries reflect engineering and working skills of quarrymen and miners and were often carefully adjusted to local geological conditions to minimize physical efforts and expenditures and to maximize effectiveness [127,128]. Mining and mining-related heritage is also presented and analysed in the wider spatial context, as a key factor contributing to the identity of a region, linking to the theme of cultural landscapes addressed more comprehensively in Section 4. Selected examples include former areas of tin extraction in Cornwall [86,129], copper and lead mining in southern Spain [130], and roofing slate exploitation in eastern Czechia [131]. There are also publications, which focus on specific types of stone resources and address various associated issues such as geological context, techniques of extraction, past and present use, as well as potential geoheritage value [103,132–135].

The cultural dimension of mining and quarrying has received an increasing recognition from the World Heritage Committee of UNESCO and currently more than 20 former sites of natural resource exploitation are listed as World Heritage properties in countries such as Great Britain, Germany, Czechia, Sweden, Spain, Poland, and Chile. Among the most recent examples is the Erzgebirge/Krušnohoří Mining Region along the Czech/German border, inscribed in 2019 and consisting of 22 key localities on both sides of the border (Figure 4). Each of these, in turn, encompasses several more sites of interest. The area was a mining region since the 12th to the 20th century and was particularly thriving in the 15th and 16th centuries when most of silver available in Europe came from the Erzgebirge mines.

Figure 4. Mining heritage of Erzgebirge/Krušnohoří Mining Region—a transboundary UNESCO World Heritage property in Czechia/Germany. (**a**) evidence of exploitation of ore-bearing veins near the town of Horní Blatna, Czechia, (**b**) waste heaps and terrain depressions after collapsed shafts on Mt. Mědnik, Czechia, (**c**) remnants of a large tin mine in Geyer, Germany, (**d**) preserved traditional buildings of an old mining town, Horní Blatna, Czechia (all photographs by the authors).

With the contemporary increasing interest in geotourism, which includes mining heritage attractions, more and more quarries are being developed for tourists and offer educational opportunities. Reflecting this trend, various recent publications explore possibilities to use mining heritage as a foundation for geotourism development [136–148], although it was also pointed out that tourist visits to mining sites are hundreds of years old [149]. Various means of conveying information are employed at currently accessible sites, from individual information panels through discovery routes within individual quarries to longer trails linking adjacent quarries or mining heritage sites [150]. Conversion of a former underground mine into a tourist site usually requires a lot more effort and observation of the legal framework that controls the maintenance of underground mine spaces. Nonetheless, old mines of coal, gold, silver, ores, salt, flint, and chert, are increasingly popular visitation sites, and guided tours typically combine information about geological foundations, history of discovery, mining operations, and the use of a given resource in the past and at present [151]. Besides geotourism, Prosser [152] has recently shown the potential of quarrying to make connections with the local communities, emphasizing extraction of stone resources since antiquity as part of local history, the use of stone in buildings well-known to the locals, including dwelling houses in which they still live, conversion to local recreation grounds, and various educational initiatives implemented within former quarries.

Exploration of mining heritage extends to include industrial activities, which do not necessarily produce lasting visible evidence and do not create 'mining landscapes' but contribute to the cultural history of the region. This is the case of petroleum industry [146,153–155].

6. Cultural Heritage Dimension of Natural Geophysical Disasters

Natural processes, particularly catastrophic ones, often interfere with cultural heritage, damaging the latter or destroying it completely (Figure 5). Recent examples of substantial loss of cultural heritage include large-scale destruction of UNESCO World Heritage properties in Kathmandu, Nepal, due to the Gorkha earthquake in April 2015, an earlier damage of another UNESCO site in Bam, Iran, caused by an earthquake in 2003, and the consequences of the 2008 Wenchuan earthquake in Sichuan, China. However, ongoing exogenic processes may also pose a serious threat to sites of cultural significance, as shown by examples of heritage-listed religious sites of Hinduism affected by fluvial erosion and channel shifts of the Brahmaputra river in India [156], recurrent landslides, slow-moving slope deformations and gully erosion affecting historical settlements and structures in Italy [157,158], or historical buildings in the town of Calatayud in Spain, suffering from gypsum dissolution and land subsidence [159]. The ruined medieval church in the village of Trzęsacz in NW Poland, perched atop a coastal cliff (Figure 5a), is another example of a historical site that suffered from geomorphic processes, but one can also argue that its significance actually increased, being now also important for coastal geomorphology as a benchmark of long-term erosion [160]. In a similar way, the significance of the landslide-affected town of Bagnoregio in Italy has risen, especially after the opening of a dedicated landslide museum in 2012 [158]. Less dramatic in the long-term are consequences of river floods, commemorated by flood marks found on buildings located next to river channels (Figure 5b) [161–163].

Figure 5. Natural hazards and disasters in the cultural heritage context. (**a**) remnants of a medieval church in Trzęsacz, northern Poland, destroyed due to sea cliff retreat, (**b**) flood marks in the town of Eibelstadt on the River Main, Germany, (**c**) ruins of Herculaneum, Italy, excavated from beneath lahar and pyroclastic surge deposits from the AD79 eruption of Mt. Vesuvius, (**d**) ruins of a church in San Juan Parangaricutiro, Mexico, destroyed by a lava flow from the Parícutin volcano in 1944 (all photographs by the authors).

Of special interest are localities where everyday life in ancient times was suddenly interrupted by natural events such as volcanic eruptions, including lava flows or pyroclastic deposits overwhelming buildings, or earthquakes. These sites, if uncovered by archaeologists, become windows to the past of unprecedented importance, but they also show complex relationships between people and nature and can be used for geo-education [164,165]. The ruins of Pompeii and Herculaneum continue to attract interest of scientists exploring the history of their destruction by the 79AD eruption of Mt. Vesuvius (Figure 5c) [166,167]. Likewise, the site of Akrotiri on the island of Santorini has become famous due to its preservation after the catastrophic explosion of Thera volcano around 1600 BC [168]. Perhaps less known are the ruins of the pre-Columbian ceremonial site of Cuicuilco in Mexico City, partly buried by lava flows some 1670 ± 35 years BP, in this case probably long after abandonment by the local population [31]. There are also various examples of ancient structures destroyed by earthquakes, particularly in the Mediterranean world [169–171].

For more recent natural disasters, one can argue that the very fact of partial destruction turned a place, which would be fairly ordinary otherwise, into a site of mixed, natural-cultural importance. This is the case of the former village of San Juan Parangaricutiro in Mexico, overwhelmed in 1944 by a lava flow fed by the Parícutin volcano [172]. The sole vestige of the town is the ruined church, partly filled by lava, with the tower rising above the rugged lava surface (Figure 5d). The church itself was of minor architectural significance, but after the eruption the site has become a tourist attraction, whereas for the local population it is also a pilgrimage site and a sign of supernatural protection of the church. Similarly, towns and villages abandoned following strong earthquakes, recurrent landslides, and flood disasters, turning into 'ghost towns' if they were not totally demolished, may also become localities of combined geoheritage and cultural significance as testaments of the power of natural forces [173].

This field of inquiry is also closely related to geo-mythology and other intangible cultural heritage, as will be discussed in Section 10.

7. Geoheritage and the History of Science

A substantial part of geoheritage is linked with the history of Earth sciences, which is another part of cultural history. Of particular importance are localities where significant observations and discoveries were made, so that these sites have become benchmarks for subsequent studies. These discoveries are in turn connected with the names of people, who were later recognized, as geosciences developed, as the leading representatives of academia, influential researchers, and founders of 'schools of thought', which may have lasted for many decades. In these cases, relevant geosites do not only inform about specific rocks, fossils or structures, but also provide opportunities to recall the life and contribution of an eminent scientific figure. In addition, the further back in time we go, the more often authors of these observations were not highly specialized geologists, but rather people of broad interests and multiple skills, including poetry and art, and well-travelled.

These discoveries are usually recalled by erecting simple commemorating plaques or interpretation panels at the sites of concern. The former, however, add a cultural dimension to a geosite (e.g., localities visited by Johann Wolfgang Goethe in the sandstone tablelands of Central Europe [174]), but do not help to better understand the significance of a site. By contrast, the latter inform about the nature of the scientific achievement and sometimes about the persons themselves, as in the case of Hutton's Section in Edinburgh, Scotland, where James Hutton demonstrated the intrusive origin of a dolerite dyke. In fact, England and Scotland are areas where many important discoveries were made during the early stages of Earth science development, and these are brought back to attention through both more specialist publications aimed at professionals and efforts to explain the sites themselves to the general public [175,176]. The history, scientific content, and educational role of geological collections in museums is another recurrent theme in recent publications, illustrating another dimension of geoheritage and history of science linkages [177–180]. For example, Hose [177] summarized the main characteristics of leading geological museums in England, France, Italy, and Germany, whereas Vicedo et al. [181] emphasized the value of movable palaeontological heritage collected in the Museum of Natural History in Barcelona.

In the city of Prague, Czechia, an educational trail was set up in the valleys of Dalejské and Prokopské údolí, where numerous limestone quarries expose a marine succession of Ordovician, Silurian, and Devonian age, including some globally recognized stratotypes, most associated with the work of Joachim Barrande in the 19th century. At several stops, panels inform about the significance of these sites for the development of geology, whereas the quarries themselves are well exposed and easily available for viewing. Another recent study from Czechia recalls the contribution of J.W. Goethe to the Plutonism/Neptunism debate at the turn of the 19th century and his efforts to explain the origin of Komorná hůrka cinder cone [182]. The site itself preserves the original adit dug to test the hypothesis of volcanic versus marine origin of basalt, whereas interpretation panels put it into wider regional and scientific context. In the Alps, in turn, especially in Switzerland, numerous geosites are highly valued not only for their intrinsic value, but also for their importance for the advancements in the theory of mountain glaciation and mountain building [183]. The ruins of Serapeo in Pozzuoli, Italy, with perforations imposed by burrowing marine molluscs, played an important role in the understanding of cyclic sea-level changes and were immortalized on the cover of Lyell's 'Principles of Geology', first printed in 1830 [184]. However, not all these important localities have on-site interpretation facilities, and the awareness of their significance is then limited to a rather narrow circle of professionals.

It is worth noting that advances in geosciences and the means to commemorate them are not limited to the Western world. Figure 6 shows a panel from Yandangshan UNESCO Global Geopark in southern China that recalls how a Chinese scholar Shen Kuo, living in the 11th century, explained the role of running water in eroding the rocks and the land.

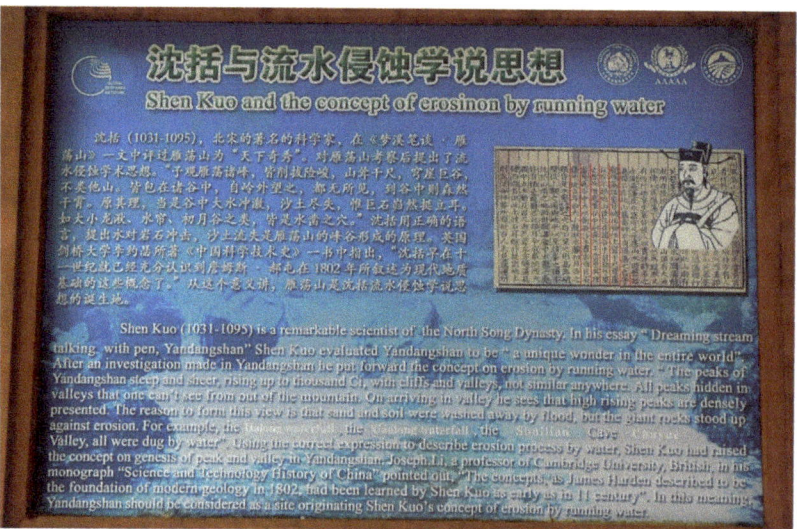

Figure 6. Interpretation panel in Yandangshan UNESCO Global Geopark, China, referring to the history of scientific inquiry in the field of geosciences (photograph by the authors).

8. Geoheritage and Early Tourism

Tourism can be defined and analysed from various perspectives, but no one doubts that it is an inseparable part of culture. Whereas the extremely broad subject of contemporary geotourism, addressed in an ever-growing number of publications from all around the world, is beyond the scope of this review as it does not yet qualify as part of "cultural heritage", early tourism with its tangible evidence and intangible legacy is certainly of heritage value.

In Europe, linkages between geoheritage and early tourism are best expressed in the context of "Grand Tours" of the 17th to 19th centuries. These travels are understood as long journeys undertaken by representatives of contemporaneous aristocratic elites, especially from western and northern Europe, with the Italian peninsula as the ultimate destination. The purpose was to become acquainted with the rich cultural heritage of Italy, traced back to Greco-Roman times, fulfilled through visits to famous historical cities (Venice, Florence, Rome) and sites of archaeological importance (e.g., Pompeii). However, many Grand Tour practitioners travelling across the Alps and the Italian peninsula did not ignore impressive natural features such as the glaciers and high-mountain scenery of the Alps [171]. In Italy itself, active volcanoes such as Mt. Vesuvius, Mt. Etna and the Aeolian Islands aroused curiosity, and so did related geothermal phenomena such as those at Campi Flegrei near Naples, and the interest can be traced back into antiquity [19]. Sometimes it is not straightforward to draw a clear dividing line between early tourism and the development of science (see Section 7). A recent project aimed to refresh the memory about the long journey of J.W. Goethe across Italy in the late 18th century and to highlight the aptness of his observations about geology and landscapes [185]. Thus, the travelling Goethe was as much a person undertaking the Grand Tour and a geotourist, as he was a naturalist, whose efforts to understand the environment are documented in his travel memoirs.

In a different geographical context, Migoń [174] reviewed the history of tourism in the Sudetes range in Central Europe, noting an interest of 18th century travellers in sandstone rock cities and the evidence of amazement in these unique rock formations recorded in travel diaries. Likewise, the nearby Karkonosze Mountains and their surroundings attracted tourists offering vistas of unusual geomorphic formations such as waterfalls, glacial cirques (not interpreted as such in those times), non-karstic fissure caves, and granite tors [101,174]. In the 19th century, the Sudetes, like many other mountains and uplands in the contemporaneous German lands, were renowned for their sites of geological and geomorphological interest. In Britain, the Romantic period between the late 18th and mid-19th centuries saw a surge of interest in natural landscapes too, as extensively documented for the wild Lake District in northern England [186] and other parts of England [187]. Gordon and Baker [188], in turn, drew attention to the "tourism of awe" that typified the 18th and 19th century visits to the Scottish Highlands, emphasizing ubiquitous appreciation of wild Highland scenery, with waterfalls, glacial troughs, and outcrops of columnar jointing in basalts. A related theme is early geotourism provisions, such as purposefully designed guidebooks [189] and outdoor geological models [174].

9. Geoheritage, Art and Literature

Linkages between geoheritage and art are not among the most popular topics and were explored in a rather limited number of publications, even though opportunities of thematic studies are multiple [190,191], but this may change, following the recent plea of Motte and McInnes [192] that "it is stimulating to continue the adventure of bridging sciences and art". Nevertheless, various studies have been recently published, addressing subjects such as natural landscape features as a source of inspiration for artists, or the use of old paintings to inform attempts to decipher landform change in the past few centuries. In this context, art presentations of sites that were subsequently destroyed, either by natural forces or due to human activities, are of particular importance. A relevant example is the site of the White and Pink Terraces in the North Island of New Zealand, once acclaimed a 'wonder of the world', but irreversibly annihilated through the explosion of the Tarawera volcano in 1886 [193]. In the Alps, the pre-20th century artistic imagery is now a most useful source of information about past extents of glaciers, used in conjunction with other lines of evidence [194]. In a similar vein, analysis of old paintings, postcards, and engravings of the coastal scenery in Brittany, France, helped understanding the patterns of coastal change and may inform current coastal management [192]. Pullin [195] offered an insightful analysis of the life and work of Eugene von Guérard, a 19th century landscape painter of German origin, who began his career in Germany, but spent many years in Australia and New Zealand, portraying various volcanic features with a very high degree of accuracy. The common use of sites of geological or geomorphological significance as motifs for landscape paintings was also emphasized for the Scottish Highlands [188], the Isle of Wight in southern England [196], the chalk cliffs of the island of Rügen in Germany [191], and badland landscapes of Italy [197]. Landscape paintings from the period of the Renaissance (15th–16th century) were analysed by Nesci and Borchia [198], who aimed to see if the landscape backdrops were imaginary or real sceneries, confirming the latter. Further examples and relevant references were provided in a brief, but informative overview of the subject by Gordon [190].

Art other than painting is rarely presented in the context of geoheritage. One novel aspect is exploration of the content of municipal heraldry in Portugal, with the emphasis on geological and geomorphological elements demonstrating close links with the landscape and the sense of place underpinned by physical features [199]. The appearance of landscapes of geoheritage value in movies and related opportunities for geotourism has also been recently addressed [200].

Natural landscapes and individual curious landforms have often been inspirational to poets and writers, especially in the Romantic period of the 19th century, so that one can find various direct and indirect references to diverse geological and geomorphological phenomena [190,191]. Examples of this kind demonstrate the cultural value of geoheritage and add value to localities otherwise important for scientific reasons.

10. Geoheritage and Intangible Cultural Heritage

The last two decades have seen an increasing awareness of the contribution that intangible cultural heritage offers to better appreciate and understand geoheritage. As for the other themes, this one too appears in various forms such as preservation of memory of natural geophysical disasters and their consequences, explanation of natural processes by indigenous communities unfamiliar with the scientific foundation of geoheritage, traditions related to various aspects of natural heritage, and reconstructed geo-mythology.

Geo-mythology issues were reviewed and illustrated by diverse examples from around the world in an edited work [201], which highlighted its importance for geological sciences. Various authors demonstrated that mythical events embedded in the collective memory of indigenous communities and related place names may in fact inform about catastrophic geophysical events such as earthquakes [202–204], tsunamis [205,206], or dramatic environmental changes due to climate change and ice expansion [207]. The most recent additions to the subject are the scholarly monograph by Burbery [208] and a review of folk tales related to submerged lands due to sudden or gradual geophysical change by Nunn [209]. However, possible linkages between the content of tales and real geological events have to be explored with extreme care due to possibilities of various recent 'contaminations', as demonstrated by referencing to Aboriginal stories about 'falls from the sky', which could be interpreted as eye-witnessed meteorite impacts but were in fact inconsistent with the timing of the origin of craters [210]. The iconic Mount Olympus in Greece was also presented in the context of ancient mythology versus modern geology [211], whereas Khoshraftar and Farsani [212] argued that geo-mythological associations at cultural heritage sites may increase interest among visitors, which in turn can be used to offer more comprehensive interpretation of geosites.

In other cases, folk tales and associated traditions and rituals are not related to real events from the geological past but simply reflect beliefs of the local population passed on from the era prior to scientific discoveries in the field of geosciences. Exploration of such traditions in the geoheritage context seems to be an emerging theme, as demonstrated by recent contributions from the Bohemian Massif in Czechia [213], the Ardennes in Belgium [214] and Lithuania [17]. Geomorphological features subject to folk explanation include distinctively shaped rock outcrops, erratic boulders, dead-ice kettles, meander loops, and caves.

Apart from oral traditions passed on from one generation to another, geoheritage connections may be also found in traditional tattooing, specific phrases used in local languages, and cultural activities [215].

11. Geoheritage—Added Value at Cultural Heritage Sites

In Sections 2 and 4, we presented geoheritage sites, which show an additional value associated with cultural heritage. However, the relationship can be also in reverse, in that cultural values are considered as superior, but this should not lead to the neglect of geodiversity and geoheritage aspects at these sites. This issue can be demonstrated at many UNESCO World Heritage sites, which were inscribed solely in the recognition of their cultural value but may contain interesting landforms and rock outcrops [216]. Selected recent examples of this kind, other than those referred to above, include:

- The artificial caves of Elephanta off the Bombay coast in India, with numerous paintings and sculptures, excavated in multi-layered basaltic lava flows. A recent publication [217] drew attention to the paucity of geological information about the site and aimed to redress the balance.
- The archaeological area of Aksum in Ethiopia, known for the famous stelae, subterranean necropolises and ruined buildings. These historical monuments are integrated with the erosional scenery of exposed syenite plugs, with impressive rock cliffs, talus-covered slopes and boulder blankets (Figure 7a) [218].
- The rock island of Mont-Saint-Michel in Normandy, France [219], made famous by the Benedictine Abbey erected on the top, but being also a first-class example of rock-controlled relief amidst megatidal flats.
- The Cultural Landscape of Sintra in Portugal, where various elements of the 19th century Romantic architecture, including gardens and parks, are incorporated into granite scenery with impressive crags and boulder fields (Figure 7b), giving way to sea cliffs and shore platforms [220].
- The archaeological site of Petra in Jordan, with numerous elaborate structures dated to the turn of BC/AD times, half-built and half-carved into striking red sandstone outcrops. Geomorphology provides a magnificent setting to the ancient city, with high cliffs, rock platforms, extremely narrow gorges, and a variety of selective weathering features [221].
- The Thingvellir area in Iceland, inscribed in recognition of its cultural significance for the history of parliamentary culture, but at the same time being "perhaps the best place on this planet to understand the process of rupturing of the crust in response to the pulling forces of plate movements" [222]. The site contains superb examples of open fractures, trenches and grabens, waterfalls and lava outcrops (Figure 7c).

It is also appropriate to mention rock art sites, which often sit amidst spectacular scenery, but the focus of research tends to be on the content and interpretation of paintings and engravings. Again, the UNESCO World Heritage list provides examples (e.g., Tadrart Acacus, Libya; Serra da Capivara, Brazil—Figure 7d) and there are many less known sites on all continents. Nonetheless, a few recent publications analysed the wider geological and geomorphological context of such localities [223,224] and consideration of environmental history may significantly help in interpretation of rock art and explanation of its current state [225]. Similarly, caves renowned for rock art or other archaeological discoveries are examples of mixed cultural-natural heritage and hence, geomorphosites, even though their core values are associated with the importance for the history of mankind [7,226,227].

Figure 7. Geoheritage context of selected UNESCO World Heritage cultural properties. (**a**) the archaeological area of Aksum, Ethiopia, with flat-topped syenite plugs, pediments and intriguing boulder fields, (**b**) granite residual hills and boulders in the cultural landscape of Sintra, Portugal, (**c**) graben in the lava plateau at Thingvellir, Iceland, testifies to recent crustal extension at the plate boundary, (**d**) magnificent sandstone and conglomerate cliffs provide the geomorphic context for the rock art site of Serra da Capivara, Brazil (all photographs by the authors).

12. Systematic Approach versus Multiple Linkages

The review above shows that the interface of geoheritage and cultural heritage is very broad, and many specific themes are explored, even though some appear more often addressed than others. In addition, in tracing recent literature one can easily observe that the range of such themes increases, and this trend is likely to continue in response to the expansion of the UNESCO Global Geopark Network, which emphasizes geoheritage—cultural heritage linkages, new international initiatives such as World Geodiversity Day, and a growing awareness among scientists that inter- and transdisciplinary studies have considerable merit. Therefore, it is tempting to offer a provisional systematic classification that would facilitate navigation across this increasingly complex field of inquiry. This, however, seems a challenging task as approaches to classification can be different, being not mutually exclusive (Table 1).

Perhaps the most obvious classification for geoscientists would be one that makes the type of geoheritage central, paying less attention to the kind of relationships with cultural heritage (Table 1, column A). This stance reflects the internal division of Earth sciences and defines the possible contribution that petrologists, palaeontologists, structural geologists and volcanologists, geomorphologists or hydrologists can have. However, some sub-disciplines may fit into several categories, especially volcanology, whose linkages with cultural heritage are often explored. Cultural dimension may be shown by specific types of volcanic rocks used for various purposes, volcanoes as singular landforms, volcanic landscapes, geothermal phenomena, as well as information about past eruptions. In this approach, the classification system may be easily expanded by adding sub-categories, for example defining types of landforms more specifically (fluvial, mass movement-related, volcanic, karst).

Table 1. Various approaches (A–G) with their underlying factors important to classification of themes at the geoheritage—cultural heritage interface.

A	B	C	D	E	F	G
Type of geoheritage	Type of human activity	Spatial scale	Temporal scale	Nature of evidence	Core values	Principal context
Rocks/stones/minerals Fossils Geological structures Landforms Landscapes Springs and other hydrological phenomena Soils	Building construction Farming Mining, quarrying and industrial architecture Tourism and travel Science and education Art Craft	Cultural landscape Urban (town, city) Rural (village) Individual site/object Subterranean sites	Prehistory Antiquity Medieval Modern era	Tangible • in situ • ex situ Intangible • language • traditions, customs • myths and beliefs • history	Geoheritage as core value Geoheritage as additional value Equal standing	Geo-conservation Geotourism Geo-education Raising awareness

An opposite way involves the type of human activity as the basis of a classification system (Table 1, column B), as in each case the nature of relationships with geoheritage is different and includes exploitation and re-use (e.g., mineral resources and building stones), adaptation to natural conditions (town planning, farming in difficult terrains), adjustment to societal needs (adaptation of caves for visitors), or recording and interpretation (education, art).

The next two approaches emphasize the spatial and temporal scale, respectively. Considering space (Table 1, column C), one can categorize inquiries according to the size of an area at which the relationships are examined. The broadest scale is represented by cultural landscapes, where protracted human use of natural resources resulted in the origin of a distinctive scenery, whose appearance, both at large and in detail, is underpinned by geology and geomorphology. Terraced wine-growing landscapes and mining districts are relevant examples. Urban geoheritage follows, usually encompassing large areas, with several specific themes that are addressed. The main ones are adjustment of urban design to natural landforms and the use of several characteristic building stones across a city. However, large cities may be also considered as cultural landscapes, as exemplified by the UNESCO World Heritage-listed Rio de Janeiro. Rural heritage comes next, although again these small inhabited spatial units may be part of much wider cultural landscapes. Defining the boundaries between these categories in numerical terms seems thus neither possible nor helpful. The decreasing spatial scale is completed with individual objects (if the focus is on anthropic constructs such as buildings), or sites (landforms, rock outcrops). A category of its own is made by subterranean sites, not visible from the ground and not fitting any of the above categories. In fact, the size of subterranean sites may vary broadly, from small caves and single adits to extensive underground systems of quarries, mines and dwelling spaces. Classification in respect to timeline is another possibility, allowing for easier identification of potential partners in scientific endeavour (archaeologists, historians, art historians), but is unlikely to be universal. The proposal shown in Table 1 (column D) is obviously Mediterranean/Europocentric and reflects major phases of civilization development and historical breakthroughs experienced in this part of the world, but it will not be applicable to other continents. For instance, it will have little relevance to East Asia, nor to Australia and Oceania.

The mutual relationships between geo- and cultural heritage may also be examined from the perspective of tangibility/intangibility of evidence (Table 1, column E). The former is more explored and can be further divided into inquiries focused on in situ objects of interest (landforms, sites of exploitation of natural resources, hydrological phenomena etc.) and those, where objects of value are preserved ex situ (e.g., mineral and fossil collections). The latter, increasingly appreciated in the last few decades, may also warrant further subdivision depending on their type (language, mythology, customs etc.).

The review of publications has also shown that the position of geoheritage values in respect to cultural heritage varies (Table 1, column F). In some studies, geoheritage and geodiversity are central, whereas cultural associations are considered as added values that enhance the significance of a site. This approach is particularly common in valorization of geosites in the context of tourist use, underlined by an assumption that the diversity of values makes the site more attractive and allows for implementation of various innovative tools and concepts in geo-education. In other cases, however, a given site is primarily known for its cultural heritage values and the efforts of geoscientists are directed to show its less known face and perhaps to achieve some sort of balance in perception of significance, in response to the general limited awareness of geoheritage and geodiversity. Equal standing is another option, as demonstrated by UNESCO World Heritage mixed properties, where neither cultural nor natural values are superior in respect to one another. Finally, the main purpose and context of studies is emphasized, acknowledging that they often have a clearly defined end-user (Table 1, column G). Thus, some have a distinct geoconservation slant, aiming to deliver practical solutions how to save geological, geomorphological, mining or building stone heritage that may face various threats arising from ongoing development and deteriorating state of environment. Many publications explore the linkages in the context of geotourism and its possible enhancement, particularly with reference to geoparks and towns/cities, whereas others focus on opportunities to use these cross-disciplinary relationships in formal and informal education, both in- and outdoor. There are also studies, which are not addressed to the specific users at the time they are presented, but simply contribute to the enlargement of our knowledge and understanding of the environment and history.

One may thus ask if classifications in this particular context are useful at all. One observation clearly emerging from the literature review is that various themes and fields of inquiry are overlapping and interlinked. For example, the frequently addressed issues of mining and quarrying heritage, distinguished as a separate theme in this paper (Section 5), are closely connected with cultural landscapes and urban geoheritage due to use of quarried stone, overlap with the history of science, and may have links with early tourism (visitations of famous ancient quarries, such as in Carrara, Italy) and art, being recorded on paintings. Mining culture may also influence intangible local and regional heritage. Likewise, rock art is relevant to the geomorphological heritage, as rock art sites are usually associated with distinctive landforms, and history of geoscience, giving insights into past environmental changes. Geomythology and other expressions of intangible cultural heritage are linked with past geophysical disasters and may help to better understand the latter. For all these reasons, we also present an alternative way to show the wide thematic spectrum within the broadly defined geoheritage—cultural heritage interface, using a mind map (Figure 8). It shows the main and the less explored themes identified in this paper, also indicating some specific areas of interest, as well as including the most evident linkages between them. However, it should not be considered as an exhaustive, definitive proposal and more themes and connections can be added, reflecting the variety of subjects, approaches, and cultural contexts.

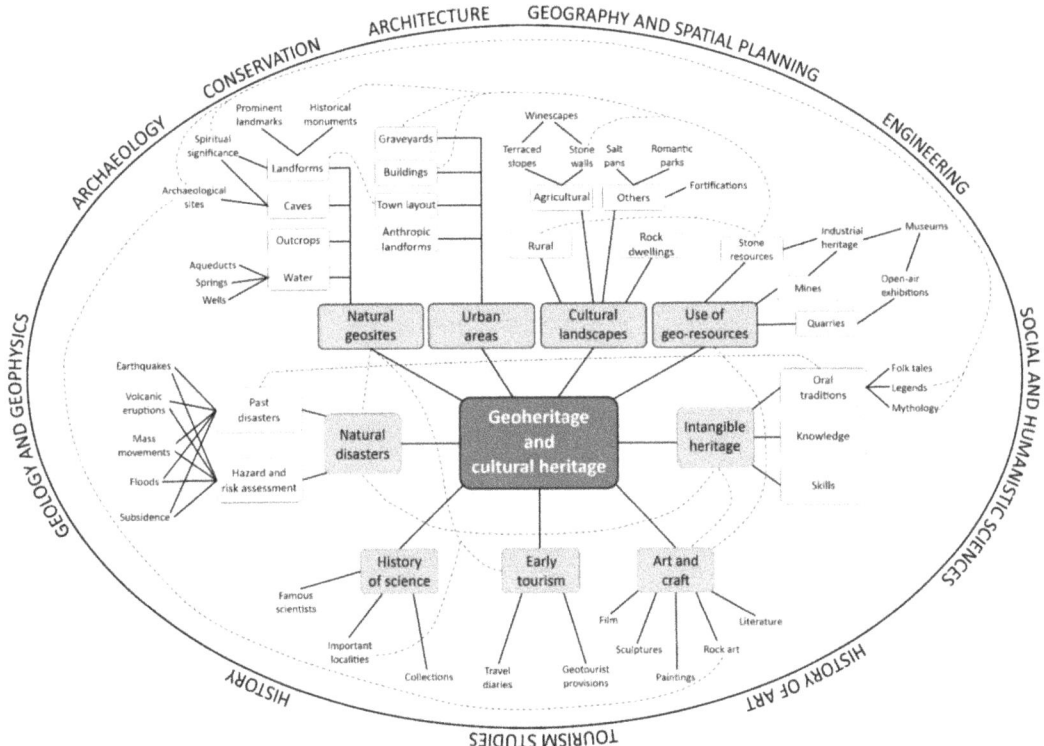

Figure 8. Themes explored at the geoheritage—cultural heritage interface, divided into main (framed by thicker lines) and secondary ones, and the most evident connections between specific subjects (dashed lines). The outer ellipse shows sciences whose fields of interest overlap with geoheritage studies.

13. Conclusions

Two main conclusions emerging from this review are: (1) the cultural relevance of geoheritage is widely recognized and has become a very popular subject of studies, which explore various aspects of this interface, resulting in a broad range of themes and specific topics; (2) among these different themes some generate much more attention than others, resulting in general thematic imbalance. Among the most frequently addressed subjects are geoheritage in urban space, with particular focus on heritage stones and their use in historical building construction as well as on recognition of geosites and their accessibility, and mining/quarrying heritage, along with associated industrial architecture. The reasons are likely complex but may include both the urgent need for such studies, driven by conservation priorities, as well as the response to plans to increase or diversify tourist visitation. In each case, preferential funding may be available. Urban space is also an obvious place to explore for scientists based in respective cities. Apparently much less explored, at least by people with a geoscience background who publish in geoscience journals, are themes more linked with history such as history of scientific discoveries, of early tourism, and the whole sphere of art. The latter may seem a subject without significant implications for geosciences themselves, but this is not true. Old landscape paintings were shown to be valuable sources of information about landform change during historical times and may inform environmental management strategies. The literature survey also revealed a very uneven regional coverage. The vast majority of studies focused on relationships between geoheritage and cultural heritage came from Europe and the subject seems to grow in popularity in South America (specifically Brazil), whereas the heritage of Asian and African countries, as well as of Australia and Oceania, appears much less explored in this context, at least within leading international journals focused on geoheritage and geoconservation.

Recent years have seen an intensification of outreach activities among geoscience practitioners, arising from a growing realization that geoheritage and geodiversity do not enjoy sufficient recognition in society and their importance is often ignored [228,229]. Within this unfortunate situation, there is poor understanding of the crucial underpinning of many cultural values, both tangible and intangible, by geoheritage and geodiversity elements. Notably, whereas procedures of geosite assessment for education and geotourism purposes often include cultural elements as an added value, elevating the significance of a site and making it more appealing to the general public [230–233], cultural heritage is often assessed without due attention paid to its landscape context. One type of evidence supporting this statement is the inscription of many UNESCO World Heritage cultural properties, which undoubtedly owe their unique characteristics to the natural setting (e.g., rock art sites, sacred mountains, mining heritage), but the setting itself is rarely highlighted. Consequences of this bias are a lack of awareness of geoheritage values among visitors, even at sites with spectacular geoheritage, as recently demonstrated via a visitors' survey at Meteora in Greece [234], or in the sandstone rock cities in the Bohemian Paradise UNESCO Global Geopark [235]. Therefore, a key research priority is action to foster understanding of geoheritage as a foundation value for cultural heritage, which can be achieved by publications exploring geoheritage and geodiversity of sites and areas recognized mainly for their cultural values. Parallel to that could be practical involvement in communication and promotion of geoheritage at cultural heritage sites to achieve more integrated interpretation offered to visitors. On the other hand, links with cultural heritage are worth underlining at primarily geoheritage sites too, as they may relate more immediately to popular experience and interests. Likewise, studies of under-researched themes, such as geoheritage in art and geoheritage, and geodiversity-related intangible cultural heritage also emerge as priorities.

Author Contributions: Both authors have equally contributed to the manuscript. All authors have read and agreed to the published version of the manuscript.

Funding: This research did not receive any specific funding.

Acknowledgments: We are grateful to three reviewers for undertaking the review, particularly to Reviewer no. 1 for suggestions how to strengthen the paper, additional bibliographic hints, and all linguistic corrections.

Conflicts of Interest: The authors declare no conflict of interest.

References

1. Brockx, M.; Semeniuk, V. Geoheritage and geoconservation—History, definition, scope and scale. *J. R. Soc. West. Aust.* **2007**, *90*, 53–87.
2. Gordon, J.E. Geoheritage, geotourism and the cultural landscape: Enhancing the visitor experience and promoting geoconservation. *Geosciences* **2018**, *8*, 136. [CrossRef]
3. Reynard, E.; Giusti, C. The landscape and the cultural value of geoheritage. In *Geoheritage. Assessment, Protection, and Management*; Reynard, E., Brilha, J., Eds.; Elsevier: Amsterdam, The Netherlands, 2018; pp. 147–166.
4. Olson, K.; Dowling, R. Geotourism and cultural heritage. *Geoconserv. Res.* **2018**, *1*, 37–41. [CrossRef]
5. Pásková, M.; Zelenka, J.; Ogasawara, T.; Zavala, B.; Astete, I. The ABC concept—Value added to the Earth heritage interpretation? *Geoheritage* **2021**, *13*, 38. [CrossRef]
6. Reynard, E. Geomorphosites: Definitions and characteristics. In *Geomorphosites*; Reynard, E., Coratza, P., Regolini-Bissig, G., Eds.; Pfeil Verlag: Munich, Germany, 2009; pp. 9–20.
7. Hobléa, F. Karstic geomorphosites: Managing subterranean natural-cultural heritage sites. In *Geomorphosites*; Reynard, E., Coratza, P., Regolini-Bissig, G., Eds.; Pfeil Verlag: Munich, Germany, 2009; pp. 189–200.
8. Altunel, E.; D'Andria, F. Pamukkale travertines: A natural and cultural monument in the World Heritage List. In *Landscapes and Landforms of Turkey*; Kuzucuoğlu, C., Çiner, A., Kazancı, N., Eds.; Springer: Cham, Switzerland, 2019; pp. 219–229.
9. Lazaridis, G. Caves in the conglomerates of the Meteora geosite (Greece). *Cave Karst Sci.* **2020**, *47*, 6–10.
10. Rassios, A.E.; Ghikas, D.; Dilek, Y.; Vamvaka, A.; Batsi, A.; Koutsovitis, P. Meteora: A billion years of geological history in Greece to create a World Heritage Site. *Geoheritage* **2020**, *12*, 1–16. [CrossRef]

11. Kiernan, K. Human impacts on geodiversity and associated natural values of bedrock hills in the Mekong Delta. *Geoheritage* **2010**, *2*, 101–122. [CrossRef]
12. Kiernan, K. Impacts of war on geodiversity and geoheritage: Case studies of karst caves from northern Laos. *Geoheritage* **2012**, *4*, 225–247. [CrossRef]
13. Sardella, R.; Iurino, D.A.; Mecozzi, B.; Sigari, D.; Bona, F.; Bellucci, L.; Coltorti, M.; Conti, J.; Lembo, G.; Muttillo, B.; et al. Grotta Romanelli (Lecce, Southern Italy) between past and future: New studies and perspectives for an archaeo-geosite symbol of the Palaeolithic in Europe. *Geoheritage* **2019**, *11*, 1413–1432. [CrossRef]
14. Tronkov, D.; Sinnyovsky, D. Belogradchik Rocks, Bulgaria: Geological setting, genesis and geoconservation value. *Geoheritage* **2012**, *4*, 153–164. [CrossRef]
15. Carreras, J.; Druguet, E.; Siddoway, C.S. Geological heritage beyond natural spaces: The Red Rocks Amphitheatre (Morrison, CO, USA), an example of syncretism between urban development and geoconservation. *Geoheritage* **2012**, *4*, 205–212. [CrossRef]
16. Górska-Zabielska, M. The most valuable erratic boulders in the Wielkopolska region of western Poland and their potential to promote geotourism. *Geo J. Tour. Geosites* **2020**, *13*, 694–714. [CrossRef]
17. Pukelytė, V.; Baltrūnas, V.; Karmaza, B. Geoheritage as a source and carrier of culture, Lithuania. *Geoheritage* **2022**, *14*, 8. [CrossRef]
18. Migoń, P.; Michniewicz, A.; Różycka, M. Granite tors of the Waldviertel region in Lower Austria. In *Landscapes and Landforms of Austria*; Embleton-Hamann, C., Ed.; Springer: Cham, Switzerland, 2022.
19. Costa, A.; Di Vito, M.A.; Ricciardi, G.P.; Smith, V.C.; Talamo, P. The long and intertwined record of humans and the Campi Flegrei volcano (Italy). *Bull. Volcanol.* **2022**, *84*, 5. [CrossRef]
20. Megerle, H. Cultural values of geomorphosites within the Geopark Swabian Alb, Germany. *Collect. EDYTEM* **2013**, *15*, 149–154. [CrossRef]
21. Megerle, H.E. Calcerous tufa as invaluable geotopes endangered by (over-)tourism: A case study in the UNESCO Global Geopark Swabian Alb, Germany. *Geosciences* **2021**, *11*, 198. [CrossRef]
22. Thornbush, M.J.; Allen, C.D. (Eds.) *Urban Geomorphology. Landforms and Processes in Cities*; Elsevier: Amsterdam, The Netherlands, 2018.
23. Reynard, E.; Pica, A.; Coratza, P. Urban geomorphological heritage. An Overview. *Quaest. Geogr.* **2017**, *36*, 7–20. [CrossRef]
24. Vaz, T.; Zêzere, J.L. The urban geomorphological landscape of Lisbon. In *Landscapes and Landforms of Portugal*; Vieira, G., Zêzere, J.L., Mora, C., Eds.; Springer: Cham, Switzerland, 2020; pp. 295–303.
25. Pelfini, M.; Brandolini, F.; D'Archi, S.; Pellegrini, L.; Bollati, I. *Papia civitas gloriosa*: Urban geomorphology for a thematic itinerary on geocultural heritage in Pavia (Central Po Plain, N Italy). *J. Maps* **2021**, *17*, 42–50. [CrossRef]
26. Fernandes, N.F.; Tupinambá, M.; Mello, C.L.; Peixoto, M.N. Rio de Janeiro: A metropolis between granite-gneiss massifs. In *Geomorphological Landscapes of the World*; Migoń, P., Ed.; Springer: Dordrecht, The Netherlands, 2010; pp. 89–100.
27. Silva, T.M.; Ferrari, A.L.; Tupinambá, M.; Fernandes, N. The Guanabara Bay, a giant body of water surrounded by mountains in the Rio de Janeiro metropolitan area. In *Landscapes and Landforms of Brazil*; Vieira, B.C., Salgado, A.A.R., Santos, L.J.C., Eds.; Springer: Dordrecht, The Netherlands, 2015; pp. 389–399.
28. Pica, A.; Vergari, F.; Fredi, P.; Del Monte, M. The Aeterna Urbs geomorphological heritage (Rome, Italy). *Geoheritage* **2015**, *8*, 31–42. [CrossRef]
29. Faccini, F.; Giardino, M.; Paliaga, G.; Perotti, L.; Brandolini, P. Urban geomorphology of Genoa old city (Italy). *J. Maps* **2021**, *17*, 51–64. [CrossRef]
30. Pica, A.; Reynard, E.; Grangier, L.; Kaiser, C.; Ghiraldi, L.; Perotti, L.; Del Monte, M. GeoGuides, urban geotourism offer powered by mobile application technology. *Geoheritage* **2018**, *10*, 311–326. [CrossRef]
31. Palacio-Prieto, J.L. Geoheritage within cities: Urban geosites in Mexico City. *Geoheritage* **2015**, *7*, 365–373. [CrossRef]
32. Melelli, L. "Perugia Upside-Down": A multimedia exhibition in Umbria (Central Italy) for improving geoheritage and geotourism in urban areas. *Resources* **2019**, *8*, 148. [CrossRef]
33. Molewski, P. Anthropogenic degradation of dunes within a city: A disappearing feature of the cultural landscape of Toruń (Poland). *J. Maps* **2021**, *17*, 162–169. [CrossRef]
34. Kubalíková, L.; Kirchner, K.; Kuda, F.; Bajer, A. Assessment of urban geotourism resources: An example of two geocultural sites in Brno, Czech Republic. *Geoheritage* **2020**, *12*, 7. [CrossRef]
35. Németh, K.; Gravis, I.; Németh, B. Dilemma of geoconservation of monogenetic volcanic sites under fast urbanization and infrastructure developments with special relevance to the Auckland Volcanic Field, New Zealand. *Sustainability* **2021**, *13*, 6549. [CrossRef]
36. Keller, B. Lake Lucerne and its spectacular landscape. In *Landscapes and Landforms of Switzerland*; Reynard, E., Ed.; Springer: Cham, Switzerland, 2021; pp. 305–323.
37. Del Lama, E.A.; de la Corte Bacci, D.; Martins, L.; Motta Garcia, M.G.; Dehira, L.K. Urban geotourism and the old centre of São Paulo City, Brazil. *Geoheritage* **2015**, *7*, 147–164. [CrossRef]
38. Da Silva, C.M. Urban geodiversity and decorative arts: The curious case of the "rudist tiles" of Lisbon (Portugal). *Geoheritage* **2019**, *11*, 151–163. [CrossRef]
39. Pacheco, M.; Cachão, M. Urban geology of Lisbon: The importance of the National Palace of Ajuda (Lisbon, Portugal). *Geoheritage* **2021**, *13*, 84. [CrossRef]

40. Zwoliński, Z.; Hildebrandt-Radke, I.; Mazurek, M.; Makohonienko, M. Existing and proposed urban geosites values resulting from geodiversity of Poznań City. *Quaest. Geogr.* **2017**, *36*, 125–149. [CrossRef]
41. Vegas, J.; Díez-Herrero, A. An assessment method for urban geoheritage as a model for environmental awareness and geotourism (Segovia, Spain). *Geoheritage* **2021**, *13*, 27. [CrossRef]
42. Erikstad, L.; Nakrem, H.A.; Markussen, J.A. Protected geosites in an urban area of Norway, inventories, values, and management. *Geoheritage* **2018**, *10*, 219–229. [CrossRef]
43. Moradipour, F.; Moghimi, E.; Beglou, M.J.; Yamani, M. Assessment of urban geomorphological heritage for urban geotourism development in Khorramabad City, Iran. *Geoheritage* **2020**, *12*, 40. [CrossRef]
44. Kubalíková, L. Cultural ecosystem services of geodiversity: A case study from Stránská skála (Brno, Czech Republic). *Land* **2020**, *9*, 105. [CrossRef]
45. Tičar, J.; Komac, B.; Zorn, M.; Ferk, M.; Hrvatin, M.; Ciglič, R. From urban geodiversity to geoheritage: The case of Ljubljana (Slovenia). *Quaest. Geogr.* **2017**, *36*, 37–50. [CrossRef]
46. Fio Firi, K.; Maričić, A. Usage of the natural stones in the City of Zagreb (Croatia) and its geotouristical aspect. *Geoheritage* **2020**, *12*, 62. [CrossRef]
47. Górska-Zabielska, M.; Zabielski, R. Potential values of urban geotourism development in a small polish town (Pruszków, Central Mazovia, Poland). *Quaest. Geogr.* **2017**, *36*, 75–86. [CrossRef]
48. Kubalíková, L.; Drápela, E.; Kirchner, K.; Bajer, A.; Balková, M.; Kuda, F. Urban geotourism development and geoconservation: Is it possible to find a balance? *Environ. Sci. Policy* **2021**, *121*, 1–10. [CrossRef]
49. Borghi, A.; d'Atri, A.; Martire, L.; Castelli, D.; Costa, E.; Dino, G.; Favero-Longo, S.E.; Ferrando, S.; Gallo, L.M.; Giardino, M.; et al. Fragments of the Western Alpine Chain as historic ornamental stones in Turin (Italy): Enhancement of urban geological heritage through geotourism. *Geoheritage* **2014**, *6*, 41–55. [CrossRef]
50. De Wever, P.; Baudin, F.; Pereira, D.; Cornée, A.; Egoroff, G.; Page, K. The importance of geosites and heritage stones in cities—A review. *Geoheritage* **2017**, *9*, 561–575. [CrossRef]
51. Corbí, H.; Martínez-Martínez, J.; Martin-Rojas, I. Linking geological and architectural heritage in a singular geosite: Nueva Tabarca Island (SE Spain). *Geoheritage* **2019**, *11*, 703–716. [CrossRef]
52. Freire-Lista, D.M.; Fort, R. Historical city centres and traditional building stones as heritage: Barrio de las Letras, Madrid (Spain). *Geoheritage* **2019**, *11*, 71–85. [CrossRef]
53. Lezzerini, M.; Pagnotta, S.; Legnaioli, S.; Palleschi, V. Walking in the streets of Pisa to discover the stones used in the Middle Ages. *Geoheritage* **2019**, *11*, 1631–1641. [CrossRef]
54. Pereira, D.; Perez-Castro, P. Art museums: A good context for outreach activities on natural stones and heritage. *Geoheritage* **2019**, *11*, 125–132. [CrossRef]
55. Wolniewicz, P. Bringing the history of the Earth to the public by using storytelling and fossils from decorative stones of the city of Poznań, Poland. *Geoheritage* **2019**, *11*, 1827–1837. [CrossRef]
56. Key, M.M.; Lieber, S.B.; Teagle, R.J. An historical geoarchaeological approach to sourcing an eighteenth century building stone: Use of Aquia Creek Sandstone in Christ Church, Lancaster County, VA, USA. *Geoheritage* **2020**, *12*, 4. [CrossRef]
57. Dreesen, R.; Poty, E.; Mottequin, B.; Marion, J.-M.; Denayer, J. An exceptional Lower Carboniferous historical heritage stone from Belgium, the 'Pierre de Meuse'. *Geoheritage* **2021**, *13*, 100. [CrossRef]
58. Kubalíková, L.; Zapletalová, D. Geo-cultural aspects of building stone extracted within Brno City (Czech Republic): A bridge between natural and cultural heritage. *Geoheritage* **2021**, *13*, 78. [CrossRef]
59. Santi, P.; Tramontana, M.; Tonelli, G.; Renzulli, A.; Veneri, F. The historic centre of Urbino, UNESCO World Heritage (Marche region, Italy): An urban-geological itinerary across the building and ornamental stones. *Geoheritage* **2021**, *13*, 86. [CrossRef]
60. Woodcock, N.H.; Furness, E.N. Quantifying the history of building stone use in a heritage city: Cambridge, UK, 1040–2020. *Geoheritage* **2021**, *13*, 12. [CrossRef]
61. Oyarzun, R.; Duque, J.F.M.; Barrenechea, J.F.; López García, J.A. Gossans, slates, and the red and black hamlets of Segovia (Spain): Interrelated geological and architectural features. *Geoheritage* **2018**, *10*, 109–121. [CrossRef]
62. Kullberg, J.C.; Prego, A. The historical importance and architectonic relevance of the "extinct" Arrábida Breccia. *Geoheritage* **2019**, *11*, 87–111. [CrossRef]
63. Damas Mollá, L.; Uriarte, J.A.; Zabaleta, A.; Aranburu, A.; García Garmilla, F.; Sagarna, M.; Bodego, A.; Clemente, J.A.; Morales, T.; Antigüedad, I. Red Ereño: An ornamental and construction limestone of international significance from the Basque country (northern Spain). *Geoheritage* **2021**, *13*, 2. [CrossRef]
64. Cárdenes, V.; López-Piñeiro, S.; Ruiz de Argandoña, V.G. The relevance of the green phyllites of Lugo (Spain) in the architectonical heritage: An exceptional roofing slate resource. *Geoheritage* **2021**, *13*, 11. [CrossRef]
65. Kaur, G.; Makki, M.F.; Avasia, R.K.; Bhusari, B.; Duraiswami, R.A.; Pandit, M.K.; Fareeduddin; Baskar, R.; Kad, S. The Late Cretaceous-Paleogene Deccan Traps: A potential global heritage stone province from India. *Geoheritage* **2019**, *11*, 973–989. [CrossRef]
66. Garg, S.; Kaur, P.; Pandit, M.; Fareeduddin; Gurmeet, K.; Kamboj, A.; Thakur, S.N. Makrana marble: A popular heritage stone resource from NW India. *Geoheritage* **2019**, *11*, 909–925. [CrossRef]
67. Garg, S.; Agarwal, P.; Ranawat, P.S.; Kaur, P.; Singh, A.; Saini, J.; Pandit, K.M.; Acharya, K.; Kaur, G. Rajnagar marble: A prominent heritage stone from Rajasthan, NW India. *Geoheritage* **2022**, *14*, 4. [CrossRef]

68. Kaur, G.; Ahuja, A.; Thakur, S.N.; Pandit, M.; Duraiswami, R.; Singh, A.; Kaur, P.; Saini, J.; Goswami, R.G.; Prakash, J.; et al. Jodhpur Sandstone: An architectonic heritage stone from India. *Geoheritage* **2020**, *12*, 16. [CrossRef]
69. Kaur, G.; Kaur, P.; Ahuja, A.; Singh, A.; Saini, J.; Agarwal, A.; Bhargava, O.N.; Pandit, M.; Giri, R.G.; Acharya, K.; et al. Jaisalmer golden limestone: A heritage stone resource from the desert of Western India. *Geoheritage* **2020**, *12*, 53. [CrossRef]
70. Kaur, G.; Bhargava, O.N.; Ruiz de Argandoña, V.G.; Thakur, S.N.; Singh, A.; Saini, J.; Kaur, P.; Sharma, U.; Garg, U.; Singh, J.J.; et al. Proterozoic slates from Chamba and Kangra: A heritage stone resource from Himachal Pradesh, India. *Geoheritage* **2020**, *12*, 79. [CrossRef]
71. Kaur, G.; Agarwal, P.; Garg, S.; Kaur, P.; Saini, J.; Singh, A.; Pandit, M.; Acharya, K.; Roopra, V.S.; Bhargava, O.N.; et al. The Alwar quartzite built architectural heritage of North India: A case for global heritage stone resource designation. *Geoheritage* **2021**, *13*, 55. [CrossRef]
72. Bone, D.A. Historic building stones and their distribution in the churches and chapels of West Sussex, England. *Proc. Geol. Assoc.* **2016**, *127*, 53–77. [CrossRef]
73. Ito, T.; Ichizawa, Y. Castellation incorporating geology and geography: Tenth–sixteenth century castles on chert of a Jurassic Accretionary Complex in Central Japan. *Geoheritage* **2022**, *14*, 17. [CrossRef]
74. Szadkowska, K.; Szadkowski, M.; Tarka, R. Inventory and assessment of the geoheritage of the Sudetic Foreland Geopark (South-Western Poland). *Geoheritage* **2022**, *14*, 24. [CrossRef]
75. Baucon, A.; Piazza, M.; Cabella, R.; Bonci, M.C.; Capponi, L.; Neto de Carvalho, C.; Briguglio, A. Buildings that 'speak': Ichnological geoheritage in 1930s buildings in Piazza della Vittoria (Genova, Italy). *Geoheritage* **2020**, *12*, 70. [CrossRef]
76. Francischini, H.; Fernandes, M.A.; Kunzler, J.; Rodrigues, R.; Leonardi, G.; Carvalho, I.D.S. The ichnological record of Araraquara sidewalks: History, conservation, and perspectives from this urban paleontological heritage of southeastern Brazil. *Geoheritage* **2020**, *12*, 50. [CrossRef]
77. Polck, M.A.d.R.; de Medeiros, M.A.M; de Araújo-Júnior, H.I. Geodiversity in urban cultural spaces of Rio de Janeiro city: Revealing the geoscientific knowledge with emphasis on the fossil content. *Geoheritage* **2020**, *12*, 47. [CrossRef]
78. Simón-Porcar, G.; Martínez-Graña, A.; Simón, J.L.; González-Delgado, J.A.; Legoinha, P. Ordovician ichnofossils and popular architecture in Monsagro (Salamanca, Spain): Ethnopaleontology in the service of rural development. *Geoheritage* **2020**, *12*, 76. [CrossRef]
79. Valentino, D.; Borghi, A.; d'Atri, A.; Gambino, F.; Martire, L.; Perotti, L.; Vaggelli, G. STONE Pietre Egizie: A free mobile application for promoting the scientific research on ornamental stones of Museo Egizio of Torino, Italy. *Geoheritage* **2020**, *12*, 61. [CrossRef]
80. Richards, S.J.; Newsome, D.; Simpson, G. Architectural geoheritage, engaging the observer and the geotourism potential of the Lighthouse Hotel Rock Wall, Bunbury, Western Australia. *Geoheritage* **2020**, *12*, 75. [CrossRef]
81. Del Lama, E.A. Potential for urban geotourism: Churches and cemeteries. *Geoheritage* **2019**, *11*, 717–728. [CrossRef]
82. Pereira, L.S.; do Nascimento, M.A.L.; Mantesso-Neto, V. Geotouristic trail in the Senhor da Boa Sentença Cemetery, João Pessoa, State of Paraíba (PB), Northeastern Brazil. *Geoheritage* **2019**, *11*, 1133–1149. [CrossRef]
83. Melelli, L.; Silvani, F.; Ercoli, M.; Pauselli, C.; Tosi, G.; Radicioni, F. Urban geology for the enhancement of the hypogean geosites: The Perugia underground (Central Italy). *Geoheritage* **2021**, *13*, 18. [CrossRef]
84. Ramalho, E.C.; Marrero-Diaz, R.; Leitão, M.; Dias, R.; Ramada, A.; Pinto, C. Alfama springs, Lisbon, Portugal: Cultural geoheritage throughout the centuries. *Geoheritage* **2020**, *12*, 74. [CrossRef]
85. Gordon, J.E. Rediscovering a sense of wonder: Geoheritage, geotourism and cultural landscape experiences. *Geoheritage* **2012**, *4*, 65–77. [CrossRef]
86. Knight, J.; Harrison, S. A land history of men': The intersection of geomorphology, culture and heritage in Cornwall, southwest England'. *Appl. Geogr.* **2013**, *42*, 186–194. [CrossRef]
87. Wells, E.C. Cultural soilscapes. In *Function of Soils for Human Societies and the Environment*; Frossard, E., Blum, W.E.H., Warkentin, B.P., Eds.; Geological Society, Special Publications: London, UK, 2006; Volume 266, pp. 125–132.
88. Szepesi, J.; Harangi, S.; Ésik, Z.; Novák, T.J.; Lukács, R.; Soós, I. Volcanic geoheritage and geotourism perspectives in Hungary: A case of an UNESCO World Heritage Site, Tokaj Wine Region Historic Cultural Landscape, Hungary. *Geoheritage* **2017**, *9*, 329–349. [CrossRef]
89. Pijet-Migoń, E.; Migoń, P. Linking wine culture and geoheritage—Missing opportunities at European UNESCO World Heritage sites and in UNESCO Global Geoparks? A survey of web-based resources. *Geoheritage* **2021**, *13*, 71. [CrossRef]
90. Reynard, E.; Estoppey, E. The Lavaux World Heritage terraced vineyard. In *Landscapes and Landforms of Switzerland*; Reynard, E., Ed.; Springer: Cham, Switzerland, 2021; pp. 111–122.
91. Pereira, S. The terraced slopes of the Douro Valley. In *Landscapes and Landforms of Portugal*; Vieira, G., Zêzere, J.L., Mora, C., Eds.; Springer: Cham, Switzerland, 2020; pp. 151–162.
92. Lugeri, F.R.; Amadio, V.; Bagnaia, R.; Cardilo, A.; Lugeri, N. Landscape and wine production areas: A geomorphological heritage. *Geoheritage* **2011**, *3*, 221–232. [CrossRef]
93. Magliulo, P.; Di Lisio, A.; Sisto, M.; Valente, A. Geotouristic enhancement of high-quality wine production areas: Examples from Sannio and Irpinia landscapes (Campania Region, Southern Italy). *Geoheritage* **2020**, *12*, 18. [CrossRef]
94. Meadows, M.E. The Cape Winelands. In *Landscapes and Landforms of South Africa*; Grab, S., Knight, J., Eds.; Springer: Cham, Switzerland, 2015; pp. 103–109.

95. Amato, V.; Valletta, M. Wine landscapes of Italy. In *Landscapes and Landforms of Italy*; Soldati, M., Marchetti, M., Eds.; Springer: Cham, Switzerland, 2017; pp. 523–536.
96. Collier, M. Field boundary stone walls as exemplars of "novel" ecosystems. *Landsc. Res.* **2013**, *38*, 141–150. [CrossRef]
97. Duma, P.; Latocha, A.; Łuczak, A.; Piekalski, J. Stone walls as a characteristic feature of the cultural landscape of the Izera Mountains, southwestern Poland. *Int. J. Hist. Archaeol.* **2020**, *24*, 22–43. [CrossRef]
98. Rosendahl, S.; Marçal Gonçalves, M. Joining geotourism with cultural tourism: A good blend. *J. Tour. Herit. Res.* **2019**, *2*, 252–275.
99. Cyffka, B.; Bock, M. Degradation of field terraces in the Maltese Islands—Reasons, processes and effects. *Geogr. Fis. Dinam. Quat.* **2008**, *31*, 119–128.
100. Tsermegas, I.; Dłużewski, M.; Biejat, K.; Szynkiewicz, A. Function of agricultural terraces in Mediterranean conditions—Selected examples from the island of Ikaria (the Southern Sporades, Greece). *Misc. Geogr.* **2011**, *15*, 65–78. [CrossRef]
101. Migoń, P.; Latocha, A. Enhancement of cultural landscape by geomorphology. A study of granite parklands in the West Sudetes, SW Poland. *Geogr. Fis. Dinam. Quat.* **2008**, *31*, 195–203.
102. Cárdenes, V.; Ponce de León, M.; Rodríguez, X.A.; Rubio-Ordoñez, A. Roofing slate industry in Spain: History, geology, and geoheritage. *Geoheritage* **2019**, *11*, 19–34. [CrossRef]
103. Ballesteros, D.; Caldevilla, P.; Vila, R.; Barros, X.C.; Alemparte, M. Linking geoheritage and traditional architecture for mitigating depopulation in rural areas: The Palaeozoic Villages Route (Courel Mountains UNESCO Global Geopark, Spain). *Geoheritage* **2021**, *13*, 63. [CrossRef]
104. Çiner, A.; Aydar, E. A fascinating gift from volcanoes: The Fairy Chimneys and underground cities of Cappadocia. In *Landscapes and Landforms of Turkey*; Kuzucuoğlu, C., Çiner, A., Kazancı, N., Eds.; Springer: Cham, Switzerland, 2019; pp. 535–549.
105. Adamovič, J.; Mikuláš, R.; Cílek, V. *Atlas Pískovcových Skalnych Měst České a Slovenské Republiky: Geologie a Geomorfologie*; Academia: Prague, Czech Republic, 2010.
106. Bruno, D.E.; Perrotta, P. A geotouristic proposal for Amendolara territory (northern Ionic sector of Calabria, Italy). *Geoheritage* **2012**, *4*, 139–151. [CrossRef]
107. Bentivenga, M.; Capece, A.; Guglielmi, P.; Martorano, S.; Napoleone, D.; Palladino, G.; De Luca, V. The San Giorgio Lucano anthropic cave complex (Basilicata, Southern Italy): A geosite to protect and enhance. *Geoheritage* **2019**, *11*, 1509–1519. [CrossRef]
108. Boukhchim, N.; Fraj, T.B.; Reynard, E. Lateral and "vertico-lateral" cave dwellings in Haddej and Guermessa: Characteristic geocultural heritage of southeast Tunisia. *Geoheritage* **2018**, *10*, 575–590. [CrossRef]
109. Gamkrelidze, I.; Okrostsvaridze, A.; Koiava, K.; Maisadze, F. Potential Geoparks of Georgia. In *Geotourism Potential of Georgia, the Caucasus*; Springer: Cham, Switzerland, 2021; pp. 57–81. [CrossRef]
110. Coratza, P.; Gauci, R.; Schembri; Soldati, M.; Tonelli, C. Bridging natural and cultural values of sites with outstanding scenery: Evidence from Gozo, Maltese Islands. *Geoheritage* **2016**, *8*, 91–103. [CrossRef]
111. Gauci, R.; Schembri, J.A.; Inkpen, R. Traditional use of shore platforms: A study of the artisanal management of salinas on the Maltese Islands (Central Mediterranean). *SAGE Open* **2017**, *7*, 1–16. [CrossRef]
112. Gauci, R.; Inkpen, R. The physical characteristics of limestone shore platforms on the Maltese Islands and their neglected contribution to coastal land use development. In *Landscapes and Landforms of Malta*; Gauci, R., Schembri, J.A., Eds.; Springer: Cham, Switzerland, 2019; pp. 343–356.
113. Iranzo-García, E.; Kortekaas, K.H.; López, E.R. Inland salinas in Spain: Classification, characterisation, and reflections on unique cultural landscapes and geoheritage. *Geoheritage* **2021**, *13*, 24. [CrossRef]
114. Schembri, J.A.; Spiteri, S.C. By gentlemen for gentlemen—Ria coastal landforms and the fortified imprints of Valletta and its harbours. In *Landscapes and Landforms of Malta*; Gauci, R., Schembri, J.A., Eds.; Springer: Cham, Switzerland, 2019; pp. 69–78.
115. Burt, T.; Tucker, M. The geomorphology of the Whin Sill. In *Landscapes and Landforms of England and Wales*; Goudie, A., Migoń, P., Eds.; Springer: Cham, Switzerland, 2020; pp. 515–530.
116. Kausch, B.; Maquil, R. Landscapes and landforms of the Luxembourg Sandstone, Grand-Duchy of Luxembourg. In *Landscapes and Landforms of Belgium and Luxembourg*; Demoulin, A., Ed.; Springer: Cham, Switzerland, 2018; pp. 43–62.
117. Ren, F. The interplay between Taoist philosophy, Danxia Landscape and human beings—'Tao follows nature'. In *Sandstone Landscapes. Diversity, Ecology and Conservation, Proceedings of the 3rd International Conference on Sandstone Landscapes, Kudowa-Zdrój, Poland, 25–28 April 2012*; Migoń, P., Kasprzak, M., Eds.; Department of Geography and Regional Development, University of Wrocław: Wrocław, Poland, 2013; pp. 159–162.
118. Kiernan, K. Landforms as sacred places: Implications for geodiversity and geoheritage. *Geoheritage* **2015**, *7*, 177–193. [CrossRef]
119. López-García, J.A.; Oyarzun, R.; López Andrés, S.; Manteca Martínez, J.I. Scientific, educational, and environmental considerations regarding mine sites and geoheritage: A perspective from SE Spain. *Geoheritage* **2011**, *3*, 267–275. [CrossRef]
120. Mateos, R.M.; Durán, J.J.; Robledo, P.A. Marès quarries on the Majorcan coast (Spain) as geological heritage sites. *Geoheritage* **2011**, *3*, 41–54. [CrossRef]
121. Todaro, S. The potential geosite of the "Libeccio Antico" quarries: A sedimentological and stratigraphic characterisation of ornamental stone from Mt Cocuccio, Custonaci Marble District, Sicily. *Geoheritage* **2019**, *11*, 809–820. [CrossRef]
122. Gioncada, A.; Pitzalis, E.; Cioni, R.; Fulignati, P.; Lezzerini, M.; Mundula, F.; Funedda, A. The volcanic and mining geoheritage of San Pietro Island (Sulcis, Sardinia, Italy): The potential for geosite valorization. *Geoheritage* **2019**, *11*, 1567–1581. [CrossRef]
123. Ruban, D.A.; Sallam, E.S.; Khater, T.M.; Ermolaev, V.A. Golden Triangle geosites: Preliminary geoheritage assessment in a geologically rich area of eastern Egypt. *Geoheritage* **2021**, *13*, 54. [CrossRef]

124. Beretić, N.; Đukanović, Z.; Cecchini, A. Geotourism as a development tool of the Geo-mining Park in Sardinia. *Geoheritage* **2019**, *11*, 1689–1704. [CrossRef]
125. Costa, E.; Dino, G.A.; Benna, P.; Rossetti, P. The Traversella mining site as Piemonte geosite. *Geoheritage* **2019**, *11*, 55–70. [CrossRef]
126. Matías, R.; Llamas, B. Analysis using LIDAR and photointerpretation of Las Murias-Los Tallares (Castrocontrigo, León-Spain): One of the biggest Roman gold mines to use the "Peines" system. *Geoheritage* **2019**, *11*, 381–397. [CrossRef]
127. Pérez-Aguilar, A.; Juliani, C.; de Barros, E.J.; Magalhães de Andrade, M.R.; de Oliveira, E.S.; Braga, D.d.A.; Santos, R.O. Archaeological gold mining structures from colonial period present in Guarulhos and Mairiporã, São Paulo State, Brazil. *Geoheritage* **2013**, *5*, 87–105. [CrossRef]
128. Abdel-Maksoud, K.M.; Emam, M.A. Hidden geology in Ancient Egypt. *Geoheritage* **2019**, *11*, 897–907. [CrossRef]
129. Lorenc, M.W.; Cocks, A. Inscribing a landscape: The Cornish Mining World Heritage Site. *Geoturystyka* **2008**, *1*, 27–40.
130. Peréz Sánchez, A.A.; Lorenc, M.W. The cultural landscape of the Linares-La Carolina mining district (Spain). *Geoturystyka* **2008**, *3*, 13–26. [CrossRef]
131. Lenart, J. The Nízký Jeseník—Highland with abandoned deep mines. In *Landscapes and Landforms of the Czech Republic*; Pánek, T., Hradecký, J., Eds.; Springer: Cham, Switzerland, 2016; pp. 305–317.
132. Gallala, W.; Younes, A.; Ouazaa, N.L.; Hadjzobir, S. Roman millstones of Carthage (Tunisia): A geoarchaeological study using petrological and geochemical methods. *Geoheritage* **2018**, *10*, 673–686. [CrossRef]
133. Careddu, N.; Grillo, S.M. Sardinian basalt—An ancient georesource still en vogue. *Geoheritage* **2019**, *11*, 35–45. [CrossRef]
134. Careddu, N.; Grillo, S.M. "Trachytes" from Sardinia: Geoheritage and current use. *Sustainability* **2019**, *11*, 3706. [CrossRef]
135. Lapuente Mercadal, M.P.; Cuchí Oterino, J.A.; Blanc, P.; Brilli, M. Louvie-Soubiron marble: Heritage stone in the French Pyrenean Ossau Valley—first evidence of the Roman Trans-Pyrenean use. *Geoheritage* **2021**, *13*, 17. [CrossRef]
136. Stefano, M.; Paolo, S. Abandoned quarries and geotourism: An opportunity for the Salento Quarry District (Apulia, Southern Italy). *Geoheritage* **2017**, *9*, 463–477. [CrossRef]
137. Baczyńska, E.; Lorenc, M.W.; Kaźmierczak, U. The landscape attractiveness of abandoned quarries. *Geoheritage* **2018**, *10*, 271–285. [CrossRef]
138. AlRayyan, K.; Hamarneh, C.; Sukkar, H.; Ghaith, A.; Abu-Jaber, N. From abandoned mines to a labyrinth of knowledge: A conceptual design for a geoheritage park museum in Jordan. *Geoheritage* **2019**, *11*, 257–270. [CrossRef]
139. Gajek, G.; Zgłobicki, W.; Kołodyńska-Gawrysiak, R. Geoeducational value of quarries located within the Małopolska Vistula River Gap (E Poland). *Geoheritage* **2019**, *11*, 1335–1351. [CrossRef]
140. Kazancı, N.; Suludere, Y.; Özgüneylioğlu, A.; Mülazımoğlu, N.S.; Şaroğlu, F.; Mengi, H.; Boyraz-Aslan, S.; Gürbüz, E.; Onur Yücel, T.; Ersöz, M.; et al. Mining heritage and relevant geosites as possible instruments for sustainable development of miner towns in Turkey. *Geoheritage* **2019**, *11*, 1267–1276. [CrossRef]
141. Marengo, A.; Borghi, A.; Bittarello, E.; Costa, E. Touristic fruition of the disused quarry of Busca Onyx: Problematics and strategies. *Geoheritage* **2019**, *11*, 47–54. [CrossRef]
142. Brzezińska-Wójcik, T.; Skowronek, E. Tangible heritage of the historical stonework centre in Brusno Stare in the Roztocze Area (SE Poland) as an opportunity for the development of geotourism. *Geoheritage* **2020**, *12*, 10. [CrossRef]
143. Martínez, A.M.D.; Timarán, F.P. Evaluation of candidate sites in a proposal for sustainable development: "The Gold Route", Nariño, Colombia. *Geoheritage* **2020**, *12*, 56. [CrossRef]
144. Ruiz Pulpón, Á.R.; Cañizares Ruiz, M.C. Enhancing the territorial heritage of declining rural areas in Spain: Towards integrating top-down and bottom-up approaches. *Land* **2020**, *9*, 216. [CrossRef]
145. Milu, V. Preliminary assessment of the geological and mining heritage of the Golden Quadrilateral (Metaliferi Mountains, Romania) as a potential geotourism destination. *Sustainability* **2021**, *13*, 10114. [CrossRef]
146. Reyes, C.A.R.; Amorocho-Parra, R.; Villarreal-Jaimes, C.A.; Meza-Ortíz, J.A.; Castellanos-Alarcón, O.M.; Madero-Pinzon, H.D.; Casadiego-Quintero, E.; Carvajal-Díaz, J.D. Geotourism in regions with influence from the oil industry: A study case of the Middle Magdalena Valley Basin (Colombia). *Geoheritage* **2021**, *13*, 107. [CrossRef]
147. Hellqvist, M. Teaching sustainability in geoscience field education at Falun Mine World Heritage Site in Sweden. *Geoheritage* **2019**, *11*, 1785–1798. [CrossRef]
148. Cavalcanti, J.A.D.; da Silva, M.S.; Schobbenhaus, C.; de Mota Lima, H. Geo-mining heritages of the Mariana Anticline Region, southeast of Quadrilátero Ferrífero-MG, Brazil: Qualitative and quantitative assessment of Chico Rei and Passagem mines. *Geoheritage* **2021**, *13*, 98. [CrossRef]
149. Rybár, P.; Hronček, P. Mining tourism and the search for its origins. *Geotourism/Geoturystyka* **2017**, *50–51*, 27–66. [CrossRef]
150. Wrede, V.; Mügge-Bartolović, V. GeoRoute Ruhr—A network of geotrails in the Ruhr Area National GeoPark, Germany. *Geoheritage* **2012**, *4*, 109–114. [CrossRef]
151. Farsani, N.T.; Esfahani, M.A.G.; Shokrizadeh, M. Understanding tourists' satisfaction and motivation regarding mining geotours (case study: Isfahan, Iran). *Geoheritage* **2019**, *11*, 681–688. [CrossRef]
152. Prosser, C.D. Geoconservation, quarrying and mining: Opportunities and challenges illustrated through working in partnership with the mineral extraction industry in England. *Geoheritage* **2018**, *10*, 259–270. [CrossRef]
153. Radwański, A.B. The oil and ozokerite mine in Boryslav and historical monuments of petroleum and salt industries in its vicinity (Fore-Carpathian region, Ukraine). *Geoturystyka* **2009**, *3*, 35–44.

154. Radwański, A.B. The Ignacy Łukasiewicz Memorial Museum of Oil and Gas Industry in Bóbrka and historical monuments of petroleum and salt industries in the vicinity of Krosno (the Polish Outer Carpathians). *Geoturystyka* **2009**, *3*, 35–44. [CrossRef]
155. Briševac, Z.; Maričić, A.; Brkić, V. Croatian geoheritage sites with the best-case study analyses regarding former mining and petroleum activities. *Geoheritage* **2021**, *13*, 95. [CrossRef]
156. Roy, N.; Pandey, B.W.; Rani, U. Protecting the vanishing geo-cultural heritage of India: Case study of Majuli Island in Assam. *Int. J. Geoheritage Parks* **2020**, *8*, 18–30. [CrossRef]
157. Tosatti, G. Slope instability affecting the Canossa geosite (Northern Apennines, Italy). *Geogr. Fis. Dinam. Quat.* **2008**, *31*, 239–246.
158. Margottini, C.; Di Buduo, G. The geological and landslides museum of Civita di Bagnoregio (Central Italy). *Landslides* **2017**, *14*, 435–445. [CrossRef]
159. Gutiérrez, F.; Cooper, A.H. Evaporite dissolution subsidence in the historical city of Calatayud, Spain: Damage appraisal and prevention. *Nat. Haz.* **2002**, *25*, 259–288. [CrossRef]
160. Łabuz, T. Morfodynamika i tempo erozji klifu w Trzęsaczu (1997–2017). *Landf. Anal.* **2017**, *34*, 29–50. [CrossRef]
161. Munzar, J.; Deutsch, M.; Elleder, L.; Ondracek, S.; Kallabova, E.; Hradek, M. Historical floods in Central Europe and their documentation by means of floodmarks and other epigraphical monuments. *Morav. Geogr. Rep.* **2006**, *14*, 26–44.
162. Herget, J. Am Anfang war die Sintflut. In *Hochwasserkatastrophen in der Geschichte*; WBG: Darmstadt, Germany, 2012.
163. Pekárová, P.; Halmová, D.; Bačová Mitková, V.; Miklánek, P.; Pekár, J.; Škoda, P. Historic flood marks and flood frequency analysis of the Danube River at Bratislava, Slovakia. *J. Hydrol. Hydromech.* **2013**, *61*, 326–333. [CrossRef]
164. Coratza, P.; De Waele, J. Geomorphosites and natural hazards: Teaching the importance of geomorphology in society. *Geoheritage* **2012**, *4*, 195–203. [CrossRef]
165. Migoń, P.; Pijet-Migoń, E. Natural disasters, geotourism and geo-education. *Geoheritage* **2019**, *11*, 629–640. [CrossRef]
166. Wallace, A. Presenting Pompeii: Steps towards reconciling conservation and tourism at an ancient site. *Pap. Inst. Archeol.* **2012**, *22*, 115–136.
167. Scandone, R.; Giacomelli, L. Vesuvius, Pompei, Herculaneum: A lesson in natural history. *J. Res. Didact. Geogr.* **2014**, *2*, 33–41.
168. Gorokhovich, Y. Santorini, Eruption. In *Encyclopedia of Natural Hazards*, Bobrowsky, P., Ed.; Springer: Dordrecht, The Netherlands, 2013; pp. 884–895.
169. Hancock, P.L.; Chalmers, R.M.L.; Altunel, E.; Çakir, Z.; Becher-Hancock, A. Creation and destruction of travertine monumental stone by earthquake faulting at Hierapolis, Turkey. *Geol. Soc. Lond. Spec. Publ.* **2000**, *171*, 1–14. [CrossRef]
170. Silva, P.G.; Borja, F.; Zazo, C.; Goy, J.L.; Bardajı, T.; De Luque, L.; Lario, J.; Dabrio, C.J. Archaeoseismic record at the ancient Roman City of Baelo Claudia (Cádiz, south Spain). *Tectonophysics* **2005**, *408*, 129–146. [CrossRef]
171. Guidoboni, E.; Muggia, A.; Marconi, C.; Boschi, E. A case study in archaeoseismology. The collapses of the Selinunte temples (southwestern Sicily): Two earthquakes identified. *Bull. Seismol. Soc. Amer.* **2002**, *92*, 2961–2982. [CrossRef]
172. Alcantara-Ayala, I. Parícutin volcano: To the other side. In *Geomorphological Landscapes of the World*; Migoń, P., Ed.; Springer: Dordrecht, The Netherlands, 2010; pp. 59–68.
173. Gizzi, F.T.; Bentivenga, M.; Lasaponara, R.; Danese, M.; Potenza, M.R.; Sileo, M.; Masini, N. Natural hazards, human factors, and "ghost towns": A multi-level approach. *Geoheritage* **2019**, *11*, 1533–1565. [CrossRef]
174. Migoń, P. Rediscovering geoheritage, reinventing geotourism—200 years of experience from the Sudetes, Central Europe. In *Appreciating Physical Landscapes*; Special Publications; Hose, T.A., Ed.; Geological Society: London, UK, 2016; Volume 417, pp. 215–228.
175. Hose, T.A. Geological inquiry in Britain and England: A brief history. In *Geoheritage and Geotourism. A European Perspective*; Hose, T.A., Ed.; Boydell Press: Woodbridge, UK, 2016; pp. 31–53.
176. Munt, M.C. Geoheritage case study: The Isle of Wight, England. In *Geoheritage and Geotourism. A European Perspective*; Hose, T.A., Ed.; Boydell Press: Woodbridge, UK, 2016; pp. 195–204.
177. Hose, T.A. Museums and geoheritage in Britain and England. In *Geoheritage and Geotourism. A European Perspective*; Hose, T.A., Ed.; Boydell Press: Woodbridge, UK, 2016; pp. 55–79.
178. De Wever, P.; Guiraud, M. Geoheritage and museums. In *Geoheritage. Assessment, Protection, and Management*; Reynard, E., Brilha, J., Eds.; Elsevier: Amsterdam, The Netherlands, 2018; pp. 129–145.
179. Borghi, A.; Vignetta, E.; Ghignone, S.; Gallo, M.; Vaggelli, G. The "Stella Polare" expedition (1899–1900): Study and enhancement of the rock collection. *Geoheritage* **2020**, *12*, 33. [CrossRef]
180. De Lima, J.T.M.; de Souza Carvalho, I. Geological or cultural heritage? The ex situ scientific collections as a remnant of nature and culture. *Geoheritage* **2020**, *12*, 3. [CrossRef]
181. Vicedo, V.; Ozkaya de Juanas, S.; Fernández-Lluch, D.; Batlles, A.; Ballester, M. Enhancing the educational value of palaeontological heritage: The didactic collection of the Museu de Ciències Naturals de Barcelona (Catalonia, Spain) and its strategic framework. *Geoheritage* **2021**, *13*, 19. [CrossRef]
182. Rapprich, V.; Valenta, J.; Brož, M.; Kadlecová, E.; van Wyk de Vries, B.; Petronis, M.; Rojík, P. A crucial site in the argument between Neptunists and Plutonists: Reopening of the historical adit in the Komorní hůrka (Kammerbühl) volcano after 180 years. *Geoheritage* **2019**, *11*, 347–358. [CrossRef]
183. Reynard, E. Geoheritage case study: Canton Valais, Switzerland. In *Geoheritage and Geotourism. A European Perspective*; Hose, T.A., Ed.; Boydell Press: Woodbridge, UK, 2016; pp. 279–290.

184. Aucelli, P.P.C.; Brancaccio, L.; Cinque, A. Vesuvius and Campi Flegrei: Volcanic history, landforms and impact on settlements. In *Landscapes and Landforms of Italy*; Soldati, M., Marchetti, M., Eds.; Springer: Cham, Switzerland, 2017; pp. 389–398.
185. Coratza, P.; Panizza, M. Goethe's Italian journey and the geological landscape. In *Landscapes and Landforms of Italy*; Soldati, M., Marchetti, M., Eds.; Springer: Cham, Switzerland, 2017; pp. 511–521.
186. Hose, T.A. Geotourism in Britain and Europe: Historical and modern perspectives. In *Geoheritage and Geotourism. A European Perspective*; Hose, T.A., Ed.; Boydell Press: Woodbridge, UK, 2016; pp. 153–172.
187. Hose, T.A. Towards a history of geotourism: Definitions, antecedents and the future. In *The History of Geoconservation*; Special Publications; Burek, C.V., Prosser, C.D., Eds.; Geological Society: London, UK, 2008; Volume 300, pp. 37–60.
188. Gordon, J.E.; Baker, M. Appreciating geology and the physical landscape in Scotland: From tourism of awe to experiential re-engagement. In *Appreciating Physical Landscapes*; Special Publications; Hose, T.A., Ed.; Geological Society: London, UK, 2016; Volume 417, pp. 25–40.
189. Hose, T.A. Geoheritage in the field. In *Geoheritage and Geotourism. A European Perspective*; Hose, T.A., Ed.; Boydell Press: Woodbridge, UK, 2016; pp. 101–127.
190. Gordon, J.E. Geotourism and cultural heritage. In *Handbook of Geotourism*; Dowling, R., Newsome, D., Eds.; Edward Elgar: Cheltenham, UK, 2018; pp. 61–75.
191. Chylińska, D. The role of the picturesque in geotourism and iconic geotourist landscapes. *Geoheritage* **2019**, *11*, 531–543. [CrossRef]
192. Motte, E.; McInnes, R. Using artistic imagery to improve understanding of coastal landscape. Changes on the Rance Estuary (French Channel Coast). *Geoheritage* **2019**, *11*, 961–972. [CrossRef]
193. Conly, G. *Tarawera. The Destruction of the Pink and White Terraces*; Grantham House: Wellington, New Zealand, 1985.
194. Zumbühl, H.J.; Nussbaumer, S.U.; Wipf, A. Top of Europe: The Finsteraarhorn—Jungfrau glacier landscape. In *Landscapes and Landforms of Switzerland*; Reynard, E., Ed.; Springer: Cham, Switzerland, 2021; pp. 217–233.
195. Pullin, R. The artist as geotourist: Eugene von Guérard and the seminal sites of early volcanic research in Europe and Australia. In *Appreciating Physical Landscapes*; Special Publications; Hose, T.A., Ed.; Geological Society: London, UK, 2016; Volume 417, pp. 59–70.
196. Moore, R. The Isle of Wight Undercliff. In *Landscapes and Landforms of England and Wales*; Goudie, A., Migoń, P., Eds.; Springer: Cham, Switzerland, 2020; pp. 145–167.
197. Sabato, L.; Tropeano, M.; Festa, V.; Longhitano, S.G.; dell'Olio, M. Following writings and paintings by Carlo Levi to promote geology within the "Matera-Basilicata 2019, European Capital of Culture" events (Matera, Grassano, Aliano—Southern Italy). *Geoheritage* **2019**, *11*, 329–346. [CrossRef]
198. Nesci, O.; Borchia, R. Landscapes and landforms of the Duchy of Urbino in Italian renaissance paintings. In *Landscapes and Landforms of Italy*; Soldati, M., Marchetti, M., Eds.; Springer: Cham, Switzerland, 2017; pp. 257–269.
199. Da Silva, C.M. Geodiversity and sense of place: Local identity geological elements in Portuguese municipal heraldry. *Geoheritage* **2019**, *11*, 949–960. [CrossRef]
200. González-Delgado, J.Á.; Martínez-Graña, A.; Holgado, M.; Gonzalo, J.C.; Legoinha, P. Augmented reality as a tool for promoting the tourist value of the geological heritage around natural filming locations: A case study in "Sad Hill" (The Good, the Bad and the Ugly Movie, Burgos, Spain). *Geoheritage* **2020**, *12*, 34. [CrossRef]
201. Vitaliano, D. Geomythology: Geological origins of myths and legends. In *Myth and Geology*; Special Publications; Piccardi, L., Masse, W.B., Eds.; Geological Society: London, UK, 2007; Volume 273, pp. 1–7.
202. Trifonov, V.G. The Bible and geology: Destruction of Sodom and Gomorrah. In *Myth and Geology*; Special Publications; Piccardi, L., Masse, W.B., Eds.; Geological Society: London, UK, 2007; Volume 273.
203. Piccardi, L. The AD 60 Denizli Basin earthquake and the apparition of Archangel Michael at Colossae (Aegean Turkey). In *Myth and Geology*; Special Publications; Piccardi, L., Masse, W.B., Eds.; Geological Society: London, UK, 2007; Volume 273, pp. 95–105.
204. Mörner, N.-A. The Fenris Wolf in the Nordic Asa creed in the light of palaeoseismics. In *Myth and Geology*; Special Publications; Piccardi, L., Masse, W.B., Eds.; Geological Society: London, UK, 2007; Volume 273, pp. 143–163.
205. Nunn, P.D.; Pastorizo, M.R. Geological histories and geohazard potential of Pacific Islands illuminated by myths. In *Myth and Geology*; Special Publications; Piccardi, L., Masse, W.B., Eds.; Geological Society: London, UK, 2007; Volume 273, pp. 117–119.
206. Bryant, E.; Walsh, G.; Abbott, D. Cosmogenic mega-tsunami in the Australia region: Are they supported by Aboriginal and Maori legends. In *Myth and Geology*; Special Publications; Piccardi, L., Masse, W.B., Eds.; Geological Society: London, UK, 2007; Volume 273, pp. 203–214.
207. Berger, W.H. On the discovery of the ice age: Science and myth. In *Myth and Geology*; Special Publications; Piccardi, L., Masse, W.B., Eds.; Geological Society: London, UK, 2007; Volume 273, pp. 271–278.
208. Burbery, T.J. *Geomythology. How Common Stories Reflect Earth Events*; Routledge: New York, NY, USA, 2021.
209. Nunn, P. *Worlds in Shadow. Submerged Lands in Science, Memory and Myth*; Bloomsberry Sigma: London, UK, 2021.
210. Hamacher, D.W.; Goldsmith, J. Aboriginal oral traditions of Australian impact craters. *J. Astronom. Hist. Herit.* **2013**, *16*, 295–311.
211. Rassios, A.E.; Krikeli, A.; Dilek, Y.; Ghikas, C.; Batsi, A.; Koutsovitis, P.; Hua, J. The geoheritage of Mount Olympus: Ancient mythology and modern geology. *Geoheritage* **2022**, *14*, 15. [CrossRef]
212. Khoshraftar, R.; Farsani, N.T. Geomythology: An approach for attracting geotourists (case study: Takht-e Soleymān—Takab World Heritage Sites). *Geoheritage* **2019**, *11*, 1879–1888. [CrossRef]

213. Kirchner, K.; Kubalíková, L. Geomythology: An useful tool for geoconservation and geotourism purposes. In *Public Recreation and Landscape Protection—With Man Hand in Hand*; Fialová, J., Pernicová, D., Eds.; Czech Society of Landscape Engineers and Department of Landscape Management Faculty: Brno, Czech Republic, 2015; pp. 68–74.
214. Goemaere, E.; Millier, C.; Declercq, P.Y.; Fronteau, G.; Dreesen, R. Legends of the Ardennes Massif, a cross-border intangible geo-cultural heritage (Belgium, Luxembourg, France, Germany). *Geoheritage* 2021, *13*, 28. [CrossRef]
215. Fepuleai, A.; Weber, E.; Németh, K.; Muliaina, T.; Iese, W. Eruption styles of Samoan volcanoes represented in tattooing, language and cultural activities of the indigenous people. *Geoheritage* 2017, *9*, 395–411. [CrossRef]
216. Migoń, P. Geoheritage and World Heritage sites. In *Geoheritage. Assessment, Protection, and Management*; Reynard, E., Brilha, J., Eds.; Elsevier: Amsterdam, The Netherlands, 2018; pp. 237–249.
217. Sheth, H.; Samant, H.; Patel, V.; D'Souza, J. The volcanic heritage of the Elephanta Caves, Deccan Traps, western India. *Geoheritage* 2017, *9*, 359–372. [CrossRef]
218. Ferrari, G.; Ciampalini, R.; Billi, P.; Migon, P. Geomorphology of the archaeological area of Aksum. In *Landscapes and Landforms of Ethiopia*; Billi, P., Ed.; Springer: Dordrecht, The Netherlands, 2015; pp. 147–161.
219. Bonnot-Courtois, C.; Walter-Simonnet, A.-V.; Baltzer, A. The Mont-Saint-Michel Bay: An exceptional megatidal environment influenced by natural evolution and man-made modifications. In *Landscapes and Landforms of France*; Fort, M., André, M.-F., Eds.; Springer: Dordrecht, The Netherlands, 2013; pp. 41–51.
220. Kullberg, M.C.; Kullberg, J.C. Landforms and geology of the Sierra de Sintra and its surroundings. In *Landscapes and Landforms of Portugal*; Vieira, G., Zêzere, J.L., Mora, C., Eds.; Springer: Cham, Switzerland, 2020; pp. 251–264.
221. Franchi, R.; Savelli, D.; Colosi, F.; Drapp, P.; Gabrielli, R.; Moretti, E.; Peloso, D. Petra and Beida (Jordan): Two adjacent archaeological sites up to an exploitation of geomorphology-related topics for a cultural and touristic development. *Mem. Descr. Carta Geol. D'It.* 2009, *87*, 77–90.
222. Gudmunsson, A. *The Glorious Geology of Iceland's Golden Circle*; Springer: Cham, Switzerland, 2017.
223. Zerboni, A.; Perego, A.; Cremaschi, M. Geomorphological map of the Tadrart Acacus Massif and the Erg Uan Kasa (Libyan Central Sahara). *J. Maps* 2015, *11*, 772–787. [CrossRef]
224. Abioui, M.; M'Barki, L.; Benssaou, M.; Ezaidi, A.; El Kamali, N. Rock art conservation and geotourism: A practical example from Foum Chenna engravings site, Morocco. *Geoconservation Res.* 2020, *2*, 1–11. [CrossRef]
225. Tansem, K.; Storemyr, P. Red-coated rocks on the seashore: The esthetics and geology of prehistoric rock art in Alta, Arctic Norway. *Geoarcheology* 2021, *36*, 314–334. [CrossRef]
226. Biot, V. Les grottes ornées: Des géosites culturels à la resource territoriale. *Collect. EDYTEM* 2013, *15*, 119–126. [CrossRef]
227. Ortega, A.I.; Benito-Calvo, A.; Pérez-González, A.; Carbonell, E.; Bermúdez de Castro, J.M.; Arsuaga, J.L. Atapuerca karst and its palaeoanthropological sites. In *Landscapes and Landforms of Spain*; Gutiérrez, F., Gutiérrez, M., Eds.; Springer: Dordrecht, The Netherlands, 2014; pp. 101–110.
228. Gray, M. *Geodiversity. Valuing and Conserving Abiotic Nature*; Wiley Blackwell: Chichester, UK, 2013.
229. Gray, M. Geodiversity, geoheritage and geoconservation for society. *Int. J. Geoheritage Parks* 2019, *7*, 226–236. [CrossRef]
230. Reynard, E.; Fontana, G.; Kozlik, L.; Scapozza, C. A method for assessing scientific and additional values of geomorphosites. *Geogr. Helv.* 2007, *62*, 148–158. [CrossRef]
231. Reynard, E.; Perret, A.; Bussard, J.; Grangier, L.; Martin, S. Integrated approach for the inventory and management of geomorphological heritage at the regional scale. *Geoheritage* 2016, *8*, 43–60. [CrossRef]
232. Erhartič, B. Geomorphosite assessment. *Acta Geogr. Slov.* 2010, *50*, 295–319. [CrossRef]
233. Kubalíková, L. Geomorphosite assessment for geotourism purposes. *Czech J. Tour.* 2013, *2*, 80–104. [CrossRef]
234. Georgousis, E.; Savelides, S.; Mosios, S.; Holokolos, M.-V.; Drinia, H. The need for geoethical awareness: The importance of geoenvironmental education in geoheritage understanding in the case of Meteora geomorphes, Greece. *Sustainability* 2021, *13*, 6626. [CrossRef]
235. Drápela, E.; Boháč, A.; Böhm, H.; Zágoršek, K. Motivation and preferences of visitors in the Bohemian Paradise UNESCO Global Geopark. *Geosciences* 2021, *11*, 116. [CrossRef]

Article

Bridges as Geoheritage Viewpoints in the Western Caucasus

Anna V. Mikhailenko [1], Vladimir A. Ermolaev [2] and Dmitry A. Ruban [3,4,*]

1. Department of Physical Geography, Ecology, and Nature Protection, Institute of Earth Sciences, Southern Federal University, Zorge Street 40, 344090 Rostov-on-Don, Russia; avmihaylenko@sfedu.ru
2. Department of Commodity Science and Expertise, Plekhanov Russian University of Economics, Stremyanny Lane 36, 117997 Moscow, Russia; ermolaevvla@rambler.ru
3. K.G. Razumovsky Moscow State University of Technologies and Management (The First Cossack University), Zemlyanoy Val Street 73, 109004 Moscow, Russia
4. Department of Organization and Technologies of Service Activities, Higher School of Business, Southern Federal University, 23-ja Linija Street 43, 344019 Rostov-on-Don, Russia
* Correspondence: ruban-d@mail.ru

Citation: Mikhailenko, A.V.; Ermolaev, V.A.; Ruban, D.A. Bridges as Geoheritage Viewpoints in the Western Caucasus. *Geosciences* 2021, 11, 377. https://doi.org/10.3390/geosciences11090377

Academic Editors: Hara Drinia, Panagiotis Voudouris, Assimina Antonarakou and Jesus Martinez-Frias

Received: 8 August 2021
Accepted: 6 September 2021
Published: 7 September 2021

Publisher's Note: MDPI stays neutral with regard to jurisdictional claims in published maps and institutional affiliations.

Copyright: © 2021 by the authors. Licensee MDPI, Basel, Switzerland. This article is an open access article distributed under the terms and conditions of the Creative Commons Attribution (CC BY) license (https://creativecommons.org/licenses/by/4.0/).

Abstract: Distant observation of unique geological and geomorphological features facilitates comprehension and tourism of these important resources. Bridges offer an opportunity for such observation, and the idea of bridge-based geoheritage viewpoints is proposed. In the geologically-rich area of the Western Caucasus (southwestern Russia), eleven bridges were assessed semiquantitatively with the newly proposed approach. The results indicated their different but moderate utility as geoheritage viewpoints. The utility of two bridges is high. Bridges differ not only by the quality of the views they offer but also by their accessibility. Mandatory permissions and entrance fees reduce this property in several cases. Although the study area is somewhat specific due to the relatively large number of bridges and their utility, similar situations can be found in other geographical localities. Bridge-based geoheritage viewpoints are important to geotourism development, and, particularly, they contribute to establishing optimal and comfortable routes.

Keywords: geosite; geotourism; Mountainous Adygeya; scenery; tourism

1. Introduction

Geoheritage and geotourism studies have intensified in the past decade [1–7]. Aside from documentation of hundreds (if not thousands) of new geosites, new dimensions of geoheritage diversity have been revealed. In particular, it has been realized that points suitable for the comfortable observation of unique geological and geomorphological features are important elements of geoheritage landscapes, and the most valuable of them can be judged as true geosites (even if they do not expose any unique features). The idea of viewpoint geosites was proposed by Fuertes-Gutiérrez and Fernández-Martínez [8] and Palacio [9] and then developed and conceptualized by Migoń and Pijet-Migoń [10], with some subsequent additions by Mikhailenko and Ruban [11]. There were also several other works, which considered viewpoints in relation to geoheritage management in different parts of the world [12–17].

Fuertes-Gutiérrez and Fernández-Martínez [7] and Migoń and Pijet-Migoń [9] noted that viewpoint geosites can differ significantly, and they can be either natural and artificial. One can imagine many objects, standing on which offers views of unique geological and geomorphological features and panoramas of geoheritage landscapes. Evidently, the practical importance of such objects is outstanding because they facilitate inventory and monitoring of geosites for the purposes of geoconservation, as well as enhancing comprehension of geoheritage by visitors and provide emotional satisfaction [7,17]. Therefore, establishing a diversity of viewpoint geosites and paying attention to their particular types are crucial research tasks. Field investigations in the Western Caucasus—a large mountainous domain in southwestern Russia—have shown the importance of numerous bridges for the distant viewing of geoheritage features.

The objective of the present paper is to characterize the bridge-based geoheritage viewpoints of a particular geologically-rich area of the Western Caucasus. This area is known as Mountainous Adygeya, and it lies near the border of the Republic of Adygeya and the Krasnodar Region (Figure 1). Terminological and methodological solutions are also offered in this paper. More generally, the latter aims to promote bridges as important elements of geoheritage landscapes facilitating efficient management of these landscapes.

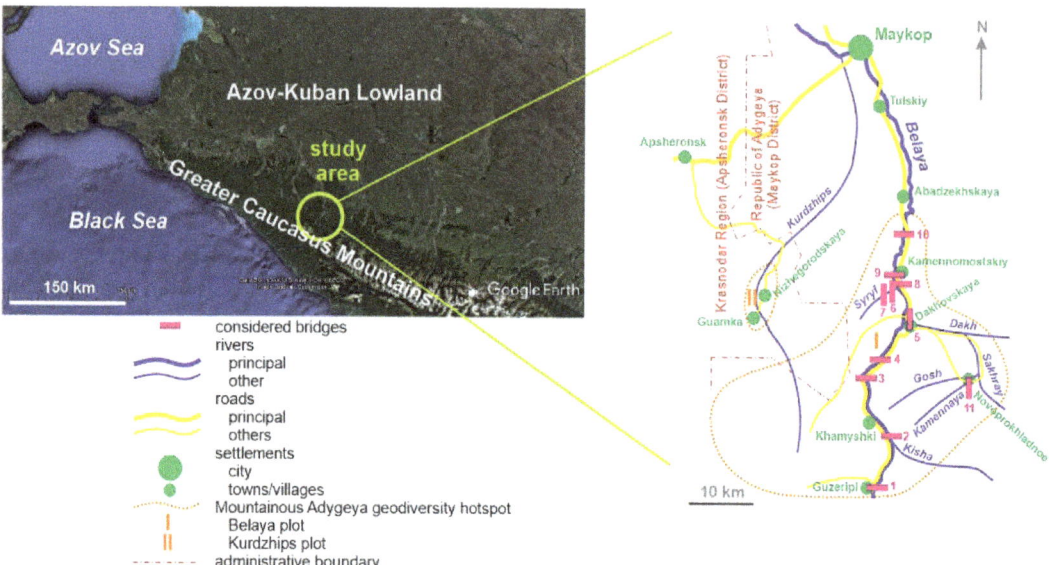

Figure 1. Location of the study area and its bridges. Inserted image was generated from the Google Earth engine.

2. Methodological Remarks

2.1. Study Area

The study area is situated in the Western Caucasus, which is the western segment of the Greater Caucasus mountain chain and the related late Cenozoic orogen (Figure 1). The general geographical and geological setting of this large domain was characterized, particularly, by Adamia et al. [18], Frolova [19], Kaban et al. [20], Lurie et al. [21], Rantsman [22], Van Hinsbergen et al. [23], and Viginsky [24]. More precisely, this area corresponds to the Mountainous Adygeya geosiversity hotspot boasting numerous and diverse unique geological and geomorphological features [25,26]. Administratively, this area belongs to the western and central parts of the Maykop District of the Republic of Adygeya and the eastern part of the Apsheronsk District of the Krasnodar Region. Geographical and geological characteristics of this area can be found in the works by Lozovoy [27], Rostovtsev et al. [28], and Ruban [29], and these characteristics are described briefly below.

The study area is dominated by mountains with a height from 500–700 m to >2500 m. Mountain ranges are generally short (<10 km), and many of them are cuesta-type ranges (sensu [30–32]). The southwestern part of the study area is occupied by the Lagonaki Highland with a height of >1800 m. The climate is temperate with rather mild winters and rather hot summers. The annual rainfall reaches 700 mm, and the Lagonaki Highland is one of the wettest places in Russia, with an annual rainfall of up to 3000 mm and more. Winters are characterized by strong snowfalls. Deep river valleys either cross mountain ranges and cut narrow canyons and gorges, or, in contrast, they stretch along ranges and form wide valleys with well-developed terraces. The principal river is the Belaya River, which is a left tributary of the larger Kuban River (one of the largest rivers of the Russian South). The other rivers and streams are elements of the dense drainage network of the Belaya River. The study area is covered by dense vegetation, including deciduous, coniferous, and mixed forests, meadows (both Alpine and riverine), and plots of true steppe (grassland). The human settlement is not dense: the population is <20,000 in the area of ~150 km^2, and it is concentrated in a few towns and villages (no detailed statistics are available). Nonetheless, there is a rather developed road infrastructure (Figure 1). Notably, Mountainous Adygeya is one of the most important tourist destinations of the Russian South [33], attracting up to 0.5 million of visitors annually [34] (this is approximated, and the true tourist flows may be stronger—detailed statistics are absent).

Geologically, the study area is diverse, and it is dominated by the Mesozoic deposits accumulated in the tropical Caucasian Sea, which was a marginal semienclosed sea of the Tethys Ocean. The most widespread rocks are Early–Middle Jurassic shales and Late Jurassic limestones and dolostones. Precambrian metamorphics and Late Paleozoic granitoids crop out in the central part of the area, and the Early–Middle Permian molassic sequence is exposed in its southern part. Cretaceous siliciclastics and carbonates crop out in the northern part of the area. The Hercynian, Cimmerian, and Alpine phases of tectonic deformations resulted in highly-complex folding and faulting. Fifteen geosites (some of them are also geomorphosites sensu [35–37]) represent unique geological and geomorphological features of Mountainous Adygeya [25,26], some of which are ranked nationally and even globally.

The material for the present study was obtained in the course of field investigations in Mountainous Adygeya, and the majority of observations were made during the field campaign in summer 2021. A total of eleven bridges were visited and examined in regard to their utility for distant viewing and comprehension of the local geoheritage landscapes (Figure 1). This material is used to develop and test the approach explained below.

2.2. Terminology and Approach Proposal

A bridge not only connects two points/places divided by a topographic low or any other natural/artificial barrier (river, road, etc.), but it is also a relatively high point offering views of the surrounding landscape into two opposite directions (from both sides of bridges). Consequently, bridges are among potential viewpoints for distant sometimes panoramic viewing of geoheritage. Indeed, such a function works if unique geological objects are available near a given bridge and can principally be visible from there (for instance, if they are not masked by vegetation or located too far away to be recognized).

Relating bridges to geoheritage requires certain terminological justifications. Fuertes-Gutiérrez and Fernández-Martínez [8], Palacio [9], and Migoń and Pijet-Migoń [10] argued the importance of viewpoints, and the term "viewpoint geosite" has been coined. The specialists broadly agreed that such sites are characterized by duality (observation point and observable object), and they are very important for geoheritage comprehension. An attentive reading of the noted works [8–10] implies the existence of two categories of such sites, namely "ordinary" viewpoints allowing observation of distant geoheritage features and viewpoints allowing distant features to be recognized as really unique. Apparently, the only latter can be judged as true geosites. The situation is even more complicated because an "ordinary" viewpoint may provide exceptional opportunity to observe one distant geosite, several geosites, or even the entire geoheritage landscape (unique features in their broad geological and nongeological contexts). It would be wrong to not link such viewpoints to geosites. Regarding the above, it is sensible to specify geoheritage viewpoints as a broad category, and viewpoint geosites as its subcategory. The former embraces all points from which unique features are visible, and the latter are the most important of them. Viewpoint geosites themselves may or may not have some heritage value (cf. [10]), and they are something in between physical geosites (i.e., geosites with intrinsic value) and the so-called "virtual" geosites (sensu [5,38]). Technically, all geoheritage viewpoints are of utmost importance, as they facilitate geoconservation and geotourism (see above). They can be nonheritage sites, viewpoint geosites, or particular elements of large linear or areal geosites. When bridges make unique features visible, these are bridge-based geoheritage viewpoints, and this provisional term is employed in this study. Bridges can be added to the other sorts of manmade viewpoints distinguished by Migoń and Pijet-Migoń [10]. It should be added that some bridges constructed for tourism are located in places with panoramic views and high aesthetic properties.

Field excursions in geologically-rich areas allow easy identification of bridge-based geoheritage viewpoints. Their number depends on the drainage network density and the development of socioeconomical, transport, and touristic infrastructure of a given territory. However, this simple identification is not enough. The utility of some sites is larger than that of the others, and, thus, their assessment is necessary. Migoń and Pijet-Migoń [10] proposed a set of criteria for assessment of viewpoint geosites, and this can be used to develop a semiquantitative scoring-based approach for the assessment of bridge-based geoheritage viewpoints. The other developments of geosite assessment, which often consider panoramic viewing (e.g., see review in [39]), are also taken into account. Two remarks are necessary. First, this approach emphasizes the "technical" properties not the uniqueness of the observable features, because not all geoheritage viewpoints are viewpoint geosites (see above), and this uniqueness may or may not be understood distantly. Second, bridges have some specific properties, which need to be taken into account. In other words, bridge-based geoheritage viewpoints cannot be accessed exactly as viewpoint geosites or any other geosites.

The criteria and the related scores proposed for the semiquantitative assessment of bridge-based geoheritage viewpoints are summarized in Table 1, and several clarifications are provided below. First, if a given bridge is wide, it cannot offer a 360° panorama, but it provides two views (for instance, two 180° panoramas) from each side. Second, unique geological and geomorphological objects may not be visible from any side of a given bridge due to the curvature of slopes, dense vegetation, shadows, and constructions, or even may not exist. Third, the bridges accessible by only cars or trains are less valuable than the bridges accessible by only pedestrians because the flow of cars or trains cannot stop to allow observation of distant features. Fourth, in cases of required permissions or entrance fees, bridges lose a significant part of their accessibility, as permissions are not always easy to obtain and not all visitors are ready to pay for "just viewing". Fifth, there are bridges that are challenging for some (if not many) visitors to walk along. This is the case with too old or damaged bridges. In Mountainous Adygeya, there are several rather long (up to 200 m and more) hanging bridges. Although they are accessible to pedestrians, some visitors are not prepared (more psychologically than physically) to walk along them due to their swinging or feel discomfort standing in the center for "lazy" viewing. Moreover, bridge swinging complicates taking photos. Indeed, this is a serious limitation to accessibility.

Each given bridge can easily be assessed by the proposed criteria (Table 1), and, the total score indicates its relative utility. As for the latter, it is proposed tentatively to differentiate utility into three grades, i.e., low, high, and moderate utility. Indeed, the set of criteria and the related scores are provisional and can be justified in the course of further investigations. Nonetheless, they reflect the diversity of situations one can face in reality, and many of these situations were encountered during field investigations in the study area.

Table 1. Criteria proposed for semiquantitative assessment of bridge-based geoheritage viewpoints.

Criterion	Grade	Score
Panoramas and other views (P)	360° panorama	50
	120–180° panoramas from two sides	40
	120–180° panorama from one side and restricted view from another side	30
	120–180° panorama from only one side	25
	Restricted views on two sides	20
	Restricted view from only one side	10
Visibility of unique geological/geomorphological features (V)	Excellent (all details are visible)	30
	Mixed (some features are visible better than the others)	20
	Poor (too general a view)	10
Diversity of visible unique geological/geomorphological features (D)	>10 features	30
	4–10 features	20
	1–3 features	10
Accessibility (A)	By cars (or other transport) and pedestrians	30
	By pedestrians only	25
	By cars (or other transport) only	15
	By only prepared pedestrians (e.g., in the case of hanging bridges)	7
	Permission required (PE)	−25 (difficult to obtain) −5 (easy to obtain)
	Entrance fee (EF)	−5
Special constructions for comfortable observation (S)	Present	15
	Absent	0
Geological value of bridge itself (G)	Associated geoheritage	30
	Stone heritage	20
	Absent	0
Cultural value of bridge itself (C)	Present	15
	Absent	0
TOTAL SCORE (total utility)	Maximum	200
	Minimum	35
	High utility	>120
	Moderate utility	80–120
	Low utility	<80

3. Results

The application of the proposed approach to the eleven bridges of Mountainous Adygeya enabled characterizing them individually and generally. This information is presented below. Additionally, one bridge was selected to demonstrate how the approach worked in detail.

3.1. General Characteristics

Eleven bridges were judged as geoheritage viewpoints in the study area (Figures 1 and 2, Table 2). Importantly, these are found in different parts of the Mountainous Adygeya geodiversity hotspot. Many of them are bridges along the principal road connecting Maykop and Guzeripl. This road stretches through the area from the north to the south, and it crosses the Belaya River and its tributaries several times. Some bridges have been constructed for the local needs, and some are elements of the local touristic infrastructure. Three general types of bridges are distinguished, namely capital constructions (usually made of concrete), light constructions (metallic and woody, often mixed), and hanging constructions (usually metallic with woody pavement) (Table 2).

Figure 2. Selected bridges of the study area. IDs correspond to Figure 1 and Table 2.

Table 2. Semiquantitative assessment of the bridge-based geoheritage viewpoints of the Mountainous Adygeya geodiversity hotspot.

ID	Locality (River)	Type	Criteria (See Table 1 For Abbreviations and Scoring System)							
			P	V	D	A	S	G	C	TOTAL
1	Guzeripl (Belaya)	Capital	30	20	10	30	0	0	0	90
2	Kisha (Belaya)	Capital	20	30	10	30 − 25PE = 5	0	30	0	95
3	Sibirka (Belaya)	Capital	40	30	10	30	15	0	0	125
4	Belaya Rechka (Belaya)	Hanging	50	10	10	7	0	0	0	77
5	Dakhovskaya (Dakh)	Capital	25	10	10	30	15	0	15	105
6	Rufabgo—inner 1 (Syryf)	Light	20	20	10	25 − 5EF = 20	15	0	0	85
7	Rufabgo—inner 2 (Syryf)	Light	20	20	10	25 − 5EF = 20	15	0	0	85
8	Rufabgo—entrance (Belaya)	Light	40	30	20	25 − 5EF = 20	15	0	0	125
9	Kamennomostskiy (Belaya)	Hanging	50	30	10	7	0	0	0	97
10	Polkovnitskaya (Belaya)	Hanging	50	10	10	7 − 5PE = 2	0	0	0	72
11	Novoprokhladnoe (Kamennaya)	Light	20	20	10	25	0	0	15	90

Notes: IDs correspond to Figures 1 and 2; PE—permission required; EF—entrance fee (see Table 1).

The considered bridges differ by their view and age, although all connect river banks (Figure 2). One bridge was constructed at the beginning of the 20th century with old technologies (#5 on Figures 1 and 2); now, this is a double bridge consisting of an old bridge and the parallel modern bridge. The majority of the bridges were constructed in Soviet times, with common maintenance in postSoviet times (for instance, #9 on Figures 1 and 2). Several bridges were constructed as elements of the tourist infrastructure in the mid-2000s (for instance, #4, #7, and #8 on Figures 1 and 2). Importantly, all these bridges are high and long, and they offer spectacular views of the geoheritage landscapes.

The properties of the bridge-based geoheritage viewpoints differed substantially (Table 2). First of all, the bridges provided different possibilities for viewing geoheritage. Interestingly, the hanging bridges (for instance, #4 on Figures 1 and 2) were ideal objects offering 360° panoramas because they were narrow. In several cases (for instance, #7 on Figures 1 and 2), the bridges crossed narrow river valleys with rather steep slopes and dense vegetations and shadows, as a result of which the views were restricted. The visibility of unique features also differed (Table 2). It was excellent in many cases, when visitors could see details of the geological and geomorphological objects. However, it was lower in many other cases. For instance, the bridge over the Dakh River (#5 on Figure 1) offered an excellent opportunity to enjoy the 180° panorama of the cuesta-type range. However, its scarp with outcrops of the Late Jurassic carbonates was visible from such a distance as a very narrow yellow strip, which could not be understood correctly without specific knowledge. Moreover, the frontal view of the cuesta scarp complicated interpretation of this landform.

The diversity of the visible features was generally low (Table 2). With one exception (see example below), the number of these features did not exceed three. The most striking difference was linked to accessibility (Table 2). It was low in the case of the hanging bridges, but it was even lower when permissions are required. For instance, the bridge over the Belaya River near the mouth of the Kisha River (#2 on Figure 1) offered a spectacular view of the red-colored Early–Middle Permian molassic sequence, which is a legacy of the Hercynian orogeny in the Greater Caucasus. However, this bridge was closed, as private property. Nonetheless, more than half of the considered bridges were relatively well accessible, with scores of 20 or more. It was very important that several bridges had spaces for comfortable observations of the geoheritage (Table 2), and some of them (#6–8 on Figures 1 and 2) were especially constructed to allow tourists to enjoy the views of the local landscapes. There was one bridge with intrinsic geoheritage value (#2 on Figure 1). When this bridge and the nearby road were maintained, huge clasts of Late Jurassic reefal limestones were used. As a result, excellent specimens of ancient corals have been found near the foundation of this bridge. Finding similar specimens in natural outcrops would be a challenging task. Finally, there were two bridges with cultural value (Table 2). One of them was the old bridge over the Dakh River (#5 on Figures 1 and 2), which seemed to be a true architectural heritage of Dakhovskaya village. It was built in the beginning of the 20th century with some old technologies (for instance, eggs were added to cement), and it is one of the local symbols; it was employed by the Russian film industry.

Despite the above-mentioned differences, the total utility of the bridge-based geoheritage viewpoints of Mountainous Adygeya does not differ strikingly and varies within 72–125 (most commonly, within 80–100) (Table 2). According to the proposed grades (Table 1), two bridges had low utility (close to the upper limit of the grade), seven bridges had moderate utility, and two other bridges had high utility (close to the lower limit of the grade). Generally, this indicates that these viewpoints can contribute to geoconservation and geotourism. Their contribution may be judged significant because of their wide distribution rather than by their outstanding utility, which is high in the only two cases.

3.2. Case Study

The bridge at the entrance to the Rufabgo touristic attraction (#8 on Figures 1 and 2) received the highest scores (Table 2), and it was chosen as a representative example for detailed characteristics. This light metallic bridge was constructed at the beginning of the 2000s as a private initiative to connect the banks of the Belaya River and, thus, to offer the shortest way to the Rufabgo waterfalls (one of the most known attractions of the Russian South with dozens and even hundreds of visitors every day) from the principal road. This bridge is located in the Khadzhokh canyon, which is a large object with an unprecedented concentration of unique geological and geomorphological features; this is a proven geosite [40]. The length of the bridge was ~50 m, its relative height exceeded 10 m, and the width of the bridge was ~2 m. Standing in the middle did not allow one to comprehend the unique features at the valley's bottom, and, thus, 180° panoramic views were offered from each side of the bridge (Figure 3). As it was inside the canyon, the distance from the unique features was not large, and their visibility was excellent.

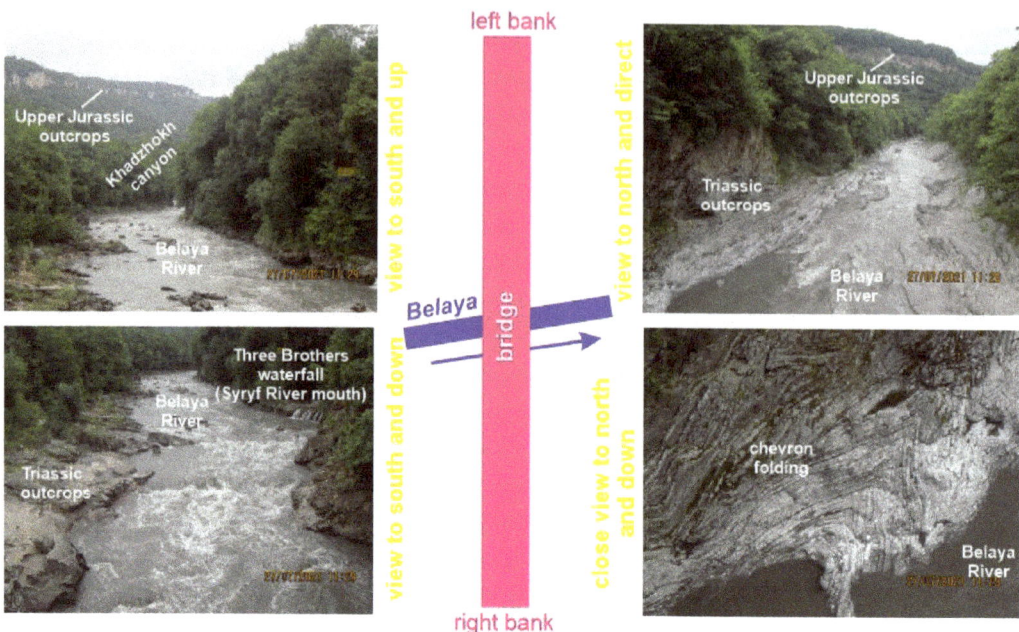

Figure 3. The entrance bridge to the Rufabgo tourist attraction.

The unique features included the Khadzhokh canyon itself, the Triassic outcrops stretching along the Belaya River (Mountainous Adygeya boasts one of the most complete sections of the entire Triassic in Russia), chevron folding of the Triassic rocks, the Late Jurassic carbonates cropped out in the upper part of the canyon, and, finally, the small Three Brothers waterfall (this is the so-called "hanging mouth" of the Syryf River, and it is the smallest of the Rufabgo waterfalls). These features are not only interesting scientifically and diverse but also very spectacular. For instance, chevron folds are structural features, the formation of which is linked to specific kinematics [41,42]. The lengthy outcrops of the Anisian (Middle Triassic) layered limestones along the Belaya River exhibit chevron folding, which creates a pattern of outstanding aesthetic value (Figure 3). According to Gaetani et al. [43], these structures formed in the second half of the Anisian stage when the area experienced significant Cimmerian deformations due to plate tectonic reorganizations at the southern Eurasian margin. The Greater Caucasus is understood as a Galatian terrane derived from Gondwana in the midPaleozoic, attached to the Proto-Alpine area (somewhere near the Carnic Alps) in the Late Paleozoic and then moved to its present position in the Triassic [44]. Regarding this scenario, the formation of chevron folds can be attributed to the phase when this terrane shifted eastwards, which caused unavoidable active contact with the other tectonic blocks.

The bridge was excellently accessible to pedestrians, and it could accommodate up to several dozen people simultaneously. However, walking along the bridge required paying a fee of 500 RUR (~7 USD), which is neither expensive nor inexpensive, and this fact restricted slightly this bridge's accessibility. In contrast, some construction peculiarities contributed to its value. First, the bridge was constructed so to not only connect the river banks but to allow viewing far along the canyon, which was not complicated by slope curvature, vegetation, or shadows. Second, the bridge had small balconies on both sides, which were specially designed to serve as comfortable observation points.

Despite the intrinsic diversity of the entire geosite, its geometry was complex and did not allow observation of even half of the unique features from any single place [40]. The considered bridge-based geoheritage viewpoint was the only place where so many features could be viewed simultaneously and with so much clarity (Figure 3). Moreover, this bridge enabled comprehension of the essence of the canyon from within. Therefore, the importance of this bridge was outstanding, and it had no natural or artificial analogues in this geosite. The high total scores for the utility of this viewpoint (Table 2) imply its potential is fully realized.

4. Discussion and Conclusions

The more or less significant utility and the wide distribution of the bridge-based geoheritage viewpoints in Mountainous Adygeya established by the semiquantitative assessment (Table 2) imply their general importance to this geodiversity hotspot. A question for study is how common is the situation in which viewpoints are so important. In the study area, the utility of the bridge-based geoheritage viewpoints was determined by the dense drainage network, the subparallel positions of the principal river and the principal road, and the exposure of unique features along the rivers. Such situations seem to be common to well-precipitated mountain domains with more or less developed road infrastructure. Domains of this kind can be found in Europe, North America, Southeast Asia, and many other parts of the world. If so, bridge-based geoheritage viewpoints have universal value, although their utility can vary depending on the peculiarities of each particular territory.

Geoheritage viewpoints are not equal to viewpoint geosites (sensu [10,11]), and, thus, the spatial relations between the bridge-based geoheritage viewpoints and the geosites in Mountain Adygeya should be clarified. Although geoheritage mapping in the study area is still in progress, the preliminary information on geoheritage distribution (for instance, it is partly summarized in [26]) allows such a clarification. From the eleven bridges considered for the present study, ten items are elements of the larger geosites, and only the Dakh bridge occurs individually (Table 3). The latter demonstrates moderate utility (Table 2), and it has significant intrinsic cultural value (see above). Consequently, it appears logical to recognize this bridge as a cultural heritage site, and not as a viewpoint geosite. Nonetheless, this site serves well as an "ordinary" geoheritage viewpoint.

Table 3. Spatial relations of the bridge-based geoheritage viewpoints to the geosites of the Mountainous Adygeya geodiversity hotspot.

ID	Key Visible Features	Geosite (Nomenclature after [26])	Approximate Representation of the Geosite Uniqueness in the View from Bridge
1	Early Jurassic shales	Molchepa locality	20%
2	Permian red molasse	Khamyshki section	30%
3	Late Paleozoic granites, pseudo-karst, gorge	Granite gorge	30%
4	Early–Middle Jurassic shales, folds	Syuk valley and locality	<10%
5	Late Jurassic carbonates, cuesta range	not attributed to any geosite	
6	Triassic rocks, chevron folds, waterfall, canyons, Late Jurassic carbonates	Khadzhokh canyon system and Rufabgo waterfalls	<10%
7			10%
8			25%
9	Late Jurassic sabkha deposits	Kamennomostskiy variegated rocks	50%
10	Lower Cretaceous deposits, including Aptian green glauconitic sandstones	Polkovnitskaya valley	<10%
11	Early–Middle Jurassic shales	Sakhray canyon	<10%

Notes: IDs correspond to Figures 1 and 2 and Table 2.

Evidently, all geoheritage viewpoints are especially important to geotourism development, and the bridges are not excluded. First, distant views solve the problem of the accessibility of the far-located features. Second, such views facilitate understanding the unique features in their broad natural context. This is a kind of shift from viewing any particular unique feature to the observation of geoheritage landscape. Third, panoramic viewing itself is enjoyable to tourists. This may also stimulate senses and, thus, contribute to the destination sensescape [45]. Fourth, bridges are notable constructions, visiting which may be interesting to tourists for nongeological reasons (hanging bridges make excursions adventurous). Fifth, bridges are related to roads and touristic infrastructure, i.e., they simplify paying attention to geological and geomorphological features from the common nongeotourist routes. If so, bridges would strengthen significantly the potential of roadside geotourism [46,47]. More generally, bridge-related geoheritage viewpoints facilitate geotourist activities, diversify geotourists' experience, and contribute to the integration of geological and nongeological tourism. This finding has two practical implications. First, geoheritage inventory and geotourism planning need to pay attention to the locally available bridges. The proposed approach of semiquantitative assessment can help in both inventory and planning. Second, bridges can be especially constructed to facilitate geotourism development on geologically-rich territories and in geoparks.

Conclusively, the evidence from the Western Caucasus indicates the bridge-based geoheritage viewpoints as very useful objects. In the study area, all eleven bridges demonstrate a certain utility (chiefly moderate) for the observation of many unique geological and geomorphological features. The approach proposed for the semiquantitative assessment of such viewpoints enables distinguishing bridge-based geoheritage viewpoints with various properties, and it was tested successfully. Nonetheless, this approach is still too tentative, and broad discussion among experts in geoconservation and geotourism is necessary in order to justify and universalize it. For instance, the carrying capacity of bridges, availability of artificial light, presence of signs explaining viewpoint opportunities, etc., may also be taken into account. For those bridge-based geoheritage viewpoints, which are viewpoint geosites, the geological uniqueness should be examined, and the relation of the unique features of bridges themselves require some specific interpretations. This seems to be a vast field for investigation and discussion. After further refinement, the approach can be demanded by practitioners because bridges provide unique opportunities for strengthening geotourism programs and geopark management.

Author Contributions: Conceptualization, V.A.E. and D.A.R.; methodology, V.A.E. and D.A.R.; investigation, A.V.M. and D.A.R.; writing—original draft preparation, D.A.R.; writing—review and editing, A.V.M., V.A.E. and D.A.R. All authors have read and agreed to the published version of the manuscript.

Funding: This research received no external funding.

Data Availability Statement: Not applicable.

Acknowledgments: The authors gratefully thank the editors for this special volume for their kind invitation to contribute and all three reviewers for their helpful recommendations. N.V. Ruban (Russia) is thanked for her field assistance, and G.I. Doludenko (Russia) is thanked for his driving support.

Conflicts of Interest: The authors declare no conflict of interest.

References

1. Crofts, R.; Tormey, D.; Gordon, J.E. Introducing New Guidelines on Geoheritage Conservation in Protected and Conserved Areas. *Geoheritage* **2021**, *13*, 33. [CrossRef]
2. Dowling, R.; Newsome, D. (Eds.) *Handbook of Geotourism*; Edward Elgar: Cheltenham, UK, 2018.
3. Henriques, M.H.; Canales, M.L.; García-Frank, A.; Gomez-Heras, M. Accessible Geoparks in Iberia: A Challenge to Promote Geotourism and Education for Sustainable Development. *Geoheritage* **2019**, *11*, 471–484. [CrossRef]
4. Kubalíková, L.; Bajer, A.; Balková, M. Brief notes on geodiversity and geoheritage perception by the lay public. *Geosciences* **2021**, *11*, 54. [CrossRef]
5. Pasquaré Mariotto, F.; Bonali, F.L. Virtual geosites as innovative tools for geoheritage popularization: A case study from Eastern Iceland. *Geosciences* **2021**, *11*, 149. [CrossRef]
6. Reynard, E.; Brilha, J. (Eds.) *Geoheritage: Assessment, Protection, and Management*; Elsevier: Amsterdam, The Netherlands, 2018.
7. Santangelo, N.; Valente, E. Geoheritage and Geotourism resources. *Resources* **2020**, *9*, 80. [CrossRef]
8. Fuertes-Gutiérrez, I.; Fernández-Martínez, E. Geosites Inventory in the Leon Province (Northwestern Spain): A Tool to Introduce Geoheritage into Regional Environmental Management. *Geoheritage* **2010**, *2*, 57–75. [CrossRef]

9. Palacio, J. Viewpoints and geological heritage. Uses in tourism and education. In *Towards the Balanced Management and Conservation of the Geological Heritage in the New Millenium*; Barettino, D., Vallejo, M., Gallego, E., Eds.; Sociedad Geológica de España: Madrid, Spain, 1999; pp. 378–384.
10. Migoń, P.; Pijet-Migoń, E. Viewpoint geosites—Values, conservation and management issues. *Proc. Geol. Assoc.* **2017**, *128*, 511–522. [CrossRef]
11. Mikhailenko, A.V.; Ruban, D.A. Environment of viewpoint geosites: Evidence from the Western Caucasus. *Land* **2019**, *8*, 93. [CrossRef]
12. Habibi, T.; Ponedelnik, A.A.; Yashalova, N.N.; Ruban, D.A. Urban geoheritage complexity: Evidence of a unique natural resource from Shiraz city in Iran. *Resour. Policy* **2018**, *59*, 85–94. [CrossRef]
13. Henriques, M.H.; dos Reis, R.P. Storytelling the Geoheritage of Viana do Castelo (NW Portugal). *Geoheritage* **2021**, *13*, 46. [CrossRef]
14. Mariño, J.; Cueva, K.; Thouret, J.-C.; Arias, C.; Finizola, A.; Antoine, R.; Delcher, E.; Fauchard, C.; Donnadieu, F.; Labazuy, P.; et al. Multidisciplinary Study of the Impacts of the 1600 CE Huaynaputina Eruption and a Project for Geosites and Geo-touristic Attractions. *Geoheritage* **2021**, *13*, 64. [CrossRef]
15. Migoń, P.; Różycka, M. When Individual Geosites Matter Less—Challenges to Communicate Landscape Evolution of a Complex Morphostructure (Orlické–Bystrzyckie Mountains Block, Czechia/Poland, Central Europe). *Geosciences* **2021**, *11*, 100. [CrossRef]
16. Paungya, N.; Singtuen, V.; Won-In, K. The preliminary geotourism study in Phetcahbun Province, Thailand. *GeoJ. Tour. Geosites* **2020**, *31*, 1057–1067. [CrossRef]
17. Tessema, G.A.; Poesen, J.; Verstraeten, G.; Van Rompaey, A.; Van Der Borg, J. The scenic beauty of geosites and its relation to their scientific value and geoscience knowledge of tourists: A case study from southeastern Spain. *Land* **2021**, *10*, 460. [CrossRef]
18. Adamia, S.; Zakariadze, G.; Chkhotua, T.; Sadradze, N.; Tsereteli, N.; Chabukiani, A.; Gventsadze, A. Geology of the Caucasus: A review. *Turk. J. Earth Sci.* **2011**, *20*, 489–544.
19. Frolova, M. The landscapes of the Caucasus in Russian geography: Between the scientific model and the sociocultural representation. *Cuad. Geogr.* **2006**, *38*, 7–29.
20. Kaban, M.K.; Gvishiani, A.; Sidorov, R.; Oshchenko, A.; Krasnoperov, R.I. Structure and density of sedimentary basins in the southern part of the east-European platform and surrounding area. *Appl. Sci.* **2021**, *11*, 512. [CrossRef]
21. Lurie, P.M.; Panov, V.D.; Panova, S.V. Cryosphere of the Greater Caucasus. *Sustain. Dev. Mt. Territ.* **2019**, *11*, 182–190. [CrossRef]
22. Rantsman, E.Y. Morphostructural subdivision and some problems of geodynamics of the Great Caucasus. *Geomorfologiya* **1985**, *1*, 3–16.
23. Van Hinsbergen, D.J.J.; Torsvik, T.H.; Schmid, S.M.; Matenco, L.C.; Maffione, M.; Vissers, R.L.M.; Gürer, D.; Spakman, W. Orogenic architecture of the Mediterranean region and kinematic reconstruction of its tectonic evolution since the Triassic. *Gondwana Res.* **2020**, *81*, 79–229. [CrossRef]
24. Viginsky, V.A. Main stages of relief evolution and topographic steps of the west Great Caucasus and adjacent foothills. *Geomorfologiya* **1986**, *2*, 44–53.
25. Ruban, D.A. Quantification of geodiversity and its loss. *Proc. Geol. Assoc.* **2010**, *121*, 326–333. [CrossRef]
26. Mikhailenko, A.V.; Ruban, D.A.; Ermolaev, V.A. Accessibility of geoheritage sites—A methodological proposal. *Heritage* **2021**, *4*, 1080–1091. [CrossRef]
27. Lozovoj, S.P. *The Lagonaki Highland*; Krasnodarskoe knizhnoe izdatel'stvo: Krasnodar, USSR, 1984. (In Russian)
28. Rostovtsev, K.O.; Agaev, V.B.; Azarian, N.R.; Babaev, R.G.; Beznosov, N.V.; Hassanov, N.A.; Zesashvili, V.I.; Lomize, M.G.; Paitschadze, T.A.; Panov, D.I.; et al. *The Jurassic of the Caucasus*; Nauka: St. Petersburg, Russia, 1992. (In Russian)
29. Ruban, D.A. *Mountains Ranges and Summits of the Northeastern Periphery of the Lagonaki Highland*; DGTU-Print: Rostov-on-Don, Russia, 2020. (In Russian)
30. Davis, W.M. The Drainage of Cuestas. *Proc. Geol. Assoc.* **1899**, *16*, 75–93. [CrossRef]
31. Duszyński, F.; Migoń, P.; Strzelecki, M.C. Escarpment retreat in sedimentary tablelands and cuesta landscapes—Landforms, mechanisms and patterns. *Earth-Sci. Rev.* **2019**, *196*, 1028. [CrossRef]
32. Peterek, A.; Schröder, B. Geomorphologie evolution of the cuesta landscapes around the Northern Franconian Alb-Review and synthesis. *Z. Geomorphol.* **2010**, *54*, 305–345. [CrossRef]
33. Ivlieva, O.V.; Shmytkova, A.V.; Sukhov, R.I.; Kushnir, K.V.; Grigorenko, T.N. Assessing the tourist and recreational potential in the South of Russia. *E3S Web Conf.* **2020**, *208*, 05013. [CrossRef]
34. Ruban, D.A.; Mikhailenko, A.V.; Zorina, S.O.; Yashalova, N.N. Geoheritage Resource of a Small Town: Evidence from Southwestern Russia. *Geoheritage* **2021**, *13*, 82. [CrossRef]
35. Panizza, M. Geomorphosites: Concepts, methods and examples of geomorphological survey. *Chin. Sci. Bull.* **2001**, *46*, 4–6. [CrossRef]
36. Reynard, E.; Coratza, P.; Giusti, C. Geomorphosites and Geotourism. *Geoheritage* **2011**, *3*, 129–130. [CrossRef]
37. Reynard, E.; Coratza, P.; Hobléa, F. Current Research on Geomorphosites. *Geoheritage* **2016**, *8*, 1–3. [CrossRef]
38. Pasquaré Mariotto, F.; Antoniou, V.; Drymoni, K.; Bonali, F.L.; Nomikou, P.; Fallati, L.; Karatzaferis, O.; Vlasopoulos, O. Virtual geosite communication through a webgis platform: A case study from Santorini island (Greece). *App. Sci.* **2021**, *11*, 5466. [CrossRef]

39. Štrba, L.; Kršák, B.; Sidor, C. Some Comments to Geosite Assessment, Visitors, and Geotourism Sustainability. *Sustainability* **2018**, *10*, 2589. [CrossRef]
40. Mikhailenko, A.V.; Ruban, D.A.; Ermolaev, V.A. The Khadzhokh Canyon System-An Important Geosite of the Western Caucasus. *Geosciences* **2020**, *10*, 181.
41. Bastida, F.; Aller, J.; Toimil, N.C.; Lisle, R.J.; Bobillo-Ares, N.C. Some considerations on the kinematics of chevron folds. *J. Struct. Geol.* **2007**, *29*, 1185–1200. [CrossRef]
42. Wu, Y.; Eckert, A.; Liu, X.; Obrist-Farner, J. The role of flexural slip during the development of multilayer chevron folds. *Tectonophysics* **2019**, *753*, 124–145. [CrossRef]
43. Gaetani, M.; Garzanti, E.; Polino, R.; Kiricko, Y.; Korsakhov, S.; Cirilli, S.; Nicora, A.; Rettori, R.; Larghi, C.; Bucefal Palliani, R. Stratigraphic evidence for Cimmerian events in NW Caucasus (Russia). *Bull. Soc. Géol. Fr.* **2005**, *176*, 283–299. [CrossRef]
44. Ruban, D.A. The Greater Caucasus—A Galatian or Hanseatic terrane? Comment on "The formation of Pangea" by G.M. Stampfli, C. Hochard, C. Vérard, C. Wilhem and J. von Raumer [Tectonophysics 593 (2013) 1–19]. *Tectonophysics* **2013**, *608*, 1442–1444. [CrossRef]
45. Buzova, D.; Sanz-Blas, S.; Cervera-Taulet, A. "Sensing" the destination: Development of the destination sensescape index. *Tour. Manag.* **2021**, *87*, 104362.
46. Ranjbaran, M.; Sotohian, F. Development of Haraz Road geotourism as a key to increasing tourism industry and promoting geoconservation. *Geopersia* **2021**, *11*, 61–79.
47. Štrba, L.; Baláž, B.; Lukác, M. Roadside geotourism—An alternative approach to geotourism. *e-Rev. Tour. Res.* **2016**, *13*, 598–609.

Article

Inventory and Assessment of the Geomorphosites in Central Cyclades, Greece: The Case of Paros and Naxos Islands

Niki Evelpidou *, Anna Karkani, Maria Tzouxanioti, Evangelos Spyrou, Alexandros Petropoulos and Lida Lakidi

Faculty of Geology and Geoenvironment, National and Kapodistrian University of Athens, 15784 Zografou, Greece; ekarkani@geol.uoa.gr (A.K.); mtzouxanioti@geol.uoa.gr (M.T.); evspyrou@geol.uoa.gr (E.S.); alexpetrop@geol.uoa.gr (A.P.); leda.lak@gmail.com (L.L.)
* Correspondence: evelpidou@geol.uoa.gr

Citation: Evelpidou, N.; Karkani, A.; Tzouxanioti, M.; Spyrou, E.; Petropoulos, A.; Lakidi, L. Inventory and Assessment of the Geomorphosites in Central Cyclades, Greece: The Case of Paros and Naxos Islands. *Geosciences* **2021**, *11*, 512. https://doi.org/10.3390/geosciences11120512

Academic Editors: Deodato Tapete and Jesus Martinez-Frias

Received: 15 November 2021
Accepted: 10 December 2021
Published: 14 December 2021

Publisher's Note: MDPI stays neutral with regard to jurisdictional claims in published maps and institutional affiliations.

Copyright: © 2021 by the authors. Licensee MDPI, Basel, Switzerland. This article is an open access article distributed under the terms and conditions of the Creative Commons Attribution (CC BY) license (https:// creativecommons.org/licenses/by/ 4.0/).

Abstract: The Cycladic landscape is characterized by landforms of natural beauty and rarity. Landforms resulting from differential erosion, weathering, tectonics, drainage network, sea level changes, and depositional processes can contribute to the development of geotourism in the area. This can be achieved by supporting conservation, protection and promotion of the geo-environment and nature, educating students, residents, and visitors. The aim of this work is to develop an inventory of the main geomorphosites of Paros and Naxos islands by assessing their scientific and additional values, using qualitative and quantitative criteria. Our results show that, besides the high scientific interest of the 75 geomorphosites, most are also characterized by a high ecological value and can potentially lead to a significant increase in the islands' tourism. The results of this work aim at raise awareness on the geomorphological heritage of central Cyclades and provide a basis for their promotion, protection, and management.

Keywords: geomorphological heritage; geomorphological synthesis; geocultural sites; geoheritage assessment; Aegean

1. Introduction

During the last decades, there have been several attempts for the promotion and preservation of the geomorphological and geological heritage in several regions [1–3]. Interest in geoconservation and geomorphological heritage dates back to the 1990s [4]. Geomorphological heritage refers to the total of geomorphosites of an area [5] and can be a witness to climate change, tectonic evolution and the related changes in the history of life at the surface of the Earth. Geomorphosites are areas of particular geomorphological interest, and several authors state that they can reveal part of the Earth's history regarding, for instance, the palaeoclimate, the palaeogeography, the palaeoecology, etc. [5–7]. Geomorphosites usually include landforms, but several authors [7,8] state that any part of the surface of Earth can be considered as a geomorphosite, as long as it contributes to the knowledge and/or comprehension of the Earth's history. The conditions under which the geomorphosites have formed, i.e., geological, geomorphological, tectonic, climatic etc., can aid in the knowledge and comprehension of geology and Earth history.

Geomorphosites are usually not only characterized by geomorphological interest, but by geological, ecological, environmental, cultural and/or archaeological, etc. as well, whereas the aesthetic and/or socioeconomic value is also of paramount significance [5,8–10]. What is more, a landform cannot be considered as a geomorphosite if its interest is limited to geomorphology. Panizza [5] has therefore suggested that the term "geomorphosite" can have two different definitions. The first one has already been mentioned and only includes the geomorphological interest, while the second one also includes the other aforementioned values (scientific, cultural, economic, etc.).

The assessment of the geomorphological heritage has been developing since the 1990s, and attention is paid to both geotourism, geomorphological–cultural heritage, etc., and the environmental impact [9,11,12]. Yet, although the quantitative assessment of each value follows certain criteria according to the methodology of Reynard et al. [1], the assessment is subjective and is relevant to the assessors' experience [1,13–15]. The assessment of geomorphological heritage and geoheritage in general and the evaluation of individual sites of interest can aid in its promotion, as well as its preservation. This means that an area of geomorphological interest has been greatly studied and it has been assessed regarding its scientific, cultural, aesthetic, etc. value, which is more likely to be promoted, resulting in the attraction of more geotourists and contributing to the economy of the local society, and what is more, it can also be preserved against degeneration due to time, weather conditions, vandalisms by the massive tourism or even by the locals. Geotourism is of paramount importance concerning an area's economic development, as it regards both the primary tourism and secondary tourism forms [16,17]. Several research groups have developed qualitative and quantitative methods for the assessment of geomorphosites (Reynard et al., 2016b) [18–20]. De Lima et al. [21] proposed four criteria to be considered when conducting an inventory: the topic, the value, the scale, and the use. The proposal was accepted by many scientists, but no universal consensus exists on the assessment methods [22]. The research on geomorphosites is currently developed at geoheritage assessment and inventories and geoheritage management at specific geomorphological environments and through the collaboration with other related sectors [23].

The mapping of the geomorphosites is another aspect. However, it has gained much less interest than their evaluation itself [24]. For instance, the corresponding framework, as well as global symbology and other similar mapping aspects have yet to develop [24,25]. Carton et al. [26,27] suggested a differentiation between the map depiction that concerns specialists (e.g., geomorphologists) and non-specialists. Coratza and Regolini-Bissig (2009) have also suggested geomorphosite mapping methods.

One of the countries in Europe that has great potential on geoheritage is Greece. Its interest regarding geomorphology is intense, as it belongs to the active Greek orogenetic arc [28], which shows its unique geological, hydrological and geomorphological features [29]. The complex geological and geomorphological setting and evolution, the great variety in climatic conditions and the numerous islands and indented coastline make Greece a region with a great diversity of natural sceneries [30].

Since 1937, natural areas of specific geological and ecological importance have been placed under special protection. Several geological heritage sites of international value are nationally protected areas and belong to Geoparks. The first two geoparks in Greece were the Lesbos Petrified Forest and the Psiloritis Geopark in Crete [30]. Today there are four more geoparks, Chelmos–Vouraikos UNESCO Global Geopark in the Peloponnese, Vikos–Aoos UNESCO Global Geopark in Epirus, Sitia UNESCO Global Geopark in Crete and Grevena–Kozani UNESCO Global Geopark in Macedonia [31].

In Greece, not many actions have been taken for the assessment of geomorphosites, and this work aims at filling this gap by giving prominence to the geomorphological heritage. More specifically, the two areas of interest are Naxos and Paros islands, Cyclades (Figure 1). These islands have been selected because they are directly related to the active Greek orogenetic arc, whereas there are sufficient data related to their geomorphology. They belong to the Cyclades and are part of the Cycladic plateau, whose morphology has been influenced by climatic conditions since Miocene times and the geomorphology presents some particular characteristics [32].

In the present work, a total of 75 geomorphosites on the islands of Naxos and Paros were selected and assessed according to the methodology proposed by Reynard et al. [1]. In particular, their scientific value was initially graded, referring to integrity, rarity, representativity and palaeogcographical value. Additional assessed values include ecology, economy, culture and aesthetics [1]. All values were synthesized, and thus the final assessment of the geomorphosites was conducted.

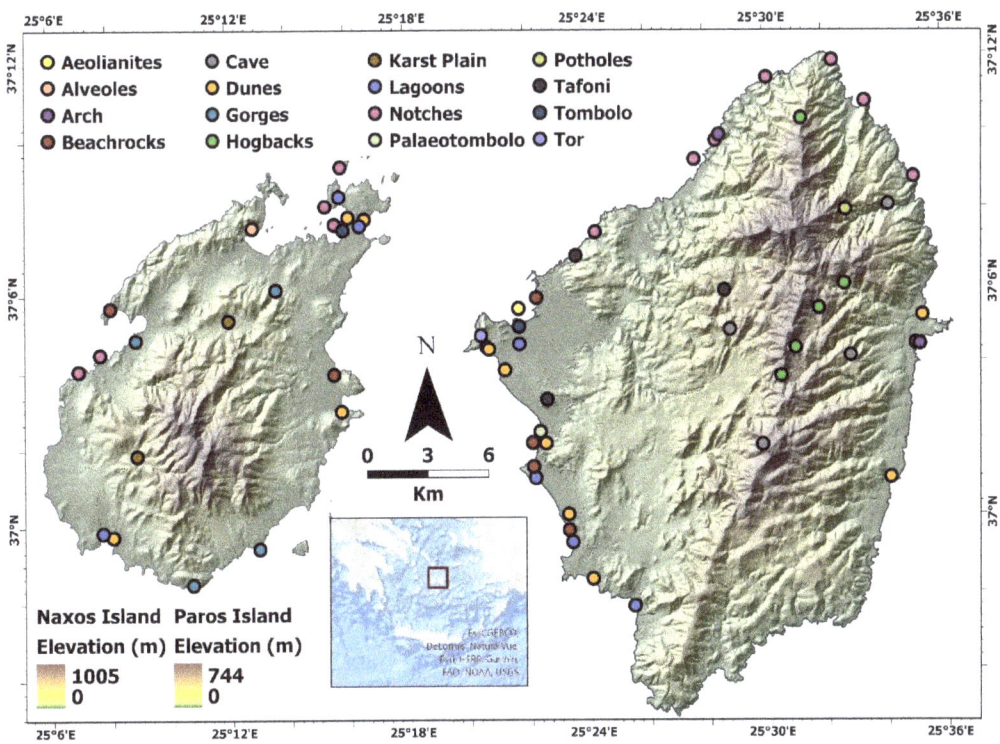

Figure 1. Naxos and Paros islands. The selected geomorphosites have been classified and are shown in the map.

2. Study Area

The study area is located in the central Aegean Sea and includes Naxos and Paros islands, both of which belong to the Cyclades complex, Aegean Sea, Greece. The Cyclades consist of about 33 main Islands, the largest being Naxos, while Paros is the third largest. Naxos covers an area of 430 Km^2, while its coastal zone extends for 148 Km. The island is mountainous, with a central mountain range trending N-S. The maximum altitude is 1001 m, the top of Zeus. Paros is located to the west of Naxos. It has an elliptical shape with a total area of 196 Km^2. The territory of the island is mountainous in the center and flat on the coast. The altitude increases toward the interior of the island and specifically, increases in the central and southern part. The highest point of the island is the peak of Aghios Ilias with an altitude of 771 m [33].

Lithologically, Naxos consists of migmatized gneissic rocks containing marbles, metapelites and amphibolites, while a sequence of schists, gneisses, marbles and metavolcanics surround the migmatite dome [34]. Above the metamorphic series, Miocene–Pliocene sediments are found [35,36], mainly deposited in a fluvial, partly marine environment. The lithology of Paros is dominated by marbles, gneiss and schists, above which Neogene and Quaternary deposits are found [33].

Geomorphological Regime

The islands of the Cyclades form the so-called Cycladic plateau, whose altitudes generally increase eastward (the eastern Cycladic islands are larger in area and higher in altitudes than the western ones). The largest part of the Cycladic plateau is currently located beneath sea level and at an average depth of 100 m. A deep N-S tectonic graben segregates the westernmost islands (e.g., Kea and Kythnos) from the rest of the Cyclades.

The marginal slopes of the plateau are located at a depth of approximately 200 m and generally coincide with the continental slope (Figure 2).

The Aegean Sea is located in the convergence zone of the African and Eurasian plates [37]. Naxos and Paros islands are located in the center of the Aegean microplate (which is part of the Eurasian plate) and are relatively seismically inactive, although the regions surrounding the Cyclades are seismically active [37]. During the Eocene, tectonic compression took place, which was succeeded by an extensional regime during Oligocene or Miocene [38–40]. The Cycladic region is characterized by a relatively thin continental crust of about 25–26 Km, which is owed to two processes during the Cenozoic and Neogene. During the Cenozoic gravitational collapse of the Aegean crust took place due to the southward retreat of a subduction front, while during the Neogene, the extrusion of the Anatolian block westward took place. These procedures were due to the extensional tectonic regime at the center of the Aegean plate, as the area is behind the modern volcanic arc of Aegean (e.g., [41,42]).

During the Quaternary, active tectonics played a significant role in the geomorphology of the Cyclades, but during Late Quaternary, eustasy also played a determining role. During the Last Glacial Maximum (18 Ka BP), most of the Cyclades were part of one single island and were gradually segregated when the sea-level rose. Paros and Naxos were the last islands that were separated, 8 Ka BP (Figure 2).

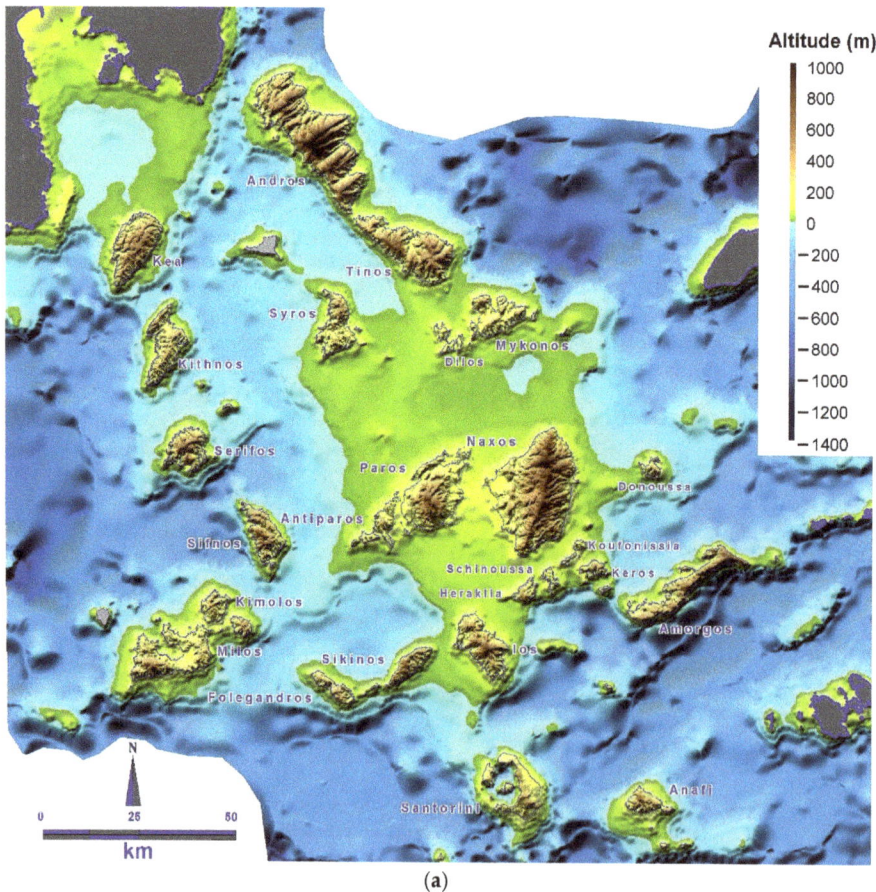

(a)

Figure 2. *Cont.*

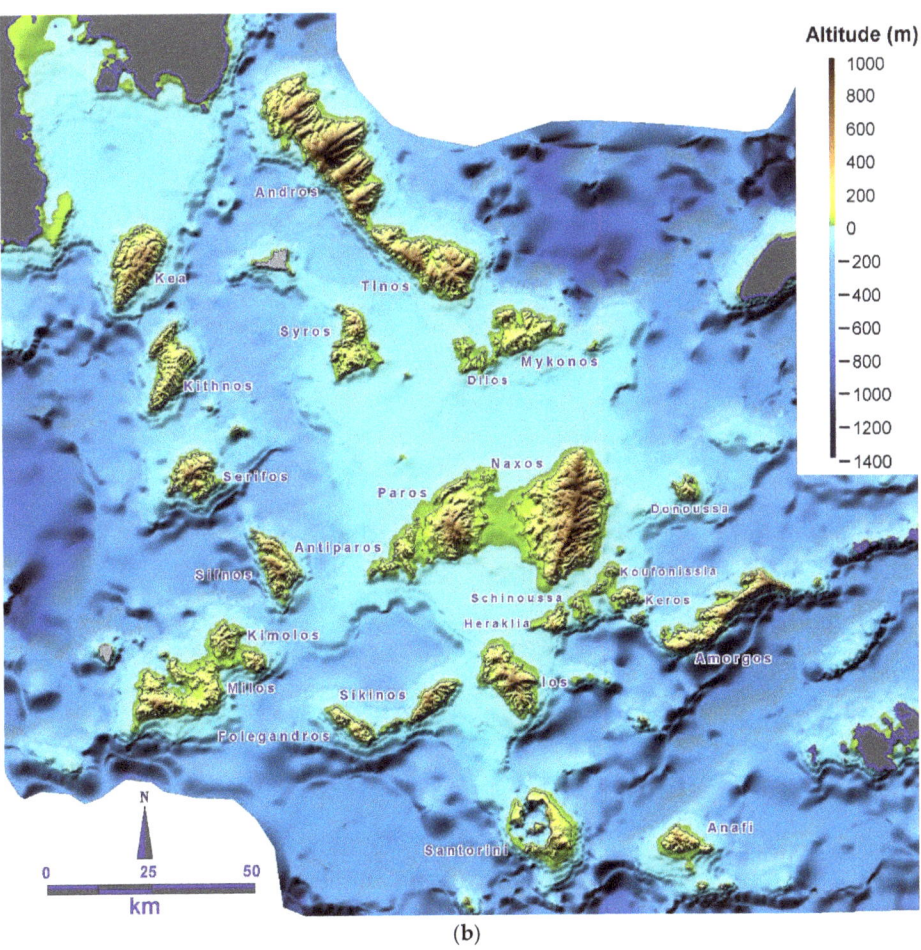

(b)

Figure 2. *Cont.*

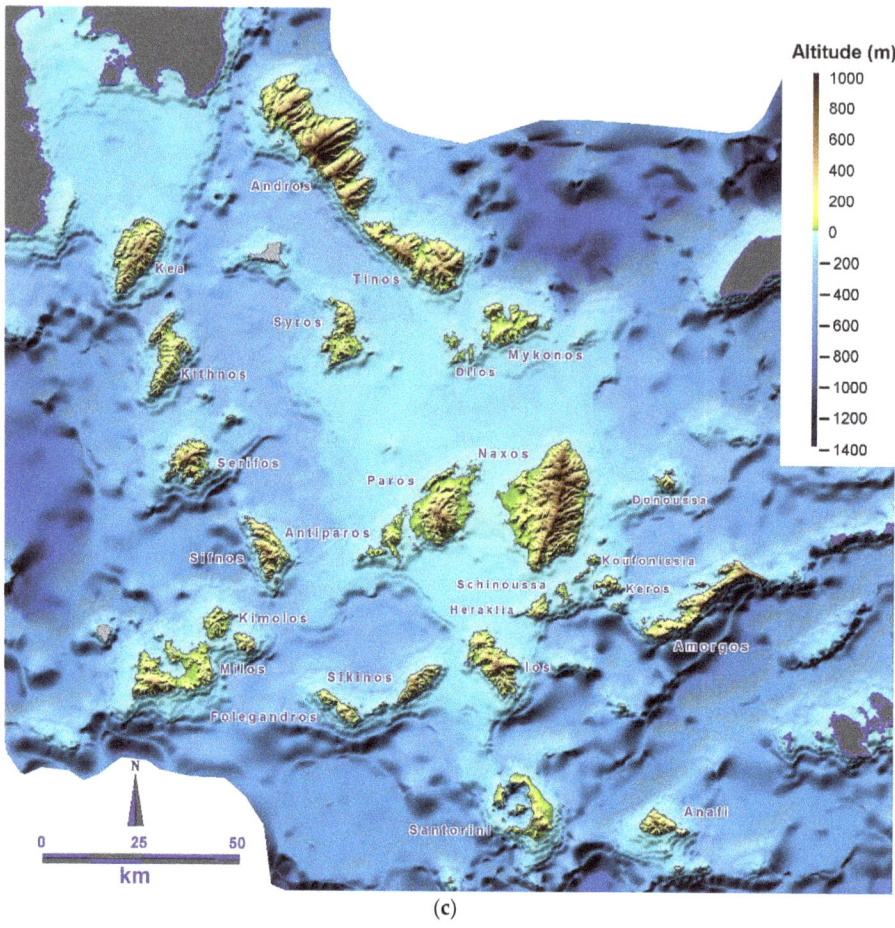

Figure 2. Representation of the Cyclades region as it was 18,000 years ago and how the individual islands were gradually separated, as the region reached its present form. (**a**) Palaeogeography 18,000 years ago; (**b**) 12,500 years ago; (**c**) today.

The current geomorphological landscape of both islands is mainly owed to their lithology, tectonics, drainage system and climate. Most of their landscape is a result of the combination of differential erosion due to frequent lithological alterations and exogenic processes. When it comes to tectonics, in certain areas, planation surfaces can be observed, indicating periods of tectonic stability and uplift. Additionally, downward erosion is notable in both islands, indicating an intense tectonic uplift status and wetter climate periods in the past.

As far as lithology is concerned, differential erosion has played a crucial role to the present-day geomorphology. Typical examples of landforms controlled by differential erosion are hogbacks, which dominate in the mainland, due to the frequent alterations, especially in the center of the islands, of marble, schists and gneisses. Gneisses are more vulnerable to erosion than marbles, thus, upon their erosion, the marbles protrude, revealing the previous relief of the area.

The existence of granodiorite has led to the creation of a unique relief form, due to its weathering. Tafoni are typical landforms owed to the weathering of granodiorites and other crystalline rocks. They consist of cavernous holes resembling honeycombs and are created due to both aeolian erosion processes and chemical weathering [43]. Smaller tafone-type cavities are usually referred to as alveoles [44]. Tafone-type landforms can be found in the western part of Naxos on granodiorite and the north part of Paros on granite. Erosion of the granodiorite produces sediments that end up in the coastal zone, hence the long and wide beaches of western Naxos, which are characterized by a sand composition that is similar to that of the granodiorite. Depositional processes have also born an impact on the landscape of Paros and Naxos. There are several plains, most of which are yet small in area. The plain of Naxos is the largest one in the Cyclades and covers a significant area compared to other plains of the Cyclades.

The western part of Naxos and the eastern part of Paros are characterized by long, sandy beaches, which is quite rare for the Cycladic landscape. In Naxos and Paros, this regime is owed to the lithological structure of the islands and the dominant weathering and erosional processes. The coasts of the islands are mainly steep and rocky, and to a lesser extent sandy, their morphology having mainly been shaped during Late Holocene [33]. They are characterized by a manifold morphology, i.e., bays, capes and several landform types. For instance, erosional landforms such as coastal caves can be found. Coastal cliffs are among the most dominant in both islands and in many cases, they have led to the formation of pocket beaches at their base. Yet, depositional processes also play an important role in shaping the coasts of the islands. Besides pocket beaches, for instance, sandy beaches are also common, in some of which dunes and/or coastal lagoons can be found [36,45,46]. Some rare depositional landforms include tombolos. The submarine landforms of Naxos and Paros, such as notches [47] and beachrocks [48,49], are quite common and generally reveal tectonic subsidence events and the consequent relative sea-level rise, as well as the evolution of the coastal zone [50–52] (see Supplementary Material). Beachrocks also act in a protective way for the coasts, as part of the wave energy is consumed on them.

3. Materials and Methods

The data used in the present study have derived from several field trips in the two islands during the last decades. They concern information about geomorphosites of the study area and they were used for their inventory, classification, and evaluation. The inventory of geomorphosites was developed in three stages. Initially, the geomorphological characteristics of the study area were identified and mapped. Consequently, the geomorphosites representing the geomorphology of the area were recorded, and finally, the ones that had scientific value were selected.

The assessment of geomorphosites follows the method proposed by Reynard et al. 2007 [1], which concerns the evaluation of geomorphosites at regional level and provides criteria for their evaluation [1,15]. These criteria are divided into five main categories, which are further divided into subcategories. The basic categories mentioned in the method are general data, descriptive data, scientific value, additional values and synthesis and were used to classify and evaluate the selected geomorphosites.

More specifically, the studied sites were named and categorized according to the method proposed by Reynard et al. [1] and examined regarding two values: scientific and additional values (ecological, cultural, economic and aesthetic value). Each geomorphosite is named with an identification code, which derives from: (a) the abbreviation of its location with capital letters, (b) the main process of its formation and (c) its number. The main processes of our study area and their codes represent the following: WEA = weathering, KAR = karstic, LIT = littoral, FLU = fluvial, EOL = aeolian, DIFR = differential erosion.

The scientific value is divided into four categories, namely: integrity, representativeness, rareness, and palaeogeographical interest [1]. Integrity is related to the state of each geomorphosite conservation. Representativeness is related to geomorphosite exemplarity. Rareness is related to the rarity of the geomorphosite depending on the reference area. Finally, palaeogeographical value is related to the importance of geomorphosite's location as a component of Earth's evolution. Each geomorphosite was graded for each of the aforementioned values, with a grade ranging from 0 to 1 [1]. Table S1 presents the assessment criteria and scoring for the scientific value, based on Reynard et al. [1] and

Bouzekraoui et al. [53]. The final scientific value derives from the average of integrity, representativeness, rareness, and palaeogeographical interest.

Regarding the additional values, they are composed of the ecological, cultural, economic, and aesthetic value [1]. The ecological value corresponds to the arithmetic mean of two elements: (i) ecological impact, which accounts for the significance of a geomorphosite for the development of a particular ecosystem, and (ii) the protection status of the geomorphosite. The aesthetic values correspond to the arithmetic mean of two elements: (i) the viewpoints of a particular geomorphosite, that is its visibility, and (ii) structure that considers the contrasts and vertical development of a landform, therefore those with color contrasts or high vertical development obtain a higher value. The cultural value is composed of four elements, religious importance, historical importance, artistic or literary importance and geohistorical importance; for this value, the element having the highest value is considered. Lastly, the economic value takes into consideration the qualitative and/or quantitative assessment of the products generated by a particular geomorphosite. Table S2 presents the assessment criteria and scoring for the additional values, based on Reynard et al. [1] and Bouzekraoui et al. [53].

The results from the geomorphological synthesis were presented on GIS. Geomorphosites, the categories of criteria and their evaluations were recorded in a database, which were subsequently incorporated into GIS. After analyzing and processing the data, several maps were created with the use of ArcGIS Pro v.2.8.3 and its modules, which depict the selected geomorphosites categorized.

4. Results and Discussion

4.1. Description of the Geomorphosites

The sites of geomorphological interest that were selected through this research are described in this section. In this research, a total of 75 geomorphosites were selected. Sites referred below may be found in the two interactive maps developed for the purposes of this paper. Specifically, the interactive map of Naxos can be found at https://arcg.is/0LzWrv1 (accessed on 13 December 2021) and for Paros at https://arcg.is/01f09a (accessed on 13 December 2021). The maps in Figures 3 and 4 depict the studied geomorphosites, as well as spider diagrams regarding their scientific and additional values. In these spider diagrams, one can observe the distribution of the individual values (scientific, aesthetic, ecological, economic and cultural) for each geomorphosite. The following link provides this information in a dashboard https://www.arcgis.com/apps/dashboards/a3cf026ba865488eb966407991b2e5b8 (accessed on 13 December 2021). For practical reasons, the selected sites were categorized according to the geomorphological environment/processes that led to their formation. Therefore, the main categories include coastal, karstic, aeolian and erosional landforms (the latter excluding landforms that belong to the previous categories).

The Supplementary Table S3 includes the detailed geomorphosites assessment, including the name of each site, location, the landform(s) that can be observed, as well as its grading regarding the individual values discussed above (integrity, representativeness, rareness, palaeogeographical interest, ecological, cultural, economic and aesthetic value) and its final grading. The coastal landforms of Naxos and Paros islands include beachrocks, coastal lagoons, tidal notches, coastal dunes, tombolos and palaeotombolos. Beachrocks are mainly found in the areas of Aghios Georgios, Plaka, Orkos, Mikri Vigla, Glyfada, Ramnos, Martselo and Tsoukalia. Tidal notches can be found in the areas of Galini, Kampos, Northern Paros, Gaidouronisi, Lageri, Koukoumvales, Parasporos, as well as in four different sited near Hilia Vrysi. Tombolos are found in Aghios Georgios and Lageri, whereas a palaeotombolo can be observed in Plaka, Naxos. Coastal lagoons can be found in Mikri Vigla, Glyfada, Aghios Prokopios, Aghios Georgios, Agiasso and Lageri.

Aeolian landforms include one aeolianite site in Manto island, Naxos, as well as coastal dunes found in the areas of Aghios Georgios, Aghia Anna, Plaka, Kastraki, Glyfada, Pyrgaki, Kanaki, Azala, Lageri, Alyki, Molos and Santa Maria. Karstic landforms in Paros and Naxos islands include one cave, two potholes and two karst plains. The former refers to the Cave of Zas mountain near Filoti, Naxos, whereas the two karst plains are located in Korakas Hill, Marathi and Moutsi. The two potholes are located in Koronos-Koronida and Moutsouna, Naxos.

Erosional landforms include gorges, hogbacks, tafonis, tors and alveoles. Four gorges were considered worthy of assessing, namely in Mavros Kavos, Pyrgos Cape, Xiropotamos River and Paroikia. Characteristic hogbacks are found in northern Naxos, whereas the most important tafoni formations are in Stelida, Moni Chrysostomou, Plaka and Kinidaros. Tor sites include the areas of Stelida and Mikri Vigla, Naxos island, whereas a characteristic alveolus site is Kolymbithres in Paros island.

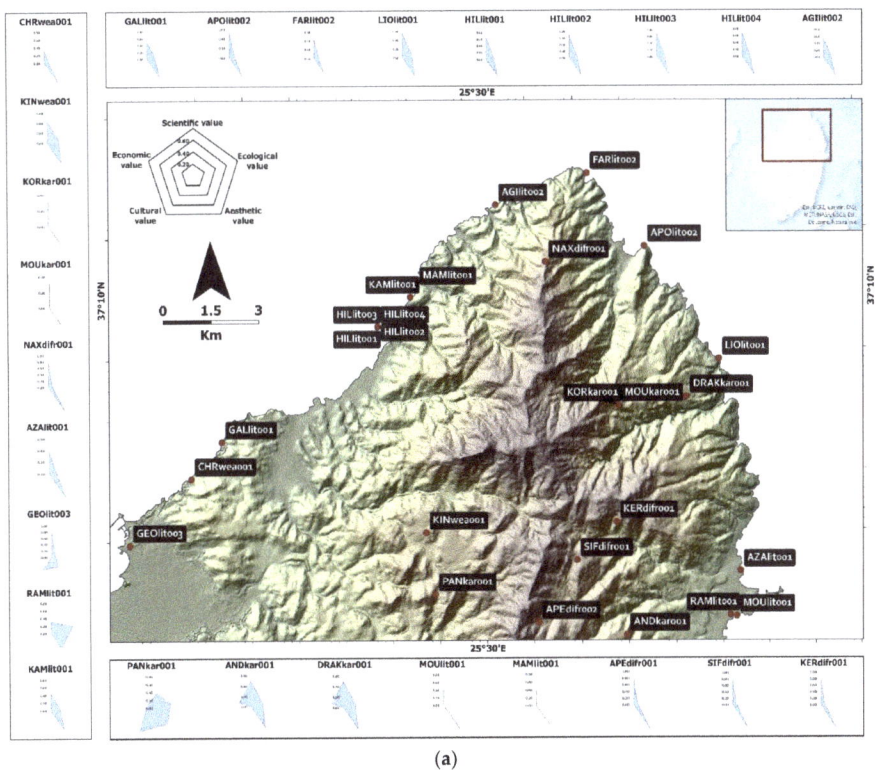

(a)

Figure 3. Cont.

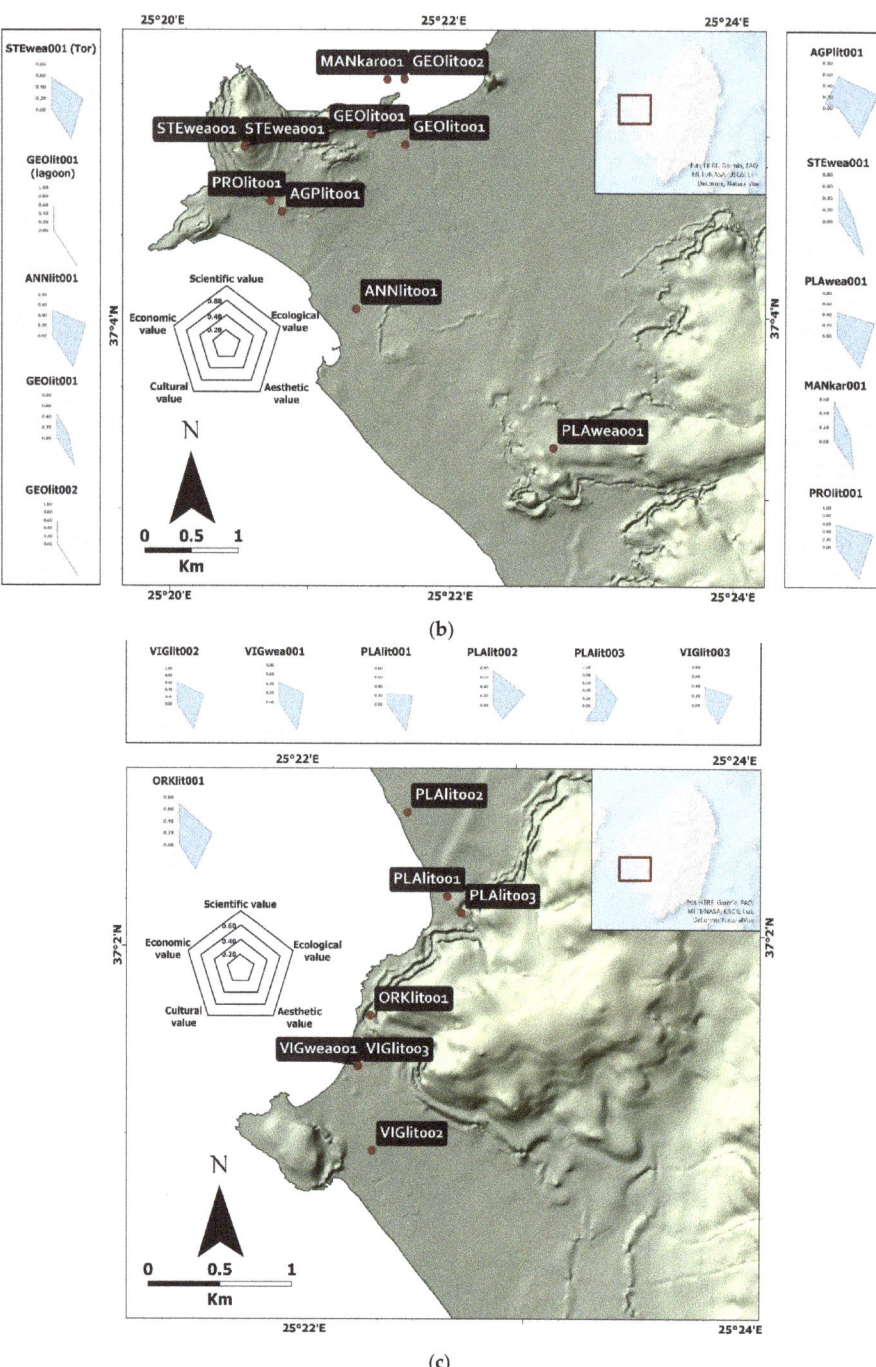

Figure 3. Cont.

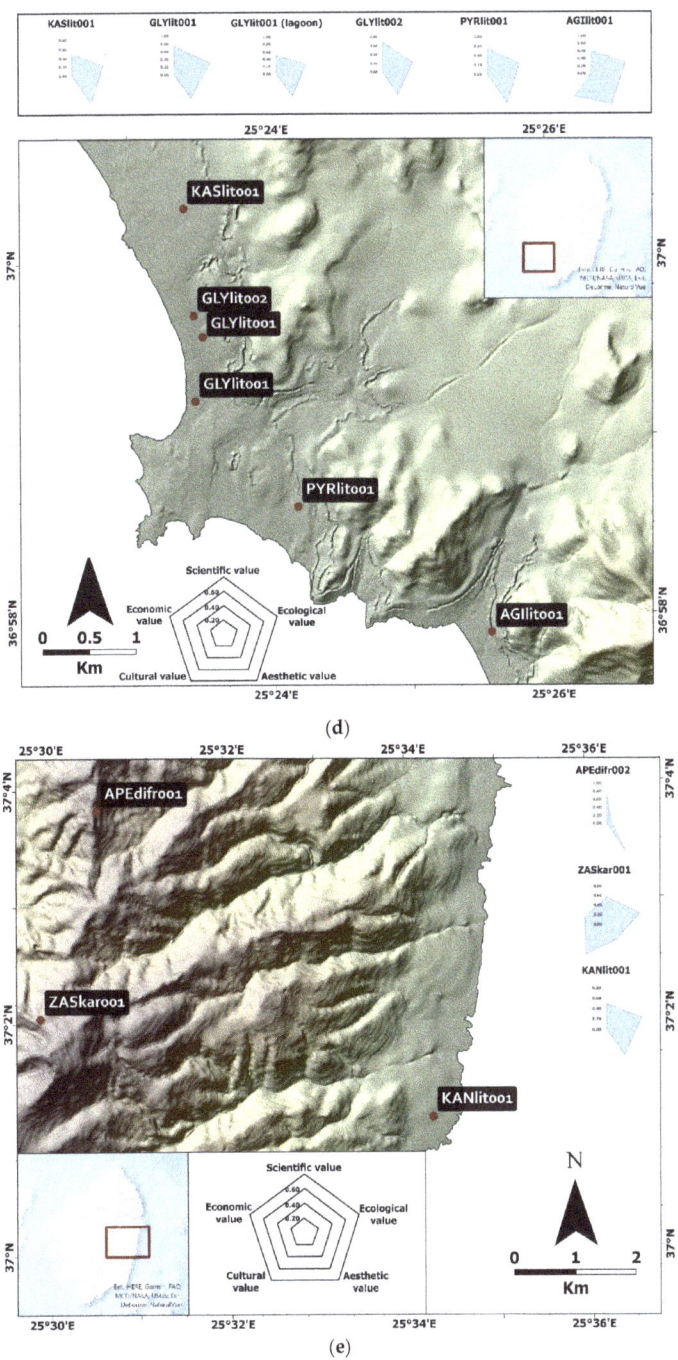

Figure 3. Terrain map of Naxos Island split into five parts (**a–e**), where the selected geomorphosites, as well as spider diagrams regarding their values are depicted. Each spider diagram consists of a sequence of equi-angular spokes, with each spoke representing one of the following variables: scientific value, economic value, ecological value, cultural value, and aesthetic value.

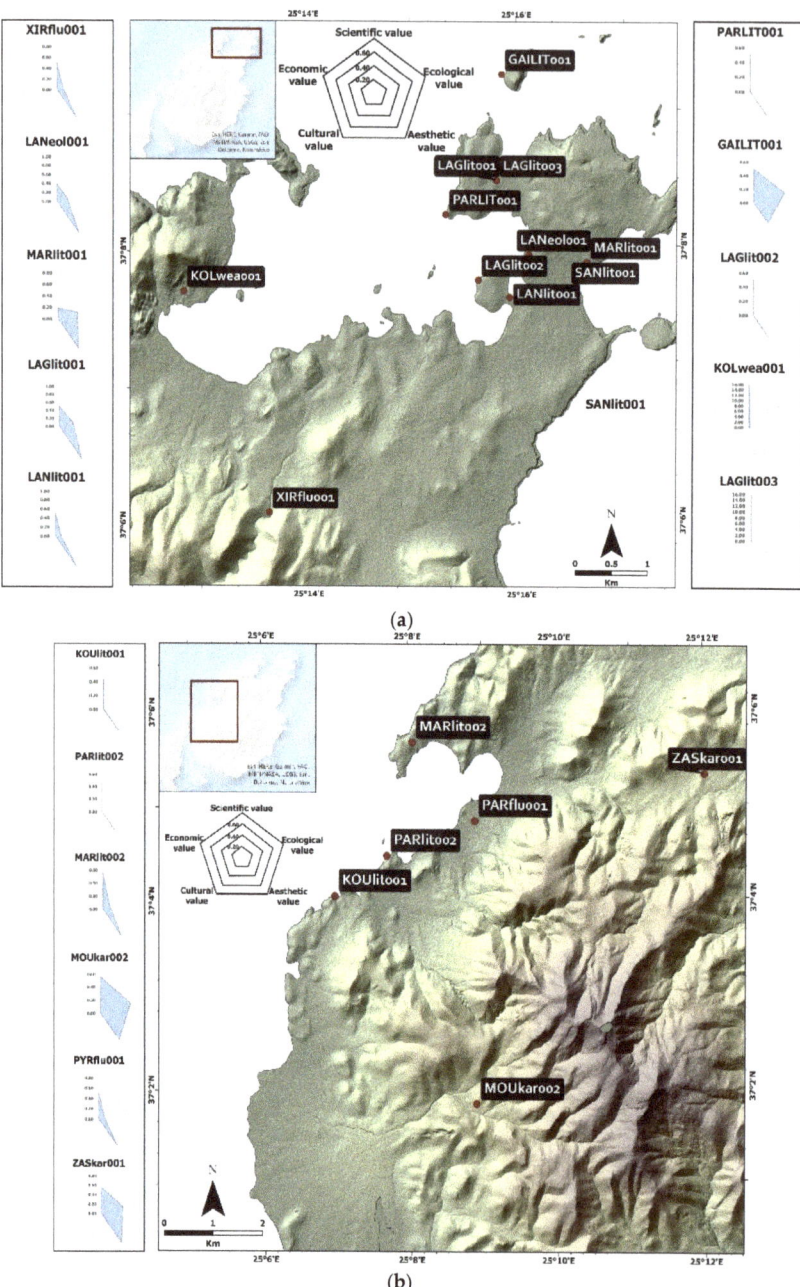

Figure 4. *Cont.*

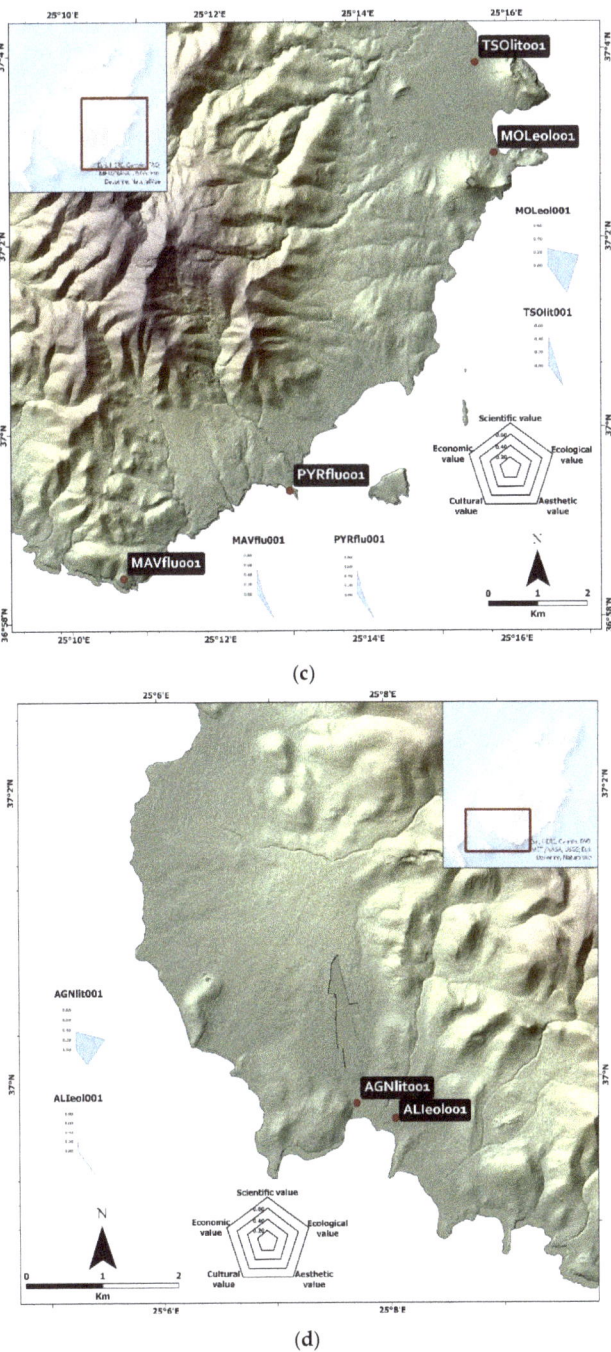

Figure 4. Terrain map of Paros Island split into four parts (**a–d**), where the selected geomorphosites, as well as spider diagrams regarding their values are depicted. Each spider diagram consists of a sequence of equi-angular spokes, with each spoke representing one of the following variables: scientific value, economic value, ecological value, cultural value, and aesthetic value.

4.2. Quantitative Assessment of the Geomorphosites

The quantitative assessment of the geomorphosites can be seen in the Supplementary Table S3. Overall, the scientific value ranges between 0.38 and 0.81. 47 out of the 75 geomorphosites are graded with at least 0.5. Integrity is high (0.75 or 1) for 61% of them and 0.5 for 30% of them. As far as representativeness is concerned, 42% of the 75 geomorphosites are rated with 0.75 or 1 and 13% with 0.5. When it comes to rareness, 36% of the geomorphosites are rare (0.75 or 1-graded) and another 36% of them medium (0.5). The palaeogeographical value is the lowest ranking among the four aspects of the scientific value, as the highest rate is 0.5 and only refers to a 12% of the total number of the geomorphosites (Figure 5 for Naxos; Figure 6 for Paros). Overall, regarding the scientific value of the studied geomorphosites, the criteria of integrity received the highest values, while the criteria of palaeogeographic value received the lowest ranking. Geomorphosites, such as hogbacks, dunes and beachrocks were among the ones receiving the highest ranking in terms of scientific value overall, while features such as tors or potholes received lower ratings (Table 1).

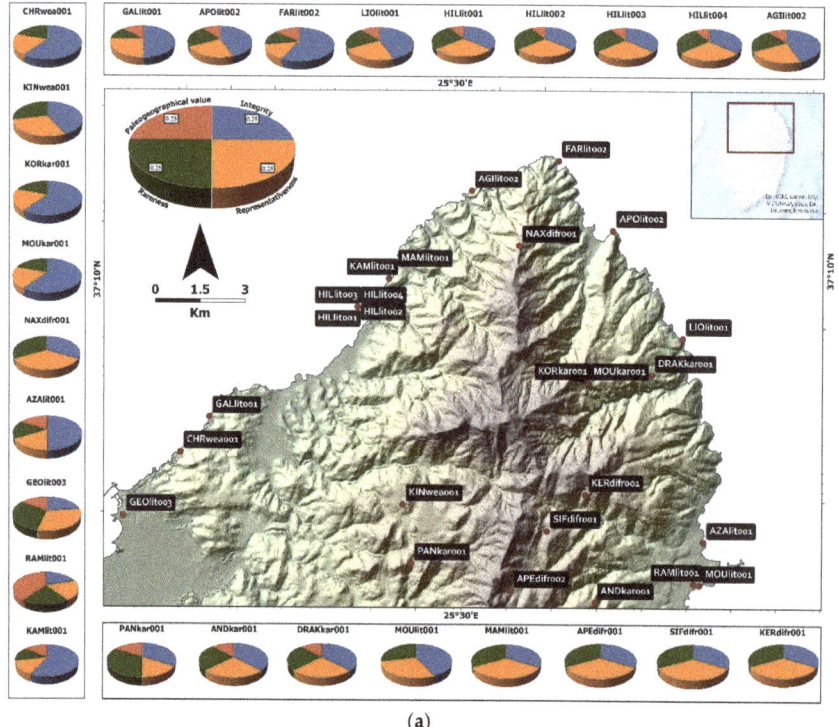

(a)

Figure 5. *Cont.*

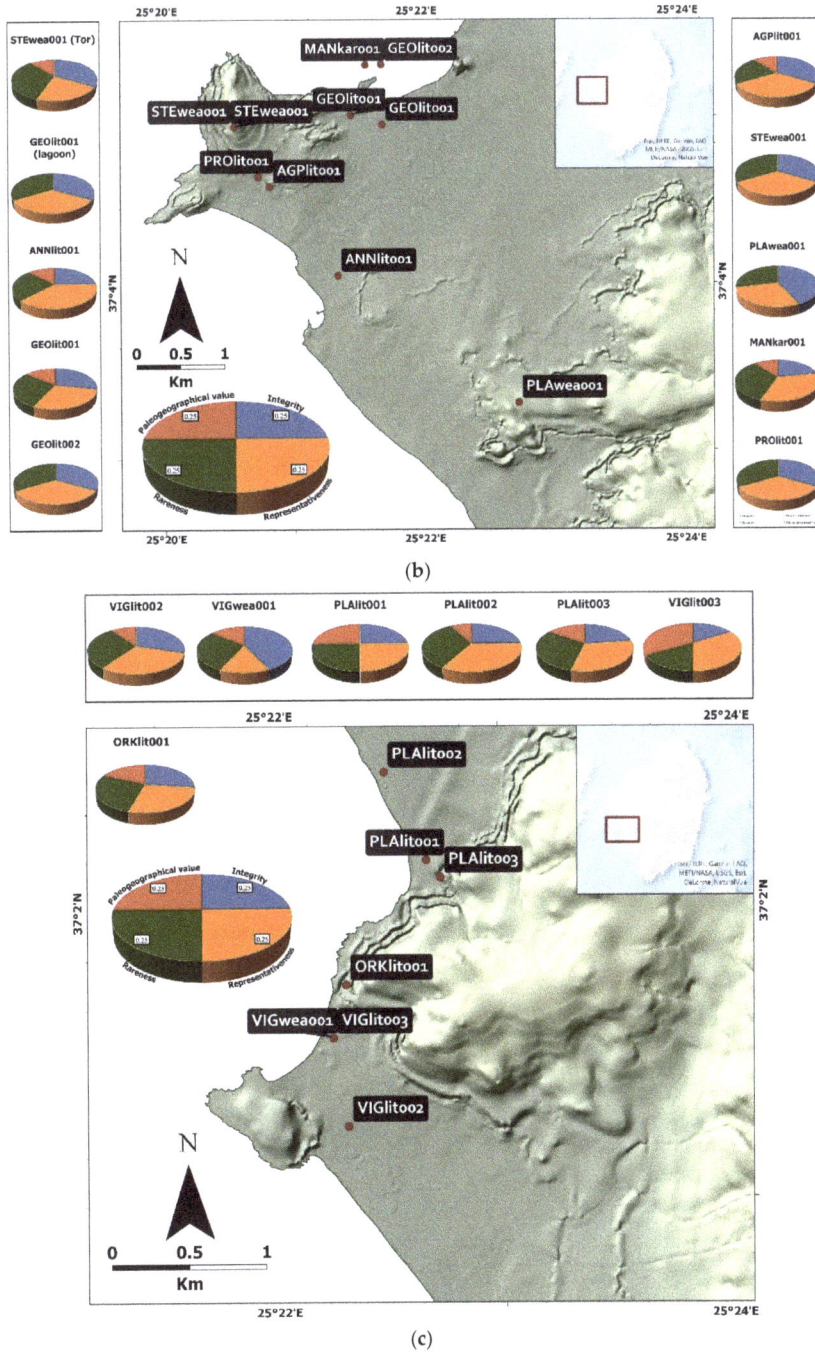

Figure 5. Cont.

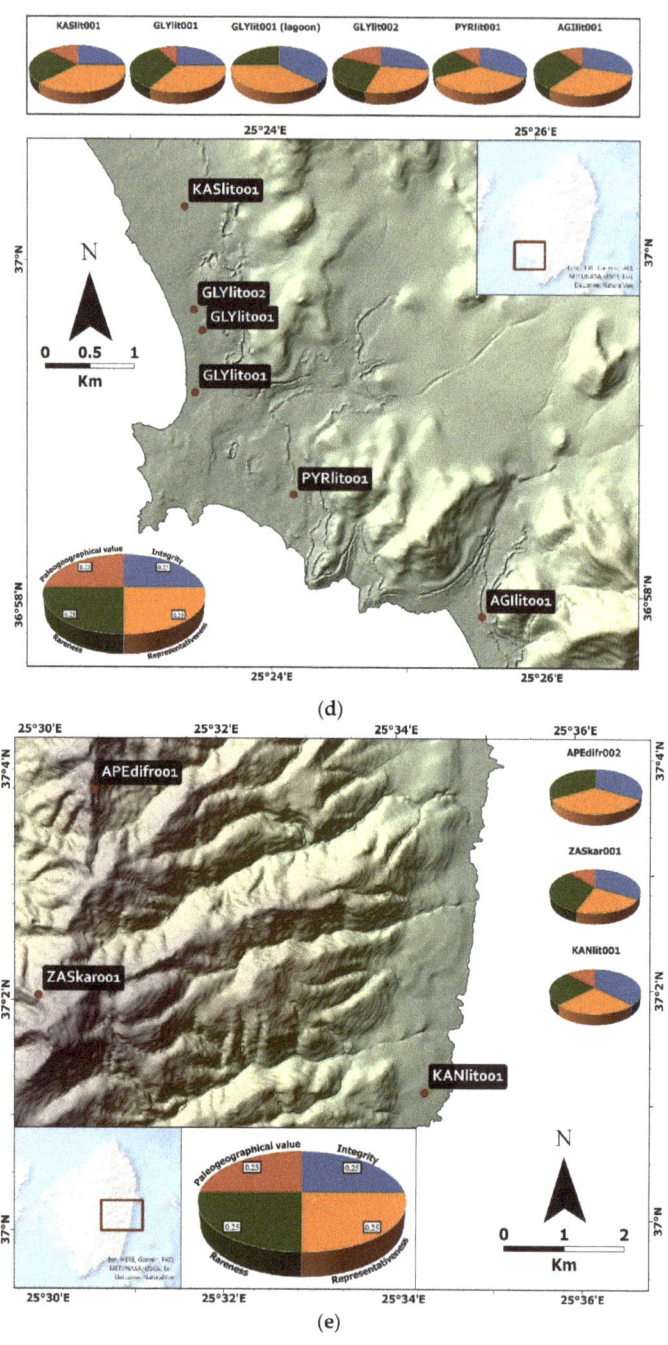

Figure 5. Map of Naxos split into five parts (**a**–**e**) showing the geomorphosites and their scientific value. The diagrams framing the map show the contribution of each parameter (integrity, representativeness, rareness, and palaeogeographical value) to the final calculation of the scientific value for each geomorphosite. The blue section of the pie chart represents the integrity, the green section represents the rareness, the orange section defines the representativeness, and the red section represents the palaeogeographical value.

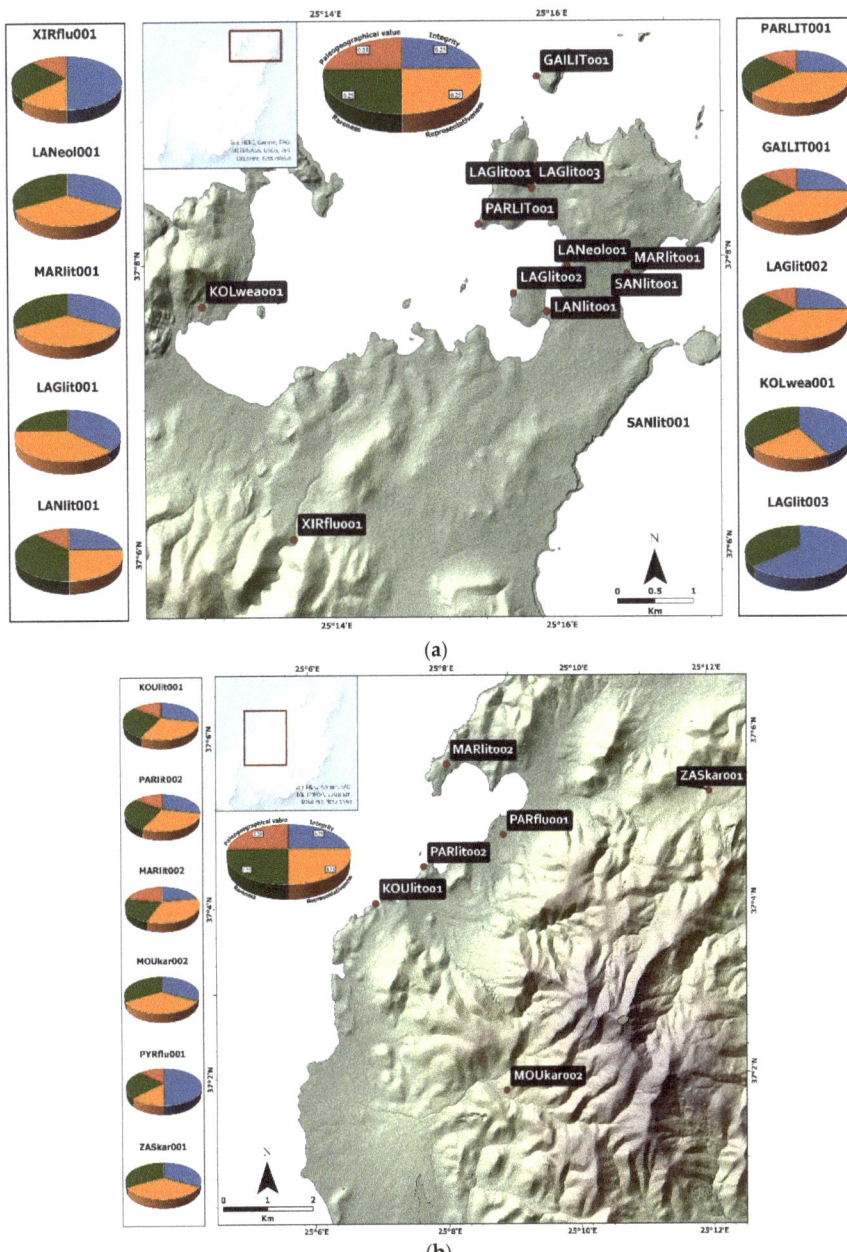

Figure 6. *Cont.*

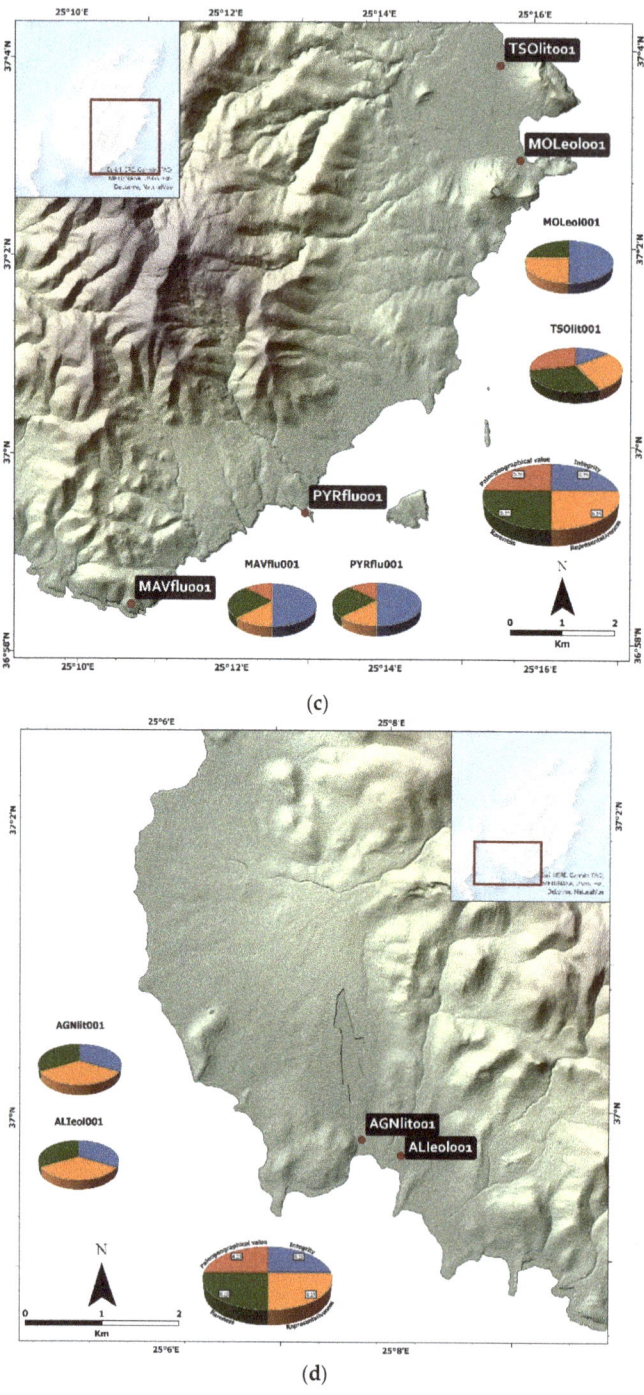

Figure 6. Map of Paros split into four parts (**a**–**d**), showing the geomorphosites and their scientific value. The blue section of the pie chart represents the integrity, the green section represents the rareness, the orange section defines the representativeness, and the red section represents the palaeogeographical value.

Table 1. Average values of the ratings for the assessed geomorphosites.

Identification Code	Scientific Value	Ecological Value	Aesthetic Value	Cultural Value	Economic Value
STEwea001	0.56	0.25	0.75	0	0
CHRwea001	0.31	0.125	0.625	0	0
PLAwea001	0.44	0.75	0.75	0	0
KINwea001	0.44	0.25	0.5	0	0
STEwea001	0.56	0.625	0.625	0	0
VIGwea001	0.44	0.625	0.75	0	0
KORkar001	0.31	0	0.25	0	0
MOUkar001	0.31	0	0.25	0	0
NAXdifr001	0.75	0.125	0.875	0	0
GEOlit001	0.44	0.25	0.625	0	0
ANNlit001	0.50	0.75	0.75	0	0
PLAlit001	0.25	0.625	0.75	0	0
KASlit001	0.50	0.75	0.75	0	0.25
GLYlit001	0.75	1	0.75	0	0.25
PYRlit001	0.56	0.75	0.75	0	0
KANlit001	0.50	0.75	0.625	0	0
AZAlit001	0.38	0.125	0.5	0	0
MANkar001	0.56	0.25	0.5	0	0
VIGlit002	0.63	0.875	0.875	0	0
GLYlit001	0.50	0.875	0.75	0	0
PROlit001	0.56	1	1	0	0
GEOlit001	0.56	0	1	0	0.25
AGIlit001	0.63	1	1	0.75	0.25
GEOlit002	0.56	0	1	0	0
PLAlit002	0.75	0.75	0.375	0	0
ZASkar001	0.56	0.75	0.375	0.75	0.5
GEOlit003	0.81	0.125	0.375	0.5	0
PLAlit003	0.81	0.625	0.5	0.5	0
ORKlit001	0.69	0.625	0.5	0	0
VIGlit003	0.38	0.625	0.5	0	0
GLYlit002	0.69	0.75	0.625	0	0
RAMlit001	0.31	0.625	0.5	0	0
GALlit001	0.50	0.25	0.625	0	0
KAMlit001	0.44	0.25	0.625	0	0
AGIlit002	0.56	0.25	0.625	0	0
FARlit002	0.44	0.125	0.375	0	0
APOlit002	0.56	0.125	0.5	0	0
LIOlit001	0.56	0.25	0.5	0	0
HILlit001	0.69	0.25	0.625	0	0
HILlit002	0.69	0.25	0.625	0	0
HILlit003	0.69	0.25	0.625	0	0
HILlit004	0.69	0.25	0.625	0	0
MAVflu001	0.50	0.125	0.625	0	0
PYRflu001	0.50	0.125	0.625	0	0
XIRflu001	0.50	0.125	0.625	0	0
PARflu001	0.50	0.125	0.625	0	0
ZASkar001	0.56	0.5	0.75	0	0
MOUkar002	0.56	0.5	0.5	0	0
LANeol001	0.38	0.25	0.875	0	0
ALIeol001	0.19	0	0.625	0	0
MOLeol001	0.25	0.5	0.5	0	0
MARlit001	0.19	0.375	0.625	0	0
LAGlit001	0.50	0.375	1	0	0
LANlit001	0.50	0.125	0.875	0	0
MARlit002	0.56	0.125	0.5	0	0
TSOlit001	0.44	0.125	0.375	0	0

Table 1. Cont.

Identification Code	Scientific Value	Ecological Value	Aesthetic Value	Cultural Value	Economic Value
PARLIT001	0.50	0	0.375	0	0
GAILIT001	0.50	0.5	0.375	0	0
LAGlit002	0.50	0	0.375	0	0
KOUlit001	0.44	0	0.375	0	0
PARlit002	0.44	0	0.375	0	0
KOLwea001	0.44	0.5	0.5	0	0
LAGlit003	0.44	0.75	0.75	0	0
SANlit001	0.25	0.625	0.5	0	0
PANkar001	0.38	0.375	0.5	0.75	0.25
ANDkar001	0.50	0.25	0.5	0	0.25
DRAKkar001	0.50	0.25	0.5	0	0.25
MOUlit001	0.44	0	0.75	0	0
MAMlit001	0.38	0	0.625	0	0
APEdifr001	0.75	0.125	0.875	0	0
APEdifr002	0.75	0.125	0.875	0	0
SIFdifr001	0.75	0.125	0.875	0	0
KERdifr001	0.75	0.125	0.875	0	0
AGPlit001	0.56	0.75	0.625	0	0.25
AGNlit001	0.38	0.625	0.375	0	0

As far as the additional values are concerned, 55% of the 75 geomorphosites are highly rated (more than 0.5), whereas 19% of them are rated with 0.5. Regarding ecology, 29% of the selected geomorphosites are graded with more than 0.5 and 7% with 0.5. The other two aspects (economy and culture) do not seem to play an important role in the geomorphological heritage of the two islands, as there is only one geomorphosite whose economical value is 0.5 (the rest being 0-graded), two geomorphosites whose cultural value is 0.75 and another two whose cultural value is 0.5 (the rest being 0-graded) (Figure 7 for Naxos; Figure 8 for Paros). It should however be noted that no quantitative data were available for a direct estimation of the economic value of the studied geomorphosites.

The aesthetic and the ecological value vary depending on the type of geomorphosite. The aesthetic value maintains high values in the majority of geomorphosites, compared to the ecological value which fluctuates from 0 to 0.875. Unlike the aesthetic value and the ecological value, the cultural value appears only in lagoon (e.g., AGIlit001), beachrock (e.g., GEOlit003, PLAlit003) and cave (e.g., ZASkar001, PANkar001) geomorphosites.

The highest-ranking geomorphosites include the beachrocks in Aghios Georgios, Naxos and the tidal notches NW of Koukoumavles, Paros, the scientific value reaching 0.81. The additional values in these cases are relatively low (generally between 0.1 and 0.6). Besides them, there exist another 11 geomorphosites (18% in total) whose scientific value exceeds 0.6. Regarding their additional values, they are characterized by a high ecological and aesthetic value. A total of 39 geomorphosites (65%) are characterized by medium value, between 0.4 and 0.6. These geomorphosites are generally characterized by a high aesthetic value (more than 0.5), whereas 14 of them are also of high ecological value. A total of 11 geomorphosites (16%) are of low scientific value (less than 0.4), most of which are of high aesthetic value and only three of high ecological value.

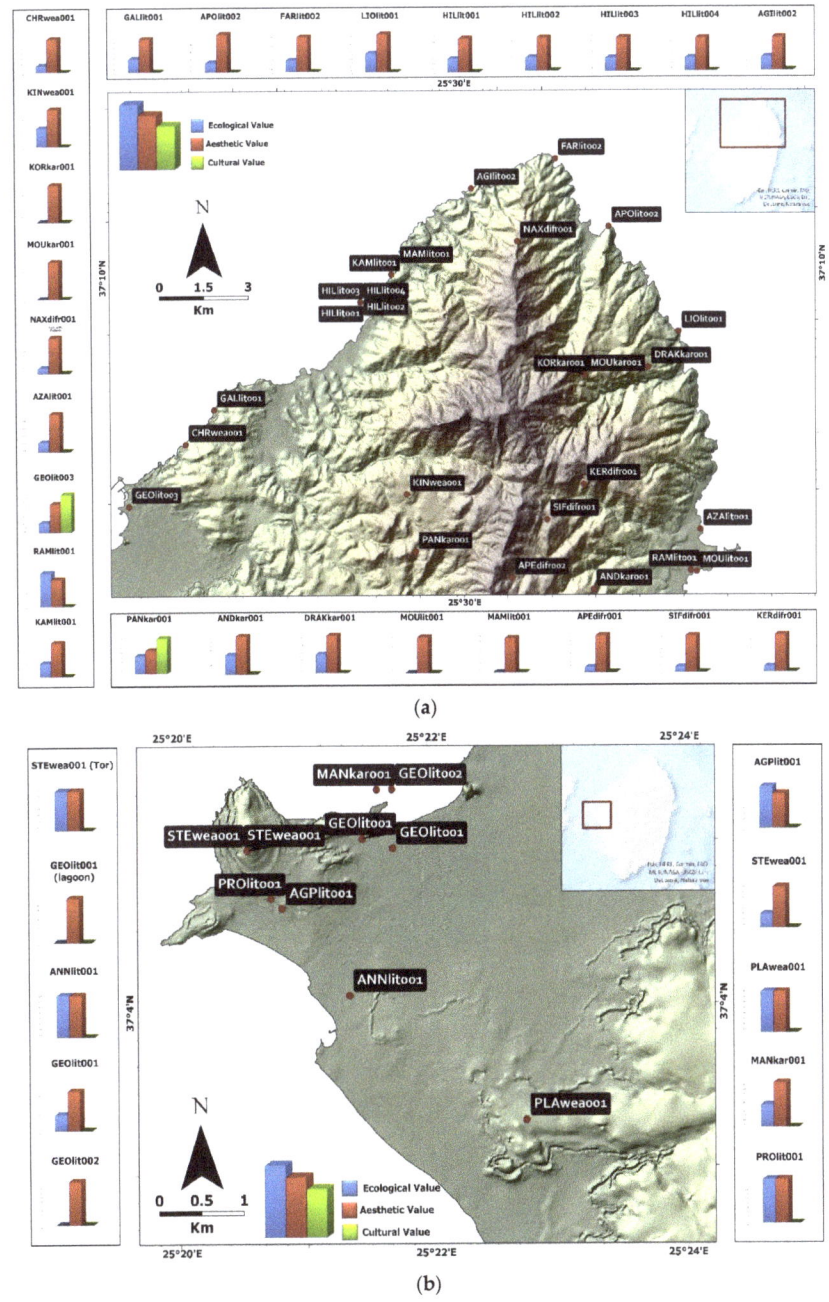

Figure 7. Cont.

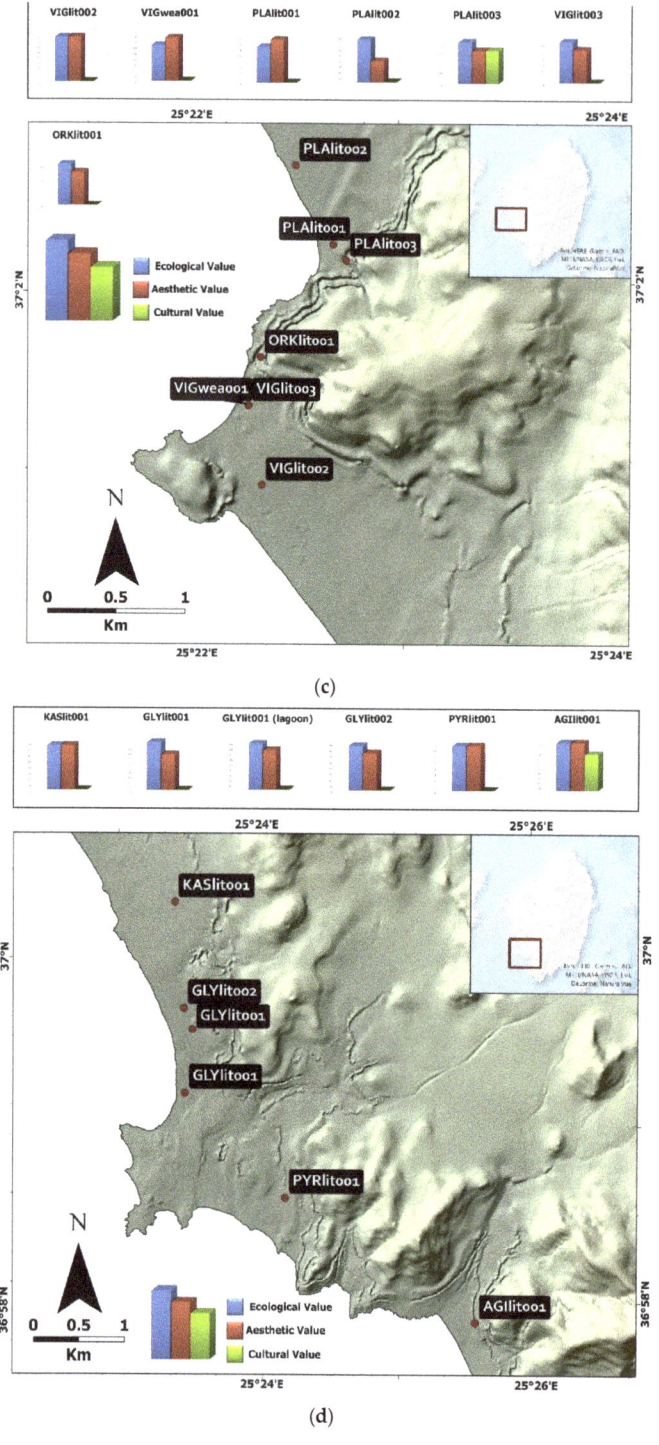

Figure 7. Cont.

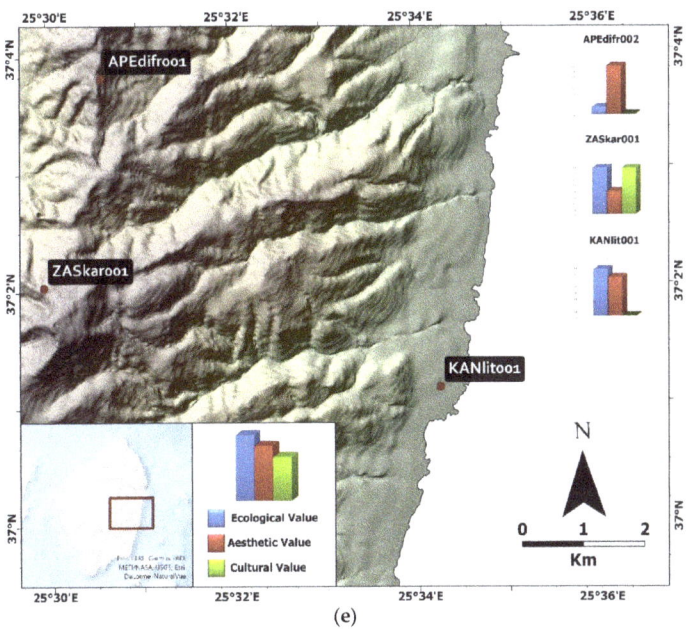

Figure 7. Map of Naxos split into five parts (**a**–**e**), showing the geomorphosites and their additional values. The diagrams framing the map show the relationship between ecological, aesthetic, and cultural value for each geomorphosite. The blue columns represent the ecological value, the red columns represent the aesthetic value, and the green columns represent the cultural value.

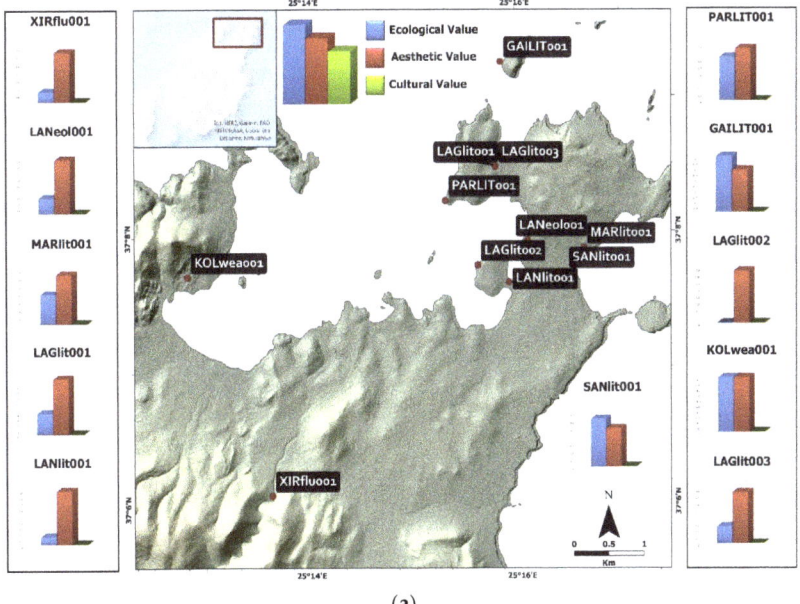

Figure 8. *Cont.*

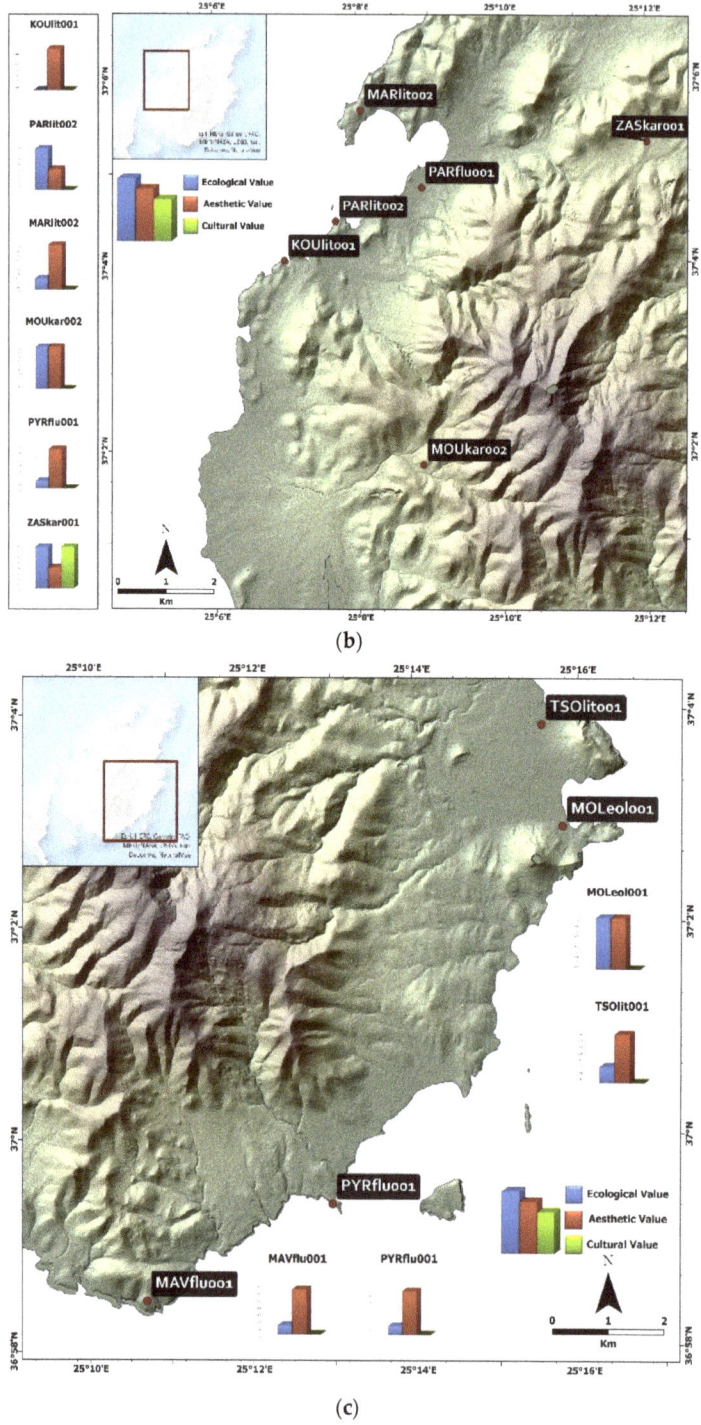

Figure 8. Cont.

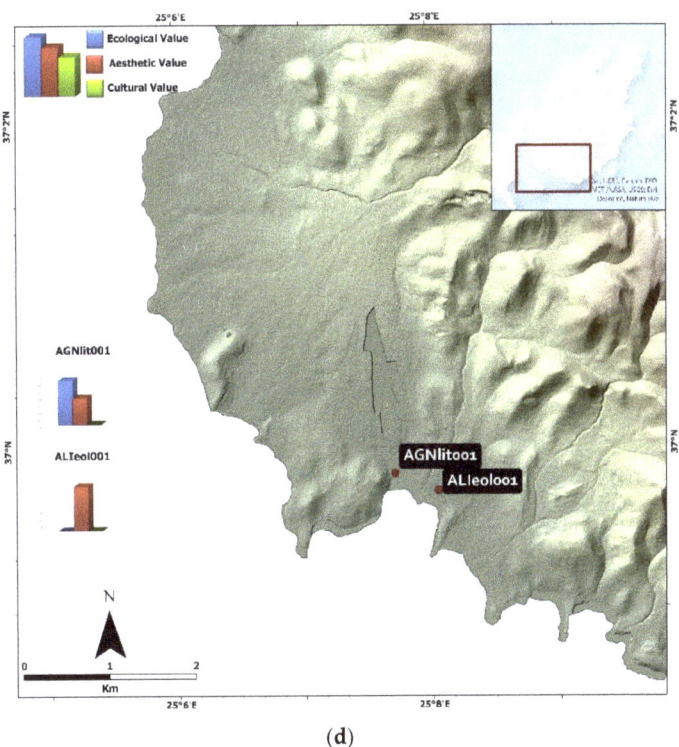

(d)

Figure 8. Map of Paros split into four parts (**a–d**), showing the geomorphosites and their additional values. The diagrams framing the map show the relationship between ecological, aesthetic, and cultural value for each geomorphosite. The blue columns represent the ecological value, the red columns represent the aesthetic value, and the green columns represent the cultural value.

4.3. The Geotouristic Potential of Paros and Naxos Islands

Both islands are characterized by special geomorphological and environmental features. Presently, both islands receive most tourists during the summer period and while certain geomorphosites are known and visited, such as the cave of Zas (ZASkar001) or Cave of Panagia (PANkar001), other landforms are either visited, but their particular value is not acknowledged due to lack of information, or they remain completely unknown. Overall, the geotouristic potential of both islands is poorly exploited at present. For the case of Paros, recent efforts include the development of an environmental-cultural park on the north part of the island. However, its most prominent actions are related to cultural activities primarily during the high summer season. In the case of Naxos, the island is visited by many international tourists wishing to experience quitter vacations and more natural landscapes. While both islands have all those elements that can attract the interest of geotourists, they have not exploited this possibility yet, possibly because they accept a lot of tourism in the summer months anyway and feel that they are covered or possibly because they put all their efforts in the summer tourism and lack the energy to expand the touristic season. The effort to develop geotourism needs organization and the right decisions from decision makers and the local society in order to open the tourist product and attract visitors at different times and not only during the summer, hence reducing the intensification of tourism in the summer months, which has reached its limits and often leads to negative comments. Both islands still lack a clear plan for the development of geotourism and the promotion of geotouristic features and activities.

The above call for better promotion and information to the public and perspective tourists on the geomorphological heritage of both islands; such actions could include the inclusion of geomorphosites to tourist guides, the design of geo-routes, the development of StoryMaps, etc. The promotion of geomorphological heritage in the area will clearly contribute to the enhancement of tourism, supporting the environmental and geomorphological beauties of the island and enhancing geotourism and eco-tourism. It is important to emphasize the difference in quality and time of arrival for this category of tourists, which will extend the touristic season, but at the same time will respect the heritage of the islands. The promotion of geomorphological heritage will provide the appropriate tools for informing the public about environmental pressures, their effects, and the need for protection. It will also provide reliable proposals and solutions for geoenvironmental problems and, on the other hand, will promote and promote the concept of geo-conservation.

New generations of tourists are composed of travelers who need nature, they are more educated and looking for alternative tourist destinations. In the list of the most frequent touristic destinations, geological and geomorphological landscapes often stand out, replacing other classic tourist destinations of the past. Among the most famous geotouristic destinations are Yellowstone, the Grand Canyon, Yosemite Valley Park and the Great Barrier Reef [54].

Such destinations, with proper management, contribute to the sustainable development of the areas, i.e., a development that meets the needs of current generations, without compromising the rights and needs of future generations. It is worth mentioning that tourism, due to the intensifying competition, is being modernized and enriched by developing new products. In the global tourism market and especially in the European, in recent years, alternative forms of tourism development are being promoted, especially those that are compatible with the directions of protection of the natural and cultural environment.

5. Conclusions

A total of 75 geomorphosites were selected, mapped and assessed in this research, namely 22 in Paros and 53 in Naxos island, Cyclades, Greece. They were assessed such that their promotion, as well as their geoconservation, can be rendered possible. The assessment of the geomorphosites followed the assessment method proposed by Reynard et al. [1]. Based on field work and the consequent assessment of the visited areas as potential geomorphosites, we have concluded that the islands of Naxos and Paros are of vast geomorphological heritage that is worth conserving and promoting and can lead to the further economic development of the two regions. Both islands are abundant in unique coastal landforms great sea-level and palaeogeography indicators, as well as landforms created in other environments. Besides the high scientific interest of the 63 geomorphosites, most of them are also characterized by a high ecological value and can potentially lead to a significant increase in the islands' tourism, not in the sense of massive tourism, which usually leads to degradation, but qualitive tourism, for instance geotourism and ecotourism. The promotion and management of these geomorphosites can contribute to the development of the local economy, given that both islands are highly popular and therefore do not lack the corresponding facilities for hosting geotourism.

Supplementary Materials: The following are available online, Table S1: Geomorphosites database. Geomorphological data for Naxos island are available at https://arcg.is/0LzWrv1 (accessed on 13 December 2021) and for Paros at https://arcg.is/01f09a (accessed on 13 December 2021).

Author Contributions: Conceptualization, N.E. and A.K.; methodology, N.E., A.K., M.T. and E.S.; investigation, N.E., A.K., M.T., E.S., A.P. and L.L.; writing—original draft preparation, N.E., A.K., E.S. and A.P.; writing—review and editing, N.E., A.K., M.T., E.S. and A.P.; visualization, M.T.; supervision, N.E. All authors have read and agreed to the published version of the manuscript.

Funding: This research received no external funding.

Data Availability Statement: The data used for this work can be found at the Supplementary Material.

Conflicts of Interest: The authors declare no conflict of interest.

References

1. Reynard, E.; Fontana, G.; Kozlik, L.; Scapozza, C. A method for assessing "scientific" and "additional values" of geomorphosites. *Geogr. Helv.* **2007**, *62*, 148–158. [CrossRef]
2. Brilha, J.B. *Património Geológico e Geoconservação: A Conservação da Natureza na sua Vertente Geológica*; Palimage: Coimbra, Portugal, 2005.
3. Gray, M. *Geodiversity: Valuing and Conserving Abiotic Nature*; Wiley: Oxford, UK, 2004.
4. Martini, G. (Ed.) *Actes du Premier Symposium International sur la Protection du Patrimoine Géologique, Digne-les-Bains, 11–16 June 1991*; Société Géologique de France: Paris, France, 1994.
5. Panizza, M. Geomorphosites: Concepts, methods and example of geomorphological survey. *Chin. Sci. Bull.* **2001**, *46*, 4–5. [CrossRef]
6. Panizza, M.; Piacente, S. Geomorphological assets evaluation. *Z. Geomorphol.* **1993**, *87*, 13–18.
7. Grandgirard, V. Géomorphologie, Protection de la Nature et Gestion du Paysage. Ph.D. Thesis, Faculté des Sciences, Université de Fribourg, Fribourg, Switzerland, 1997. (In French).
8. Reynard, E.; Panizza, M. Geomorphosites: Definition, assessment and mapping. *Géomorphologie Relief Process. Environ.* **2005**, *11*, 177–180. [CrossRef]
9. Panizza, M.; Piacente, S. *Geomorfologia Culturale*; Pitagora: Bologna, Italy, 2003.
10. Fassoulas, C.; Zouros, N. Evaluating the influence of Greek geoparks to the local communities. *Bull. Geol. Soc. Greece* **2017**, *43*, 896. [CrossRef]
11. Rivas, V.; Rix, K.; Francés, E.; Cendrero, A.; Brunsden, D. Geomorphological indicators for environmental impact assessment: Consumable and non-consumable geomorphological resources. *Geomorphology* **1997**, *18*, 169–182. [CrossRef]
12. Cendrero, A.; Panizza, M. Geomorphology and environmental impact assessment: An introduction. *Suppl. Geogr. Fis. Din. Quat.* **1999**, *3*, 167–172.
13. Bruschi, V.M.; Cendrero, A. Geosite evaluation: Can we measure intangible values? *Il Quat.* **2005**, *18*, 293–306.
14. Bruschi, V.M.; Cendrero, A. Direct and parametric methods for the assessment of geosites and geomorphosites. In *Geomorphosites*; Reynard, E., Coratza, P., Regolini-Bissig, G., Eds.; Pfeil: München, Germany, 2009; pp. 73–88.
15. Reynard, E.; Perret, A.; Bussard, J.; Grangier, L.; Martin, S. Integrated approach for the inventory and management of geomorphological heritage at the regional scale. *Geoheritage* **2016**, *8*, 43–60. [CrossRef]
16. Pralong, J.-P. *Géotourisme et Utilisation de Sites Naturels D'intérêt Pour les Sciences de la Terre. Les Régions de Crans-Montana Sierre (Valais, Alpes Suisses) et Chamonix-Mont-Blanc (Haute-Savoie, Alpes françaises)*; Travaux et Recherches; Institut de Géographie: Lausanne, Switzerland, 2006; Volume 32, 224p.
17. Reynard, E.; Holzmann, C.; Guex, D. Géomorphologie et tourisme: Quelles relations? In *Géomorphologie et Tourisme*; Travaux et Recherches; Reynard, E., Holzmann, C., Guex, D., Summermatter, N., Eds.; Institut de Géographie: Lausanne, Switzerland, 2003; Volume 24, pp. 1–10.
18. Grandgirard, V. L'évaluation des géotopes. *Geol. Insubrica* **1999**, *4*, 59–66.
19. Brilha, J. Inventory and quantitative assessment of geosites and geodiversity sites: A review. *Geoheritage* **2016**, *8*, 119–134. [CrossRef]
20. Giusti, C.; Calvet, M. The inventory of French geomorphosites and the problem of nested scales and landscape complexity. *Geomorphol. Relief Process. Environ.* **2010**, *2*, 223–244. [CrossRef]
21. de Lima, F.F.; Brilha, J.B.; Salamuni, E. Inventorying geological heritage in large territories: A methodological proposal applied to Brazil. *Geoheritage* **2010**, *2*, 91–99. [CrossRef]
22. Mucivuna, V.C.; Reynard, E.; da Glória Motta Garcia, M. Geomorphosites assessment methods: Comparative analysis and typology. *Geoheritage* **2019**, *111*, 1799–1815. [CrossRef]
23. Reynard, E.; Coratza, P.; Hobléa, F. Current Research on Geomorphosites. *Geoheritage* **2016**, *8*, 1–3. [CrossRef]
24. Reynard, E. Scientific research and tourist promotion of geomorphological heritage. *Geogr. Fis. Din. Quat.* **2008**, *31*, 225–230.
25. Bissig, G. Mapping geomorphosites: An analysis of geotourist maps. *Geotourism/Geoturystyka* **2008**, *14*, 3. [CrossRef]
26. Carton, A.; Coratza, P.; Marchetti, M. Guidelines for geomorphological sites mapping: Examples from Italy. *Géomorphologie* **2005**, *3*, 209–218. [CrossRef]
27. Carton, A.; Coratza, P.; Marchetti, M. Nota preliminare sulla cartografia dei geomorfositi. In *La Memoria della Terra, la Terra della Memoria*; Piacente, S., Poli, G., Eds.; L'Inchiostroblu: Bologna, Italy, 2003; pp. 114–120.
28. Papanikolaou, D.I. *Geology of Greece*; Patakis: Athens, Greece, 2015.
29. Goldsworthy, M.; Jackson, J. Active normal fault evolution in Greece revealed by geomorphology and drainage patterns. *J. Geol. Soc.* **2020**, *157*, 967–981. [CrossRef]
30. Zouros, N.C. Geomorphosite assessment and management in protected areas of Greece. *Geogr. Helv.* **2007**, *62*, 169–180. [CrossRef]
31. Hellenic Geoparks Forum. Available online: http://www.hellenicgeoparks.gr/?page_id=46 (accessed on 20 September 2021).
32. Gaki-Papanastassiou, K.; Evelpidou, N.; Maroukian, H.; Vassilopoulos, A. *Palaeogeographic Evolution of the Cyclades Islands (Greece) during the Holocene*; Springer: Dordrecht, The Netherlands, 2009; pp. 297–304.

33. Evelpidou, N. Geological and Geomorphological Observations in Paros Island (Cyclades) Using Photo Interpretation and GIS Methods. Master's Thesis, Faculty of Geology and Geoenvironment, National and Kapodistrian University of Athens, Athens, Greece, 1997.
34. Pe-Piper, G.; Kotopouli, C.N.; Piper, D.J. Granitoid rocks of Naxos, Greece: Regional geology and petrology. *Geol. J.* **1997**, *32*, 153–171. [CrossRef]
35. Angelier, J.; Lyberis, N.; Le Pichon, X.; Barrier, E.; Huchon, P. The tectonic development of the Hellenic arc and the Sea of the Crete: A synthesis (Mediterranean). *Tectonophysics* **1982**, *86*, 159–196. [CrossRef]
36. Jansen, J.B.H. *Geological Map of Greece, Island of Naxos (1:50,000)*; Institute for Geology and Mineral Resources: Athens, Greece, 1973.
37. Bargnesi, E.A.; Stockli, D.F.; Mancktelow, N.; Soukis, K. Miocene core complex development and coeval supradetachment basin evolution of Paros, Greece, insights from (U–Th)/He thermochronometry. *Tectonophysics* **2013**, *595*, 165–182. [CrossRef]
38. Gautier, P.; Brun, J.P.; Moriceau, R.; Sokoutis, D.; Martinod, J.; Jolivet, L. Timing, kinematics and cause of Aegean extension: A scenario based on a comparison with simple analogue experiments. *Tectonophysics* **1999**, *315*, 31–72. [CrossRef]
39. Gautier, P.; Brun, J.P. Crustal-scale geometry and kinematics of late-orogenic extension in the central Aegean (Cyclades and Ewia Island). *Tectonophysics* **1994**, *238*, 399–424. [CrossRef]
40. Jolivet, L. A comparison of geodetic and finite strain pattern in the Aegean, geodynamic implications. *Earth Planet. Sci. Lett.* **2001**, *187*, 95–104. [CrossRef]
41. Tirel, C.; Gueydan, F.; Tiberi, C.; Brun, J.P. Aegean crustal thickness inferred from gravity inversion. Geodynamical implications. *Earth Planet. Sci. Lett.* **2004**, *228*, 267–280. [CrossRef]
42. Zhu, L.; Mitchell, B.J.; Akyol, N.; Cemen, I.; Kekovali, K. Crustal thickness variations in the Aegean region and implications for the extension of continental crust. *J. Geophys. Res.* **2006**, *111*. [CrossRef]
43. Soukis, K.; Koufosotiri, E.; Stournaras, G. *Special Landforms on Tinos Island: Spheroidal Weathering "TAFONI" Forms*, 3rd ed.; International Scientific Symposium of Protected Areas and Natural Monuments: Mytilini, Greece, 1998. (In Greek)
44. Theodoropoulos, D. Honeycomb weathering phenomena (TAFONI) on Tinos Island. *Ann. Géologiques Pays Helléniques* **1975**, *26*, 149–158. (In Greek)
45. Cordier, S.; Schlüchter, M.-L.; Evelpidou, N.; Pavlopoulos, K.; Bouchet, M.; Frechen, M. Morphology and OSL-420 based geochronology of the Holocene coastal dunes fields of Naxos Island (Cyclades, Greece): Preliminary 421 results. In Proceedings of the XVIII INQUA Congress, Bern, Switzerland, 21–27 July 2011.
46. Evelpidou, N.; Melini, D.; Pirazzoli, P.; Vassilopoulos, A. Evidence of a recent rapid subsidence in the S–E 417 Cyclades (Greece): An effect of the 1956 Amorgos earthquake? *Cont. Shelf Res.* **2012**, *39–40*, 27–40. [CrossRef]
47. Evelpidou, N.; Melini, D.; Pirazzoli, P.A.; Vassilopoulos, A. Evidence of repeated late Holocene rapid subsidence 439 in the SE Cyclades (Greece) deduced from submerged notches. *Int. J. Earth Sci.* **2013**, *103*, 381–395. [CrossRef]
48. Karkani, A.; Evelpidou, N.; Vacchi, M.; Morhange, C.; Tsukamoto, S.; Frechen, M.; Maroukian, H. Tracking 427 shoreline evolution in central Cyclades (Greece) using beachrocks. *Mar. Geol.* **2017**, *388*, 25–37. [CrossRef]
49. Saitis, G.; Koutsopoulou, E.; Karkani, A.; Anastasatou, M.; Stamatakis, M.; Gatou, M.-A.; Evelpidou, N. A multi- analytical study of beachrock formation in Naxos and Paros Islands, Aegean Sea, Greece and their palaeoenvironmental significance. *Z. Geomorphol.* **2021**, *63*, 19–32. [CrossRef]
50. Karkani, A. Study of the geomorphological and environmental evolution of the coastal zone of Central Cyclades. Ph.D. Thesis, National and Kapodistrian University of Athens, Athens, Greece, 2017.
51. Sakellariou, D.; Galanidou, N. Pleistocene submerged landscapes and Palaeolithic archaeology in the tectonically 451 active Aegean region. *Geol. Soc. Lond. Spec. Publ.* **2016**, *411*, 145–178. [CrossRef]
52. Desruelles, S.; Fouache, É.; Ciner, A.; Dalongeville, R.; Pavlopoulos, K.; Kosun, E.; Coquinot, Y.; Potdevin, J.-L. 447 Beachrocks and sea level changes since Middle Holocene: Comparison between the insular group of Mykonos– 448 Delos–Rhenia (Cyclades, Greece) and the southern coast of Turkey. *Glob. Planet. Chang.* **2009**, *66*, 19–33. [CrossRef]
53. Bouzekraoui, H.; Barakat, A.; Touhami, F.; Mouaddine, A.; El Youssi, M. Inventory and assessment of geomorphosites for geotourism development: A case study of Aït Bou Oulli valley (Central High-Atlas, Morocco). *Area* **2018**, *50*, 331–343. [CrossRef]
54. Gray, M. "Simply the best": The search for the world's top geotourism destinations. In *The Geotourism Industry in the 21st Century*; Sadry, B.N., Ed.; Apple Academic Press: Boca Raton, FL, USA, 2020; pp. 207–226, ISBN 9780429292798.

Article

How Academics and the Public Experienced Immersive Virtual Reality for Geo-Education

Fabio L. Bonali [1,2,*], Elena Russo [1], Fabio Vitello [3], Varvara Antoniou [4], Fabio Marchese [1,†], Luca Fallati [1], Valentina Bracchi [1], Noemi Corti [1], Alessandra Savini [1], Malcolm Whitworth [5], Kyriaki Drymoni [1], Federico Pasquaré Mariotto [6], Paraskevi Nomikou [4], Eva Sciacca [7], Sofia Bressan [1], Susanna Falsaperla [8], Danilo Reitano [8], Benjamin van Wyk de Vries [9], Mel Krokos [10], Giuliana Panieri [11], Mathew Alexander Stiller-Reeve [12,13], Giuseppe Vizzari [14], Ugo Becciani [7] and Alessandro Tibaldi [1,2]

1. Department of Earth and Environmental Sciences, University of Milan Bicocca, Piazza della Scienza 1-4, 20126 Milan, Italy; elena.russo@unimib.it (E.R.); fabio.marchese@unimib.it (F.M.); luca.fallati@unimib.it (L.F.); valentina.bracchi@unimib.it (V.B.); n.corti3@campus.unimib.it (N.C.); alessandra.savini@unimib.it (A.S.); kyriaki.drymoni@unimib.it (K.D.); s.bressan2@campus.unimib.it (S.B.); alessandro.tibaldi@unimib.it (A.T.)
2. CRUST-Interuniversity Center for 3D Seismotectonics with Territorial Applications, 66100 Chieti, Italy
3. INAF-Istituto di Radioastronomia, via Gobetti 101, 40129 Bologna, Italy; fabio.vitello@inaf.it
4. Department of Geology and Geoenvironment, National and Kapodistrian University of Athens, 15784 Athens, Greece; vantoniou@geol.uoa.gr (V.A.); evinom@geol.uoa.gr (P.N.)
5. School of the Environment, Geography and Geosciences, University of Portsmouth, Portsmouth PL01 3QL, UK; malcolm.whitworth@port.ac.uk
6. Department of Human and Innovation Sciences, Insubria University, Via S. Abbondio 12, 22100 Como, Italy; pas.mariotto@uninsubria.it
7. INAF-Osservatorio Astrofisico di Catania, Via Santa Sofia 78, 95123 Catania, Italy; eva.sciacca@inaf.it (E.S.); ugo.becciani@inaf.it (U.B.)
8. Istituto Nazionale di Geofisica e Vulcanologia-Osservatorio Etneo, Piazza Roma 2, 95125 Catania, Italy; susanna.falsaperla@ingv.it (S.F.); danilo.reitano@ingv.it (D.R.)
9. Laboratoire Magmas et Volcans, Observatoire du Physique du Globed e Clermont, IRD, UMR6524-CNRS, Université Clermont Auvergne, 63170 Aubiere, France; ben.vanwyk@uca.fr
10. School of Creative Technologies, University of Portsmouth, Portsmouth PO1 2DJ, UK; mel.krokos@port.ac.uk
11. CAGE, Center for Arctic Gas Hydrate, Environment and Climate, The Arctic University of Tromsø, 9019 Tromso, Norway; giuliana.panieri@uit.no
12. Department of Informatics, Systems and Communication, University of Milan Bicocca, Viale Sarca 336/14, 20126 Milan, Italy; mathew@stillerreeve.no
13. Konsulent Stiller-Reeve, 5281 Valestrandsfossen, Norway
14. Center for Climate and Energy Transformation, University of Bergen, 5020 Bergen, Norway; giuseppe.vizzari@unimib.it
* Correspondence: fabio.bonali@unimib.it; Tel.: +39-0264482015
† Now at Red Sea Research Center, King Abdullah University of Science and Technology, Thuwal 23955-6900, Saudi Arabia.

Abstract: Immersive virtual reality can potentially open up interesting geological sites to students, academics and others who may not have had the opportunity to visit such sites previously. We study how users perceive the usefulness of an immersive virtual reality approach applied to Earth Sciences teaching and communication. During nine immersive virtual reality-based events held in 2018 and 2019 in various locations (Vienna in Austria, Milan and Catania in Italy, Santorini in Greece), a large number of visitors had the opportunity to navigate, in immersive mode, across geological landscapes reconstructed by cutting-edge, unmanned aerial system-based photogrammetry techniques. The reconstructed virtual geological environments are specifically chosen virtual geosites, from Santorini (Greece), the North Volcanic Zone (Iceland), and Mt. Etna (Italy). Following the user experiences, we collected 459 questionnaires, with a large spread in participant age and cultural background. We find that the majority of respondents would be willing to repeat the immersive virtual reality experience, and importantly, most of the students and Earth Science academics who took part in the navigation confirmed the usefulness of this approach for geo-education purposes.

Citation: Bonali, F.L.; Russo, E.; Vitello, F.; Antoniou, V.; Marchese, F.; Fallati, L.; Bracchi, V.; Corti, N.; Savini, A.; Whitworth, M.; et al. How Academics and the Public Experienced Immersive Virtual Reality for Geo-Education. *Geosciences* 2022, 12, 9. https://doi.org/10.3390/geosciences12010009

Academic Editors: Hara Drinia, Panagiotis Voudouris, Assimina Antonarakou and Jesus Martinez-Frias

Received: 26 November 2021
Accepted: 21 December 2021
Published: 24 December 2021

Publisher's Note: MDPI stays neutral with regard to jurisdictional claims in published maps and institutional affiliations.

Copyright: © 2021 by the authors. Licensee MDPI, Basel, Switzerland. This article is an open access article distributed under the terms and conditions of the Creative Commons Attribution (CC BY) license (https://creativecommons.org/licenses/by/4.0/).

Keywords: immersive virtual reality; geology; photogrammetry; education; Iceland; Santorini; Etna

1. Introduction

Virtual reality (VR) is considered as a modern approach that provides 3D visualization in geological sciences, geoinformation for data collection and dissemination [1–3], as well as an immersive user experience [4,5]. Nowadays, VR landscapes can also rely on open or ad hoc-created geospatial datasets (e.g., [2]), including digital terrain and photogrammetry-derived 3D models [6], as well as bathymetric data [4]. The latter are also provided in the form of virtual outcrops and virtual geosites, and are considered as a key strategy to overcome common difficulties among students, such as visualizing three-dimensional concepts on a two-dimensional medium (for instance, a book), or on an image-based virtual tour [7–9].

A first attempt to apply VR techniques to studying geomorphological processes was made by Hilde et al. [10] and Anderson [11], who applied 3D visualization techniques to picturing ocean bottom topography by reconstructing a 3D scenario and a 3D tour. Subsequently, Anderson [12] refined the first approximation approach of Hilde et al. [10] to produce three-dimensional models of Mars using Viking Orbiter images, to study an extended period of glaciation in the Elysium Region. Other VR approaches focused on organising the so-called virtual tours of key outcrops, geolocated in a GIS platform, navigated by students by visualizing and studying 2-D images [7–9] or 3D digital outcrop models [13]. This approach has been introduced to support teaching activity in Earth Sciences, to reduce teaching costs, and to improve student learning efficiency and increase learning interest [9].

In addition, virtual tours have been useful at different stages: as a digital review before field campaigns, as a tutorial review, and as a digital asset substitute for field site inspection [8]. A mixed approach proposed a number of 3D models to be explored similarly to a virtual tour, using a PC, a tablet or a Smartphone; for example, McCaffrey et al. [14] applied this approach to petroleum geoscience, whereas Pasquaré Mariotto et al. [3] combined the use of virtual geosites with a WebGIS Platform (https://arcg.is/1e4erK0 (accessed on 26 November 2021)) to improve geoheritage communication.

VR applications have been increasingly used in recent years for geoscience education, scientific research, geoheritage communication and geotourism purposes; in fact, due to this approach, geological sites are made more accessible and available, including those that are dangerous or expensive to travel to, or have limited access. According to previous research efforts [3,15], this can be largely attributed to the possibility of making 3D models available as a free web resource, providing users with a number of virtual geological landscapes. These are the so-called virtual outcrops or virtual geosites, which are available to the public via web resources and represent, in 3D, geomorphological features and structures with photorealistic textures, based on photogrammetry techniques [3,6]. As such, virtual geosites can be an innovative tool for popularizing geoheritage in a broader audience by explaining ongoing, active geological and environmental processes [3,15,16]. Virtual geosites can also engage younger generations, who are particularly interested in highly interactive forms of communication and teaching [15]. An important advancement in the exploration of virtual outcrops and virtual geosites is the immersive virtual reality approach tailored by Tibaldi et al. [5], which has been developed in the framework of two research and innovation projects: namely the Italian Argo3D (https://argo3d.unimib.it/ (accessed on 26 November 2021)) and the EU Erasmus + 3DTeLC (http://3dtelc.lmv.uca.fr/ (accessed on 26 November 2021)) Projects.

In these projects, users can explore virtual geosites in the first person, thus experiencing the feeling of an immersive environment as if they were directly in the field. Furthermore, they can either walk or fly in photorealistic and photogrammetry-based landscapes (virtual scenarios) [6]. This possibility allows them to explore remote or inaccessible areas that would not be accessible otherwise, e.g., when vertical cliffs are present or when exploring recent volcanically and tectonically active areas.

With the final aim of raising awareness about the above-cited projects, we organized nine educational events, with the total participation of 459 attendees, belonging to different age groups and cultural backgrounds, ranging from the lay public to highly-specialised scientific community members, and involving both students and researchers in Earth Sciences.

The participants firstly experienced immersive virtual reality and afterwards provided their feedback on the experience via anonymous questionnaires, the results of which are presented in this paper. Participants were categorised in the following five groups, four of which are subgroups:

(i) All Participants;
(ii) Academics/Researchers in Earth Sciences (Academics) that include PhD students and postdocs;
(iii) MSc Students in Earth Sciences (MSc);
(iv) Middle and High School Students (Schools students);
(v) Lay Public (i.e., participants that do not belong to the other groups).

For the dissemination events, we selected four Virtual Geosites which are considered as stunning volcano–tectonic environments due to their cultural, historical and scientific value. These are: the Metaxa Mine located in the Santorini volcanic complex in Greece [17]; a rift-transform triple-junction [18]; the 1984 Krafla eruption area [19], both located along the North Volcanic Zone (NVZ) of Iceland; and the Mt Pizzillo area, situated in the NE rift of Mt Etna, Italy [20].

In this paper, we explore the impact of immersive virtual reality as applied to geological exploration aimed at geo-education, geological exploration and, potentially, geotourism in volcanic areas.

2. Materials and Methods

2.1. Immersive Virtual Reality Approach

As detailed in Choi et al. [21], VR applications can be classified as non-immersive and, more importantly, fully immersive experiences. Non-immersive VR is typically referred to as 3D visualization and displays 3D models on a computer screen and/or mobile device without using any head-mounted displays. Fully immersive VR was applied during our dissemination activities and provided users with virtual ways to explore classical geological sites, being able to fly above virtual landscapes, using a suitable VR headset and thumbsticks. Users were provided with a holistic view of a virtual geosite, which allowed them to explore specific geomorphological and volcanotectonic features at a range of different aerial scales. This approach has been fully developed for Earth Sciences by Tibaldi et al. [5], with contributions from Gerloni et al. [22] and Krokos et al. [6]. The former has introduced a navigation mechanism dedicated to explore outcrops in a fully immersive way by replicating real-world field exploration. This approach is based on an offline-based visual discovery framework developed with the Unity game engine (https://unity.com (accessed on 26 November 2021)). The application employs an Oculus Rift (https://www.oculus.com (accessed on 26 November 2021)) as a head-mounted VR device and allows earth scientists to navigate in their own immersive VR scenarios, based on photo-realistic photogrammetry outputs [6].

In our immersive virtual reality application, the landscape was built upon unmanned aerial system (UAS)-based photogrammetry techniques and is thus able to provide centimetric pixel size resolution for the resulting virtual landscape in the form of a 3D model. This has been successfully applied to models of different aerial extents and resolutions, ranging from about 50 to 1000 m (among the longest) and from 0.8 to 4 cm/pixel, respectively.

We used Agisoft Metashape (https://www.agisoft.com/ (accessed on 26 November 2021)) to process the 3D-model and generate a Wavefront OBJ Tiled model, which was designed and tested as the landscape input for our approach by Krokos et al. [6],

Tibaldi et al. [6] and Antoniou et al. [17]. The use of 3D tiled models enabled us to obtain reliable representations of geometrical features especially when the objects are oblique to the studied outcrop.

When the user starts their virtual exploration, they walk around on a "solid ground surface" (Figure 1a–c), moving with the thumbsticks on the controllers. Our immersive virtual reality system allows them to choose between three different modes to navigate the scenery, named "walk mode", "drone mode" and "flight mode"(Figure 1b). In the default "walk mode", the user can navigate the virtual geosite slowly at a virtual height of around 2 m. (Figure 1c). In the "drone mode", the user can experience the site as a radio-controlled drone, flying above the virtual landscape. In the "flight mode" (Figure 1e), the user moves to a higher elevation and looks down over the terrain. This mode allows the user to move across the site at the fastest speed.

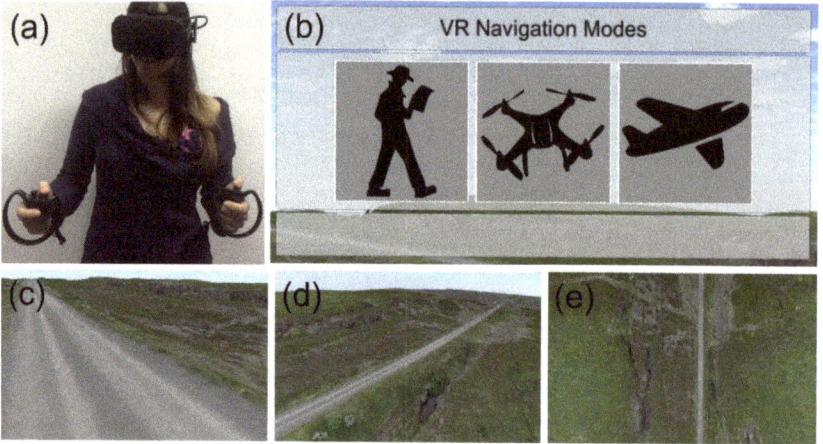

Figure 1. (**a**) User experiencing geological exploration. The VR headset and the input peripherals belong to the Oculus Rift. (**b**) The three exploration modes available in our application; (**c**) walk; (**d**) drone; and (**e**) Flight modes; road for scale.

2.2. Dissemination Events

The dissemination events described below were held between 2018 and 2019 in Italy, Greece, and Austria. Each participant was offered the possibility to explore two virtual geosites of their personal choice, each for five minutes, for a total of 10 min for each participant, during which they tried all three navigation modes.

During the virtual exploration, each user was supported by a trained staff member, who was knowledgeable about the virtual geosites and the use of the immersive tools. The aim was to offer a navigation tutorial to the user, so that they could quickly grasp the main geological and geomorphological features of the selected site. The dissemination events were as follows:

1. Title: *Field exploration using immersive virtual reality in the framework of the "B.Inclusion days"* (https://www.unimib.it/eventi/binclusion-days-2018), held at the University of Milan-Bicocca, Italy. This event was supported by the EGU 2018 Public Engagement Grant (https://www.egu.eu/news/400/egu-2018-public-engagement-grants-awarded-to-suzanne-imber-and-fabio-bonali/) and was held in collaboration with the Disability and DSA (Disabled Students' Allowance) service of the University of Milan-Bicocca (https://en.unimib.it/services/bicocca-campus/disability) (10 October 2018). During this event we collected feeback from 25 participants aged 20–56.
2. Title: *Santorini Summer School*, in the framework of the EU Erasmus + Project— Bringing the 3D-world into the classroom: a new approach to Teaching, Learning and

Communicating the science of geohazards in terrestrial and marine environments (http://3dtelc.lmv.uca.fr/). This event was held in Santorini (Thira), Greece (12–21 October 2018). On this occasion, we collected 14 feedback forms from participants aged 23–30.

3. Title: *Volcano-tectonic applications using immersive virtual reality*. This event was held in Milan, Italy, at the University of Milan-Bicocca, in the framework of a class in active tectonics and volcano tectonic settings held by Prof. Alessandro Tibaldi (https://en.unimib.it/alessandro-tibaldi) (10 January 2019). During this event we collected 10 feedback forms from participants aged 22–24.

4. Title: *Geological exploration without barriers: Shaping geological 3D virtual field-surveys for overcoming motor disabilities*, held at the University of Milan-Bicocca, Milan, Italy (16 January 2019). This event was supported by the EGU 2018 Public Engagement Grant (https://www.egu.eu/news/400/egu-2018-public-engagement-grants-awarded-to-suzanne-imber-and-fabio-bonali/) and was held in collaboration with the Disability and DSA (Disabled Students' Allowance) service of the University of Milano-Bicocca (https://en.unimib.it/services/bicocca-campus/disability). At this event we collected 37 feedback forms and the age of the participants was 19–59.

5. Title: *Geological exploration using Immersive Virtual Reality*. This event was held in Milan at the University of Milan-Bicocca (https://www.unimib.it/eventi/realta-virtuale-immersiva-esplorare-territorio) in the framework of the Digital Week event (https://www.milanodigitalweek.com/), Italy (16 March 2019). During this event we collected data from 24 participants, the age of participants ranging 20–73.

6. Title: *Primavera in Bicocca 2019* (https://www.unimib.it/sites/default/files/orientamento/programma_primavera_in_bicocca_2019.pdf) (21 March 2019). This event was held in Milan at the University of Milan-Bicocca, Italy. During this event we collected 10 feedback forms and the participants' age was 18–20.

7. Title: *Geological exploration without barriers in tour*. This event was held at the National Institute of Astrophysics (INAF), Catania, Italy and was supported by Argo3D funding (https://argo3d.unimib.it/) (30 March 2019). During this event we collected 21 questionnaires, and the age of the participants was 13–18.

8. Title: *Virtual reality for geohazards and geological studies*, held in Vienna during the EGU General Assembly 2019 (https://www.egu2019.eu/), Austria Center, Austria (7–12 April 2019). The event was supported by the EGU outreach committee (https://www.egu.eu/outreach/). On this occasion we collected 155 feedback forms and the age of the participants was 21–70.

9. Title: *3D and immersive Virtual Reality: new frontiers in geological exploration*. This event was held during the MeetMeTonight event (https://www.meetmetonight.it/), at the Natural Sciences Museum of Milan, Italy (27–28 September 2019). During this event we collected 163 feedback forms, and the age of the participants was 13–66.

2.3. Questionnaires

The aim of the questionnaires was to obtain feedback from people of different ages and backgrounds for the use of photorealistic immersive virtual reality in Earth Sciences, and to explore its applicability to teaching and studying as well as for communication purposes. The questions were categorised into three main groups: (i) general information and questions; (ii) general questions about the experience; (iii) specific questions about its use in Earth Sciences applications.

To facilitate the participants' replies, a specific procedure was designed. Firstly, the participants were asked to fill in an authorization form. For minors under the age of 18, parents were asked to fill in the authorization forms on their behalf. Once the required consent was given, the participants lined up to access the experience. Once their turn was announced, the participants were seated at the station equipped with a computer and the Oculus Rift (with headset and two controllers). In this phase, one of the staff members introduced the immersive virtual reality experience with a short tutorial to explain how to explore the VR environment, as well as providing some basic geological notions about the environment in which they would soon be immersed. During each event, after the experience, the participants were invited to fill in a Google Form questionnaire that was the same for each dissemination event. The latter was divided into three macro categories:

(1) General information aimed at collecting anonymous data such as age and job title, and then general questions. Questions: (i) Have you ever used a VR headset before? (reply, YES/NO); (ii) Do you usually go hiking/trekking? (reply, YES/NO);

(2) General questions on the experience aimed at evaluating the experience with the virtual reality just tested: (iii) Which is your favourite navigation mode? (reply: Walk, Drone or Flight mode); (iv) How satisfied are you with the portrayal of the virtual landscape? (five-level Linkert item—1 Unsatisfied, 2 Slightly satisfied, 3 Neutral, 4 Moderately satisfied, 5 Very satisfied); (v) How would you rank your experience with virtual exploration? (five-level Linkert item—1 Unsatisfied, 2 Slightly satisfied, 3 Neutral, 4 Moderately satisfied, 5 Very satisfied); (vi) Would you like to repeat the experience? (reply, YES/NO); (vii) Did you experience any form of sickness during the navigation? (reply, YES/NO), If so, which one? (Open reply)

(3) Specific questions on the potential use of Immersive Virtual Reality in Earth Sciences, aimed at assessing the perception of the users on the possible adoption of this technology in the geo-education context: (viii) How useful do you think it is as a studying/learning tool in Earth Sciences? (five-level Linkert item—1 Useless, 2 Slightly useful, 3 Neutral, 4 Moderately useful, 5 Very useful); (ix) How useful do you think it is as a teaching tool in Earth Sciences? (five-level Linkert item—1 Useless, 2 Slightly useful, 3 Neutral, 4 Moderately useful, 5 Very useful).

As a final point regarding the questionnaires, it was decided to propose a limited number of questions to the participants, focusing on acquiring a large number of replies, following a trade-off approach. In fact, we planned to reach a reasonable balance between the possibility to acquire useful feedback about the use of immersive virtual reality for geo-education and the practical plausibility of administering the form within the events. We especially focused on getting feedback on the quality, usability, and usefulness of the developed system.

We did not carry out a specific analysis of perceived usability of the developed immersive VR environment and system, because it would have required the participants to fill in a very long and time-consuming questionnaire [23].

3. The Virtual Geosites

3.1. The Metaxa Mine, Santorini (Greece)

The Metaxa Mine is in the SE part of the Santorini volcanic complex, which is part of the active South Aegean Volcanic Arc. The volcanic centres extend from Sousaki and Methana (close to Central Greece) to the Nisyros-Kos islands (at the Eastern border with Turkey), with Santorini volcano being its southernmost expression. The latter is located on the Aegean microplate and is associated with recent volcano tectonic activity in the region [24]. The activity of the volcanic arc is related to the subduction of the African plate beneath the Eurasian plate [25], but locally around Santorini Island the stress field is extensional due to the slab rollback [26] and the associated regional tectonics.

The Metaxa Mine is an iconic geosite where visitors have a chance to view and explore the Late Bronze Age Minoan era deposits. Moreover, it is an industrial heritage site as the extraction of pumice during the previous century gave considerable income opportunities to the local population. Locally, the landscape presents an asymmetrical morphology [17], surrounded by almost vertical and overhanging cliffs and covered by debris flow deposits and remnants of anthropogenic activities. Its heterogeneous floor is scattered with small hills made of covered excavation materials. Along the vertical slopes of the mine, and especially at the entrance, an excellent outcrop showing Late Bronze Age deposits [5] can be observed. The virtual geosite for this site (Figure 2) has an extent of about 520 × 400 m and a texture resolution of 0.9 cm/pix. Further details are in Antoniou et al. [17] while

3D details with annotations can be found at https://geovires.unimib.it/ (accessed on 26 November 2021), within the geosites section, and at https://arcg.is/1e4erK0 (accessed on 26 November 2021).

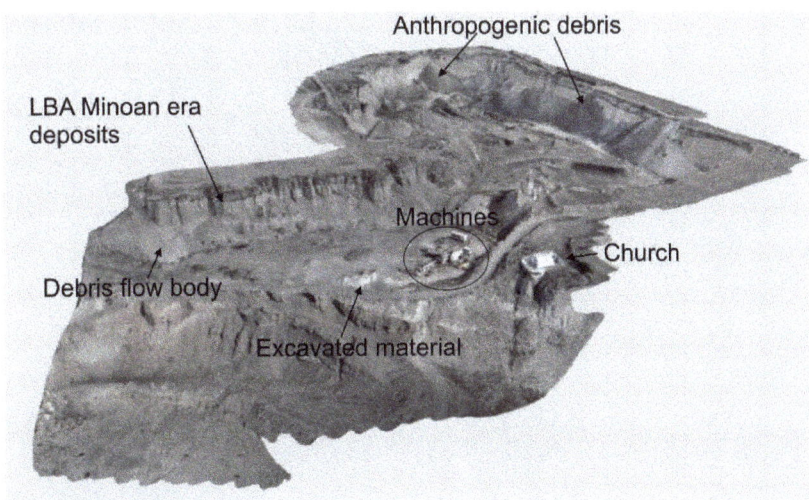

Figure 2. 3D eastern view of the Virtual Geosite of the Metaxa Mine. See the white church on the right hand side for scale.

3.2. 1984 Krafla Eruption Site, Northern Volcanic Zone (Iceland)

The virtual geosite is located within the Krafla fissure swarm in the Northern Volcanic Zone of Iceland, along the emerging part of the Mid Atlantic ridge. This unique volcano tectonic terrain is composed of active rift zones, extension fractures, faults, eruptive fissures, and basaltic volcanoes. The area is occupied by a number of volcanic craters with a N-S alignment (Figure 3), lava flows, as well as a series of extension fractures and normal faults associated with the 1984 Krafla eruption [27]. The larger cone in the foreground (350 m × 150 m) was formed in 1984, at the end of the "Krafla Fires" eruptive cycle [28]. In its central part, one can observe the pyroclastic cone flank [19] as well as two normal faults which cross the post-Latest Glacial Maximum lava flows (older than 7 ka BP; 27). In the model, a cluster of recent monogenetic volcanoes (scoria cones) is also present, as well as an older scoria cone filled by the 1984 lava flow. The virtual geosite for this site (Figure 4) has an extent of about 1010 × 680 m and a texture resolution of 2.6 cm/pix. Additional details are made available in 3D with annotations at: https://geovires.unimib.it/ (accessed on 26 November 2021), within the volcano tectonics section.

Figure 3. 3D view of the virtual geosite of the Krafla eruption site; the scale is highlighted in the figure.

Figure 4. 3D view of the rift-transform triple junction virtual geosite; the scale is highlighted in the figure.

3.3. The Rift-Transform Triple Junction, Northern Volcanic Zone (Iceland)

To the east of Husavik, in Northern Iceland, a remarkable subaerial triple-junction intersection is preserved between the Husavik-Flatey Fault dextral transform fault and rifting in the Northern Volcanic Zone (Figure 4). The geomorphological and structural landforms are visible due to the presence of a sheet of pahoehoe lavas that record numerous structural features that are displayed in fantastic detail. These features include transform and extensional fault structures, riedel fault complexes, transpressional faulting and compressional strike-slip relay ramps, as well as second-order R, R' and P shears [18], providing a unique opportunity to investigate this dynamic structural environment using a virtual reality model created from drone imagery acquired over the triple junction region for the first time. Two interacting fault systems are clearly visible in the drone data and using the VR (the drone flight mode is particularly good for viewing these structures), comprising a

NW-SE transform-affinity faults of the Husavik-Flatey Fault and the roughly N-S normal faults of the Theistareykir rifting in the volcanic zone. The site is dominated by the vertical fault scarp face that extends north-south and the Husavik-Flatey Fault that extends away from this cliff face, as well as a series of small fault scarps with evidence of opening and right lateral movement.

3.4. Mt Pizzillo Area, NE Rift of Mt. Etna (Italy)

Mt. Etna is located along the eastern coast of Sicily, Italy, and since 2013 it has been inscribed on the UNESCO World Heritage list. Its long eruptive history, starting from 600,000 years ago [29] and its continuous gas and vapor emissions along with its sporadic Strombolian activity, has produced several spectacular lava fountains and lava flows which make this volcano an important reference for volcanological studies. Etna is, therefore, an ideal site to test immersive virtual reality and design a virtual survey of geological and structural features, such as its dense network of eruptive fissures forming the so-called NE, S and W Rifts. In particular, the area of investigation is located in the northern sector of the volcano: it belongs to the NE Rift, a group of eruptive fissures that mark a zone of frequent, shallow volcanic intrusions. The rough terrain is affected by active deformation (approximately 2 cm/yr [30]) and is covered by cinder cones and historical lava flows. The virtual geosite (Figure 5), which has an aerial extent of about 230 × 200 m and a texture resolution of 2 cm/pix, is an area located within Mt Etna's NE rift, on the northern flank of the volcano. Here, the users can observe a swarm of NE-SW extension fractures with centimetric dilation (up to about 70 cm) and several piercing points, which can be clearly observed along the fracture walls. The two widest main fractures gradually transit to normal faults, dipping towards each other. The faults can be clearly recognized on the flank of the pyroclastic cone of historical age [29]. The primary fracture located on the right is next to a short normal fault dipping to the left. By following the trace of the fault where it reaches the highest point on the flank, it is possible to see its partitioning into two faults that dip towards each other and form a graben structure. Further details are made available in 3D with annotations at: www.geovires.unimib.it, within the volcano tectonics section.

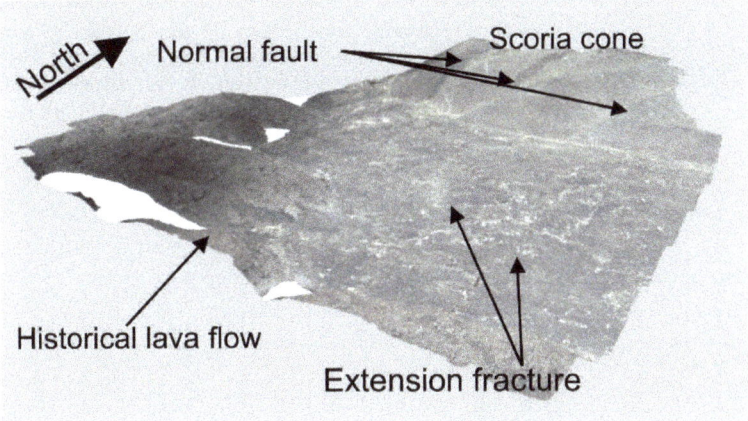

Figure 5. 3D view of the virtual geosite of Mt Pizzillo, along the NE Rift, Mt Etna.

4. Feedback from the Users

In this section we show the results acquired from 459 questionnaires from participants, aged 13–73. We focus both on the entire dataset of participants and on the following categories:

(i) Academics/researchers in Earth Sciences (A/RES) (n = 144, age 25 ÷ 70);
(ii) MSc students in Earth Sciences (MScSES) (n = 35, age 21 ÷ 30);

(iii) middle and high school students (MHSS) (n = 104, age 13 ÷ 20);
(iv) Lay public (LP) (not related to the previous categories, n = 176, age 19 ÷ 73) (Figure 6). It is worth noting that 10 out of 459 users are affected by a physical disability that would hinder the hiking or trekking needed to visit these geosites in real life.

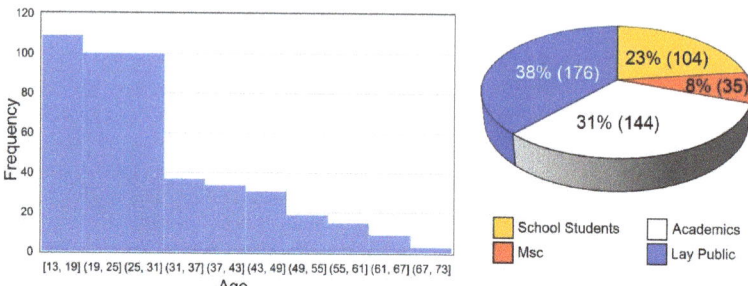

Figure 6. Graphs showing the results related to the General questions provided to the participants after their experience. To the left, relative frequency histogram displaying the age of the participants, grouped each 6 years, and, to the right, Pie chart showing the frequency of each category.

4.1. General Questions

The preliminary questions allowed us to identify the most frequent age groups that participated in the experience during our events. The histogram in Figure 6 shows that the users mainly belonged to two main groups: 13–31 years old, followed by the range between 31 and 49 years old, while older ages were less represented. For their backgrounds, the graph in Figure 6 shows that most of the users belonged to the lay public (38%), followed by academics (31%), school students (23%) and finally MSc class (age 21–30) (8%).

4.1.1. Have You Ever Used a VR Headset Before?

As part of their feedback, all participants were asked about their familiarity with the VR equipment before the event. The results are displayed in Figure 7 and show that the majority of the users (54%) had not used similar equipment before. This observation, however, changes if we consider each category singularly. In particular, most students (52%) ranging from 13 to 20 years old (school students) had already used a VR headset before. On the other hand, lay public and academics show similar percentages: most (55% in the former case and 56% in the latter) had not used a VR headset before the dissemination event.

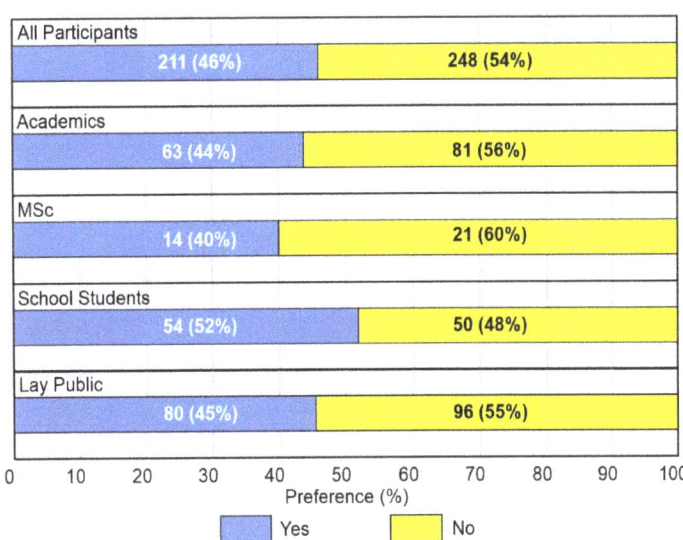

Figure 7. Bar charts showing the users' familiarity with VR equipment before the dissemination events. The results represent both the total number of participants and the background of each group, respectively.

4.1.2. Do You Usually Go Hiking/Trekking?

Figure 8 shows the habits of the participants with respect to hiking/trekking: most users (72% of all participants) enjoyed the direct contact with nature and thus were used to going hiking or trekking. This percentage increased when considering both academics and MSc (84% in the first case and 83% in the latter). Finally, school students (59%) and lay public (68%) also reached quite high percentages. This was a fundamental question to compare replies from two different datasets of participants that were, or were not, addicted to field excursions, thus with different levels of expertise in landscape exploration.

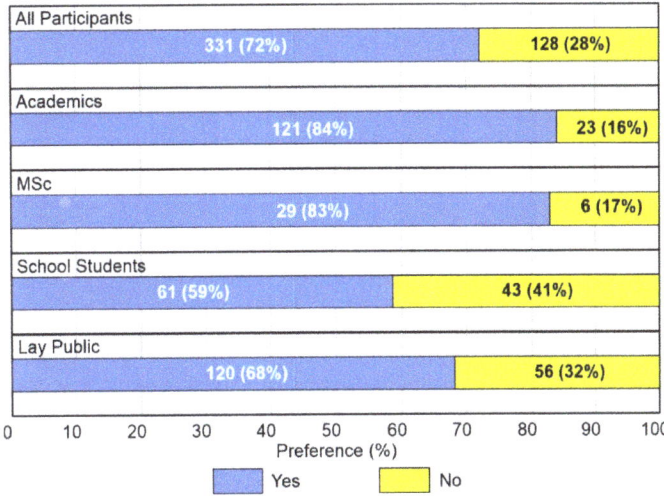

Figure 8. Bar charts showing the habits of the participants with regard to hiking/trekking. The results represent both the total number of participants and the background of each group respectively.

4.2. General Questions on Immersive Virtual Reality

4.2.1. Which Is Your Favourite Navigation Mode?

For most participants, as displayed in Figure 9, the most popular navigation mode during immersive virtual reality exploration was the drone. This was clearly shown both by considering all participants (69%) and the single categories (for which the percentage ranged from 67% to 71%). Between the other two modes, the walk and flight ones, users mostly chose the walk mode (20–21%). A discrepancy, however, was found in the results provided by the MSc group, who showed a slight preference for the flight mode (17%) over the walk mode (14%).

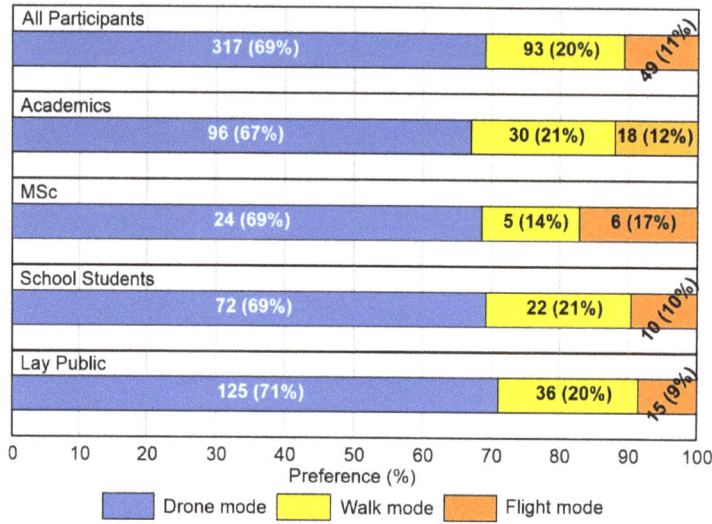

Figure 9. Bar charts showing the preferred navigation mode. The results represent both the total number of participants and the background of each group respectively.

4.2.2. How Satisfied Are You with the Portrayal of the Virtual Landscape?

The users' satisfaction on the portrayal of the immersive VR environments showed that most declared themselves to be moderately (49%) to very satisfied (38%) (Figure 10), with the academics as the most satisfied (47%), followed by the lay public and MSc. The data for school students showed the lowest satisfaction rate for rank 5, even though the majority responded 4 and 5.

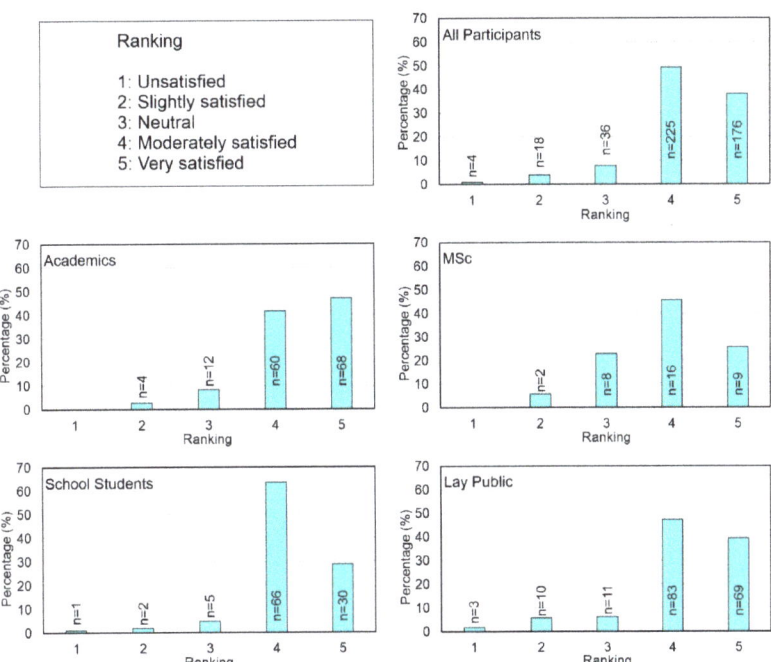

Figure 10. Bar charts showing the portrayal feedback of the Immersive Virtual Reality exploration. The results represent both the total number of participants and the background of each group respectively.

4.2.3. How Would You Rank Your Experience with Virtual Exploration?

More feedback was collected about the general satisfaction of the participants with the immersive virtual reality exploration. Most users were moderately to very satisfied with their experience (Figure 11). The highest percentages were shown by academics and the lay public. Based on our statistical analysis, school students were a little bit less satisfied when compared with the other categories. In fact, 18% declared themselves to be basically neutral.

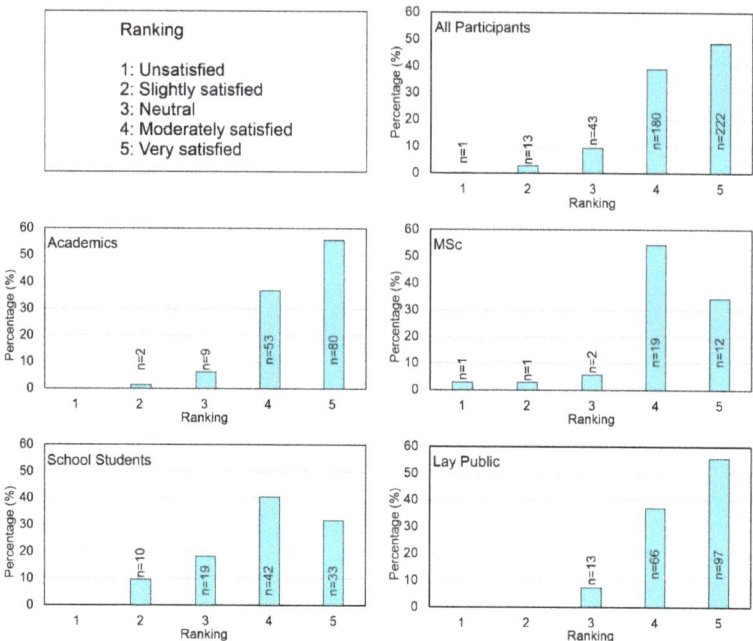

Figure 11. Bar charts showing the experience of the participants during their immersive virtual reality exploration. The results represent both the total number of participants and the background of each group respectively.

4.2.4. Would You like to Repeat the Experience?

Based on our results, a high percentage of the users wanted to repeat the experience (Figure 12). In fact, an average of 97% of all participants definitely wanted to repeat the experience; The school students category scored 100%.

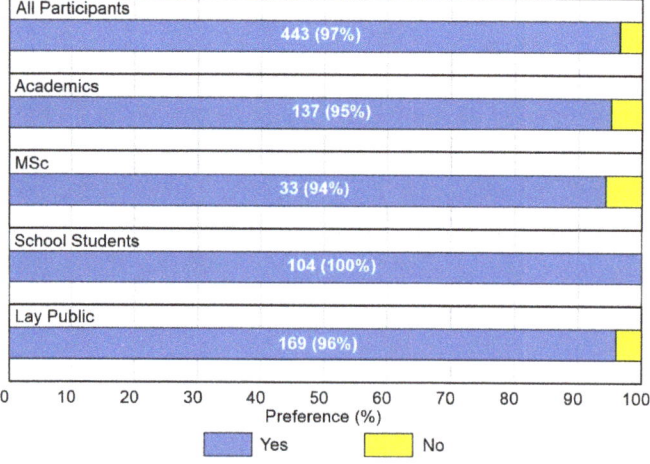

Figure 12. Bar charts showing the participants' wish to repeat the immersive virtual reality exploration experience. The results represent both the total number of participants and the background of each group respectively.

4.2.5. Did You Experience Any Form of Physical Sickness during the Navigation? If so, Which One?

Short-term sickness was mentioned by 26% of the participants (Figure 13). This varied across the different groups, with academics being the most impacted, and MSc and school students being the least impacted. The overall 119 sickness effects reported consisted of: slight dizziness (65%), slight headache (22%), slight nausea (10%) and disorientation (3%). None resulted in an interruption of the experience, suggesting that these effects are minor in most cases. Regarding this point, we are aware that there are more thorough, systematic, and validated ways of evaluating the feeling of sickness due to the usage of VR technologies (in particular the Simulator Sickness Questionnaire—SSQ—introduced by Kennedy et al. [31]), but we only wanted to have an at-a-glance indication about the fact that some adverse effects were perceived by the users, and also to evaluate the real need of performing additional experiments on this point.

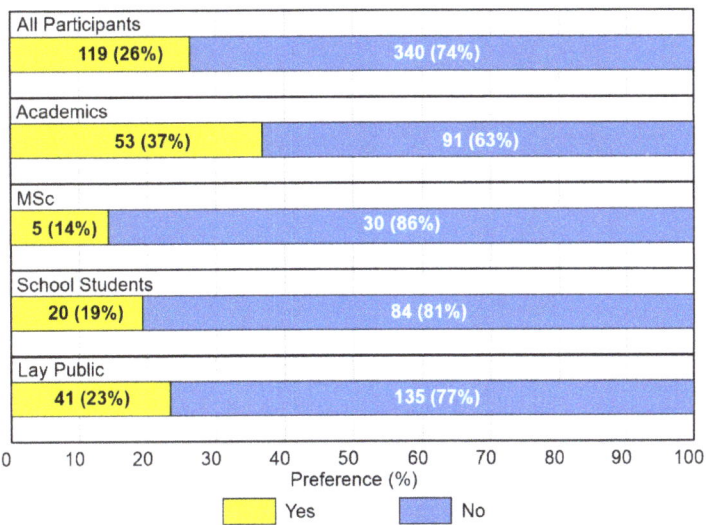

Figure 13. Bar charts showing possible side effects associated with the immersive virtual reality experience. The results represent both the total number of participants and the background of each group respectively.

4.3. Teaching

4.3.1. How Useful Do You Think Immersive Virtual Reality Is as a Studying/Learning Tool in Earth Sciences?

Figure 14 shows the effectiveness of immersive virtual reality as a learning tool. All participants believed that it could be very useful (59%) followed by moderately useful (30%). When considering each category, the perception of the users remained very similar. In particular, the lay public was found to be the most enthusiastic, as 67% found it 'very useful'. The other categories recognized the potential of this technology as a learning instrument. In particular, 56% of the school students and 57% of the MSc agreed with this scenario. Academics were less fond of it: 52% provided a very positive opinion, but 13% remained neutral.

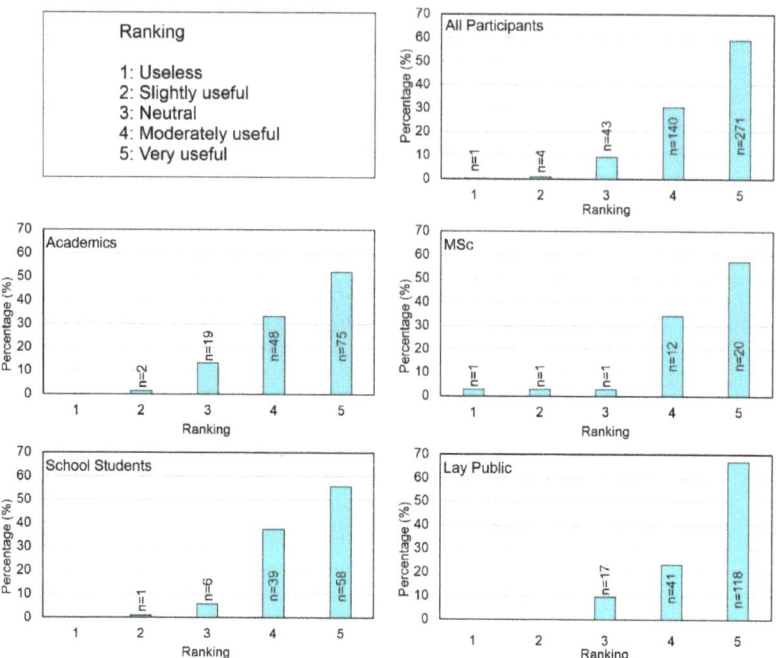

Figure 14. Bar charts showing the possible efficiency of immersive virtual reality as a learning tool in Earth Sciences. The results represent both the total number of participants and the background of each group respectively.

4.3.2. How Useful Do You Think Immersive Virtual Reality Is as a Teaching Tool in Earth Sciences?

The great majority of the users (68%) declared that immersive virtual reality can be effectively used as a geo-education tool (Figure 15). In detail, academics and MSc showed similar results (70% and 69%, respectively) and found this possibility very promising. In contrast, school students only believed that it was 'very useful' at 52%, offering a general positive opinion when the 'moderately useful' replies, totalling 95%, were considered.

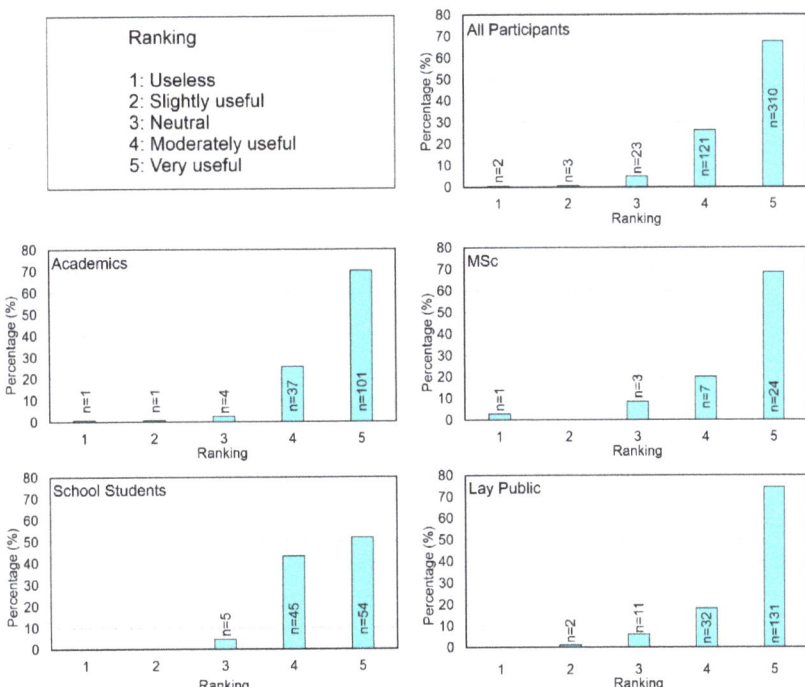

Figure 15. Chart bars showing the possible efficiency of immersive virtual reality as a geo-education tool. The results represent both the total number of participants and the background of each group respectively.

5. Discussion

Overall, the results of our initiative of using immersive virtual reality as a tool to carry out dissemination events on geo-education has been very positive. We held nine events, in three different countries, both in Italian and English languages, engaging a total of 459 international participants from various backgrounds. All dissemination events were carried out with the same format, virtual geosites and holistic view, aimed at understanding the impact of VR experiences on geosciences. Such events also involved people affected by temporal, or permanent, motor disabilities, who could explore "in person" inaccessible sites, proving how this technology could provide equality when dealing with geological field work and exploration. Thanks to the feedback given by all participants collected during these events, we provided a quantitative assessment of the immersive virtual reality experience, to evaluate its applicability as a tool for virtual geological exploration, with a particular focus on geo-education and communication. This represents an additional value, since previous attempts to evaluate VR techniques applied to Earth Sciences were mainly qualitative.

Generally speaking, there is a general agreement in the literature, mainly by practice and qualitative evaluations, that VR techniques are useful for teaching, as well as for learning in Earth Sciences (e.g., [7–9]). Over the last decade, several low-cost solutions have been made available to enable a quick and easy immersive virtual reality experience, both in terms of hardware and software solutions [32–34], providing teachers, students, and researchers with innovative tools to be applied in Earth Sciences. So far, in 2002, Kaiser et al. [32] applied immersive VR technology to improve mine planning, thanks to a collaboration among scientists with different areas of expertise. Recently, Kinsland and Borst [35] successfully used such technology to visualize geological data as 3D objects to be managed by scientists. As regards geo-education and research activity, McNamee and Bogert [36] started using immersive virtual reality through a commercial application that allows users to fly and walk in a virtual scenario, in order to address two of the main problems in teaching geological processes in an introduction-level physical geology course: describing dimensions and scale. A slightly different approach has been used by Visneskie et al. [37], where the immersive experience was conducted with a set of 360-VR-Videos organized as a virtual tour to be used with mobile headsets. At a more advanced level, Tibaldi et al. [5], Antoniou et al. [17] and Caravaca et al. [38] designed a system where the user is capable of moving within a georeferenced and scaled 3D scene that can be explored simulating a real field trip.

Following this general view that VR tools are useful for geo-education and research activities in Earth Sciences, we aim to contribute to the discussion with a new relevant quantitative evaluation on the immersive virtual reality system developed by Tibaldi et al. [5] and many of the authors of the present paper, and successively applied for use in research activity by Antoniou et al. [17].

In evaluating the replies to the questionnaires, the first important output was that most participants would be glad to repeat the experience. This held for at least 97% of the respondents (all participants), with 100% positive replies provided by school students (Figure 12). More in detail, 96% of the participants, who had never used a VR headset before, would be willing to repeat the experience, whereas for those who had already tried a VR headset, the percentage rose to 98%.

For the exploration mode, most participants preferred to fly in "drone mode", around and above the main geological objects within the virtual geosites (Figure 9), by a percentage always greater than 67%, with the highest being reached by the lay public (71%), and a greater appreciation from participants that had already used VR (75%), when compared to those who had experienced it for the first time (64%). In addition, when considering if users were used to hiking or not, the drone mode reached a percentage of 68% and 71%, respectively, confirming that this was the best choice for navigation.

Another important aspect pertains to the type of virtual scenario we have proposed: again, in this case, the majority of participants ranked with high marks (4–5) the portrayal of the virtual landscape (87%), which is based on drone-captured pictures. This appreciation rate came especially from the lay public (86%), academics (89%), and school students (92%); on the other hand, the percentage of MSc was 71% (Figure 10).

The above results confirm the general appreciation for a photogrammetry-based virtual scenario. The results acquired from school students are the most interesting. Indeed, they show the lowest satisfaction for rank 5, even though the majority responded 4 and 5. We think that this could be due to the fact that school students are far more used to gaming and hyper real visualisations than other categories: as a consequence, their expectations might probably be higher. An improvement that can be made to the virtual landscape, for a better application to geo-education, is the possibility of adding everyday scaled objects to the scene, to help the user better perceive dimensions and distances. This approach could help overcome distance underestimation problems in VR compared to reality [39,40].

For the overall evaluation of the exploration of volcanic areas through the system, with the support of a geo-expert, the majority of the participants ranked with high value (4–5) the experience (87%), reaching a percentage of 93% among the lay public, 92% among academics, 88% among MSc, whereas the percentage plummeted to 72% for school students (Figure 11).

For the use of immersive virtual reality as a tool for studying and teaching Earth Sciences, it is worth noting that these questions have a different importance for academics than for both the student categories (MSc and school students). For the former group, both teaching and studying dimensions are relevant: in terms of studying activities, it is interpreted as an application to study landscapes, and thus as an application for research purposes as they are allowed to make direct observations. For MSc and school students, it is the studying/learning question that has a greater importance, since it would be applied as a tool for advanced learning activity. Regarding the results of these questions, as shown in Figures 14 and 15, there is a general appreciation from all groups, with percentages in the range 85–96%: particularly, immersive virtual reality is considered as a great study instrument by 85–93% of the sample; the geo-education potential of it is highly regarded by 89–96% of the sample; in this latter case, the highest value (5) dominates the replies (Figure 15).

Some participants suffered from sickness effects, but none stopped the experience, and among those affected by sickness effects, only 8% of this subset would not like to repeat the experience, corresponding to about 2% of the total, suggesting that educational VR applications still need to be slightly improved. The latter point could be further explored through more detailed analyses on perceived adverse effects and system usability, following Vizzari [41]'s approach. It is noted that in order to achieve this, it is necessary to focus on a smaller number of participants, in ad hoc events lasting longer, with specific tasks being carried out, and that this is out of the scope of the present work. From a technical point of view, the use of VR headsets capable of working with a frame per second value higher than 90 can help reduce sickness effects [42]. In addition to that, it can be suggested that the user repeat the experience a few minutes after the first test. In fact, based on our personal experience, the more frequently immersive virtual reality systems are used, the more comfortable users become with the VR experience.

Overall, our results demonstrate and confirm the great potential of immersive virtual reality for geo-education and as a tool for geoscience communication to citizens, with possible applications for geotourism. Students, academics, and non-academics can explore virtual geosites in person, acquiring geological knowledge without temporal or spatial limitations, potentially reducing travel costs and carbon emissions [43], and decreasing unnecessary energy consumption.

The most important and innovative benefits we consider regarding the use of Immersive VR can be summed up as follows:

(i) It has a great potential to help users learn geology in a more interactive way, enhancing interest and improving learning efficiency, even if geological field trips are still crucial for a better understanding of Earth Science processes;
(ii) It can open up the possibility of studying virtual geosites in person for people affected by motor disabilities; it is worth noting that all participants belonging to this category ranked all questions with values greater than 4 and wished to repeat the experience in the future;
(iii) It implies a relevant cost cutting, especially for students, since geological outcrops can be brought into the lab;
(iv) It can be considered as an approach to reduce carbon emissions, due to the decrease in travelling needs for many people;
(v) It allows researchers to virtually travel to key geological spots and do science even in abnormal times, such as the COVID-19 pandemic.

Furthermore, immersive virtual reality can be also used by students to examine the geology field route beforehand, so as to make the trip easier and more efficient. It allows users to carry out tasks that could be difficult in the real world due to constraints and restrictions, such as cost, scheduling, or location. Immersive virtual reality can be an alternative plan for field trip teaching in case the actual field trip cannot be conducted, or before scheduling a field trip to better plan the survey of the area.

We wish to highlight that both VR and immersive VR tools are especially crucial nowadays to overcome limitations due to travel restrictions and lockdown periods aimed at containing the COVID-19 pandemic, which have especially affected Earth Science teaching and research activities. In fact, teaching usually entails field activities with the involvement of large student groups and research activities often need field surveying in foreign countries. For both, the use of VR can also reduce the carbon imprint of education and research.

6. Conclusions

Our work has been focused on the quantitative evaluation of the participation of different types of people in immersive virtual reality experiences dedicated to Earth Sciences, especially volcanic areas. Our quantitative survey has involved a total of 459 questionnaires, filled in by participants at nine events between 2018 and 2019. The geological sites selected for the experience are located in Greece (Santorini), North Iceland and Italy (Mt. Etna).

The analysis of the questionnaires, based on nine questions, has enabled us to shed light on how useful this approach is for geo-education purposes. The majority of the respondents have shown a high degree of interest in the immersive experience. They particularly appreciated the opportunity to fly like a drone over the displayed geological objects. Another key outcome of the survey is the respondents' major satisfaction with the portrayal of the environment, made possible by cutting-edge, unmanned aerial system (UAS)-based photogrammetry techniques.

Importantly, 97% of the participants wished to repeat the experience.

With regard to the geo-education potential of this technique, most participants ranked with high value (4–5) the experience (94%), reaching a percentage of 96% among academics in Earth Sciences. Therefore, this study confirms that both students and academics see VR as a useful tool for geo-education.

Furthermore, our data shows the important role that immersive VR might have as a tool for:

(i) popularizing Earth Sciences teaching and research by making geological key areas available to the public in terms of 3D models and scientific explanations of geological processes;
(ii) including people affected by motor disabilities who would not have access to dangerous/remote areas (e.g., tectonically or volcanically active) otherwise.

Based on the above considerations, immersive VR can be also regarded as a groundbreaking tool to improve democratic ways to access information and experience, as well as to promote inclusivity and accessibility in geo-education while reducing travel, saving time, and carbon footprints.

Author Contributions: F.L.B. has tailored and managed all the events as well as organized and written the manuscript. E.R., F.V., V.A., F.M., L.F., V.B., N.C., A.S., M.W., K.D., F.P.M., P.N., E.S., S.B., S.F., D.R., B.v.W.d.V., M.K., G.P., M.A.S.-R., G.V., U.B. and A.T. have participated in the events and contributed to writing the manuscript. All authors have read and agreed to the published version of the manuscript.

Funding: This research received no external funding devoted to this specific research.

Informed Consent Statement: Not applicable.

Data Availability Statement: The dataset with the replies from the participants is available online at https://doi.org/10.5281/zenodo.5792209 (accessed on 26 November 2021).

Acknowledgments: This research has been provided in the framework of the following projects: (i) the MIUR project ACPR15T4_00098–Argo3D (http://argo3d.unimib.it/ (accessed on 26 November 2021)); (ii) 3DTeLC Erasmus + Project 2017-1-UK01-KA203-036719 (http://www.3dtelc.com (accessed on 26 November 2021)); (iii) EGU 2018 Public Engagement Grant (https://www.egu.eu/outreach/peg/ (accessed on 26 November 2021)). Agisoft Metashape is acknowledged for photogrammetric data processing. This article is also an outcome of Project MIUR–Dipartimenti di Eccellenza 2018–2022. Finally, this paper is an outcome of the Virtual Reality lab for Earth Sciences—GeoVires lab (https://geovires.unimib.it/ (accessed on 26 November 2021)). The work supports UNESCO IGCP 692 'Geoheritage for Resilience'.

Conflicts of Interest: The authors declare no conflict of interest.

References

1. Trinks, I.; Clegg, P.; McCaffrey, K.; Jones, R.; Hobbs, R.; Holdsworth, B.; Holliman, N.; Imber, J.; Waggott, S.; Wilson, R. Mapping and analysing virtual outcrops. *Visual Geosci.* **2005**, *10*, 13–19. [CrossRef]
2. Edler, D.; Keil, J.; Wiedenlübbert, T.; Sossna, M.; Kühne, O.; Dickmann, F. Immersive VR Experience of Redeveloped Post-Industrial Sites: The Example of "Zeche Holland" in Bochum-Wattenscheid. *J. Cartogr. Geogr. Inf.* **2019**, *69*, 267–284. [CrossRef]
3. Pasquaré Mariotto, F.; Bonali, F.L. Virtual Geosites as Innovative Tools for Geoheritage Popularization: A Case Study from Eastern Iceland. *Geosciences* **2021**, *11*, 149. [CrossRef]
4. Lütjens, M.; Kersten, T.; Dorschel, B.; Tschirschwitz, F. Virtual Reality in Cartography: Immersive 3D Visualization of the Arctic Clyde Inlet (Canada) Using Digital Elevation Models and Bathymetric Data. *MTI* **2019**, *3*, 9. [CrossRef]
5. Tibaldi, A.; Bonali, F.L.; Vitello, F.; Delage, E.; Nomikou, P.; Antoniou, V.; Becciani, U.; Van Wyk de Vries, B.; Krokos, M.; Whitworth, M. Real world–based immersive Virtual Reality for research, teaching and communication in volcanology. *Bull. Volcanol.* **2020**, *82*, 38. [CrossRef]
6. Krokos, M.; Bonali, F.L.; Vitello, F.; Varvara, A.; Becciani, U.; Russo, E.; Marchese, F.; Fallati, L.; Nomikou, P.; Kearl, M.; et al. Workflows for virtual reality visualisation and navigation scenarios in earth sciences. In Proceedings of the 5th International Conference on Geographical Information Systems Theory, Applications and Management, Heraklion, Crete, Greece, 3–5 May 2019; SciTePress: Setùbal, Portugal, 2019; pp. 297–304.
7. Hurst, S.D. Use of "virtual" field trips in teaching introductory geology. *Comput. Geosci.* **1998**, *24*, 653–658. [CrossRef]
8. Warne, M.; Owies, D.; McNolty, G. Exploration of a first year university multimedia module on field geology. In Proceedings of the Beyond the Comfort Zone: Proceedings of the 21st ASCILITE Conference, Perth, Australia, 5–8 December 2004; ASCILITE: Tugun, Australia, 2004; pp. 924–933.
9. Deng, C.; Zhou, Z.; Li, W.; Hou, B. A panoramic geology field trip system using image-based rendering. In Proceedings of the 2016 IEEE 40th Annual Computer Software and Applications Conference (COMPSAC), Atlanta, GA, USA, 10–14 June 2016; IEEE: Piscataway, NJ, USA; Volume 2, pp. 264–268.
10. Hilde, T.W.C.; Carlson, R.L.; Devall, P.; Moore, J.; Alleman, P.; Sonnier, C.J.; Lee, M.C.; Herrick, C.N.; Kue, C.W. [TAMU]2-Texas A&M University and topography and acoustic mapping undersea system. In Proceedings of the Oceans 91, New York, NY, USA, 1–3 October 1991; Volume 1, pp. 750–755.
11. Anderson, D.M. Seafloor mapping, imaging and characterization from [TAMU]2 side-scan sonar data sets: Database management. In Proceedings of the International Symposium on Spectral Sensing Research, Maui, HI, USA, 15–20 November 1992; Volume 2, pp. 737–742.
12. Anderson, D.M. Glacial features observed in the Elysium Region of Mars. In Proceedings of the Tenth Thematic Conference on Geologic Remote Sensing, San Antonio, TX, USA, 9–12 May 1994; pp. I-263–I-273.
13. Martínez-Graña, A.M.; Goy, J.L.; Cimarra, C.A. A virtual tour of geological heritage: Valourising geodiversity using Google Earth and QR code. *Comput. Geosci.* **2013**, *61*, 83–93. [CrossRef]
14. McCaffrey, K.J.W.; Hodgetts, D.; Howell, J.; Hunt, D.; Imber, J.; Jones, R.R.; Tomasso, M.; Thurmond, J.; Viseur, S. Virtual fieldtrips for petroleum geoscientists. In *Geological Society, London, Petroleum Geology Conference Series*; Geological Society of London: London, UK; Volume 7, pp. 19–26.
15. Mariotto, F.P.; Antoniou, V.; Drymoni, K.; Bonali, F.; Nomikou, P.; Fallati, L.; Karatzaferis, O.; Vlasopoulos, O. Virtual Geosite Communication through a WebGIS Platform: A Case Study from Santorini Island (Greece). *Appl. Sci.* **2021**, *11*, 5466. [CrossRef]
16. Chang, S.C.; Hsu, T.C.; Jong, M.S.Y. Integration of the peer assessment approach with a virtual reality design system for learning earth science. *Comput. Educ.* **2020**, *146*, 103758. [CrossRef]
17. Antoniou, V.; Bonali, F.L.; Nomikou, P.; Tibaldi, A.; Melissinos, P.; Mariotto, F.P.; Vitello, F.R.; Krokos, M.; Whitworth, M. Integrating Virtual Reality and GIS Tools for Geological Mapping, Data Collection and Analysis: An Example from the Metaxa Mine, Santorini (Greece). *Appl. Sci.* **2020**, *10*, 8317. [CrossRef]
18. Rust, D.; Whitworth, M. A unique ~12 ka subaerial record of rift-transform triple-junction tectonics, NE Iceland. *Sci. Rep.* **2019**, *9*, 9669. [CrossRef]
19. Pasquaré Mariotto, F.; Bonali, F.L.; Venturini, C. Iceland, an open-air museum for geoheritage and Earth science communication purposes. *Resources* **2020**, *9*, 14. [CrossRef]
20. Tibaldi, A.; Corti, N.; De Beni, E.; Bonali, F.L.; Falsaperla, S.; Langer, H.; Neri, M.; Cantarero, M.; Reitano, D.; Fallati, L. Mapping and evaluating kinematics and the stress and strain field at active faults and fissures: A comparison between field and drone data at the NE rift, Mt Etna (Italy). *Solid Earth* **2021**, *12*, 801–816. [CrossRef]
21. Choi, D.H.; Dailey-Hebert, A.; Estes, J.S. *Emerging Tools and Applications of Virtual Reality in Education*; Choi, D.H., Dailey-Hebert, A., Estes, J.S., Eds.; Information Science Reference: Hershey, PA, USA, 2016.
22. Gerloni, I.G.; Carchiolo, V.; Vitello, F.R.; Sciacca, E.; Becciani, U.; Costa, A.; Riggi, S.; Bonali, F.L.; Russo, E.; Fallati, L.; et al. Immersive virtual reality for earth sciences. In Proceedings of the 2018 Federated Conference on Computer Science and In-formation Systems (FedCSIS), Poznan, Poland, 9–12 September 2018; IEEE: Piscataway, NJ, USA; pp. 527–534.
23. Kalawsky, R.S. VRUSE—A computerised diagnostic tool: For usability evaluation of virtual/synthetic environment systems. *Appl. Ergon.* **1999**, *30*, 11–25. [CrossRef]
24. Browning, J.; Drymoni, K.; Gudmundsson, A. Forecasting magma-chamber rupture at Santorini volcano, Greece. *Sci. Rep.* **2015**, *5*, 15785. [CrossRef] [PubMed]

25. Le Pichon, X.; Angelier, J. The Hellenic arc and trench system: A key to the neotectonic evolution of the eastern Mediter-ranean area. *Tectonophysics* **1979**, *60*, 1–42. [CrossRef]
26. Brun, J.P.; Faccenna, C.; Gueydan, F.; Sokoutis, D.; Philippon, M.; Kydonakis, K.; Gorini, C. Effects of slab rollback acceleration on Aegean extension. *Bull. Geol. Soc.* **2017**, *50*, 5–23. [CrossRef]
27. Saemundsson, K.; Hjartarson, A.; Kaldal, I.; Sigurgeirsson, M.A.; Kristinsson, S.G.; Víkingsson, S. *Geological Map of the Northern Volcanic Zone, Iceland. Northern Part 1: 100.000*; Iceland GeoSurvey and Landsvirkjun: Reykjavik, Iceland, 2012.
28. Thordarson, T.; Larsen, G. Volcanism in Iceland in historical time: Volcano types, eruption styles and eruptive history. *J. Geodyn.* **2007**, *43*, 118–152. [CrossRef]
29. Branca, S.; Coltelli, M.; Groppelli, G.; Lentini, F. Geological map of Etna volcano, 1:50,000 scale. *Italy J. Geosci.* **2011**, *130*, 265–291. [CrossRef]
30. Tibaldi, A.; Groppelli, G. Volcano-tectonic activity along structures of the unstable NE flank of Mt. Etna (Italy) and their possible origin. *J. Volcanol. Geoth. Res.* **2002**, *115*, 277–302. [CrossRef]
31. Kennedy, R.S.; Lane, N.E.; Berbaum, K.; Lilienthal, M.G. Simulator Sickness Questionnaire: An Enhanced Method for Quantifying Simulator Sickness. *Int. J. Aviat. Psychol.* **1993**, *3*, 203–220. [CrossRef]
32. Mat, R.C.; Shariff, A.R.M.; Zulkifli, A.N.; Rahim, M.S.M.; Mahayudin, M.H. Using game engine for 3D terrain visualization of GIS data: A review. In *IOP Conference Series: Earth and Environmental Science*; IOP Publishing: Bristol, UK, 2014; Volume 20, p. 012037.
33. Murray, J.W. *Building Virtual Reality with Unity and Steam VR*; AK Peters/CRC Press: Boca Raton, FL, USA, 2017. [CrossRef]
34. Kaiser, P.K.; Henning, J.G.; Cotesta, L.; Dasys, A. Innovations in mine planning and design utilizing collaborative immersive virtual reality (CIRV). In Proceedings of the 104th CIM Annual General Meeting, Vancouver, BC, Canada, 28 April–1 May 2002; pp. 1–7.
35. Kinsland, G.L.; Borst, C.W. Visualization and interpretation of geologic data in 3D virtual reality. *Interpretation* **2015**, *3*, SX13–SX20. [CrossRef]
36. McNamee, B.D.; Bogert, K. Gaining perspective with google earth virtual reality in an introduction-level physical geology course. In Proceedings of the GSA Annual Meeting, Indianapolis, IN, USA, 4–7 November 2018.
37. Visneskie, H.; Parks, J.; Johnston, J. Learning about Ontario's Paleozoic Geology with Virtual Reality Google Expedition Tours. Available online: https://uwspace.uwaterloo.ca/bitstream/handle/10012/15693/learning_about_ontarios_paleozoic_geology.pdf?sequence=1&isAllowed=y (accessed on 26 November 2021).
38. Caravaca, G.; Le Mouélic, S.; Mangold, N.; L'Haridon, J.; Le Deit, L.; Massé, M. 3D digital outcrop model reconstruction of the Kimberley outcrop (Gale crater, Mars) and its integration into Virtual Reality for simulated geological analysis. *Planet. Space Sci.* **2020**, *182*, 104808. [CrossRef]
39. Murgia, A.; Sharkey, P.M. Estimation of Distances in Virtual Environments Using Size Constancy. *Int. J. Virtual Real.* **2009**, *8*, 67–74. [CrossRef]
40. Keil, J.; Edler, D.; O'Meara, D.; Korte, A.; Dickmann, F. Effects of Virtual Reality Locomotion Techniques on Distance Estimations. *ISPRS Int. J. Geo-Inf.* **2021**, *10*, 150. [CrossRef]
41. Vizzari, G. Virtual Reality to Study Pedestrian Wayfinding: Motivations and an Experiment on Usability. In Proceedings of the 2020 IEEE International Conference on Artificial Intelligence and Virtual Reality (AIVR), Utrecht, The Netherlands, 14–18 December 2020; IEEE: Piscataway, NJ, USA; pp. 205–208. [CrossRef]
42. Kersten, T.; Drenkhan, D.; Deggim, S. Virtual Reality Application of the Fortress Al Zubarah in Qatar Including Performance Analysis of Real-Time Visualisation. *J. Cartogr. Geogr. Inf.* **2021**, *71*, 241–251. [CrossRef]
43. Dolf, M.; Teehan, P. Reducing the carbon footprint of spectator and team travel at the University of British Columbia's varsity sports events. *Sport Manag. Rev.* **2015**, *18*, 244–255. [CrossRef]

Article

Assessment of Geological Heritage Sites and Their Significance for Geotouristic Exploitation: The Case of Lefkas, Meganisi, Kefalonia and Ithaki Islands, Ionian Sea, Greece

Evangelos Spyrou [1], Maria V. Triantaphyllou [2,*], Theodora Tsourou [2], Emmanuel Vassilakis [1], Christos Asimakopoulos [1,2], Aliki Konsolaki [1], Dimitris Markakis [1,2], Dimitra Marketou-Galari [2] and Athanasios Skentos [2]

Citation: Spyrou, E.; Triantaphyllou, M.V.; Tsourou, T.; Vassilakis, E.; Asimakopoulos, C.; Konsolaki, A.; Markakis, D.; Marketou-Galari, D.; Skentos, A. Assessment of Geological Heritage Sites and Their Significance for Geotouristic Exploitation: The Case of Lefkas, Meganisi, Kefalonia and Ithaki Islands, Ionian Sea, Greece. *Geosciences* 2022, *12*, 55. https://doi.org/10.3390/geosciences12020055

Academic Editors: Hara Drinia, Panagiotis Voudouris, Assimina Antonarakou, Luisa Sabato and Jesus Martinez-Frias

Received: 13 December 2021
Accepted: 20 January 2022
Published: 24 January 2022

Publisher's Note: MDPI stays neutral with regard to jurisdictional claims in published maps and institutional affiliations.

Copyright: © 2022 by the authors. Licensee MDPI, Basel, Switzerland. This article is an open access article distributed under the terms and conditions of the Creative Commons Attribution (CC BY) license (https://creativecommons.org/licenses/by/4.0/).

[1] Department of Geography and Climatology, Faculty of Geology and Geo-Environment, National and Kapodistrian University of Athens, Panepistimioupolis, 15784 Athens, Greece; evspyrou@geol.uoa.gr (E.S.); evasilak@geol.uoa.gr (E.V.); chrisasimako@gmail.com (C.A.); alikikons@geol.uoa.gr (A.K.); jimmark211@gmail.com (D.M.)

[2] Department of Historical Geology and Palaeontology, Faculty of Geology and Geo-Environment, National and Kapodistrian University of Athens, Panepistimioupolis, 15784 Athens, Greece; ttsourou@geol.uoa.gr (T.T.); demimarkgal@gmail.com (D.M.-G.); thanasis.skentos@aecom.com (A.S.)

* Correspondence: mtriant@geol.uoa.gr; Tel.: +30-210-727-4893

Abstract: Geological heritage or geoheritage refers to the total of geosites, i.e., areas of high geological interest in a given area. Geosites have a high potential of attracting geotourists, thus contributing to the development of the local economy. Assessing sites of geological interest can contribute to their promotion, as well as their preservation and protection. Greece's geotectonic position in the convergent zone between the African and Eurasian plates has contributed to the existence of a considerable wealth of geosites, with the particularly active geotectonic region of the Ionian Sea characterized as a geoheritage hotspot. The purpose of this study is the selection of several such sites from the islands of Lefkas, Meganisi, Kefalonia and Ithaki and their assessment regarding their scientific, environmental, cultural, economic and aesthetic value. The most representative sites for the individual disciplines of geology (e.g., geomorphology, tectonics, stratigraphy and palaeontology) have been chosen, mapped and assessed, while indicative georoutes are proposed, which could aid the island's geotouristic promotion to geologist and non-geologist future visitors.

Keywords: geosites; geoheritage; geotourism; georoutes; Ionian islands

1. Introduction

Geological heritage or geoheritage refers to the total of geological sites in a given area, often abbreviated as "geosites". Geosites are areas of high geological interest; i.e., tectonical, palaeontological, mineralogical, stratigraphical, geomorphological, palaeogeographical, etc. (see [1]). More specifically, Gray [2] has itemized the individual components of geological heritage, namely mineralogy, petrology, paleontology, geomorphology, sedimentology, tectonics, pedology and, of course, the Earth's history. Sites of geomorphological heritage are particularly referred as geomorphosites [3,4]. Kozlowski [5] adds another individual component, the superficial water (i.e., rivers, lakes etc.), whereas González-Trueba [6] also considers the sea and oceans. The value of geoheritage has been widely recognized (e.g., [7]), with many researchers (e.g., [2,8–10]) trying to promote areas of high geological interest.

Besides their scientific value in general, geosites are also characterized by an ecological/environmental, cultural, economic and aesthetic value. These values are essential for a site of interest to be considered as a geosite [11–15]. Of course, the determining factor for recognizing an area as a geosite is its scientific value, however the rest of the values are also of paramount importance when assessing the geosites and/or comparing them to others [16]. As a general term, a geosite is a geological element, which is important for the comprehension of the Earth's history and the processes that took place at the corresponding period [13,17]. Thus, geosites and the associated fossils, rocks and minerals may contribute to our knowledge regarding palaeoclimate, palaeogeography, past life, tectonics and other processes [18,19]. In addition, following the uniformitarian principle proposed by Hutton, stating that the natural laws and processes occur in the present days in the same way as they did in the past, geosites can contribute to our understanding of current geological processes.

Geosites have a high potential of attracting geotourists, thus contributing to the development of the local economy [20,21]. Geotourism is a recently developed form of tourism that mainly regards visiting areas of particular geological and/or geomorphological interest (e.g., [22–27]), contributing to the sustainable development and the prosperity of the local economy of an area rich in geosites [28–31]. What is more, geosites do not only apply to connoisseurs of the principles of geology, but to mere tourists as well; for instance, families etc. [32].

Assessing sites of geological interest regarding their scientific, environmental, cultural, economic and aesthetic value can contribute to their promotion, as well as their preservation and protection (e.g., [33,34]). Protection usually refers to corrosion due to weathering, massive tourism impacts, as well as vandalism by the locals (e.g., [14]). Additionally, not much prominence is usually given to their geological importance, as the ecological value and the biodiversity are in most cases the main goal when it comes to protection (e.g., [2,35]). In recent years, however, there have been many attempts to assess the overall value of the geosites (e.g., [36–39]).

There are two ways of is assessing a geosite. The first one is qualitative and is conducted by means of some qualitative criteria. These were introduced by Watson and Slaymaker [40] and further developed by Pena dos Reis and Henriques [41], and they are still used in some cases. The second is the quantitative one [17,36–39,42–44]. This method is more representative, objective and impartial [45] and is therefore more widely used, despite the fact that different criteria have been proposed by different authors [7,37–39,45–49]. The assessment criteria always include the issues of rarity, representativeness and integrity [50], but other criteria are also set correspondingly to the type, location and geological interest of the geosites, including for instance, palaeontological, palaeoclimatological and paleogeographical aspects, the educative value, the ecological interest etc. (e.g., [47]).

An area that contains many sites of high geological as well as ecological, cultural, economic and/or aesthetic value is often referred to as a geopark. UNESCO Global Geoparks represent geographical areas where sites and landscapes of international geological significance are managed with a holistic concept of protection, education and sustainable development [51], and have been established both for the protection of the geosites and the preservation of the geoheritage (e.g., [52]). Furthermore, geopark entities usually collaborate with each other in order to exchange experience and geoconservation/promotion or management techniques (e.g., [53,54]). Geoparks meeting certain criteria are included in UNESCO's list of global geoparks; currently, six of these geoparks are located in Greece [54].

Greece's geotectonic position, i.e., in the convergent zone between the African and Eurasian plates [55] has contributed to the existence of a considerable wealth of geosites [42,53,56–59]. It is often referred to as a "natural geological laboratory", hosting a significant number of sites of geological interest, regarding mineralogy, palaeontology, stratigraphy, geomorphology etc. [54,60].

The islands studied for their geoheritage component in this paper are Lefkas, Meganisi, Kefalonia and Ithaki, located in the Ionian Sea, western Greece (Figure 1), covering an area of 303 km^2, 22 km^2, 773 km^2 and 117 km^2, respectively. Lefkas, the northernmost among the four, took its name from "Lefkata" cape to the south and was previously called Aghia Mavra [61]. Meganisi is a small islet, administratively belonging to Lefkas Regional Unit. Its name literally means "Large Island", while [62] suggests that it corresponds to the Homeric island of Krokylea. Kefalonia represents the largest island of the Ionian Sea and the sixth largest in Greece, inhabited since ancient times. Ithaki Island, administratively connected to Kefalonia, is also inhabited since the antiquity and thought to be the mythical island of Odysseus, even though [62] states that Lefkas is actually the Homeric Ithaca. Recently, an extended study aims to prove that Paliki peninsula on the island of Kefalonia is the land of Odysseus (e.g., [63]; Odysseus Unbound Project).

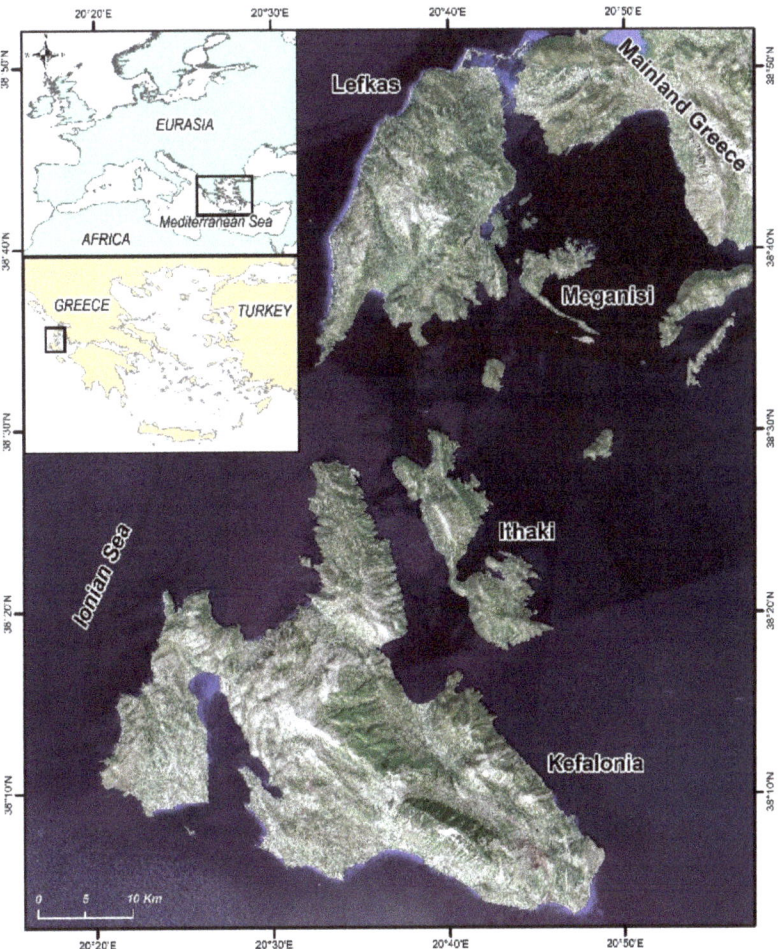

Figure 1. Map of the study area and location of the considered Ionian islands (Lefkas, Meganisi, Kefalonia, Ithaki).

So far, considerable efforts have been given for the establishment of the Kefalonia-Ithaki Geopark, highlighting numerous geotopes on the islands [64], with an end target to be included in the UNESCO geopark list. The initiative is supported by a large consortium of stakeholders, including the Region of Ionian islands.

In the present study we prefer to keep the term geosites referring to sites that combine natural geoscientific monuments with aesthetic, naturalistic, cultural, historical, touristic and educational values (e.g., [14,65]), instead of geotopes, which are defined as the smallest geographic unit with prominent geological features (e.g., [65]). Thus, we aim to expand our geoheritage knowledge for the Ionian islands by documenting and assessing selected geosites (mostly geomorphosites, in cases exhibiting paleontological, tectonic or mineralogical interest), located not only on the islands of Kefalonia and Ithaki but on Lefkas and Meganisi as well, while adding some more sites to the existing lists. Based on our assessment, indicative georoutes are designed and introduced, particularly highlighting inter-island and thematic geotourism.

2. Study Area, Geomorphological and Geological Setting

The geomorphology of the islands is generally mountainous, with steep slopes and many coastal cliffs. Lefkas is characterized by a mountainous repousse with the highest peak being Stavrota (1182 m). Other high peaks include Elati (1126 m), Ai Lias (1014 m) and Mega Oros (1012 m). Furthermore, there are many places indicating deep erosion (deep V-shaped valleys, such as in Roupakias, Episkopos and the gorge of Melissa), as seen from satellite images from Google Earth. Deep erosion is mostly owed to its intense tectonic activity [66]. The coastline primarily comprises steep cliffs, but there are many sandy beaches, however, in front of the steep cliffs. Meganisi is also mountainous, although it is characterized by mild terrain, gentle slopes and low altitudes (its highest summit barely exceeds 300 m). The largest part of its coast is cliffy, consisting of many successive beaches and coves, cartographically resembling rias, especially in its northeastern part. Kefalonia and Ithaki are also mountainous and characterized by deep erosion, with few plains located near the coastal zones. According to Koumantakis [67], there are plains surfaces formed by processes of erosion and deposition, which compose terraces of different height. The majority of the coasts are rocky, but, in the west and southwest, there are coasts with yellow, fine-grained sand that forms dunes [68]. About 2/3 of the total length of the coastline in Kefalonia are subject to coastal erosion, resulting in the creation of various landforms, such as marine terraces, tidal notches, beachrocks and aeolianites [69]. Regarding the karstic landforms, which are abundant in both islands, they are a result of the intense chemical erosion in carbonate rocks, with the karstification preventing the development of the hydrographic network, thus resulting in the formation of many hydrological basins that cover small areas [67].

The islands' geotectonic position at the transition of the convergence zone of the Greek arc to the collision zone of the Adria [55,70] is confirmed by the clockwise submarine downthrown fault-block of Kefalonia. Kefalonia Island is located on the tectonic front of the Hellenic thrust and represents the active plate boundary between the European and African plates, which is characterized by an oceanic and continental subduction [71]. Due to their geotectonic location, all studied islands are marked by intense seismicity [70]. The geological structure of the islands (Figures 2 and 3) includes post-alpine formations and the sedimentary alpine Paxos and Ionian geotectonic units that belong to the external Hellenides [55,72,73].

The tectonic structure is driven by deforming episodes of a compressive type, while the neotectonic forms are represented mainly by faults that cut all the formations of the islands [74,75]. The lithological formations are almost exclusively carbonate rocks [67,70,75–77]. The Paxos unit, represented mostly by microcrystalline limestones of Cretaceous age, covers Lefkata Peninsula and the area of Moutlou in Lefkas (Figure 2) [78] and almost all of Kefalonia (Figure 3) [73]. The Ionian unit, which is thrusted over the Paxos unit, covers all of Lefkas except for the Lefkata peninsula, Meganisi and Ithaki and the area of Sami at southeastern Kefalonia [55], including mainly carbonates of Triassic-Jurassic age (Pantokrator neritic limestones, Ammonitic Rosso beds, Vigla pelagic limestones, brecciated Senonian limestones [79] and limited flysch and molassic occurrences [80–82]. The stratigraphic column is completed with Pliocene–Quaternary post-alpine formations that lay unconformably above the alpine basement [67,75,82–86].

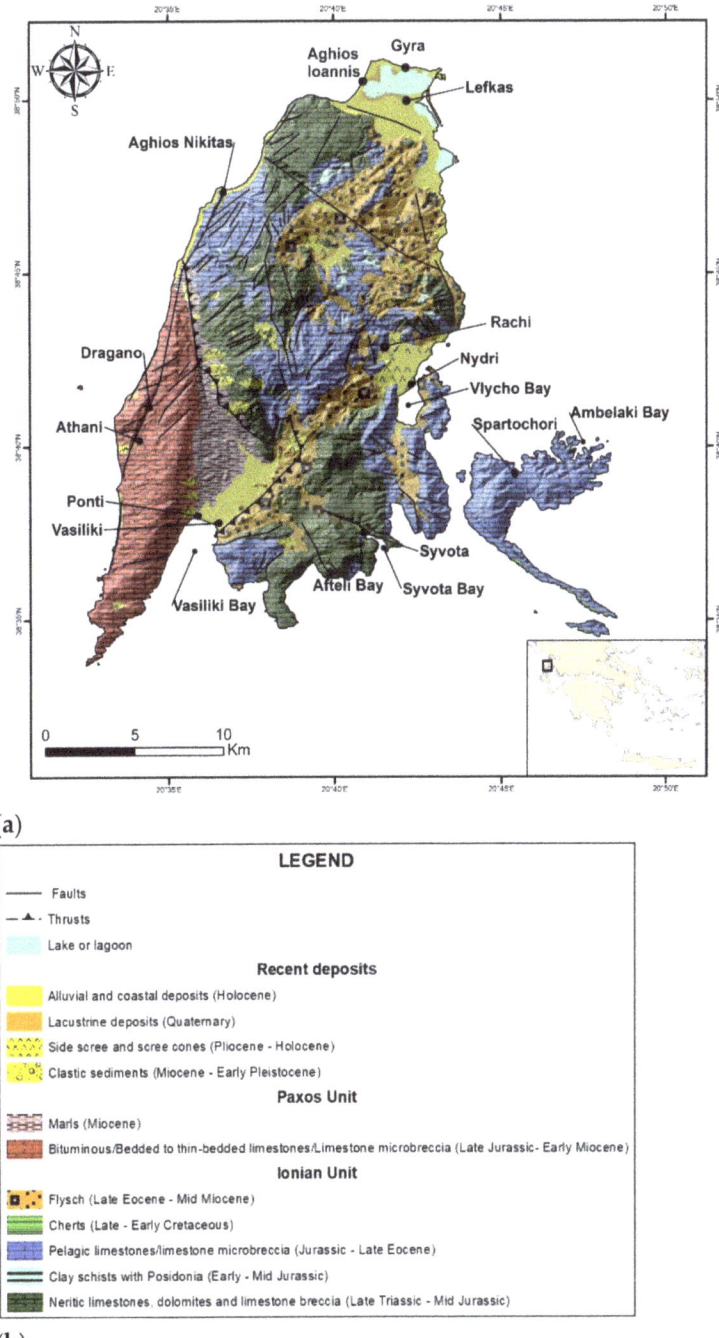

Figure 2. (a) Geological map of Lefkas and Meganisi islands, showing the distribution of their geotectonic units and the main faults and thrusts. (b) Legend of the geological formations.

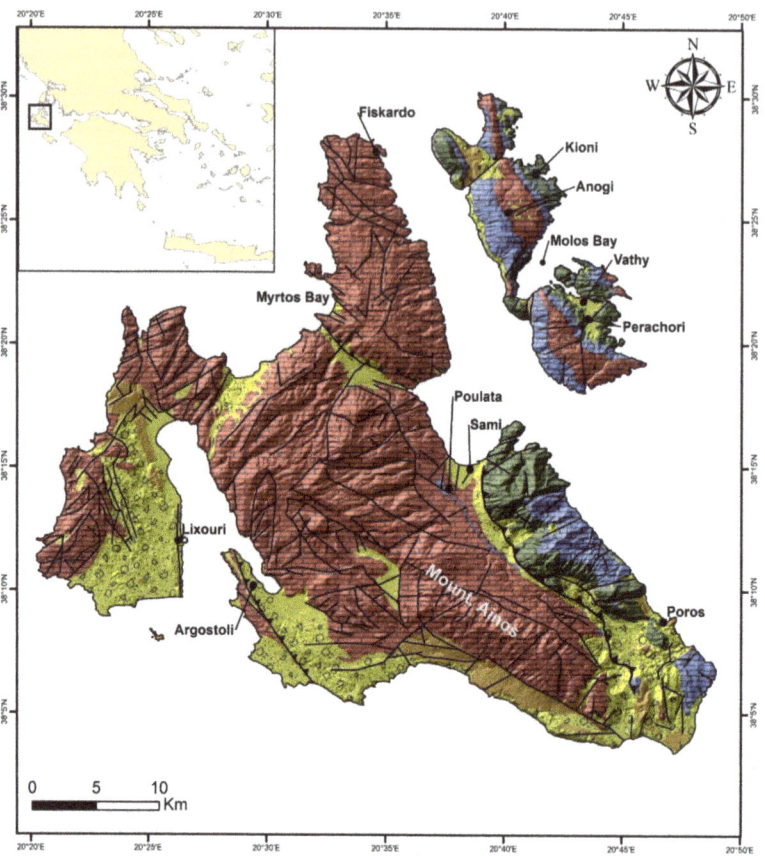

Figure 3. Geological map of Kefalonia and Ithaki island, showing the distribution of the geotectonic units and the main faults and thrusts. For the legend of the geological formations see Figure 2b.

3. Methodology

This paper is mainly based on bibliographical references for the study areas. Initially, papers and other authorities have been studied regarding geoheritage and its assessment, as well means of promotion and/or geoconservation. Several sites already proposed as potential geosites (e.g., [59,64]) have been evaluated in the present study, while further sites of interest have been spotted in the literature and graded according to the evaluation criteria applied (Table 1). In addition, some geosites have been documented and assessed upon in situ visits, which took place in Lefkas-Meganisi in 2020 (geosites L02–03, L06–07, L09–10, L12–15, M01; E. Spyrou), Kefalonia in 2013–2015, 2018–2019 (geosites K01, K03–04, K06–07, K09–11, K13–18, also including cave terrestrial laser scanning, marine terraces sampling etc.; D. Marketou-Galari, M. Triantaphyllou, E. Vassilakis, A. Konsolaki) and Ithaki in 2005 (I03–06; M. Triantaphyllou) (for geosites coding see Tables 2–4).

Geosites have been individually assessed when it comes to their scientific, environmental, cultural, economic and aesthetic value. The method used in this paper is the one proposed by Skentos [42], Skentos and Triantaphyllou [43]. More specifically, the criteria include geological history, representativeness, geodiversity, rarity, conservation, education, history—archaeology, religion, visibility, relief differentiation, accessibility, touristic infrastructure and ecological value (Table 1), applying a combination of the schemes proposed by Grandgigard [87], Rivas et al. [36], Theodosiou et al. [59], Zouros [48], Reynard et al. [47] and Fassoulas et al. [7]. Each geosite is evaluated for each criterion individually, with grading values ranging from 1 to 5. For instance, regarding geological history, values 1 to 5 are given to geosites of minor, moderate, major significance locally, moderate significance regionally and major significance regionally. When it comes to conservation, a very poorly preserved geosite is graded with 1, whereas a totally conserved one with 5, etc. For each geosite, the grades regarding the aforementioned values have been averaged, thus giving each geosite its final and overall grade. According to Skentos [42], the final grade assesses the geosites as follows: a geosite graded with >3.99 points is of global interest/significance, one graded between 3.99 to 3.50 points is of national interest/significance and one graded with less than 3.49 points is of local significance.

Selected geosites from each island have been used for the creation of indicative georoutes in a way that anyone interested can follow them and enjoy their natural beauty. All data are imported in the G.I.S. ArcMap 10.4 software, and maps have been created including the assessed geosites and the designed georoutes. Additionally, satellite images from Google Earth Pro are auxiliary used for distance estimations.

Table 1. Evaluation criteria for geosite assessment according to Skentos [42].

	1	2	3	4	5
Geological History	Small participation at local level	Moderate participation at local level	Great participation at local level	Moderate participation at regional level	Great participation at regional level
Representativeness	Not at all	Low	Medium	High	Unique
Geodiversity	1	<3	<5	<10	>10
Rarity	>20	>10	>5	>2	Unique
Conservation	Totally damaged	Low	Medium	High	Intact
Education	Not at all	-	Medium	-	High
History—Archaeology	Not at all	Existing-Low importance	Minor importance	Moderate importance	Great importance—Geohistoric site
Religion	Not at all	Existing-Low importance	Minor importance	Moderate importance	Great importance—Geohistoric site
Visibility	1	2	3	4	>4
Landscape Differentiation	Not at all	Low	Medium	High	Very high
Accessibility	Not accessible	Low	Medium	High	Very high
Tourist Infrastructure	Not at all	Low	Medium	-	High
Ecological Value	Not at all	-	Medium	-	High

Table 2. Lefkas island geosites assessment, following the evaluation criteria of Table 1.

Lefkas Island	Stavros Cave	Gyra Spit	Lefkas Lagoon	Asvotypa	Alabaster Cave	Alexander Saltpans	Melissa Gorge	Drymonas Fault	Dimosaris Waterfalls	Roupakias Gorge	Aghios Nikitas Fault	Katoulou River	Egremoni Beach	Moutlou/Porto Katsiki Beach	Ponti-Vasiliki Beach	Maradochori Lake	Choirospilia
Code name	L01	L02	L03	L04	L05	L06	L07	L08	L09	L10	L11	L12	L13	L14	L15	L16	L17
X coordinate	20.683	20.703	20.701	20.685	20.687	20.723	20.675	20.627	20.685	20.611	20.572	20.650	20.558	20.550	20.603	20.640	20.656
Y coordinate	38.8435	38.8492	38.8409	38.8267	38.8139	38.8048	38.7968	38.7710	38.7262	38.6767	38.6825	38.6711	38.6375	38.6012	38.6299	38.6462	38.6206
Geological history	3	4	3	3	3	2	4	4	2	2	4	2	3	4	3	3	3
Representativity	2	4	4	3	3	3	4	3	3	2	4	1	3	4	3	3	3
Geodiversity	2	4	2	2	3	1	3	1	2	3	1	3	4	4	3	2	3
Rarity	2	3	3	2	4	4	2	3	4	2	3	2	4	5	2	2	2
Conservation	3	4	4	4	3	4	4	5	5	5	5	4	5	4	4	5	3
Education	1	1	2	1	2	1	2	3	2	2	3	4	1	1	1	3	4
History-Archaeology	2	1	1	2	1	1	3	1	1	1	1	1	1	1	1	1	1
Religion	1	1	1	1	1	1	1	1	1	1	1	1	1	1	1	1	3
Visibility	4	5	5	4	2	5	4	3	4	4	3	4	5	5	5	3	3
Relief differentiation	3	2	1	2	2	1	4	5	4	4	5	2	3	5	3	1	1
Accessibility	3	4	4	3	3	4	3	3	3	3	2	4	3	4	4	3	3
Touristic facilities	1	3	1	1	3	1	3	1	4	2	1	2	5	5	5	1	2
Ecological value	3	3	5	3	4	3	4	2	4	3	2	3	5	5	4	5	3
Total score	2.31	3.00	2.77	2.46	2.62	2.54	3.15	2.69	3.08	2.62	2.62	2.54	3.69	3.92	3.00	2.54	2.54

Table 3. Kefalonia island geosites assessment, following the evaluation criteria of Table 1.

Kefalonia Island	Myrtos Beach	Neochori Tidal Notches	Melissani Cave	Karavomylos	Aghioi Theodoroi Cave	Aggalaki Cave	Drogarati Cave	Avythos Lake	Ionian Thrust	Poros Gorge	Mount Ainos	Valsamata Polje	Argostoli Ponors	Livadi Wetland	Paliki Marine Terraces	Platia Ammos Beach	Kounopetra	Xi Beach
Code name	K01	K02	K03	K04	K05	K06	K07	K08	K09	K10	K11	K12	K13	K14	K15	K16	K17	K18
X coordinate	20.536	20.625	20.623	20.614	20.617	20.623	20.628	20.711	20.732	20.771	20.673	20.587	20.474	20.429	20.358	20.356	20.388	20.415
Y coordinate	38.3430	38.3354	38.2570	38.2551	38.2356	38.2365	38.2278	38.1713	38.1544	38.1500	38.1372	38.1671	38.1643	38.2825	38.2146	38.2187	38.1538	38.1605
Geological history	4	3	5	2	3	3	5	1	4	4	4	4	5	2	4	4	1	3
Representativity	3	3	4	4	3	3	4	3	3	2	4	2	5	4	3	3	3	3
Geodiversity	3	2	5	3	3	3	5	1	3	5	4	5	4	2	3	3	1	3
Rarity	2	2	5	5	3	4	5	2	3	2	4	4	5	4	2	2	4	3
Conservation	5	5	5	5	4	3	5	5	4	5	4	5	5	4	4	4	1	5
Education	4	3	3	3	1	2	3	1	1	2	5	1	3	2	1	1	1	1
History-Archaeology	2	1	4	2	2	3	2	1	1	2	4	2	2	1	1	1	1	2
Religion	1	1	1	1	1	1	1	1	1	2	4	1	2	1	1	1	1	1
Visibility	5	1	4	4	2	2	2	2	3	4	4	4	2	5	4	3	4	5
Relief differentiation	5	2	3	1	5	2	2	2	3	3	5	2	1	2	3	4	1	5
Accessibility	4	1	4	5	3	3	4	4	3	3	4	4	4	4	3	2	4	5
Touristic facilities	5	2	5	4	3	3	5	1	3	2	4	2	5	3	2	5	2	5
Ecological value	4	2	4	4	4	4	4	3	3	4	5	4	4	5	3	4	1	4
Total score	3.62	2.23	4.00	3.31	2.85	2.77	3.62	1.85	2.77	3.08	4.08	3.00	3.54	3.23	2.62	2.92	1.92	3.46

Table 4. Ithaki and Meganisi islands geosites assessment, following the evaluation criteria of Table 1.

	Ithaki Island						Meganisi Island	
	Loizos Cave	Anogi Monoliths	Vathy Tidal Notches	Nymphs Cave	Sarakiniko Beach	Vathy	Aghios Ioannis Lagoon	Papanikolis Cave
Code Name	I01	I02	I03	I04	I05	I06	M01	M02
X coordinate	20.638	20.677	20.699	20.706	20.753	20.728	20.442	20.621
Y coordinate	38.4392	38.4213	38.3766	38.3601	38.3662	38.3971	38.3918	38.2542
Geological history	3	4	3	4	5	2	2	3
Representativity	2	5	3	4	3	2	3	3
Geodiversity	3	4	2	3	3	1	1	3
Rarity	2	5	3	4	2	3	3	4
Conservation	2	5	5	4	4	5	5	4
Education	1	1	1	1	5	2	1	1
History-Archaeology	4	2	1	5	2	3	1	4
Religion	4	1	1	4	1	2	1	1
Visibility	3	5	1	2	3	3	3	2
Relief differentiation	2	4	2	2	3	3	2	3
Accessibility	3	4	1	4	4	4	2	2
Touristic facilities	3	3	2	4	4	5	1	2
Ecological value	4	2	3	4	4	3	4	3
Total score	2.77	3.46	2.15	3.46	3.31	2.92	2.23	2.69

4. Results and Discussion

4.1. Distribution and Documentation of Geosites

In the present study a total of 43 geosites, namely 17 from Lefkas, 18 from Kefallonia, 5 from Ithaki and 2 from Meganisi have been selected and described for each island individually. All geosites have been provided with distinct code names, namely L01–L17 for Lefkas, M01–M02 for Meganisi, K01–K18 for Kefalonia and I01–I06 for Ithaki and have been accordingly GeoReferenced (Figure 4).

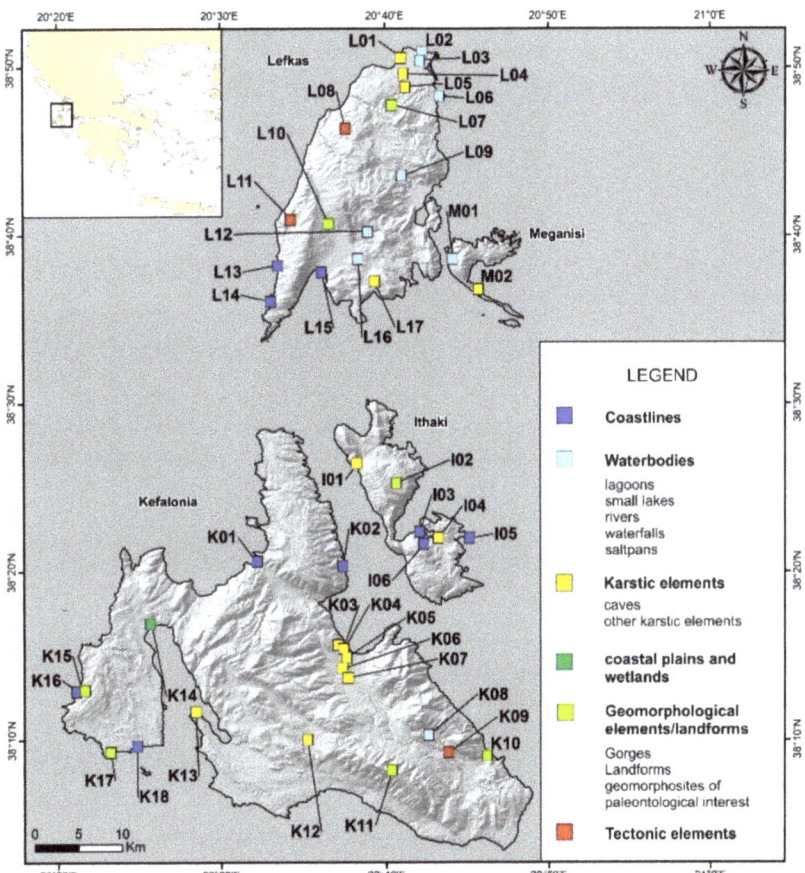

Figure 4. Location of selected geosites (mostly geomorphosites, in cases exhibiting paleontological, tectonic or mineralogical interest) on all four investigated islands. Note the different coding for each considered island.

4.1.1. Lefkas Island

Rivers and Waterfalls: Dimosaris river flows through the pelagic limestones of Ionian unit at the village of Rachi and forms estuaries at Nydri beach (Figures 2 and 4). One sizable waterfall and at least two smaller ones (L09) are developed along the river flow. The large one located upstream, forms a river lake ideal for swimming (Figure 5a,b). *Katourlou* river (L12) passes through the village of Syvros (Figures 2 and 4), and displays several riverbed landforms, such as step, riffle and pool sequences (Figure 5c), ending up to its estuaries in Vasiliki beach. Its springs near the village (Kerasia Springs) form a waterfall and are located between the Ionian carbonates and younger molassic sediments, as shown in Figures 2 and 4. It appears to be the ideal place for geoeducation, especially the understanding the hydrogeological parameters and processes and the formation of various landforms within riverbeds.

Figure 5. Selected geosites on Lefkas island, for coding see Figure 4: (**a,b**) *Dimosaris* waterfalls (L09); (**c**) *Katourlou* river (L12); (**d,e**) *Lefkas* lagoon (L03); (**f**) *Alexander* saltpans (L06); (**g**) estuaries at *Ponti-Vassiliki* beach (L15). Photos (**a–g**) courtesy E. Spyrou.

Gorges: Roupakias gorge (L10) represents a deep V-shaped intramountainous valley as a result of differential erosion, arguably due to tectonic uplift at the area of the Ionian thrust [74,88]. With a maximum altitude of 400 m difference from watershed to riverbed, it displays a notable waterfall. *Melissa* gorge (L07) is another deep V-shaped valley within the limestones of the Ionian unit, with altitude difference from watershed to riverbed that in places exceeds 80 m (as estimated from Google Earth satellite images). In the area, one can also find watermills and old stone bridges [89].

Lagoons, small lakes and saltpans: Lefkas lagoon (L03) is the largest lagoon of the Ionian Islands (Figure 5d,e), separated from the Ionian Sea by the narrow sandy *Gyra* spit (L02), which is a large sandy barrier landform in the northernmost part of Lefkas (Figure 6a,b), ranging from the area of Gyra to Aghios Ioannis (Figures 2 and 4). The sediment of the sand strip comes mainly from landslides due to earthquakes on the west coast of the island [90,91] (May et al., 2012a,b), while white and coarse beachrocks can be found at the mean sea-level, along the strip's coastline ([91] and own observations). Moreover, palaeo-tsunami deposits have been located, probably associated with the 365 A.D. earthquake in Crete [91–93]. The wave activity is intense in the area, but its maximum intensity is usually observed at the western part, namely in the Aghios Ioannis area. For this reason, the beachrocks in Aghios Ioannis are eroded to a greater extent, exhibiting a glassy texture (Figure 6c,d). *Maradochori* lake (L16) is a small circular lake formed within a doline at Maradochori plain (Figures 2 and 4). The site is of high ecological value, hosting many species of animals and hydrophile plants [94]. *Alexander* saltpans (L06) at Karyotes (Figures 2 and 4), comprises a location protected from the waves and southerly winds [95], where salt mining began in the 17th century by the Enetians [61]. The Alexander saltpans stopped functioning in 1988, being today abandoned, although they have been characterized as a protected industrial museum and a site of Natura 2000 network (Figure 5f), representing a significant element of cultural heritage [96].

Figure 6. Selected geosites on Lefkas island, for coding see Figure 4: (**a**) *Gyra* spit (L02); (**b**) beach rocks at *Gyra* (L02); (**c**) beach rocks at *Ag. Ioannis* (L02); (**d**) beach rocks at *Ag. Ioannis* (L02), detail of the glassy texture; (**e**) *Moutlou/Porto Katsiki* beach (L14); (**f**) *Egremnoi* beach (L13). Selected geosites on Meganisi island, for coding see Figure 4. (**g,h**) *Aghios Ioannis* lagoon (M01). Photos (**a–h**) courtesy E. Spyrou.

Coastlines: *Moutlou/Porto Katsiki* (L14) and *Egremnoi* (L13) are both sandy to gravelly beaches in front of an almost vertical cliff, consisting of microcrystalline bedded limestones of the Paxos unit (Figures 2 and 4). They are considered to be amongst the most famous beaches of Lefkas, as well as amongst its most beautiful ones (Figure 6e). The same as all of the southwestern part of Lefkas, they are often subject to landslides [97–99], most of which occur up to a few months after major earthquakes [66]. These geosites are suitable for geoeducation, particularly for the understanding tectonic movements, landslide processes in relation to geology and coastal erosion processes in relation to seismicity. *Ponti-Vasiliki* beach (L15) is a sandy beach in front of Ponti and Vasiliki settlements (Figures 2 and 4). It exhibits six estuaries (Figure 5g), whose streams are characterized by scant, yet existing flow at the end of the wet season, including the river formed by the confluence of the Roupakias and Katourlou waterways. This beach is ideal for understanding delta river formation and subsequent interaction with coastal sediment transport.

Caves: Choirospilia (L17) is a cave near the village of Evgiros (Figures 2 and 4). According to Dörpfeld [62], this is the pigsty of Eumeus, the faithful swineherd of the Homeric Odysseus. Interestingly, Dörpfeld [62] identifies the adjacent bay of Afteli as the port of disembarkation of Telemachus, the son of Odysseus, on his return from Pylos. *Choirotrypa* or *Alabaster* cave (L05) is the largest known cave of Lefkas that displays a typical stalactite decoration [57] and has been used in the past as a stable. During the beginning of the World War II (1940–1941), many residents found refuge in this cave [100]. *Asvotrypa* (L04) represents a cave near Fryni (Figures 2 and 4), and according to mythology and archaeological evidence, it was used as a place of worship for the Nymphs [100]. *Stavros* cave (L01) near Aghios Ioannis (Figures 2 and 4), has a typical stalactite decoration and an artificial entrance, created for ammunition storage during World War II [100].

Tectonic elements: Aghios Nikitas-Athani (L11) and *Drymonas* (L08) faults represent two of the main faults of the island, both located in its western part (Figures 2 and 4), with visible and accessible fault surfaces [101].

4.1.2. Meganisi Island

Lagoons: A coastal swamp in the area of *Aghios Ioannis* (M01) is separated from the sea by a sandy spit (Figure 6g,h), which is vegetated, thus presenting a stabilized piece of land. The overall geomorphological evolution of the island displays close connection with the structural and the seismotectonic evolution, also affected by the prevailing lithology and the sea activity [102].

Caves: Papanikolis cave (M02) is an impressively large coastal cave in Meganisi with very good decoration [57,103] most probably formed by coastal erosion (Figures 2 and 4). It was named after the homonymous submarine that found refuge there during World War II [103].

4.1.3. Kefalonia Island

Gorges: Poros gorge (K10) has an overall length of 4 km, spreading in a NE–SW direction near the boundaries of the homonymous settlement (Figures 3 and 4). The corrosive ability of the water has played a major role in the formation of the gorge within the limestones of the Ionian unit, as it controlled the geomorphological processes that took place in a neighboring tectonic graben, leading to its current form [104]. The area is strongly linked to mythology and archaeology, being described as the location of the ancient port of the Pronnaians [105] and associated with prehistoric evidence [106].

Small lakes, coastal plains and wetlands: Avythos (Akoli) lake (K08) at the southwestern part of the island (Figures 3 and 4), has been formed in the Pantokrator limestones of the Ionian Unit [104]. It was initially believed by the locals that the lake had no bottom, hence the names "Avythos" and "Akoli"; nevertheless, this karstic lake, located at 355 m altitude, is constantly filled by the aquifer formed in the underlying limestone beds [107]. *Livadi* coastal plain (K14) hosts the most important wetland of Kefalonia (Figure 7a,c), located in the northern part of the Argostoli Gulf at the edge of Thinia valley (Figures 3 and 4), with water supply mainly coming from springs located around the swamp. It is a typical example of a coastal swamp that hosts many rare, or even endangered species. Apart from the wide variety of reptile fauna, some of the critically endangered species that live in the wetland are *Aquila heliacal* and *Circus pygargus*. According to the hypothesis of Odysseus Unbound [63], Livadi marsh has the potential of having been a harbor in the late Myceanean age (around the 12th century BC), at the edge of a marine channel that could have turned Paliki peninsula to an island separated from the main part of Kefalonia at the time of Odysseus. Landslips from earthquakes and other major tectonic events, even tsunami backwash deposits [108,109], are expected to have filled the channel, forming the present Thinia valley, thus turning the "Paliki island" into the peninsula we see today [110].

Coastlines: *Myrtos* (K01) is one of the most spectacular beaches of Kefalonia (Figure 8a,b). It is located in the northern part of the island (Figures 3 and 4), with distinct geomorphological characteristics of very steep slopes of Paxos unit limestones due to intense uplift of tectonic origin. The coastal area is particularly vulnerable to the generation of slope failures induced by earthquakes [111], offering appropriate geoeducation potential for the understanding of landslide processes. Such example has been provided by the severe mud-flows after the Ianos Medicane (September 2020) that caused partial destruction of the downhill road to Myrtos beach area with the volume of the landslide material being estimated at 8664 m^3, calculated with quantification methodologies based on high resolution Digital Surface Models (Figure 8c,d), derived from photogrammetric image data before and after the event [112]. *Xi* (K18) beach (Akrotiri bay) is located in the southern part of Paliki peninsula (Figure 9d,e), with prominent cliffs displaying more than 100 m of blue clayey postalpine deposits of Early Pleistocene age [86,113]. Its name comes from the top view, in which it looks like the Greek letter Ξ (pronounced as "ksi"). The mineralogical composition of the clays is mostly of smectite–illite, which is suitable for cosmetological and pharmaceutical applications [114]. The blue clayey cliffs contrast impressively with the maroon color of the sand (Figure 9d), which is a result of the disintegration of the clay minerals. *Platia Ammos* beach (K16) in the western coast of the Paliki peninsula is featured by relatively steep hard-to-erode limestone cliffs. These geomorphological features are mainly driven by the tectonic regime of the island [115]. The shorelines of Kefalonia are considered ideal for the study of *Tidal Notches* (K02). Particularly, at Neochori in Fiskardo peninsula (Figures 3 and 4), the notches have an average height of 60 cm and an inward depth of about 20 cm. This type of notch is characterized by larger floor height than the roof height, which suggests a gradual sea-level rise, followed by a period of stability [116].

Figure 7. Selected geosites on Kefalonia island, for coding see Figure 4: (**a,c**) *Livadi* marsh and coastal plain (K14); the limestone outcropping in the area contains numerous rudist fragments (**b**); (**d**) *Melissani* cave (K03). Photos (**a–c**) courtesy E. Spyrou; (**d**) D. Marketou-Galari.

Other geomorphological elements: Mount Ainos (K11), located at the SE part of the island (Figures 3 and 4), is the highest mountain of Kefalonia. It displays a large biodiversity of endemic plants, including the typical flora of *Abies cephalonica* and numerous endangered bird species; thus, it has been declared since 1962 as national park and wildlife refuge. Moreover, it has been registered in the network of Natura 2000 and EU sites for the protection of the avifauna. The surficial karstic landforms on Mount Ainos create an impressive rocky landscape, while the planation surfaces formed by the combined action of erosion and dissolution reflect the gradual tectonic uplift of the island [73] (Underhill, 1989). Several planation surfaces are located at different elevations ranging from 100–1300 m [69]. At the NW part of Ainos, the geomorphosite exhibits paleontological interest, displaying fossiliferous thin- to thick-bedded limestones (Figure 9a–c; Upper Cretaceous, Paxos unit) with abundant rudist fragments [75,117,118]. This formation of Paxos unit is also present in Lefkas island [78]. The presence of rudist fragments indicate transportation from a nearby reef environment and not an in situ association. Rudist fragments are also present in the carbonate deposits at the vicinity of Livadi coastal plain (Figure 7b,c).

Kounopetra (K17), at the southern edge of Paliki peninsula (Cape Kounopetra; Figures 3 and 4), refers to a perpetually moving rock as a result of the clay composition at its base combined with the ripple of the waves. However, the ground displacement caused by recent earthquakes stopped the motion of the rock. *Paliki* marine terraces (K15; Figures 3 and 4) have been identified in the western part of Paliki peninsula [69,86,115], being very important geomorphological sea-level indices [69]. According to Gaki–Papanastasiou et al. [115], there is a Quaternary sequence of eight marine terraces, whose altitude ranges from 2 m to 440 m (Figure 10a,c). Biostratigraphic investigations based on calcareous nannofossil, indicate that the older terrace can be found in southern Paliki Peninsula, just above the Gelasian-Calabrian boundary in the Early Pleistocene [84,85]. The younger terrace of Middle Pleistocene age (nannofossil biozone MNN19f; [86]), is a geomorphosite with paleontological interest that can be located in Cape Kounopetra and Lixouri areas (Lepeda), bearing fossiliferous beds (Figure 10b), especially rich in invertebrate macrofaunal. In particular, 78 species of bivalves (Pectinidae, Cardiidae, *Glycymeris* spp., *Mytilus* spp. etc.), 87 species of gastropods, as well as some scaphopods, corals and echinoderms have been determined in the area [119].

Caves: A series of spectacular caves is formed in the area of Sami within the Paxos unit limestones. The sinkhole of *Melissani* (K03) is located about 2 km NW of Sami (Figures 3 and 4), and is characterized by a NNW–SSE main development axis. The cave displays very good decoration, significant archaeological findings and good accessibility [56,57,120]. The cave's entrance (15 × 25 m) was formed naturally, due to the collapse of a large part of the ceiling caused by the seismic activity in the region (Figure 7d). Concerning the geological interest of the cave, there are indications that it constitutes part of the wider hydrological karstic network [121] Maurin, while it is also characterized for its prominent biodiversity. A noteworthy example is the existence of an endangered species of heron in Greece, *Botaurus stellaris* [104]. According to archaeological excavations, the cave was used for worshiping purposes during ancient times (4th–3rd century B.C.) [104]. Aggalaki cave (K06) is a sinkhole (Figures 3 and 4), whose maximum depth reaches 50 m, located southeast of Poulata [104,122]. The cave interior is distributed in two chambers with intersecting axes; Vassilopoulos [104] suggests that both chambers end up in a brackish water lake, with a permeable network of faults causing the aquifer salination. The cave is relatively big (total surface category IV; 5001–7000 km^2) and bears good speleodecoration [57]. *Aghioi Theodoroi* cave (K05), located north of Aggalaki (Figures 3 and 4), is a spectacular sinkhole with dimensions of 23 × 20 m and maximum depth of 55 m, also hosting a lake with dimensions of 28 × 13 m [104] (Vassilopoulos, 2003). *Drogarati* cave (K07) interior is divided in two parts, which are developed within faulted Paleogene limestones (Figures 3 and 4). The second part is a two-level chamber with dimensions of 62 × 49 m, as it was measured after point cloud data processing, which was acquired by Terrestrial Laser Scanner survey (Figure 11a,b). The latter was combined with a point cloud derived after photogrammetric processing of aerial images of the open surface above the cave, captured by an Unmanned Aerial System and the result was quite impressive, as it is made clear how the chamber is connected to a collapsed doline (Figure 11a,b). Drogarati is considered one of the most impressive caves of Kefalonia not only due to its good speleodecoration and sufficient size [56,57,123] but also due to high level of acoustics acquired in the internal chamber, suitable for hosting numerous concerts and events [104].

Figure 8. Selected geosites on Kefalonia island, for coding see Figure 4: (**a,b**) the spectacular *Myrtos* beach (K01); (**c,d**) the main geomorphic changes at the popular *Myrtos* beach, can be clearly seen by comparing the orthoimages before (**c**) and after (**d**) the Sept. 2020 Medicane Ianos disaster (Vassilakis et al., 2021). Photos courtesy (**a**) D. Marketou-Galari and (**b**) M.V. Triantaphyllou.

Figure 9. Selected geosites on Kefalonia island, for coding see Figure 4: (**a,b**) Upper Cretaceous thin- to thick-bedded limestones with rudist fragments on *Mount Ainos* (K11); (**c**) rudist fragments on *Mount Ainos* (K11), detail of the transported and fragmented shells; (**d,e**) *Xi* beach (K18) with blue clayey cliffs and maroon-colored sand. Photos (**a–e**) courtesy M.V. Triantaphyllou, I. Mikellidou.

Figure 10. Selected geosites on Kefalonia island, for coding see Figure 4: (**a**) the Pleistocene *Paliki* marine terraces; (**c**) the youngest of *Paliki* marine terraces (K15) at Lepeda, with fossiliferous beds rich in Pectinidae (**b**). Photos (**a–c**) courtesy M.V. Triantaphyllou, D. Papanikolaou.

Other karstic elements: The *Argostoli* ponors (K13) are a typical example of the complex hydrogeological structure of the island mostly within the Paxos unit limestones (Figures 3 and 4). In 1963, Maurin and Zotl [121] poured a pigment in a ponor, which was detected after two weeks in the brackish springs of Sami and Melissani cave. They then certified the existence of an underground karstic network, formed by conduits that connect Argostoli and Sami, covering a distance of approximately 15 km. The current water supply can reach up to 0.3 m^3/sec, making it feasible to power water mills, whereas it is noteworthy that, during the past, the high velocity of inflow led to the construction of a hydroelectric plant [69]. Polje of *Valsamata* (K12) is the larger karstic landform of the region (Figures 3 and 4), with a total area of 6.4 km^2, developed in a NW–SE tectonic graben [69]. The area is of high hydrogeological interest, since the sinkholes drain the plain, supplying springs and local aquifers [124]. *Karavomylos* spring (K04) constitutes a submarine brackish spring located north of Sami (Figures 3 and 4). The construction of a wall isolates the spring from the sea, creating an artificial lake (Figure 11c,d). It is of utmost hydrogeological importance, because of the connection with the Argostoli ponors [104,107].

Tectonic elements: Ionian thrust (K09). The rocks prevailing on Kefalonia island are mainly limestones and dolomites that belong to the relative autochthon unit of Paxos (Pre-Apulian), occurring on Paliki peninsula in the west and on the major part of the central Kefalonia, and the allochthon tectonic nappe of the Ionian unit, occurring along the eastern part. The stratigraphy of Paxos unit comprises a thick shallow water carbonate platform with ages from Early Cretaceous to Early Miocene, followed by Middle Miocene clastic sequences of flysch type, while the Ionian nappe on Kefalonia is composed of the Late Triassic to Eocene carbonates (Figure 3). Overthrusting of the Ionian nappe took place during Middle-Late Miocene but compressional deformation continued also in the Late Miocene–Early Pliocene [55,73]. A small part of the Ionian unit is observed in the SE part of the island. The Ionian thrust juxtaposes the stratigraphic horizons of both units. It can be seen along the road that connects Sami and Poros (Figure 4), where carbonate rocks of the Mesozoic (Ionian unit) are thrusted over the Miocene corroded marls of Paxos unit (Figure 12a). The tectonic event of the thrust plays an important role in defining the current geomorphology of the island [125].

Figure 11. Selected geosites on Kefalonia island, for coding see Figure 4: (**a,b**) photorealistic representation of the *Drogarati* cave position (K07) under the surface, in proportional scale, produced after the combined point cloud originated from Terrestrial Laser Scanner and Unmanned Aerial System data processing; (**c,d**) *Karavomylos* spring and the artificial lake (K04). Photos courtesy (**a,b**) E. Vassilakis, (**c,d**) M. V. Triantaphyllou.

Figure 12. Selected geosites on Kefalonia island, for coding see Figure 4: (**a**) the *Ionian* thrust (K09) along the road that connects Sami and Poros. Selected geosites on Ithaki island, for coding see Figure 4; (**b**) the *Vigla Shale Member* outcrop, Ionian unit sequence, at the area of Vathy (I06). Photos courtesy (**a**) M.V. Triantaphyllou, I. Mikellidou, (**b**) M.V. Triantaphyllou.

4.1.4. Ithaki Island

Caves: Loizos or *Polis* cave is located in the northern part of Ithaca (I01), in the bay of Polis (Figures 3 and 4). Archaeological excavations were initiated by D. Loizos in 1868 and 1873, being systematically performed in between 1930–1932, with important findings consisting of pot fragments, weapons, masks and bronze tripods [126]. The cave was used for worshipping ancient gods, including Athena, Artemis, Hera, the Nymphs and even Odysseus [103]. Cave of *Nymphs* (I04) is situated to the west of Vathy settlement (Figures 3 and 4), and is characterized by both historical and archaeological interest, related to the worship of the Nymphs, while also linked to the return of Odysseus at the Homeric Ithaca [127]. The cave bears good speleodecoration and sufficient accessibility [56,57].

Landforms: Monoliths of Anogi (I02) are striking erosion products of various shapes, carved in the Upper Cretaceous limestones of Ionian unit (Figures 3 and 4). Apart from their geological importance, these landforms are related to the local culture, as they resemble statues, leading the locals to give them various names, depending on their shape and form.

Coastlines: Vathy port, located in the Gulf of Molos (I03; Figures 3 and 4), is identified by many historians as the ancient Forkyna, in which Odysseus arrived after the land of the Phaeacians. The ria-type slopes have been formed under the influence of the Quaternary eustatic movements, while the recognized *Tidal Notches* (I03), with an average height of 45 cm and inward depth of about 15 cm, indicate a gradual sea-level rise. A representative site at the NE entrance of Vathy port (cape Frygano), hosts an almost complete stratigraphic sequence of the Ionian unit, with the included Vigla Shale Member (Figure 12b) dated as late Aptian (nannofossil biozone BC21; [128]). A similar exposure is spotted at *Sarakiniko* beach, at the SE part of Ithaki island (I05).

4.2. Geosite Assessment

All presented geosites, mostly geomorposites, exhibiting paleontological, tectonic or mineralogical interest, have been assessed according to Skentos [42] evaluation criteria, graded for each criterium individually. As shown in Tables 2–4 and Figure 13, several geosites on all four islands are of local significance, with total scores less than 3.5. More specifically, in Lefkas, only 6 of the 17 assessed geosites are graded with more than three points; the highest values (>3.5) being attributed to Moutlou and Egremnoi sites.

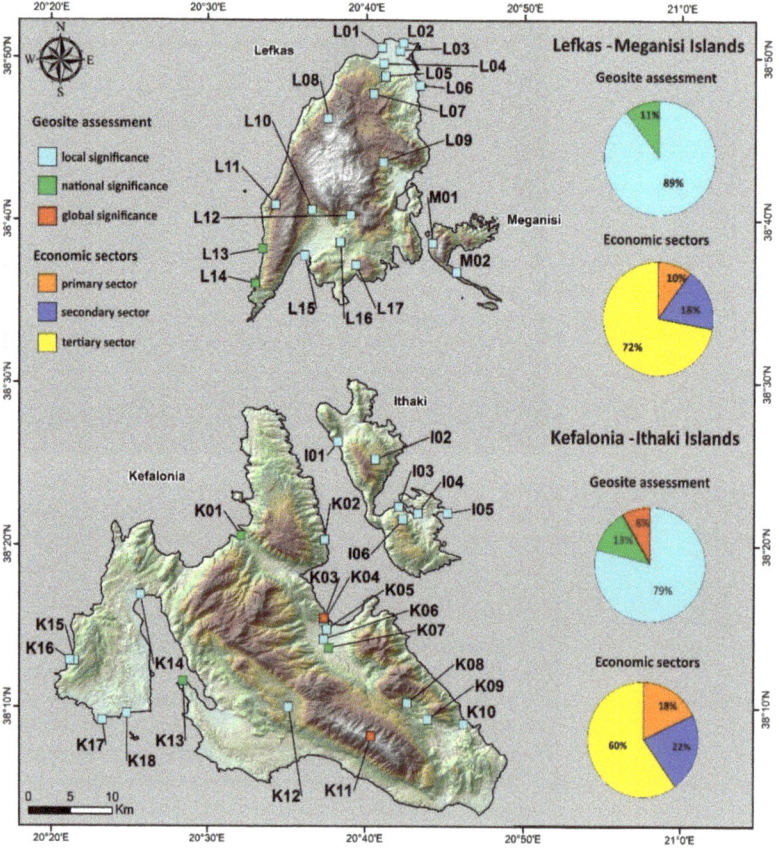

Figure 13. Categorization of the assessed geosites according to the evaluation criteria of Skentos [42], on all four investigated islands. (Data source for the distribution of the economic sectors; Hellenic Statistical Authority).

In Meganisi, both of the considered geosites are graded with less than 3 points, while in Kefalonia, 10 out of the 18 geosites (>50%) are assessed with at least 3 points. Five of them (28%) exceed 3.5 in total score, with Melissani cave and Mount Ainos scored as geosites of global interest.

Finally, in Ithaki, 3 out of 5 assessed geosites (60%), namely the cave of Nymphs, Anogi Monoliths and Sarakiniko beach surpass 3 points in total score.

Overall, the considered Ionian islands display a mid to high geotouristic activity [42]. The socioeconomic features of the Ionian islands' local economy are revealing a strong dominance of the tertiary sector (including transport and storage, information and communication, public sector, administration and services and well-established touristic infrastructure; hotels, restaurants etc.) (see Figure 13). In Lefkas–Meganisi islands, the touristic activities comprise 16.6% of the tertiary sector, while the same activities display 8.98% in Kefalonia and Ithaki; data based on the Hellenic Statistical Authority. This feature enhances significantly the geotouristic potential, particularly towards the development of inter-island georoutes and thematic geotouristic activities, e.g., "speleo-tourism".

4.3. Indicative Georoutes

Visiting Lefkas, Meganisi, Kefalonia and Ithaki geosites' unique wealth, consists of the basis of what we define here as *Georoutes*; certain feasible geotouristic routes that can be geographically outlined along the local road and path network verified in Google Earth Pro, guided and finally followed by the touristic masses that seek more autonomy through alternative touristic ways. In this line, promoting georoutes as a tool of cultural and economic growth in the regional level can significantly contribute to the education and to facilitate access and promotion of scientific culture among citizens, also spreading the awareness of protection and valorisation of their geological heritage. Thus, the interconnection of selected geosites should be considered as the basic factor for the implementation of geotouristic development and geoconservation policies at the involved Ionian islands.

The selected georoutes, presented below (Figures 14–16), are provisional examples of connecting geosites with specific kind of value. The first one involves a series of geomorphosites and points of cultural interest on Kefalonia island, aiming to raise the issue for the Paliki peninsula actually being a separate island (see Section 4.3.1). The second proposed georoute points out the potential of inter-island geotourism as the comparative advantage of the region of Ionian islands, while the third one is thematic, particularly specialized in connecting caves. Caves are considered as one of the most appreciated geotourism targets in the world; thus, a "speleo-tourism" approach would provide an advanced inter-island geotouristic potential in a complex islandic system, such as the quadruplet of Lefkas-Meganisi-Kefalonia-Ithaki. A "Speleoroute" should focus on the understanding of the karst system evolution and processes, emphasizing to caves in a manner combining adventure, science and education. It may also be designed to incorporate other forms of alternative tourism, including hiking, birdwatching, etc.

4.3.1. Paliki Georoute: Peninsula or Island?

Paliki peninsula offers an excellent georoute option for someone visiting Kefalonia (Figure 14). Paliki is located on the western side of the island, creating thusly a bay, inside which reside two towns on either side, Lixouri and Argostoli. There, one may find many landmarks of great beauty that also hold significance of either geological or cultural interest. With Lixouri as a starting point, the visitor will pass by the Paliki landscape marked by the presence of marine terraces that tell the geological history of the last 2 million years for the Paliki peninsula.

At Xi beach, the effects of coastal erosion in conjunction with the geological background can be observed, which give this geosite a "wild look" as the red sands "collide" with the gray clay, offering a unique swimming experience. Moving westwards, a second stop at Cape Kounopetra enables the discussion about seismic activity and the tectonic structure of Kefalonia island. After visiting Kipoureon monastery, established in 1759, another perfect spot for an astonishing sea view will be Platia Ammos beach. A visit to the Livadi wetland before returning back to Lixouri provides the chance to discuss the potential evidence for the Paliki peninsula actually being a separate island at the time of Odysseus, thus embodying the real Homeric Ithaca.

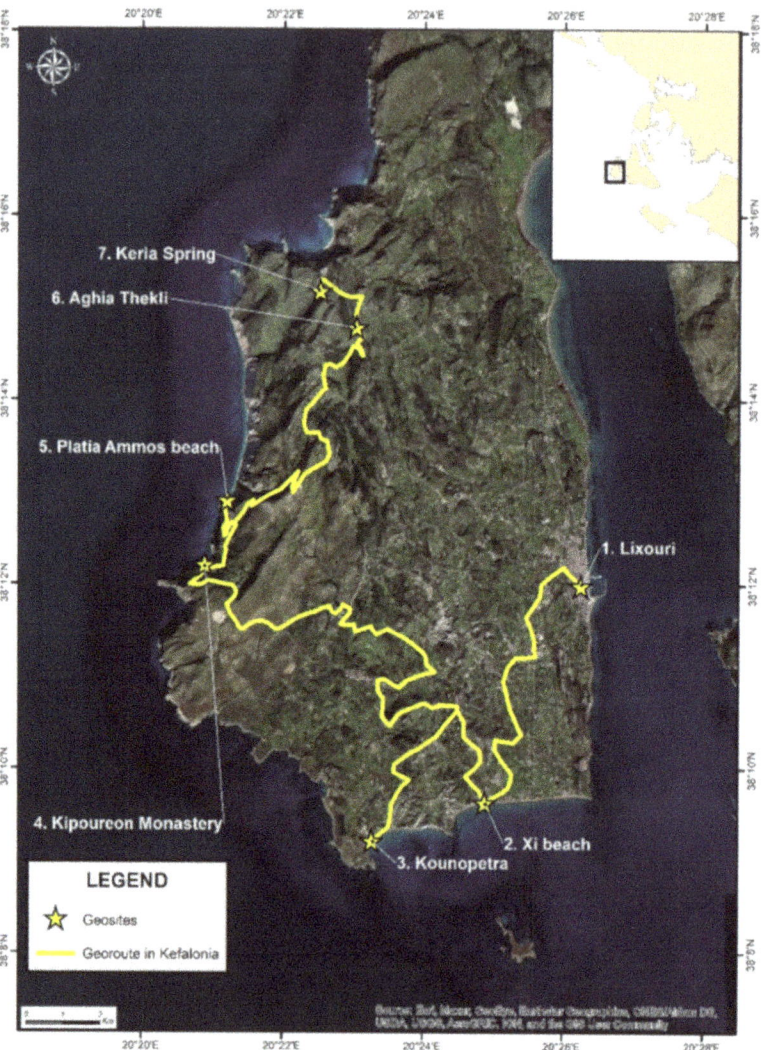

Figure 14. Designed georoute for the Paliki peninsula, Kefalonia island.

4.3.2. Inter-Island Georoutes: Lefkas and Meganisi

The geotouristic exploitation of the western part of Lefkas can be linked to the geosites of Meganisi within an inter-island georoute (Figure 15). Starting with Lefkas lagoon, the largest lagoon of the Ionian Islands, separated from the Ionian Sea by the narrow sandy Gyra spit, the visitor moving to the south can reach the Alexander saltpans in order to admire the natural landscape and the preindustrial scenery of the area. The next stop will be further to the southwest at Dimosaris waterfalls, a site that brings the visitor to a breathtaking mountainous setting.

The georoute then turns to a fully marine mode, leading the visitor to the Aghios Ioannis lagoon in Meganisi and last but not least to the impressively large coastal cave Papanikolis, with a size capable to host even the homonymous submarine during World War II.

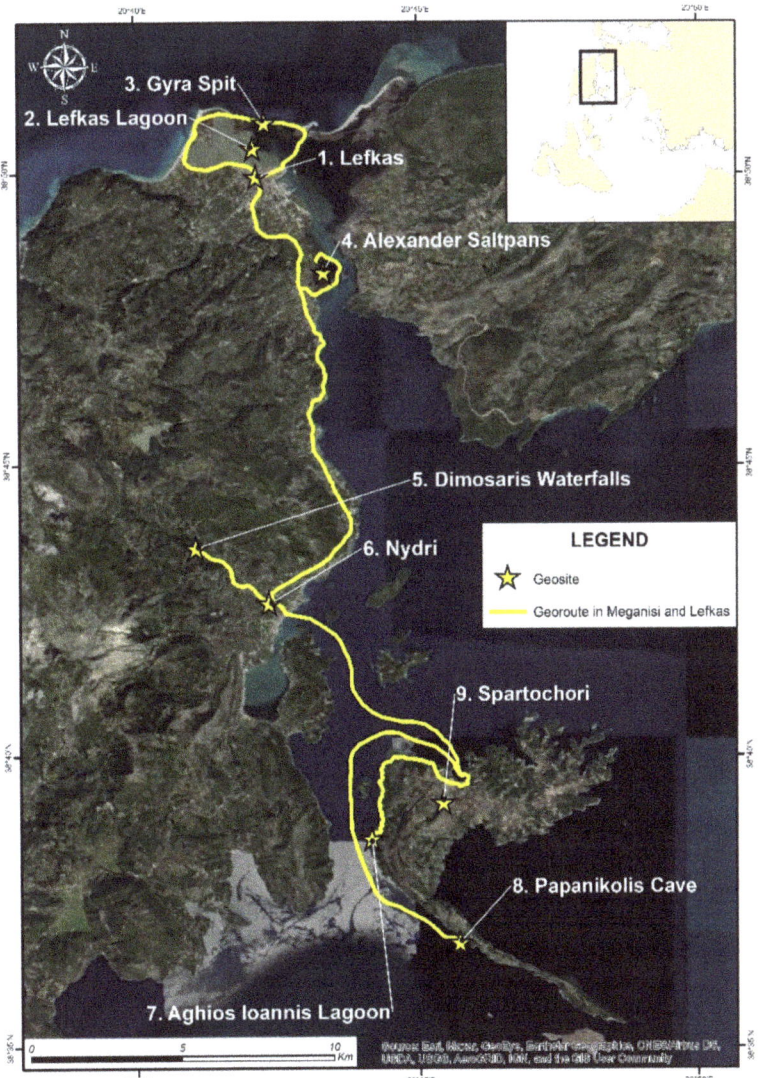

Figure 15. Lefkas and Meganisi inter-island georoute.

4.3.3. Inter-Island Georoutes: Kefalonia-Ithaki "Speleoroute"

The carbonate lithology and the intense karstic phenomena on all four investigated islands turns them to ideal destinations for cave tourism. A typical speleotouristic example is the karst system of Kefalonia, which results in a spectacular cave network, formed by processes that reveal the complex hydrogeological structure of the island.

All caves and karstic phaenomena selected for this georoute combine incomparable beauty, scientific interest and convenient access (Figure 16). The starting point is Melissani cave, a geosite of global interest at the area of Sami. Apart from its geological importance as an element belonging to an impressive underground network, which connects the Argostoli ponors to the Karavomylos spring, covering a distance of approximately 15 km, the Melissani cave also displays a large biodiversity and archaeological significance. The second stop takes place a few kilometers to the south, in Drogarati cave. The numerous speleothems that form the interior decoration of the cave offer many breathtaking sights to the visitor. The photorealistic representation of the Drogarati cave position under the surface, in proportional scale, produced after the combined point cloud originated from Terrestrial Laser Scanner and Unmanned Aerial System data processing (Figure 8c,d), provides an excellent basis for the development of a virtual inter-island thematic "Speleopark". In the same area, Aggalaki and Ag. Theodoroi caves represent stunning sinkholes, with internal brackish lakes.

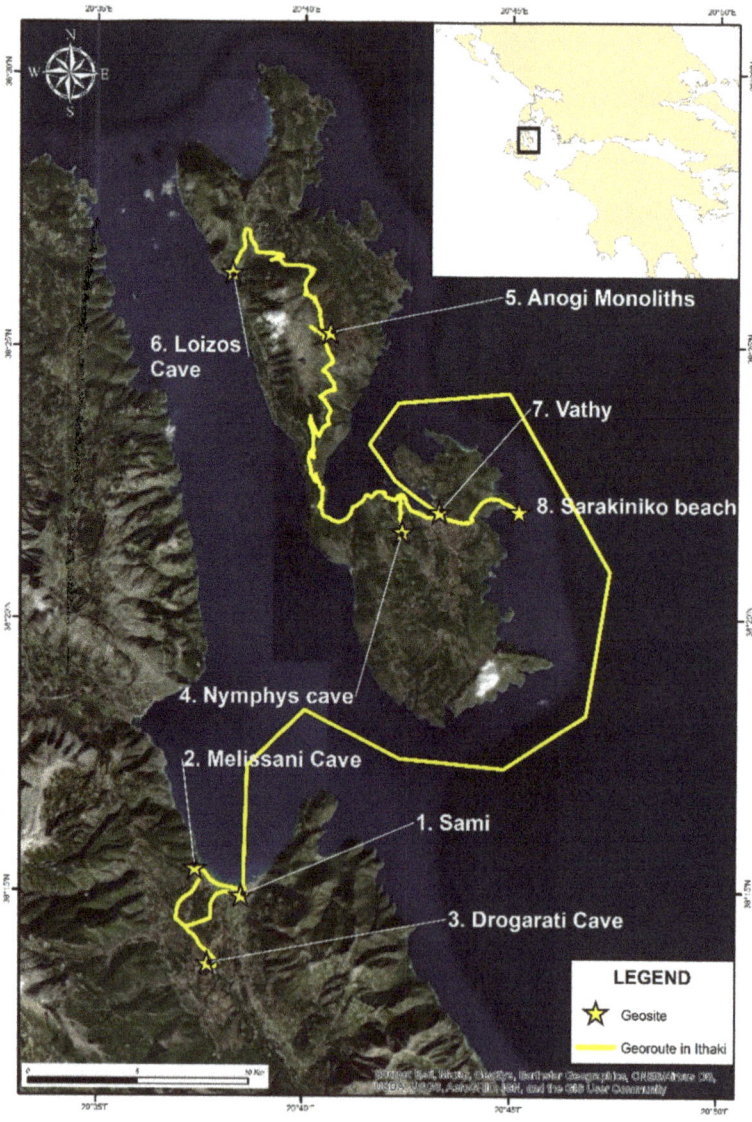

Figure 16. Kefalonia and Ithaki inter-island "Speleoroute".

The georoute continues in Ithaki, where visitors head to Vathy, a settlement located in a picturesque landscape, with a fantastic view of fjord-type slopes in the Gulf of Molos. The cave of Nymphs, right outside Vathy, will be the next stop, with significant historical and archaeological interest. A wide variety of karstic landforms can be observed in the area of the Monoliths of Anogi, making it ideal for the understanding of rock-weathering processes and impacts. Located a few kilometers to the north, Loizos cave, with substantial archaeological findings, represents the final stop of the "Speleoroute", before returning to Sami.

5. Conclusions

Through this research, we selected a total of 43 geosites (17 from Lefkas, 18 from Kefalonia, 6 from Ithaki and 2 from Meganisi). They were mapped and assessed following the criteria proposed by Skentos [42]. According to our findings, all four islands are characterized by a medium-to-high geotouristic potential, at least as far as the selected geosites are concerned. This means that they contain many sites of geological interest, in combination with other values, thus they were found worthy of promoting, as well as preserving. One of the primary goals of this study was to promote these islands as potential geotouristic destinations; thus, three different indicative georoutes were proposed, for either geologist or non-geologist future visitors.

Author Contributions: Conceptualization, E.S., M.V.T., T.T., E.V. and A.S.; methodology, M.V.T., T.T., E.V. and A.S.; investigation, E.S., M.V.T., C.A., D.M., D.M.-G. and A.S.; data curation, E.S., M.V.T., C.A., A.K., D.M., D.M.-G. and A.S.; writing—original draft preparation, E.S., M.V.T., T.T.; writing—review and editing, E.S., E.V. and A.S.; visualization, E.S., T.T., E.V. and A.K.; project administration, E.S. and M.V.T. All authors have read and agreed to the published version of the manuscript.

Funding: This research received no external funding.

Institutional Review Board Statement: Not applicable.

Informed Consent Statement: Not applicable.

Data Availability Statement: The data presented in this study are available on request from the corresponding author.

Conflicts of Interest: The authors declare no conflict of interest.

References

1. Zouros, N. Assessment, protection, and promotion of geomorphological and geological sites in the Aegean area, Greece. *Geomorphol. Relief Process. Environ.* **2005**, *11*, 227–234. [CrossRef]
2. Gray, M. *Geodiversity: Valuing and Conserving Abiotic Nature*; Wiley: Oxford, UK, 2004.
3. Ruban, D.A. Quantification of geodiversity and its loss. *Proc. Geol. Assoc.* **2010**, *121*, 326–333. [CrossRef]
4. Henriques, M.H.; dos Reis, R.P.; Brilha, J.; Mota, T. Geoconservation as an emerging geoscience. *Geoheritage* **2011**, *3*, 117–128. [CrossRef]
5. Kozłowski, S. Geodiversity. The concept and scope of geodiversity. *Prz. Geol.* **2004**, *52*, 833–837.
6. González-Trueba, J.J. *El Matizo Central de los Picos de Europa. Geomorfologia y Sus Implicaciones Geoecológjcas en la Alta Montana Cantábrica*; Universidad de Cantabria: Santander, Spain, 2007.
7. Fassoulas, C.; Mouriki, D.; Dimitriou-Nikolakis, P.; Iliopoulos, G. Quantitative Assessment of Geotopes as an Effective Tool for Geoheritage Management. *Geoheritage* **2012**, *4*, 177–193. [CrossRef]
8. Sharples, C. *Concepts and Principles of Geoconservation*; Tasmanian Parks & Wildlife Service: Tasmania, Australia, 2002; p. 79. Available online: https://nre.tas.gov.au/Documents/geoconservation.pdf (accessed on 1 September 2021).
9. Burek, C.V.; Prosser, C.D. The history of geoconservation: An introduction. *Geol. Soc. Spec. Publ.* **2008**, *300*, 1–5. [CrossRef]
10. Bruschi, V.; Cendrero, A. Direct and parametric methods for the assessment of geosites and geomorphosites. In *Geomorphosites*; Reynard, E., Coratza, P., Regolini-Bissig, G., Eds.; Verlag Friedrich Pfeil: München, Germany, 2009; pp. 73–88.
11. Panizza, M.; Piacente, S. Geomorphological assets evaluation. *Z. Geomorphol. Suppl.* **1993**, *87*, 13–18.
12. Panizza, M. Geomorphosites: Concepts, methods and examples of geomorphological survey. *Chin. Sci. Bull.* **2001**, *46*, 4–5. [CrossRef]
13. Reynard, E. Géomorphosites et paysages [Geomorphosites and landscapes]. *Geomorphol. Relief Process. Environ.* **2005**, *11*, 181–188. [CrossRef]

14. Reynard, E.; Panizza, M. Geomorphosites: Definition, assessment and mapping. *Geomorphol. Relief Process. Environ.* **2005**, *11*, 177–180. [CrossRef]
15. Fassoulas, C.; Zouros, N. Evaluating the influence of Greek Geoparks to the local communities. *Bull. Geol. Soc. Greece* **2017**, *43*, 896. [CrossRef]
16. Chingombe, W. Preliminary geomorphosites assessment along the panorama route of mpumalanga province, South Africa. *Geoj. Tour. Geosites* **2019**, *27*, 1261–1270. [CrossRef]
17. Grandgirard, V. Géomorphologie, Protection de la Nature et Gestion du Paysage. Bachelor's Thesis, Université de Fribourg, Fribourg, Switzerland, 1997.
18. Henriques, M.H.; Pena dos Reis, R. Framing the Palaeontological Heritage Within the Geological Heritage: An Integrative Vision. *Geoheritage* **2015**, *7*, 249–259. [CrossRef]
19. Brilha, J. Inventory and Quantitative Assessment of Geosites and Geodiversity Sites: A Review. *Geoheritage* **2016**, *8*, 119–134. [CrossRef]
20. Eder, F.W.; Patzak, M. Geoparks-geological attractions: A tool for public education, recreation and sustainable economic development. *Episodes* **2004**, *27*, 162–164. [CrossRef]
21. Zouros, N. The European Geoparks Network. *Episodes* **2004**, *27*, 165–171. [CrossRef]
22. Pralong, J.-P. A method for assessing tourist potential and use of geomorphological sites. *Geomorphol. Relief Process. Environ.* **2005**, *11*, 189–196. [CrossRef]
23. Emanuel, R. Scientific research and tourist promotion of geomorphological heritage. *Geogr. Fis. Din. Q.* **2008**, *31*, 225–230.
24. Dowling, R.K. Global Geotourism—An Emerging Form of Sustainable Tourism. *Czech J. Tour.* **2014**, *2*, 59–79. [CrossRef]
25. Kubalíková, L. Geomorphosite assessment for geotourism purposes. *Czech J. Tour.* **2014**, *2*, 80–104. [CrossRef]
26. Pica, A.; Fredi, P.; Del Monte, M. The ernici mountains geoheritage (Central Apennines, Italy): Assessment of the geosites for geotourism development. *Geoj. Tour. Geosites* **2014**, *14*, 193–206.
27. Ólafsdóttir, R.; Tverijonaite, E. Geotourism: A systematic literature review. *Geosciences* **2018**, *8*, 234. [CrossRef]
28. Osipova, E.; Shadie, P.; Zwahlen, C.; Osti, M.; Shi, Y.; Kormos, C.; Bertzky, B.; Murai, M.; van Merm, R.; Badman, T. *IUCN World Heritage Outlook 2: A Conservation Assessment of All Natural World Heritage Sites*; International Union for Conservation of Nature: Gland, Switzerland, 2017; ISBN 9782831718743.
29. Cappadonia, C.; Coratza, P.; Agnesi, V.; Soldati, M. Malta and sicily joined by geoheritage enhancement and geotourism within the framework of land management and development. *Geosciences* **2018**, *8*, 253. [CrossRef]
30. Dowling, R.; Newsome, D. *Handbook of Geotourism*; Edward Elgar Publishing Ltd.: Cheltenham, UK, 2018.
31. Filocamo, F.; Rosskopf, C.M.; Amato, V. A Contribution to the Understanding of the Apennine Landscapes: The Potential Role of Molise Geosites. *Geoheritage* **2019**, *11*, 1667–1688. [CrossRef]
32. Gordon, J.E. Geoheritage, geotourism and the cultural landscape: Enhancing the visitor experience and promoting geoconservation. *Geosciences* **2018**, *8*, 136. [CrossRef]
33. Comanescu, L.; Nedelea, A.; Dobre, R. Evaluation of geomorphosites in vistea valley (Fagaras Mountains-Carpathians, Romania). *Int. J. Phys. Sci.* **2011**, *6*, 1161–1168. [CrossRef]
34. Comănescu, L.; Dobre, R. Inventorying, evaluating and tourism valuating the geomorphosites from the central sector of the Ceahlău National Park. *Geoj. Tour. Geosites* **2009**, *2*, 86–96.
35. Reynard, E.; Coratza, P. Geomorphosites and geodiversity: A new domain of research. *Geogr. Helv.* **2007**, *62*, 138–139. [CrossRef]
36. Rivas, V.; Rix, K.; Francés, E.; Cendrero, A.; Brunsden, D. Geomorphological indicators for environmental impact assessment: Consumable and non-consumable geomorphological resources. *Geomorphology* **1997**, *18*, 169–182. [CrossRef]
37. Coratza, P.; Giusti, C. Methodological proposal for the assessment of the scientific quality of geomorphosites. *Alp. Mediterr. Q.* **2005**, *18*, 307–313.
38. Serrano, E.; González-Trueba, J.J. Assessment of geomorphosites in natural protected areas: The Picos de Europa National Park (Spain). *Geomorphol. Relief Process. Environ.* **2005**, *11*, 197–208. [CrossRef]
39. Pereira, P.; Pereira, D.; Caetano Alves, M.I. Geomorphosite assessment in Montesinho Natural Park (Portugal). *Geogr. Helv.* **2007**, *62*, 159–168. [CrossRef]
40. Watson, E.; Slaymaker, O. *Mid-Wales, a Survey of Geomorphological Sites*; Department of Geography, University College of Wales: Aberystwyth, UK, 1966.
41. dos Reis, R.P.; Henriques, M.H. Approaching an integrated qualification and evaluation system for geological heritage. *Geoheritage* **2009**, *1*, 1–10. [CrossRef]
42. Skentos, A. *Geosites of Greece: Record, Schematic, Geological Regime and Geotouristic Assessment*; National and Kapodistrian University of Athens: Athens, Greece, 2012.
43. Skentos, A.; Triantaphyllou, M. Registration and tourism assessment of Greek geosites using G.I.S. techniques. In Proceedings of the 32nd International Geographical Congress, IGU 2012, Cologne, Germany, 26–30 August 2012.
44. Brilha, J. Geoheritage: Inventories and evaluation. In *Geoheritage Assessment, Protection, and Management*; Elsevier: Amsterdam, The Netherlands, 2018; pp. 69–85. [CrossRef]
45. Bruschi, V.M.; Cendrero, A. Geosite evaluation; can we measure intangible values? *Alp. Mediterr. Q.* **2005**, *18*, 293–306.

46. Bonachea, J.; Bruschi, V.M.; Remondo, J.; González-Díez, A.; Salas, L.; Bertens, J.; Cendrero, A.; Otero, C.; Giusti, C.; Fabbri, A.; et al. An approach for quantifying geomorphological impacts for EIA of transportation infrastructures: A case study in northern Spain. *Geomorphology* **2005**, *66*, 95–117. [CrossRef]
47. Reynard, E.; Fontana, G.; Kozlik, L.; Pozza, C.S. A method for assessing «scientific» and «additional values» of geomorphosites. *Geogr. Helv.* **2007**, *62*, 148–158. [CrossRef]
48. Zouros, N.C. Geomorphosite assessment and management in protected areas of Greece Case study of the Lesvos island—Coastal geomorphosites. *Geogr. Helv.* **2007**, *62*, 169–180. [CrossRef]
49. Pereira, P.; Pereira, D. Methodological guidelines for geomorphosite assessment. *Geomorphol. Relief Process. Environ.* **2010**, *16*, 215–222. [CrossRef]
50. Grandgirard, V. L'évaluation des géotopes. *Geol. Insubrica* **1999**, *4*, 59–66.
51. UNESCO. UNESCO Global Geoparks (UGGp). Available online: https://en.unesco.org/global-geoparks (accessed on 13 January 2022).
52. Stoffelen, A.; Groote, P.; Meijles, E.; Weitkamp, G. Geoparks and territorial identity: A study of the spatial affinity of inhabitants with UNESCO Geopark De Hondsrug, The Netherlands. *Appl. Geogr.* **2019**, *106*, 1–10. [CrossRef]
53. Drinia, H.; Tsipra, T.; Panagiaris, G.; Patsoules, M.; Papantoniou, C.; Magganas, A. Geological heritage of syros island, cyclades complex, Greece: An assessment and geotourism perspectives. *Geosciences* **2021**, *11*, 138. [CrossRef]
54. Zafeiropoulos, G.; Drinia, H.; Antonarakou, A.; Zouros, N. From geoheritage to geoeducation, geoethics and geotourism: A critical evaluation of the Greek region. *Geosciences* **2021**, *11*, 381. [CrossRef]
55. Papanikolaou, D.I. *The Geology of Greece*; Springer Nature: Basel, Switzerland, 2021; ISBN 9783319761015.
56. Petrocheilou, A. *The Caves of Greece*; Athinon, E., Ed.; Ekdotike Athinon: Athens, Greece, 1984.
57. Triantaphyllou, M. Greek caves and touristic development. *Bull. Speleol. Soc. Greece* **1992**, *20*, 28–76.
58. Velitzelos, E.; Mountrakis, D.; Zouros, N.; Soulakelis, N. *Atlas of the Geological Monuments of the Aegean Sea*; Publications of the Ministry of the Aegean: Athens, Greece, 2002.
59. Theodosiou, E.; Fermeli, G.; Koutsouveli, A. *Our Geological Heritage*; Publ. Kaleidoskopio: Athens, Greece, 2006.
60. Zouros, N.; Valiakos, I. Geoparks management and assessment. *Bull. Geol. Soc. Greece* **2017**, *43*, 965. [CrossRef]
61. Rondoyannis, P.G. *History of the Island of Lefkas*; Lefkas, Etaeria Lefkadikon Meleton: Athens, Greece, 1980; Volume 1.
62. Dörpfeld, W. *Alt-Ithaka: Ein Beitrag zur Homer-Frage, Studien und Ausgrabungen auf der Insel Leukas-Ithaki*; Verlag Richard Uhde: Munich, Germany, 1927.
63. Underhill, J.R. Relocating Odysseus' homeland. *Nat. Geosci.* **2009**, *2*, 455–458. [CrossRef]
64. Geopark Kefalonia—Ithaca Geopark Kefalonia—Ithaca. Available online: https://kefaloniageopark.gr/en (accessed on 13 January 2022).
65. Ielenicz, M. Geotope, Geosite, Geomorphosite. *Ann. Valahia Univ. Târgoviște Geogr. Ser.* **2009**, *9*, 7–22.
66. Nikolakopoulos, K.; Kyriou, A.; Koukouvelas, I.; Zygouri, V.; Apostolopoulos, D. Combination of aerial, satellite, and UAV photogrammetry for mapping the diachronic coastline evolution: The case of Lefkada Island. *ISPRS Int. J. Geo-Inf.* **2019**, *8*, 489. [CrossRef]
67. Koumantakis, I.E. *Study of Kefalonia Underground Water Capacity, Geographical and Morphological Data, Geo-Logical Conditions and Hydrogeology*, Unpublished Technical Report. 1990.
68. Lekkas, E. Kefalonia—Ithaca islands, Programme of Devisal of the Neotectonical Map of Greece. National and Kapodistrian University of Athens: Athens, Greece, 1996.
69. Karymbalis, E.; Papanastassiou, D.; Gaki-Papanastassiou, K.; Tsanakas, K.; Maroukian, H. Geomorphological study of Cephalonia Island, Ionian Sea, Western Greece. *J. Maps* **2013**, *9*, 121–134. [CrossRef]
70. Cushing, E.M. Evolution Stucturale de la Marge Nord-Ouest Hellenique dans l'Ile de Levkas et Ses Environs (Grece Nord-Occidentale). Ph.D. Thesis, Universite de Paris-Sud, Orsay, France, 1985.
71. Papanikolaou, D. Tectonostratigraphic models of the Alpine terranes and subduction history of the Hellenides. *Tectonophysics* **2013**, *595–596*, 1–24. [CrossRef]
72. Aubouin, J.; Dercourt, J. Zone preapulienne, zone ionienne et zone du Gavrovo en Peloponnese occidental. *Bull. Société Géologique Fr.* **1962**, *S7-IV*, 785–794. [CrossRef]
73. Underhill, J.R. Late Cenozoic deformation of the Hellenide foreland, western Greece. *Geol. Soc. Am. Bull.* **1989**, *101*, 613–634. [CrossRef]
74. Lekkas, E.; Danamos, G.; Lozios, S. Neotectonic structure and neotectonic evolution of Lefkada Island. *Bull. Geol. Soc. Greece* **2001**, *34*, 157–163. [CrossRef]
75. Lekkas, E.; Danamos, G.; Maurikas, G. Geological Structure and Evolution of Kefallonia and Ithaki islands. *Bull. Geol. Soc. Greece* **2001**, *34*, 11. [CrossRef]
76. Lekkas, E. The Athens earthquake (7 September 1999): Intensity distribution and controlling factors. *Eng. Geol.* **2001**, *59*, 297–311. [CrossRef]
77. Rondoyianni, T.; Mettos, A.; Paschos, P.; Georgiou, C. *Neotectonic Map of Greece, Scale 1:100.000, Lefkada Sheet*; I.G.M.E.: Athens, Greece, 2007.
78. Bornovas, J. Géologie de l'île de Lefkade. *Geol. Geophys. Res. Spec.* **1964**, *10*, 143.

79. Karakitsios, V. The Influence of Preexisting Structure and Halokinesis on Organic Matter Preservation and Thrust System Evolution in the Ionian Basin, Northwest Greece. *Am. Assoc. Pet. Geol. Bull.* **1995**, *79*, 960–980.
80. Drinia, H.; Antonarakou, A.; Kontakiotis, G.; Tsaparas, N.; Segou, M.; Karakitsios, V. Paleo-bathymetric evolution of the Early Late Miocene deposits of the Pre-Apulian zone. *Bull. Geol. Soc. Greece* **2007**, *40*, 39–52. [CrossRef]
81. Kousis, I. *Geological and Stratigraphical Study of Meganisi, Ionian Islands*; University of Patras: Patras, Greece, 2013.
82. Triantaphyllou, M.V. Calcareous nannofossil biostratigraphy of Langhian deposits in Lefkas (Ionian Islands). *Bull. Geol. Soc. Greece* **2010**, *43*, 754–762. [CrossRef]
83. Agiadi, K.; Triantaphyllou, M.; Girone, A.; Karakitsios, V. The early Quaternary palaeobiogeography of the eastern Ionian deep-sea Teleost fauna: A novel palaeocirculation approach. *Palaeogeogr. Palaeoclimatol. Palaeoecol.* **2011**, *306*, 228–242. [CrossRef]
84. Triantaphyllou, M.V. Biostratigraphical and ecostratigraphical observations based on calcareous nannofossils, of the Eastern Mediterranean Plio-Pleistocene deposits. *GAIA* **1996**, *1*, 229.
85. Triantaphyllou, M.V.; Drinia, H.; Dermitzakis, M.D. Biostratigraphical and paleoenvironmental determination of a marine Plio/Pleistocene outcrop in Cefallinia Island (Greece). *Géologie Méditerranéenne* **1999**, *26*, 3–18. [CrossRef]
86. Papanikolaou, D.; Triantaphyllou, M. Growth folding and uplift of Lower and Middle Pleistocene marine terraces in Kephalonia: Implications to active tectonics. In Proceedings of the 4th International INQUA Meeting on Paleoseismology, Active Tectonics and Archeoseismology (PATA), Aachen, Germany, 9–15 October 2013; pp. 185–188.
87. Grandgirard, V. Methode pour la realisation d'un inventaire de geotopes geomorphologiques. *UKPIK Cah. l'Institute Geogr. l'Universite Fribg.* **1995**, *10*, 121–137.
88. Bathrellos, G.D.; Antoniou, V.E.; Skilodimou, H.D. Morphotectonic characteristics of Lefkas Island during the Quaternary (Ionian Sea, Greece). *Geol. Balc.* **2009**, *38*, 23–33.
89. Beligianni, E. *Searching for the Stone Bridges of Greece*; Livani Publications: Athens, Greece, 2008; ISBN 9601417737.
90. May, S.M.; Vött, A.; Brückner, H.; Grapmayer, R.; Handl, M.; Wennrich, V. The Lefkada barrier and beachrock system (NW Greece)—Controls on coastal evolution and the significance of extreme wave events. *Geomorphology* **2012**, *139–140*, 330–347. [CrossRef]
91. May, S.M.; Vött, A.; Brückner, H.; Smedile, A. The Gyra washover fan in the Lefkada Lagoon, NW Greece—Possible evidence of the 365 AD Crete earthquake and tsunami. *Earth Planets Sp.* **2012**, *64*, 859–874. [CrossRef]
92. Vött, A.; Brückner, H.; Brockmüller, S.; Handl, M.; May, S.M.; Gaki-Papanastassiou, K.; Herd, R.; Lang, F.; Maroukian, H.; Nelle, O.; et al. Traces of Holocene tsunamis across the Sound of Lefkada, NW Greece. *Glob. Planet. Chang.* **2009**, *66*, 112–128. [CrossRef]
93. May, S.M. Sedimentological, Geomorphological and Geochronological Studies on Holocene Tsunamis in the Lefkada—Preveza Area (NW Greece) and Their Implications for Coastal Evolution. Ph.D. Thesis, University of Cologne, Cologne, Germany, 2010.
94. Gazi, A. Landscape, Natural Environment and Settlement of Prehistoric and Ancient Lefkas (from Palaeolithic Period till the Roman Conquest) Alternative: Environmental and Cultural Prehistory of Lefkas Island. Master's Thesis, University of Edinburgh, Edinburgh, UK, 2013.
95. Santa, V. Alexandros Saltpans of Lefkas, a Historical Memory Place, in the Roads of Salt. Available online: https://aromalefkadas.gr (accessed on 10 December 2021).
96. Koukounari, E.-A.; Mikroni, S. Villaggio di Sale. Diploma Thesis, University of Ioannina, Ioannina, Greece, 2021.
97. Ganas, A.; Elias, P.; Bozionelos, G.; Papathanassiou, G.; Avallone, A.; Papastergios, A.; Valkaniotis, S.; Parcharidis, I.; Briole, P. Coseismic deformation, field observations and seismic fault of the 17 November 2015 M = 6.5, Lefkada Island, Greece earthquake. *Tectonophysics* **2016**, *687*, 210–222. [CrossRef]
98. Valkaniotis, S.; Papathanassiou, G.; Ganas, A. Mapping an earthquake-induced landslide based on UAV imagery; case study of the 2015 Okeanos landslide, Lefkada, Greece. *Eng. Geol.* **2018**, *245*, 141–152. [CrossRef]
99. Roufi, A.; Vassilakis, E.; Poulos, S. Western Lefkada Shoreline Displacement Rates Based on Photogrammetric Processing of Remote Sensing Datasets from Various Sources. In Proceedings of the 15th International Congress of the Geological Society of Greece, Athens, Greece, 22–24 May 2019; pp. 512–513.
100. Petrocheilou, A. Speleological researches in the area of Apolpaina and Aghios Ioannis, Lefkas. *Bull. Speleol. Soc. Greece* **1966**, *8*, 182–195.
101. Rondoyanni, T.; Tsiambaos, G. The active fault of Aghios Nikitas-Athani in Lefkas and the geological hazard of the island. In Proceedings of the 3rd Panhellenic Conference of Antiseismic Engineering and Engineering Seismology, Athens, Greece, 5–7 November 2008. Article 1961.
102. Gournelos, T.; Evelpidou, N.; Vassilopoulos, A.; Poulos, S. Structural control of geomorphological evolution of Meganissi Island (Ionian Sea) coastal zone and natural hazard risk detection based on fuzzy sets. In *Coastal and Marine Geospatial Technologies*; Springer: Berlin, Germany, 2010; pp. 305–314. [CrossRef]
103. Petrocheilou, A. «Loizos» Cave of Ithaca: No. sp. rec. 812A. *Bull. Speleol. Soc. Greece* **1971**, *11*, 36.
104. Vassilopoulos, A. Analysis of geomorphological and geographical data using the technology of geographical in-formation systems in Kephallenia island. *GAIA* **2003**, *12*, 227.
105. Marinatos, S. *Cephallinia, Istorikos Kai Archaeologikos Peripatos*; T.E.T. Cephallinias: Kefalonia, Greece, 1962.
106. Chatziotou, E.-M.; Stratouli, G. Drakaina Cave at Poros, Kephalonia. In Evidence for the prehistoric use of the cave and its use as a cult place during historical times. In Proceedings of the VI Panionian Congress, Thessaloniki, Greece; 2001; pp. 61–76.
107. Galiatsatou, A. Karstic Structures on Kefallinia Island. University of Aegean: Mytilene, Greece, 2017.

108. Vött, A.; Hadler, H.; Willershäuser, T.; Ntageretzis, K.; Brückner, H.; Warnecke, H.; Grootes, P.M.; Lang, F.; Nelle, O.; Sakellariou, D. Ancient harbours used as tsunami sediment traps—The case study of Krane (Cefalonia Island, Greece). In *Harbors and Harbor Cities in the Eastern Mediterranean from Antiquity to the Byzantine Period: Recent Discoveries and Current Approaches; BYZAS*; Ege Yayinlari: Istanbul, Turkey, 2014; Volume 19, pp. 743–771.
109. Tsanakas, K.; Karymbalis, E.; Cundy, A.; Gaki-Papanastasiou, K.; Papanastasiou, D.; Drinia, H.; Koskeridou, E.; Maroukian, H. Late Holocene geomorphic evolution of the Livadi coastal plain, gulf of Argostoli, Cephalonia island, western Greece. *Geogr. Fis. Din. Q.* **2019**, *42*, 43–60.
110. Underhill, J.; Styles, P.; Pavlopoulos, K.; Apostolopoulos, G. Ithaca the story continues. *Geosci. Online*. 2020, 30. Available online: https://www.geolsoc.org.uk/Geoscientist/Archive/May-2018/Ithaca-the-story-continues (accessed on 13 December 2021).
111. Mavroulis, S.; Lekkas, E. Revisiting the most destructive earthquake sequence in the recent history of greece: Environmental effects induced by the 9, 11 and 12 august 1953 Ionian sea earthquakes. *Appl. Sci.* **2021**, *11*, 8429. [CrossRef]
112. Vassilakis, E.; Konsolaki, A.; Petrakis, S.; Kotsi, E.; Fillis, C.; Lozios, S.; Lekkas, E. Quantification of Mass Movements with Structure-from-Motion Techniques. The Case of Myrtos Beach in Cephalonia, After Ianos Medicane (September 2020). *Bull. Geol. Soc. Greece Spec. Publ.* **2021**, *8*, 16–20.
113. Triantaphyllou, M.V. Quantative calcareous nannofossil biostratigraphy of Bay Akrotiri section (Cefallinia island, W. Greece). Tracing the gephyrocapsid size-trend in an early Pleistocene terrigenous sequence. *Bull. Geol. Soc. Greece* **2001**, *34*, 645–652. [CrossRef]
114. Kourlis, K. *The Use of Blue Marls in the Industry of Cosmetology*; National Technical University of Athens: Athens, Greece, 2018.
115. Gaki-Papanastassiou, K.; Karymbalis, E.; Maroukian, H.; Tsanakas, K. Geomorphic evolution of Western (Paliki) Kephalonia Island (Greece) during the Quaternary. *Bull. Geol. Soc. Greece* **2010**, *43*, 418. [CrossRef]
116. Evelpidou, N.; Karkani, A.; Kázmér, M.; Pirazzoli, P. Late Holocene shorelines deduced from tidal notches on both sides of the Ionian thrust (Greece): Fiscardo peninsula (Cephalonia) and Ithaca island. *Geol. Acta* **2016**, *14*, 13–24. [CrossRef]
117. BP Co. *The Geological Results of Petroleum Exploration in Western Greece*; Institute for Geology and Subsurface Research: Athens, Greece, 1971; Volume 10.
118. Mikellidou, I.; Patruno, S.; Triantaphyllou, M.V.; Pomoni-Papaioannou, F.; Karakitsios, V. The upper Cretaceous palaeo-slope transition: An integrated calcareous nannofossil and micro-facies approach (Ionian Islands, Preapulian Zone, Western Greece). In Proceedings of the EGU2018, Vienna, Austria, 8–13 April 2018.
119. Georgiades-Dikaioulia, E. The Neogene of Kephallinia. *Ann. Géologiques Pays Hélléniques* **1967**, *18*, 43–106.
120. Petrocheilos, I. Melissani cave. *Bull. Speleol. Soc. Greece* **1951**, *19*, 114.
121. Maurin, V.; Zotl, J. Salt water encroachment in the low altitude karst water horizons of the island of Kefalonia (Ionian Islands). *Bull. Int. Assoc. Sci. Hydrol.* **1967**, *74*, 423–438.
122. Zelilidis, A. No TitleGeological parameters affecting the stability and evolution of two caves in Kephallenia and Zante islands. *Bull. Geol. Soc. Greece* **1994**, *21*, 105–116.
123. Petrocheilou, A. Possibilites de recherche a Drogarati. *Bull. Speleol. Soc. Greece* **1968**, *9*, 179–191.
124. Papaspyridakou-Verykiou, E.; Bathrellos, G.; Skylodimou, C. Karst phenomena of Kefalonia: The physiographical evolution of Valsamata and Troianata polje. In Proceedings of the 2007 Panhellenic Geographical Conference, Athens, Greece, 4–7 October 2007; pp. 44–49.
125. Kapatsoris, A. *Tectonic Analysis of the Two Major Thrusts in the Areas of Myrtos and Aghia Kyriaki in the Northern Part of Kefalonia Island*; University of Patras: Patras, Greece, 2012.
126. Benton, S. Excavations in Ithaka, III The cave at Polis I. *Annu. Br. Sch. Athens* **1934**, *35*, 45–53. [CrossRef]
127. Giannoulidou, K. The historical caves of antiquity. *Plato* **1975**, *27*, 90–102.
128. Triantaphyllou, M.V.; Karakitsios, V.; Mantzouka, D. Calcareous Nannofossil Biostratigraphy of the Basal Part of Vigla Shale Member (Ionian Zone) in Ithaki Island; Preliminary Results. *Bull. Geol. Soc. Greece* **2006**, *39*, 126. [CrossRef]

Article

Quantitative Assessment of the Geosites of Chelmos-Vouraikos UNESCO Global Geopark (Greece)

Vasilis Golfinopoulos [1], Penelope Papadopoulou [1], Eleni Koumoutsou [2,3], Nickolas Zouros [4], Charalampos Fassoulas [5], Avraam Zelilidis [1] and George Iliopoulos [1,*]

[1] Department of Geology, University of Patras, 265 04 Patras, Greece; gkolfinopoulosv@upnet.gr (V.G.); penelpapadop@upatras.gr (P.P.); a.zelilidis@upatras.gr (A.Z.)
[2] Chelmos Vouraikos UNESCO Global Geopark, 35 Ag. Alexiou Str., 250 01 Kalavryta, Greece; koumoutsou@upatras.gr
[3] Department of Biology, University of Patras, 265 04 Patras, Greece
[4] Department of Geography, University of the Aegean, 811 00 Mytilene, Greece; nzour@aegean.gr
[5] Natural History Museum of Crete, University of Crete, 712 02 Heraklion, Greece; fassoulas@nhmc.uoc.gr
* Correspondence: iliopoulosg@upatras.gr

Citation: Golfinopoulos, V.; Papadopoulou, P.; Koumoutsou, E.; Zouros, N.; Fassoulas, C.; Zelilidis, A.; Iliopoulos, G. Quantitative Assessment of the Geosites of Chelmos-Vouraikos UNESCO Global Geopark (Greece). *Geosciences* 2022, 12, 63. https://doi.org/10.3390/geosciences12020063

Academic Editors: Jesús F. Jordá Pardo and Jesus Martinez-Frias

Received: 29 December 2021
Accepted: 26 January 2022
Published: 29 January 2022

Publisher's Note: MDPI stays neutral with regard to jurisdictional claims in published maps and institutional affiliations.

Copyright: © 2022 by the authors. Licensee MDPI, Basel, Switzerland. This article is an open access article distributed under the terms and conditions of the Creative Commons Attribution (CC BY) license (https://creativecommons.org/licenses/by/4.0/).

Abstract: The assessment of the geosites of Chelmos-Vouraikos UNESCO Global Geopark (UGGp) was carried out based on an established methodology for the evaluation of geoparks' geosites. Such assessments should be used for sustainable development and geoconservation in geoparks. The selected methodology is based on a wider range of criteria concerning the overall value of each geosite, compared to other locations. Each criterion was scored and then three indices, Vedu, Vprot and Vedu were estimated for each geosite. The application of this methodology at Chelmos-Vouraikos UGGp has produced results which not only highlight the value of each geosite, but also provide ways for their utilization. The assessment of the 40 geosites of the geopark, identified geosites with high educational and touristic value (such as Portes–Triklia and the Cave of the Lakes), while geosites with increased protection-need value (the Tectonic Graben of Kalavryta) were also highlighted. Therefore, the assessment results will be used by the geopark to plan the effective management of the geosites based on their strengths and weaknesses, and which thus will promote the geopark and will contribute to the sustainable development of the local communities. The proposed methodology uses all possible criteria for its impartial application and despite a few minor problems that have been identified, it is considered appropriate for the assessment of geosites in Geoparks. The application of such evaluation methodologies is considered crucial for the development, protection and touristic promotion of geoparks.

Keywords: geosite assessment; Chelmos-Vouraikos UGGp; geoheritage; geotourism; geoconservation

1. Introduction

UNESCO Global Geoparks (UGGps) "are single, unified geographical areas where sites and landscapes of international geological significance are managed with a holistic concept of protection, education, and sustainable development. Their bottom-up approach of combining conservation with sustainable development, while involving local communities is becoming increasingly popular" [1,2]. The concept of geoparks began to evolve in Europe 26 years ago [1,3,4]. The main goal of a geopark is the territorial development of an area with significant geoheritage and natural and cultural wealth as well, in order to promote the sustainable development of local communities through the promotion of geotourism and education [3–5]. Today, there are 169 UNESCO Global Geoparks located in 44 countries around the world [2]. In Greece to date, six areas have been designated as UNESCO Global Geoparks: Lesvos (2000), Psiloritis Natural Park (2001), Chelmos-Vouraikos (2009), Vikos–Aoos (2010), Sitia (2015) and Grevena–Kozani (2021). According to UNESCO [2], UGGps have multidimensional aims which include primarily the protection and conservation of their territorial geoheritage and additionally the cultural and environmentally sustainable development of their territories [6]. Geoscience education offered at all levels is another goal for UGGps promoting awareness about the history of the earth and sustainable development [7]. UNESCO Global Geoparks are basically about humans and about exploring and celebrating the connection between our communities and the Earth [2]. The Earth shapes human identities: it has shaped people's agriculture practices, the building materials and methods that have been used for constructing their accommodation, and even their mythology, folklore as well as folk traditions. In addition, it is the basement for all earth systems and services (natural values, biodiversity, resources, etc.) [8,9]. UNESCO Global Geoparks, therefore, engage in a range of activities to celebrate these connections. The Operational Guidelines of the UGGps define 10 focus areas in which the geoparks should develop activities [9] and for which they are evaluated every four years [10]. These areas refer to the natural resources of their territories, and their sustainable use; to geohazards and climate change; to education of local inhabitants and visitors; to natural and cultural heritage; to science and research; to local culture and the celebration of local heritage; to the empowerment of women and equality; to indigenous people and knowledge, if they exist; to the sustainable development through geotourism; and to geoconservation and safeguarding the geological value of their territory.

Over the past three decades, several methodologies for the quantitative and qualitative assessment of Earth's heritage have been developed [11–31] to serve geoconservation and geotourism needs. They aim in minimizing subjectivity in the procedure of organizing the results into an understandable and well recognizable ranking system. Unfortunately, subjectivity cannot be avoided when scoring scientific importance, the need for protection or the potential use of geosites, however, in most cases specific criteria with certain scoring systems are introduced to cope with this problem. When it comes to the preparation of management plans that should be both socially accepted and useful [18,19], the establishment of objective criteria is required [2,5]. The first methods were developed for the assessment of geomorphosites and landscape features in general [11,13,16–18,23–25,27,29], focusing mainly on their aesthetic and scientific values, while others on karst geomorphosites [22,26], or on volcanic geomorphosites [30] as well. Several other methods also induced the assessment of educational [21] and geotouristic values presenting in that way a more complete approach on the overall geosite value [12,15,20,31] Among existing methodologies, quite a few were developed primarily for the need of UGGps to achieve progress on the 10 focus areas [14,19,28]. The assessment of the geosites of a geopark not only has a scientific purpose but also aims at the management and conservation of its geological heritage at a certain territorial and legislative context and under the operational framework induced by UNESCO. Geodiversity needs to be considered in a wider view, combining sustainable development with the conservation of geoheritage [19]. Therefore, such assessment becomes a useful tool for site managers, because it highlights and quantifies the priorities needed for the proper management and protection of the geopark. It can also highlight priorities for sustainable tourism development (geotourism and educational tourism) as well as for the conservation of geosites.

The main objectives of this study are, first, the evaluation of the 40 geosites of Chelmos-Vouraikos UGGp in order to highlight their touristic and educational value as well as the need for their protection, and secondly, the evaluation of the selected methodology, through which its advantages, as well as the respective weaknesses that might need improvement, will be highlighted.

2. Geopark Outline

Chelmos-Vouraikos UGGp is located in North Peloponnese (Greece). It occupies an area of 647 km^2 with a total population of approximately 27,000. The area exhibits unique geoheritage, wonderfully combined with rich bioheritage and exceptional cultural and historical elements. To date, 40 geosites have been established in the geopark (Table 1,

Figure 1). They include unique geological formations (folds, faults, rocks and lithological formations, etc.), karstic geomorphs (caves, poljes, karstic springs), rich geomorphosites (gorges, high peaks, alpine lakes etc) and fossil sites.

Table 1. The geosites of Chelmos-Vouraikos UGGp and their categories. Tectonic (T), Lithological (L), Stratigraphical (S), Karstic (K), Geomorphological (Gm), Hydrological (H), Geotechnical (Gt), Palaeontological (P), Cultural (C).

ID	GEOSITE	CATEGORY
1	Niamata	T, L, S
2	Portes-Triklia	K, L, T
3	Mamousia-Rouskio	L, S
4	Trapeza Marine terrace	Gm, T
5	Kerpini Conglomerates	L, Gm
6	Roghi	L, S
7	Tectonic graben of Kalavryta	T, Gm
8	Agia Lavra	T
9	Xidias Lignites	L, S
10	Priolithos	L, S
11	Cave of the Lakes	K
12	Mavri Limna	Gm, T
13	Lousoi sinkhole	K
14	Aroanios Springs	K
15	Mati tou Ladona	K
16	Vesini radiolarites	L, S
17	Doxa lake	H, Gt, L
18	Solos	L
19	Tsivlos Lake	H, Gm
20	Water of Styx	Gm, S, K
21	Xerocambos breccias	Gm, L
22	Feneos sinkholes	K
23	Lousoi polje	K
24	Mavrolimni	Gm
25	Analipsi	L, K
26	Valvousi	K, Gm
27	Keramidaki	L, T
28	Mega Spilaio	Gm, L, P
29	Kastria spring	K
30	Spanolakkos	Gm, L
31	Palaeochori lignites	P, L
32	Valimi landslide	Gm
33	Pausanias Vine	C
34	Psili Korfi	Gm, T, K
35	Ntourntourvana	Gm, S, P, K
36	Chelonospilia	Gm
37	Madero	Gm, S
38	Eroded Conglomerates	Gm, L
39	"Balcony" of Styx	Gm, S
40	Tessera Elata	Gm, L

At least 485 million years (lower Palaeozoic) of geological history are "unfolded" in the area of Chelmos-Vouraikos UGGp. The alpine basement consists of three geotectonic zones, namely Tripolis Zone, Pindos Zone and the metamorphic Phyllites–Quartzites Suite which is the oldest. Post-alpine formations, whose deposition is related to the Corinth rift, overlap the alpine basement (Figure 1).

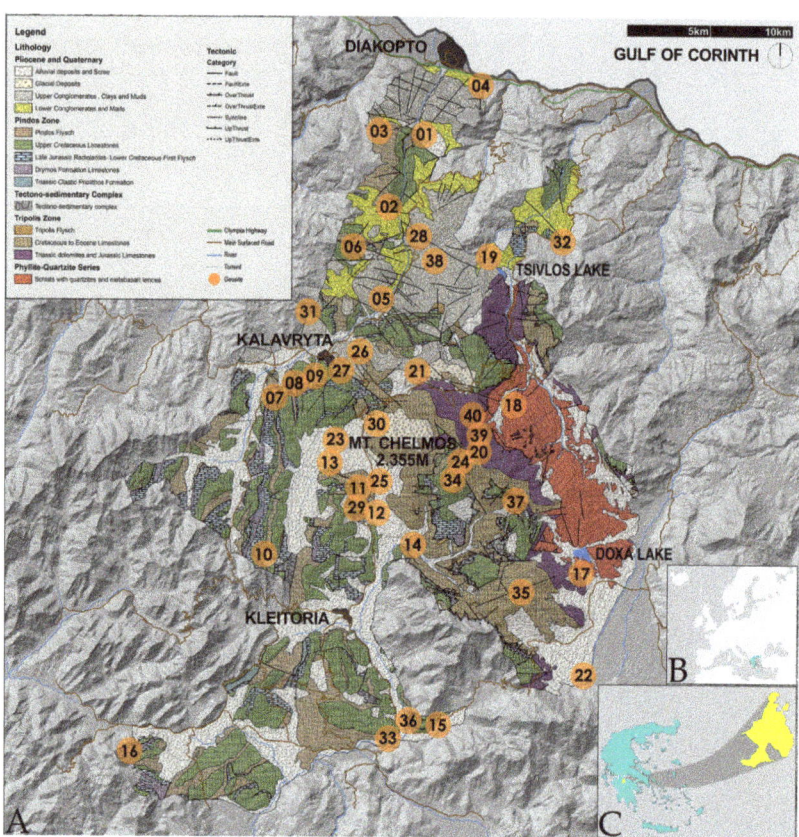

Figure 1. (**A**) Geological map of Chelmos-Vouraikos UGGp with the locations of the geosites, (**B**) Map of Europe indicating the position of Greece and (**C**) Map of Greece with the position of the study area.

Tripolis Zone consists of Mesozoic neritic limestones and dolomites [32], with a maximum visible thickness of approximately 3500 m [33]. The formations of this unit can be found at the east-southeast part of the geopark, on Chelmos Mt, along Krathis river and at the northern part of Vouraikos gorge. The basement of this zone is the Upper Palaeozoic to Lower Triassic volcano-sedimentary Tyros beds. They consist of a complex of sedimentary and volcanic rocks, which is characterized by a very low grade of metamorphosis. Carbonate sediment deposition in shallow marine environments began during the Early Mesozoic and lasted until the Late Eocene. Flysch formation followed until the end of the Oligocene. The Tripolis Zone underlies sediments of the Pindos zone and post alpine sediments. The whole sequence was overthrusted above the Phyllites–Quartzites Suite.

Pindos Zone develops mainly to the west and south part of the geopark. It consists of a Mesozoic sequence of carbonate and silicate sediments deposited in a deep-sea environment. Their thickness does not exceed 1050 m [33]. At the base of the sequence, the middle Triassic clastic Priolithos Fm is found. Drymos Fm limestones, with an Upper Triassic–Lower Jurassic age, lie on top of it. The Upper Jurassic to Lower Cretaceous Radiolarites Fm (*sensu lato*) which overlies the Drymos Fm consists of alternations of pelites, cherts (mainly radiolarites) and limestones with Calpionellids. During the Albian to Cenomanian, the First Flysch Fm was deposited. Upper Cretaceous thin-bedded limestones follow. These limestones during the Late Maastrichtian–Eocene evolved gradually into flysch [34].

The rocks of the Phyllites-quartzites Suite appear at the tectonic window of Chelmos Mt, along the Krathis River and in the Feneos plateau (polje) as well. The rocks of this Suite have been formed under high pressure/low temperature metamorphic conditions [35,36]. They consist of phyllites and quartzites that originated from a detrital sequence, whereas in some places mafic rocks are intercalated. Zircon dating methods provided an Early Palaeozoic age for these rocks [37].

Post-alpine sediments (Late Neogene–Quaternary) were deposited across a WNW-ESE direction lying parallel to the Corinth Rift system which is responsible for their formation. The total thickness of these sediments is approximately 2.8 km [38]. At the north of the study area, five major north-dipping normal faults can be found (from south and older to north and younger): Kalavryta, Kerpini–Tsivlos, Doumena, Pirgaki–Mamoussia and Helike [38]. These faults have confined a series of approximately WNW-ESE north dipping rotated fault blocks associated with the regional N-S extension of the Rift [39]. Accordingly, these fault blocks formed the half-grabens where Pliocene and Quaternary sediments were deposited unconformably on the substrate [40].

According to Ford et al. [41] and Pope et al. [42], at the top of Mount Chelmos, as well as on the surrounding edges, glaciers developed during the Middle–Late Pleistocene. Glaciofluvial brecciated deposits can be found around the mountain's high peaks along with other glacial geomorphological evidence (moraines, cirques, alpine lakes).

The area of the geopark is also valued and protected for its biodiversity and has been classified as a "National Park", managed by the Chelmos-Vouraikos Management Body and taking into consideration (a) the "IUCN Guidelines for Applying Protected Area Management Categories" and (b) the legal framework for conservation management of the Chelmos -Vouraikos National Park pertaining to four Natura 2000 sites (IUCN Management Category II) [43]. It is also rendered as a biodiversity hotspot of the Greek flora and is included in the endemism hotspot region of the mountain ranges of Northern Peloponnesus [44–46]. The natural vegetation of the Chelmos-Vouraikos UNESCO Global Geopark is the result of the interaction of various parameters and mainly of its various orographic configuration, petrological–geological composition, bioclimatic conditions and human activities that have shaped the landscape since historical times. This results in the contemporary, complex land-use fabric which includes semi-natural and natural ecosystems subject to traditional human practices, as well as ecological, historical and culturally important landscapes [10].

The people living in the area are mainly farmers and stockbreeders, who take advantage of the rich resources of their land and produce local products famous for their quality and uniqueness, such as dairy products, honey and legumes. In addition, an important driving force for the local economy is tourism. "Odontotos" rack railway runs across the steep Vouraikos gorge, passing by some of the Geopark's unique geosites, attracting thousands of tourists and giving them the opportunity to admire the incredible geomorphology of the gorge. A popular ski center is hosted at the unique glacial valleys of Chelmos Mt, while the spectacular Cave of the Lakes serves as an exceptional geomorphosite of touristic value. These are classical examples of the way that the geomorphology of the Geopark area shapes the local geotouristic character and provides the means for sustainable development. A large number of tourists is hosted every year at the numerous accommodation facilities that exist in the area.

The area played an important role during the Greek uprisings and for this reason it has a great history. This is witnessed by the numerous monuments of war atrocity which can be found scattered in the area of the geopark (e.g., Holocaust Monument at Kapi hill). The rich history is communicated to the public through the multiple museums (e.g., Holocaust Museum, Folklore-art Museums of Kleitoria and Feneos, etc.). The area has also great archaeological (e.g., archaeological sites of ancient Lousoi and ancient Kleitoria) and religious interest (e.g., historic monasteries of Mega Spilaio and Agia Lavra). One more aspect of the geopark's multidimensional value is its intangible heritage, which is wonderfully expressed through popular art especially music and poetry and mostly through the great mythological heritage. The myths and the legends that concern specific parts of the geopark are countless. Characteristic examples are the Feneos sinkholes which are connected to the mythical semi-god Hercules as are also the geosites of Portes-Triklia and Roghi.

The geopark offers a spectrum of promotional tools related to its geoheritage, especially geotrails (seven georoutes) and informative material available at the Information Center of the Management Body. Combined information regarding the geoheritage and its links to biodiversity and culture are presented on the geopark's informative signage.

3. Materials and Methods

3.1. Evaluation Methodologies for Geopark Geosites

As mentioned above, several methodologies have been developed to assess either certain types of geosites or certain needs (geoconservation, geoeducation, geotourism) [20,23–25,27,47–49]. Synthetic methodologies too, have been developed to assess in certain levels (international, national, etc.) the total values of geosites [17,20,23,24]. More specifically, Brijla [20] reviews the most common and broader applied methodologies for geoheritage assessment, proposing a new one that combines several criteria and tools of previous methods, and which can be applied for geosites and geoheritage sites and at various inventory scales (national, municipal, parks, etc.).

In addition to these, only three main methods so far have been proposed for the quantitative assessment of geoparks' geosites [14,19,28].

Zouros & Valiakos [14] proposed a quantitative method for the evaluation of the main operating elements of a geopark, which consists of five criteria with different weighting, and each criterion is subdivided into indicators. Each indicator receives a numerical value during the assessment process. The five criteria are: (1) Geology and Landscape, (2) Management Structure, (3) Interpretation and Environmental Education, (4) Geotourism, (5) Sustainable Regional Economic Development. This method represents a general assessment for geoparks, concerning the quality of the geodiversity, operation, visibility, geotourism and local development of a geopark, and thus is not limited to geosite evaluation. Actually, geosite evaluation consists only one out of the five criteria of the assessment. Hence, this assessment can be a great tool for the overall assessment of Geoparks.

The second methodology by Fassoulas et al. [19] is based on previously proposed criteria from existing evaluation methodologies [13,16,17,31]. These criteria were combined in such a way that they can be applied to all categories of geosites and can be used to evaluate all aspects of a geosite's value. This makes the methodology a useful tool for the assessment of geopark geosites, because geoparks contain a variety of different types of geosites. The criteria are the following six: (1) scientific, (2) ecological and protection, (3) cultural, (4) aesthetic, (5) economic and (6) potential for use. Each main group consists of a number of sub-criteria, and a scoring system is implemented to each criterion. Based on these criteria three individual indices are induced, the Vtour on the geotouristic value, the Vedu on the educational value and the Vprot on the conservation need. The method developed by Fassoulas et al. [19] was designed specifically for UGGps (at that time only endorsed but under close collaboration with UNESCO) as a useful tool for the geopark managers. Hence, it assesses sites in respect to the regional context of the geopark (and not the national or international) in order to highlight the local priorities that the managers should focus on for geoconservation, geoeducation or geotourism. Therefore, it uses criteria that take under consideration the multidimensional aims of UGGps and proposes that the assessment should be undertaken by a group of local and invited experts to minimize the subjectivity effect. The method has been widely considered and not only in geoparks (Google scholar citations: 188; Scopus citations: 102).

The third and most recent methodology presented by Aoulad-Sidi-Mhend et al. [28] was developed to select and classify geosites in protected areas, as well as to determine the degree of protection-need of geosites. The applied evaluation is a result of a combination of studies [17,20,31,50]. This assessment is subdivided into four parts: (1) Scientific Value,

(2) Additional Values, (3) synthesis and result of the Global Value and (4) the Degradation Risk. These parts consist of several criteria, where each criterion is scored with a numerical value. This method focuses mainly on the selection and evaluation of geosites or potential geosites in protected areas, a certainly interesting assessment for evaluating the geological heritage of such an area and contributing to the designation of new geosites.

As one of the two main and practically principal objective of this study is the evaluation of the individual geosites of an UGGp, Chelmos-Vouraikos, we consider that out of the three methods presented above for the evaluation of geoparks' geosites, the most appropriate for the evaluation of geoparks' geosites is the method proposed by Fassoulas et al. [19], which is entirely focused on geoparks. In a second step, we also look for potential weaknesses or misfunctions of the method under the modern needs of geoparks, in order to propose suggestions for its improvement [19].

3.2. Application of Selected Evaluation Methodology

Therefore, having in mind the methodology designed by Fassoulas et al. [19], a working protocol was carefully compiled (Table 2), taking into account the related published literature [10,45,51–53], especially that concerning the ecological criteria (flora and fauna). Aiming for a more accurate evaluation of the 40 geosites of Chelmos-Vouraikos UGGp, in situ inspection of each geosite took place by the research group. The geographical coordinates of each geosite were obtained. Moreover, the geosites were photographed and where permitted, rock samples were collected for the creation of the geosites database.

Table 2. Standard protocol for recording geosite characteristics [19].

NAME	LITHOLOGY
REGIONAL UNIT	CATEGORY
LONGTITUDE	TECTONIC UNIT
LATITUDE	ALTITUDE
CODE	
COMMENTS	

GEODIVERSITY	
INTEGRITY	
ECOLOGICAL IMPACTS	
PROTECTION STATUS	
ETHICS	
HISTORY & ARCHAEOLOGY	
RELIGIOUS & METAPHYSICAL	
ART & CULTURE	
VIEWPOINTS	
LANDSCAPE DIFFERENCE	
ACCESIBILITY	
VISITORS	
INTESITY OF USE	
FRAGILITY	
NATURAL RISKS	

Six main groups of criteria are considered in this method [19]: (1) scientific, (2) ecological and protection, (3) cultural, (4) aesthetic, (5) economic and (6) potential for use. Each group is sub-divided into sub-criteria (Table 3) in order to better assess the value of each geosite [19]. The scoring system for every sub-criterion ranges from 1 to 10 and five fixed scores (1, 2.5, 5, 7.5, 10) can be applied to the sub-criteria of each group (Table 3). Finally, the value of each criterion was calculated by all members of the research group as the average of the respective sub-criteria scores.

Table 3. List of criteria and description of the scoring system [19].

CRITERIA / SCORE	1	2.5	5	7.5	10
1. SCIENTIFIC					
1.1 Geological history	Single type history	Combination of at least 2 types	Combination of most types	Local story	Tells the whole local story
1.2 Representativeness	No	Low	Moderate	High	Very high
1.3 Geodiversity	<5%	25%	50%	75%	>75%
1.4 Rarity	>7	>5 <7	>3 <4	>1 <2	Unique
1.5 Integrity	Almost destroyed	Strongly deteriorated	Moderately deteriorated	Weakly deteriorated	Intact
2. ECOLOGICAL					
2.1 Ecological impact	No	Low	Moderate	High	Very high
2.2 Protection status	No protection	Limited	In spots	In large parts	Complete
3. CULTURAL					
3.1 Ethics	No	Low	Moderate	High	Very high
3.2 History	No	Low	Moderate	High	Very high
3.3 Religious	No	Low	Moderate	High	Very high
3.4 Art & Culture	No	Low	Moderate	High	Very high
4. AESTHETIC					
4.1 Viewpoints	No	1	2	3	>4
4.2 Landscape difference	No	Low	Moderate	High	Very high
5. ECONOMIC					
5.1 Visitors	<5000	>5000	>20,000	>50,000	>75,000
5.2 Attraction	No	Local	Regional	National	International
5.3 Official protection	International	National	Regional	Local	No
6. POTENTIAL FOR USE					
6.1 Intesity of use	Very intense	Intense	Moderate	Weak	No use
6.2 Impacts	Very high	High	Moderate	Low	No
6.3 Fragility	No	Low	Moderate	High	Very high
6.4 Accesibility	Close to hiking trail	Close to cobble or forest road	Close to local paved road	Close to regional road	Close to highway or town
6.5 Acceptable changes	No	Low	Moderate	High	Very high

The five scientific sub-criteria assess the scientific value of a geosite (Table 3). These are the Geological history (1.1) that depicts the contribution of a geosite to the interpretation of the overall geological history of a Geopark, the Representativeness (1.2), which addresses the status of the site as an example of the geological heritage of the Geopark, the Geodiversity (1.3), which describes the variety of the identified geological features and processes in a geosite compared to the Geopark's geodiversity, the Rarity (1.4) which concerns the uncommonness of a geosite with respect to similar geosites in the Geopark and the Integrity (1.5) which refers to the existing state of conservation of a geosite, which might be affected by human activities or/and natural processes.

The second group of sub-criteria considers the ecological value of the geosite (Table 3). Ecological impact (2.1) represents the contribution of a geosite to the development of particular ecotopes or to the existence of endemic species within this area, whereas Protection status (2.2) refers to the actual protection and conservation state of the site.

The four cultural sub-criteria highlight the participation of a geosite to the cultural heritage of the geopark (Table 3), covering all aspects of culture. Ethic importance (3.1) defines the relationship of a geosite with existing ethics or customs, Historical importance (3.2) describes the connection of the site to historical events or archaeological remains, Religious importance (3.3) concerns the religious, metaphysical or mythological value of a geosite and Art and cultural importance (3.4) assesses the presence of the geosite in the arts at a local or regional level.

The fourth group of sub-criteria identifies the aesthetic value of a geosite (Table 3). The two sub-criteria are Viewpoints (4.1) that considers the visibility of the site based on the number of viewpoints from roads or trails around the site located more than 1 km away from each other and Landscape difference (4.2), which considers the difference in shape, colour or morphology between the landscape background and the geosite.

The economic importance of a geosite is evaluated by three sub-criteria (Table 3). These are Visitors (5.1), which is based on the recorded or estimated number of visitors to the site and which should always be related to the total tourism potential of the Geopark or local region, Attraction (5.2), which considers the importance of the site as a national, regional or local attraction and Official protection (5.3), which describes the legal protection status of a geosite. It has been documented that the high protection status of an area can imply restrictions in many human activities, including economic, sometimes not even permitting the physical human presence in core zones. Therefore, it should be regarded as an economic criterion, and the higher the protection status the lower the scoring should be. In cases where legal protection represents the actual situation, scoring in sub-criterion 5.3 should be regarded as inversely proportional to the scoring of sub-criterion 2.2.

The five potential of use sub-criteria interpret the ability for possible exploitation of a geosite (Table 3). Intensity of use (6.1) indicates the present use of the site by humans, Impacts (6.2) assesses the negative effects of existing human activities on the site, Fragility (6.3) refers to the degree of resistance of a geosite's physical features with respect to potential degradation, Accessibility (6.4) describes the ease of access to the site by road or trails and Acceptable changes (6.5) considers the resistance of a geosite to changes without risking the degradation of its natural features. The last sub-criterion depends on the intensity of use and fragility of the respective geosite.

In addition, based on the values of the six criteria, three indices are calculated for each geosite, the V_{tour}, referring to the geotouristic value the V_{edu} for the educational value and the V_{prot} for the conservation/protection need [19]. The formulae for the calculation of the three indices are using differently weighed coefficients for the specific criteria that are considered for each index, as not all criteria have the same effect on each index. For the geotouristic index the aesthetic, cultural, economic and potential of use criteria are used for its calculation, with the aesthetic criterion considered as more important for this index and thus a 0.4 coefficient is used for it in the formula (0.2 for the other three criteria) [19].

$$Vtour = (0.4 \times Aesthetic) + (0.2 \times Cultural) + (0.2 \times Potential\ of\ Use) + (0.2 \times Economic)$$

For the education index, the scientific, aesthetic, cultural, and ecological criteria are used for its calculation, with the scientific criterion considered more important for it and thus a 0.4 coefficient is used for it in the formula (0.2 for the other three criteria) [19].

$$Vedu = (0.4 \times Scientific) + (0.2 \times Cultural) + (0.2 \times Aesthetic) + (0.2 \times Ecological)$$

Concerning the calculation of the protection need index, the ecological risk factor (F_{ecol} [19]) needs to be estimated first. This factor calculates the ecological risk of a geosite from the ratio of the Ecological impact score against the Protection status score of the respective geosite. Hence, the calculation of the protection need index is based on the scientific criterion and the ecological risk factor, but takes into account as well the Integrity sub-criterion using the formula:

$$Vprot = (Scientific + Fecol + (11 - Integrity))/3$$

In this way with such a quantitative methodology (scoring scale 1–10) it is much easier to prioritize and identify the geosites that are more important for their geotouristic and educational value and thus take actions to develop them accordingly, and also identify which geosites are vulnerable, facing destruction or deterioration problems, so that protection measures can be undertaken. Therefore, the Fassoulas et al. [19] method has been applied for the evaluation of the 40 geosites of Chelmos-Vouraikos UGGp in order to identify the priorities of Chelmos-Vouraikos UGGp in geoconservation, geoeducation and geotourism development [19].

4. Results

The analytic results (criteria scores) of the 40 assessed geosites of Chelmos-Vouraikos UGGp are presented in Table 4. The average score of the scientific criteria ranges from 2.1 to 9.5, while the score of ecological, aesthetic and economic criteria ranges from 1 to 10. Furthermore, the cultural criteria values do not exceed 5.9. The criteria for potential use start from 3.9 and reach up to 9.5.

The geological report and the criteria analysis for ten typical examples of geosites (geosites 2, 7, 11, 14, 19, 20, 24, 28, 35 and 38) are provided in detail in the following sections. These geosites are presented here in detail as they consist some of the most popular and characteristic geosites of the geopark (geosites 2, 7, 11, 14, 19, 28). Some of them also present great need for protection, such as geosites 7 and 14. In addition, very high scores for some of them indicate that they have not been exploited geotouristically so far (geosites 20, 24, 35), while for other geosites, such as geosite 38, the need for immediate protection is highlighted.

Table 4. Assessment of the results of the 40 geosites of Chelmos-Vouraikos UGGp.

Code	Name	Longitude	Latitude	Scientific Score	Ecology Score	Cultural Score	Aesthetic Score	Economic Score	Potential Use Score	F_{ecol}	V_{edu}	V_{prot}	V_{tour}
C 01	Niamata	22°10′23.62″ E	38°9′4.26″ N	9.5	5.5	2.0	5.5	7.0	6.7	0.1	6.4	3.5	5.3
C 02	Portes-Triklia	22°9′37.21″ E	38°6′31.92″ N	7.5	10.0	2.5	8.8	7.0	6.4	1.0	7.3	4.0	6.7
C 03	Mamousia-Rouskio	22°8′48.59″ E	38°8′56.29″ N	2.1	1.8	2.0	1.8	4.0	5.2	2.5	1.9	3.5	2.9
C 04	Trapeza Marine terrace	22°14′10.98″ E	38°10′31.17″ N	5.9	1.0	2.0	3.0	5.3	7.0	1.0	3.6	3.5	4.1
C 05	Kerpini Conglomerates	22°8′48.68″ E	38°3′21.15″ N	3.7	10.0	2.0	6.3	7.0	3.9	1.0	5.1	2.7	5.1
C 06	Roghi	22°7′47.25″ E	38°5′30.27″ N	2.9	10.0	2.4	3.0	4.0	4.9	1.0	2.4	4.6	3.5
C 07	Tectonic graben Kalavryta	22°6′30.94″ E	38°1′59.59″ N	6.5	3.0	3.3	8.8	10.0	8.2	5.0	5.6	6.7	7.8
C 08	Agia Lavra	22°4′31.05″ E	38°1′7.82″ N	7.0	1.0	5.5	3.0	5.3	7.0	1.0	4.7	3.8	4.8
C 09	Xidias Lignites	22°6′13.32″ E	38°1′20.43″ N	4.4	1.0	1.4	3.8	1.5	5.2	1.0	3.0	2.1	3.1
C 10	Priolithos	22°3′8.89′ E	37°54′53.60″ N	4.9	1.0	2.0	1.8	4.0	7.0	1.0	2.9	3.1	3.3
C 11	Cave of the Lakes	22°8′24.15″ E	37°57′34.73″ N	7.0	10.0	2.0	10.0	6.2	5.2	1.0	7.2	3.8	6.7
C 12	Mavri Limna	22°8′21.80″ E	37°57′10.19″ N	3.6	1.0	1.4	1.0	4.0	6.7	1.0	2.1	1.9	2.8
C 13	Lousoi sinkhole	22°6′45.48″ E	37°58′31.08″ N	5.2	1.0	3.0	1.0	2.0	9.5	1.0	3.1	2.4	3.3
C 14	Aroanios Springs	22°9′57.82″ E	37°56′1.34″ N	5.2	3.8	1.4	6.3	5.3	5.2	2.0	4.4	4.4	4.9
C 15	Mati tou Ladona	22°10′57.65″ E	37°50′12.29″ N	5.7	2.5	1.8	4.3	4.5	5.0	1.0	4.0	3.4	4.0
C 16	Vesini radiolarites	21°58′39.73″ E	37°49′53.02″ N	2.3	1.0	2.6	1.0	4.0	4.4	1.0	1.8	2.3	2.6
C 17	Doxa lake	22°17′13.74″ E	37°55′53.79″ N	4.9	7.5	4.6	10.0	7.5	6.2	0.5	6.4	2.1	7.7
C 18	Solos	22°14′18.30″ E	38°0′15.90″ N	4.4	1.0	2.4	1.0	2.3	6.7	1.0	2.6	2.1	2.7
C 19	Tsivlos Lake	22°13′58.24″ E	38°4′37.61″ N	7.7	7.5	1.4	5.0	7.0	6.7	0.5	5.9	3.1	5.0
C 20	Water of Styx	22°12′18.18″ E	37°59′0.32″ N	6.5	8.8	3.3	5.0	2.3	4.9	0.8	6.0	2.8	4.1
C 21	Xercambos breccias	22°11′46.87″ E	38°0′33.05″ N	3.7	6.3	1.0	3.0	7.0	5.0	1.5	3.5	4.6	3.8
C 22	Feneos sinkholes	22°17′27.78″ E	37°51′35.88″ N	4.4	1.0	3.0	1.0	4.5	5.0	1.0	2.8	2.1	2.9
C 23	Lousoi polje	22°6′10.93′ E	37°59′4.63″ N	8.5	1.0	4.0	7.5	5.0	6.2	1.0	5.9	4.3	6.0
C 24	Mavrolimni	22°12′5.15″ E	37°58′42.43″ N	6.0	10.0	1.0	7.5	1.5	6.4	1.0	6.1	2.7	4.8
C 25	Analipsi	22°8′55.55″ E	37°58′2.60″ N	5.0	1.0	2.8	1.0	4.0	7.2	0.7	3.0	2.3	3.2
C 26	Valvousi	22°8′11.26″ E	38°1′6.96″ N	3.7	6.3	1.4	7.5	1.0	5.2	0.7	4.5	3.5	4.5
C 27	Keramidaki	22°7′26.77″ E	38°1′23.46″ N	7.0	1.4	1.4	1.0	1.0	5.2	0.7	4.5	2.9	1.9
C 28	Mega Spilaio	22°10′18.90′ E	38°5′30.17″ N	6.0	7.5	5.9	8.8	9.2	6.2	2.0	6.8	4.7	7.7
C 29	Kastria spring	22°8′21.30″ E	37°57′29.16″ N	6.0	1.0	1.0	1.0	4.0	4.6	1.0	3.0	2.7	2.3
C 30	Spanolakkos	22°8′38.89′ E	37°59′43.20″ N	6.2	3.0	2.4	1.8	1.0	7.2	0.2	3.9	2.5	2.8
C 31	Palaeochori lignites	22°6′2.14″ E	38°3′20.96″ N	6.5	1.0	1.4	2.5	1.0	6.7	1.0	3.6	4.5	2.8
C 32	Valimi landslide	22°16′28.15″ E	38°5′27.46″ N	4.6	1.0	1.0	5.0	4.0	6.7	1.0	3.2	2.2	4.3
C 33	Pausanias Vine	22°9′16.89′ E	37°49′37.12″ N	4.1	7.5	3.6	1.0	2.0	7.5	0.5	4.1	2.7	3.0
C 34	Psili Korfi	22°12′3.59′ E	37°58′19.34″ N	6.0	8.8	1.4	5.5	1.5	5.4	0.8	5.5	3.4	3.9
C 35	Ntourntourvana	22°15′11.17″ E	37°54′51.72″ N	7.5	7.5	1.0	8.8	1.5	6.4	0.5	6.5	3.0	5.3
C 36	Chelonospilia	22°9′47.68″ E	37°50′13.71″ N	2.8	1.0	2.0	1.8	4.5	8.2	1.0	2.1	1.6	3.6
C 37	Madero	22°14′27.55″ E	37°57′21.59″ N	5.4	7.5	2.4	4.3	1.5	6.4	0.5	4.7	2.3	3.5
C 38	Eroded Conglomerates	22°11′4.26″ E	38°5′0.37″ N	4.2	5.5	1.0	5.0	4.5	4.0	10.0	4.0	5.9	3.9
C 39	"Balcony" of Styx	22°12′50.53″ E	37°59′57.69″ N	7.0	1.0	1.0	1.8	1.5	4.6	1.0	3.6	3.0	2.1
C 40	Tessera Elata	22°12′50.53′ E	37°59′57.69″ N	7.0	1.0	1.0	1.0	1.5	6.7	1.0	3.4	3.8	2.2

4.1. Portes–Triklia (Geosite 2)

Portes–Triklia (38°6′31.92″ N, 22°9′37.21″ E) is located in the northern part of the geopark in Vouraikos River Gorge. The area consists of Upper Cretaceous bedded limestones of Pindos zone and fluvial-torrential deposits (Figure 2A). Intense tectonic uplift along normal faults [54] has shaped the gorge in combination with rapid vertical erosion. As a result, an impressive, very narrow and deep gorge was formed during the Lower–Middle Pleistocene [54,55].

Figure 2. Representative views of the geosites described in the results. (**A**) Portes-Triklia, geosite 2, (**B**) The Water of Styx, geosite 20, (**C**) Cave of the Lakes, geosite 11, (**D**) Aroanios Springs, geosite 14, (**E**) Tectonic graben of Kalavryta, geosite 07, (**F**) Tsivlos Lake, geosite 19.

This geosite exhibits two types of geological history of the studied area: the deposition of Pindos formation and the intense erosional processes of the river (score of sub-criteria 1.1. and 1.2., 2.5 and 10 respectively). The geological features and the processes related to this geosite are the deposition of the limestones of Pindos zone, the deposition of the fluvial-torrential deposits of Vouraikos river, the high rate of uplift of the wider area combined with the intense vertical erosion of the river (score of sub-criterion 1.3. 7.5). This unique combination of features can be found in no other geosite in the geopark (score of sub-criterion 1.4. 10). The existing status of conservation from human activity and natural processes is considered slightly damaged due to the construction of the historic "Odontotos" rack railway (score of sub-criterion 1.5. 7.5).

The floristic importance of Vouraikos gorge is very high, due to the presence of rare endemic species. The avifauna of the wider area is also remarkable especially because of the high number of species that reproduce there (score of sub-criterion 2.1. 10). The state of actual protection and conservation is very high (score of sub-criterion 2.2. 10) throughout the gorge area.

The European E4 path passes also through this geosite. Its route lies along the Rack railway tracks and is used by thousands of Greeks and foreign hikers every year. The intangible heritage of Vouraikos gorge is invaluable since it is related to Hercules myths. (scores of all sub-criteria of cultural criterion are 2.5 each).

The geosite is not visible from afar mainly due to the dense vegetation but also due to the geomorphology of the area (steep gorge) (score of sub-criterion 4.1. 10) and does not differ from the background (score of sub-criterion 4.2. 7.5).

The historic "Odontotos" rack railway is cross passing this geosite attracting thousands of visitors-Greek and foreigners- each year (e.g., 159,789 visitors during 2018 according to the Hellenic Railways Organization) (score of sub-criterion 5.1. 10). Therefore, it is considered an international attraction (score of sub-criterion 5.2. 10). Vouraikos gorge is located in a NATURA A3 zone area (SCI_GR2320003 and SPA_2320013), (score of sub-criterion 5.3. 1).

No negative effects from human use have been recorded apparently because it is located in a NATURA A3 protection zone (score of sub-criteria 6.1. and 6.2. 10 each). Its physical parameters are characterized by very high fragility (score of 6.3. 10), so changes are not acceptable without the risk of degradation of the geosite (score of sub-criterion 6.5. 1). The site can be reached either from the homonymous railway stations, or by hiking across the E4 path, departing from Diakopto or Zachlorou (score of sub-criterion 6.4. 1).

4.2. Tectonic Graben of Kalavryta (Geosite 7)

The Tectonic Graben of Kalavryta (38°1′59.59″ N, 22°6′30.94″ E) is located at the central part of the study area. Kalavryta basin formed during the Early Pliocene due to the N-S extension of the area, being the oldest half-graben of the Corinth Rift (Figure 2E). During the first stages of the basin development, lakes were formed. In these lacustrine environments, layers of clays and marls were deposited, with intercalated lignite layers rich in fossil plants [54]. Alluvial fun sediments fill in the basin upwards [55].

The Kalavryta basin depicts the post alpine geological history of the study area (score of sub-criterion 1.1. 7.5). It is the oldest basin formed due to the continuous rifting process of the Corinth Gulf (score of sub-criterion 1.2. 5). Both the tectonic mechanism of the basin formation and the deposition of the oldest post-alpine formations above the alpine basement are important geological processes that have taken place (score of sub-criterion 1.3. 7.5). Thus, the tectonic graben of Kalavryta is considered unique (score of sub-criterion 1.4. 10). However, it is strongly deteriorated by human activity, since an entire town has been built on it, and has no protection status (score of sub-criterion 1.5. 2.5).

The ornithological value of the area of Kalavryta is characterized high, along with Vouraikos gorge and Chelmos Mt. (score of sub-criteria 2.1. and 2.2. 5 and 1, respectively).

Kalavryta is one of the most historic settlements in Greece, as it is inextricably linked to both the Greek War of Independence (against Ottoman Empire) of 1821 and the German occupation during World War II. Furthermore, the Folklore and Historic Museum and the Museum of the Kalavryta Holocaust (dedicated to the history of the Massacre of Kalavryta in 1943) are located in the city of Kalavryta. Finally, the local train station is the final stop of the historic "Odontotos" rack railway (score of 3.3. 10, while the other cultural sub-criteria were ranked as 1).

The tectonic graben has at least four viewpoints (score of sub-criterion 4.1.–10), and as a low relief surrounded by high peaks it is easily discernable in the landscape (score of sub-criterion 4.2. 7.5).

According to the Kalavryta Hotels Association and the Ski Center of Kalavryta, the number of visitors usually exceeds 75,000 annually (score of sub-criterion 5.1. 10), and thus is considered an international attraction (score of sub-criterion 5.2. 10) but with no official protection (score of 5.3. 10).

Even though there is very intense human activity in this area (score of sub-criterion 6.1. 10), no negative impact on the geosite has been recorded (score of sub-criterion 6.2. 10), since it has a very high degree of resilience (score of sub-criterion 6.3.–1) in relation to potential degradation (score of sub-criterion 6.4. 10). The place can be reached by car along a well-established road network (score of sub-criterion 6.5. 10).

4.3. Cave of the Lakes (Geosite 11)

The Cave of the Lakes (37°57'34.73" N, 22°8'24.15" E) is located close to Kastria village, at an altitude of 827 m. It extends to the Amolinitsa Mt, along a NW-SE trending fault. The cave develops in Cretaceous limestones of the Tripolis and Pindos zones separated by a fault. The cave is characterised by a relatively small width and a great roof height. The total length is 1950 m, and its elevation is 85 m, covering an area of 20,000 m^2. Apart from the rich speleothems, its most impressive feature is the existence of 13 successive underground lakes, which are located at different levels. These lakes were created due to slow water flow and water stagnation, resulting in the formation of calcitic walls (gours or rimstones) which continue to grow until today. Thus, two types of geological history are combined in this geosite (score of sub-criterion 1.1. 2.5 and score of sub-criterion 1.2. 7.5) and a variety of geological processes took place for the formation of its limestones (score of sub-criterion 1.3. 7.5).

Excavations that took place in the first part of the cave found rich archaeological and paleoanthropological remains showing that the cave was inhabited since 5650 BC, from the Neolithic to the Late Helladic period. The length of the touristic route in the Cave is 500 m (Figure 2C).

This geosite is unique in relation to those that have been recognized in the geopark (score of sub-criterion 1.4. 10). The interventions in the cave are considered minimum since all tourist facilities have been built under strict protection measures (score of sub-criterion 1.5. 7.5).

In the non-touristic part of the cave, one of the most important across Europe, a winter colony of bats of the species *Miniopterus schreibersi* [(18,000 individuals), has been recorded (Life Grecabat project) along with nine more bat species (score of sub-criterion 2.1. 10). It is also located within the protected area (Zone B1-SCI_GR2320009) of the Chelmos-Vouraikos National Park (score of sub-criterion 2.2. 10).

According to the myth [56–58], the daughters of the King of Tiryns Proetus, found shelter in a cave on Mount Aroania, north to Nonacris, when they were preoccupied by insanity, for punishment because they insulted the gods [59] (score of sub-criteria 3.1., 3.2., 3.3., 3.4., 1, 1, 5, 1 respectively).

This geosite is underground, it cannot be seen from anywhere except the small natural entrance point that can be seen from the road (score of sub-criterion 4.1. 10). The surface exposure does not differ from the background (score of sub-criterion 4.2. 10).

The visitors to the geosite exceeded 50,000 in 2018 according to official data (score of sub-criterion 5.1. 7.5). Thus, it is considered as a geosite of international level (score of sub-criterion 5.2. 10). It is also located within a protected area (Zone B1) of the Chelmos-Vouraikos National Park (score of sub-criterion 5.3. 1).

Today, the access to tourists is allowed under strict protection measures (score of sub-criterion 6.1. 2.5). The only negative effect is its oxidation caused by breathing (score of sub-criterion 6.2. 2.5). The degree of resistance of its physical characteristics is thus considered high (score of sub-criteria 6.3., 6.5. 10 and 1, respectively). In addition, it is only 17 km from the town of Kalavryta which facilitates access to it (score of sub-criterion 6.4. 10).

4.4. Aroanios Springs (Geosite 14)

The springs of Aroanios River (a tributary of Ladon River) are located (37°56'1.34" N, 22°9'57.82" E) near the village of Planitero, at an altitude of 600 m. The springs are supplied by waters from Chelmos Mt which are discharged underground through the sinkholes in the Loussoi polje area (geosite 23). At the northeastern part of the polje a

SE-NW oriented normal fault brings in contact Pindos zone limestones with limestones of Tripolis zone. As a result, the water springs out the Upper Cretaceous limestones all year round from 41 small fault-overflow springs (4 m^2/s of water) [60] (Figure 2D) (score of sub-criterion 1.1. 1, score of sub-criterion 1.3. 2.5). It is considered as a place with a very high degree of geoheritage representativeness (score of sub-criterion 1.2. 10). This geosite along with Geosite 15 (Mati tou Ladona) are the only geosites of this kind in the study area (score of sub-criterion 1.4.–7.5). The existing state of conservation of the geosite is characterized as moderately deteriorated, due to lumbering (score of sub-criterion 1.5. 5).

The springs are located within the most extensive floodplain forest of the geopark, dominated by plane trees (*Platanus orientalis*), which however has been infected by the "metachromatic ulcer" disease due to anthropogenic impact [61,62] (score of sub-criterion 2.1.–5). In fact, the whole forest is not protected (except at the springs) (score of sub-criterion 2.2.–2.5).

Endemic species of fishes inhabit in Aroanios River, (e.g., *Salmo magrostigma*, *Barbus peloponnesius*). The area is known for the fish hatchery facilities and traditional watermills. Furthermore, part of the E4 European path crosses this geosite (score of all cultural criteria are 1 except of 3.4. that was scored with 2.5).

The springs are not visible from any other place, due to the dense plane trees (score of sub-criterion 4.1. 7.5). This geosite fully harmonizes with the rest of the beautiful landscape (score of sub-criterion 4.2. 5).

Planitero is a popular destination for tourists all year round but especially during the summer and autumn months (score of sub-criterion 5.1. 7.5). It is considered an attraction of national importance (score of sub-criterion 5.2. 7.5). The geosite is located within the Natura B2 zone "Aroanios Springs" and thus its protection status is considered satisfactory (score of sub-criterion 5.3. 1).

Around the springs there are restaurants (score of sub-criterion 6.1. 7.5). Downstream fish farms and old watermills exist, but they do not have negative impact on the springs (score of sub-criterion 6.2. 7.5). It is considered a geosite of moderate resistance to possible degradation (score of sub-criterion 6.3. 2.5). Thus, changes are not acceptable without damage risk (score of sub-criterion 6.5. 7.5). Its accessibility is considered easy, through a regional road, less than 25 km from Kalavryta town (score of sub-criterion 6.4.–1).

4.5. Tsivlos Lake (Geosite 19)

Tsivlos Lake is located at the northeastern part of Chelmos-Vouraikos UGGp (38°4′37.61″ N, 22°13′58.24″ E). The lake was created in 1913, due to the damming of Krathis river after a large landslide, triggered by the preceding strong rainfalls that destabilized the intensely tectonized rocks of the area, blocked the riverbed [63] (Figure 2F).

Tsivlos lake depicts a unique geological history, with its formation from a landslide (score of sub-criterion 1.1. 1, score of sub-criterion 1.3. 7.5). In the geopark there is no other permanent natural lake (score of sub-criterion 1.2. 10, score of sub-criterion 1.4. 10). The geosite is kept intact, although it is used by visitors as a recreation area (score of sub-criterion 1.5. 10).

Due to the young age of the lake, the riparian vegetation is still not well developed. It is surrounded by densely forested landscapes with mixed Mediterranean coniferous trees (*Pinus halepensis*, *Pinus nigra* and *Abies cephalonica*) (score of sub-criterion 2.1. 5). The state of protection and conservation of the lake is considered very high (score of sub-criterion 2.2. 10). Two georoutes have this geosite as a starting point (score of cultural criteria are 1, except of 3.3. 2.5).

This beautiful landscape can be accessed through the local road network (Feneos–Akrata) (score of sub-criterion 4.1. 2.5), with a high differentiation from the background (score of sub-criterion 4.2. 7.5). It receives more than 75,000 visitors each year (score of sub-criterion 5.1. 10) and is thus considered an international attraction (score of sub-criterion 5.2. 10). The lake belongs to a Natura 2000 B3 protection zone of Chelmos-Vouraikos National Park (score of sub-criterion 5.3. 1).

The use by humans around the lake is characterized weak (score of sub-criterion 6.1. 7.5). The hydroelectric power station, which was built 25 years ago, close to the lake to the east, has no negative impact on the environment (score of sub-criterion 6.2. 10). Due to its high fragility (score of sub-criterion 6.3. 7.5), changes are not acceptable either inside the lake or around it (score of sub-criterion 6.5. 1). It can be reached by a detour of the local Feneos–Akrata road (score of sub-criterion 6.4 7.5).

4.6. Water of Styx (Geosite 20)

Water of Styx (or Mavroneri) waterfall (37°59′0.32″ N, 22°12′18.18″ E) is located on Mount Chelmos. It develops on thick bedded to massive Jurassic limestones of Tripolis zone, and flows down a more than 200 m high cliff, draining the eastern side of Neraidorachi limestone plateau (Figure 2B). It constitutes one of the springs of Krathis River. Near the base of the waterfall there is a small rock shelter a few meters long. The flow of the water on the limestone colors them black, thus the locals named the waterfall after this (Mavroneri = black water). Furthermore, to the northeast, following the path that leads to the geosite, two successive low-angle thrusts can be clearly seen deforming the limestone. Thus, the Water of Styx geosite presents a combination of two types of geological history (score of sub-criterion 1.1. 2.5, score of sub-criterion 1.3. 7.5). The representativeness of the geoheritage of the geopark is high (score of sub-criterion 1.2. 7.5). This geosite is one of the three waterfalls in the study area (score of sub-criterion 1.4. 5). The place is intact by both human activity and natural processes (score of sub-criterion 1.5. 10).

Concerning the ecological features of the geosite, many (at least eight) local endemic species of flora are found (score of sub-criterion 2.1. 7.5). Most of the endemic taxa belong to one of the IUCN risk categories, while *G. stygia* (named after the Styx myth and the locality) is a priority species for conservation in the EU (Annex II 92/43/EEC). The state of actual protection and conservation is very high (score of sub-criterion 2.2. 10).

This geosite presents significant intangible heritage being a site of great mythological value since it has been connected with several myths related to the ancient Greek gods (Iris, the daughter of Uranus and Tethys and the great oaths of gods and humans above Styx waters, Thetis the mother of Achilles and the famous "Achilles heel" mentioned by Homer and others as well). Due to the morphology of the landscape and the difficulty to access it, Ancient Greeks believed that the water of Styx was a source of immortality [64] (score of cultural criteria are 1, except of 3.3.–10).

The geosite, due to the black colored limestone is prominent from two points on the regional road network (score of sub-criterion 4.1 7.5). A third place is the geosite "Balcony" of Styx (geosite 39). This geosite shows a slight change in color compared to the rest of the background (score of sub-criterion 4.2 2.5). It is also prominent due to the steep morphology of the cliff.

Visitors do not exceed 5000 per year (score of sub-criterion 5.1. 1) and so it is considered a national attraction (score of sub-criterion 5.2. 5). The geosite is part of the A1 protection zone (SCI_GR2320002) of the National Park (score of sub-criterion 5.3.–1).

The intensity of use is considered weak and is related to the grazing of cattle during the summer months (score of sub-criterion 6.1. 7.5). This use causes a low degree of visual alteration of the landscape (score of sub-criterion 6.2 7.5). The degree of durability is considered very high (score of sub-criterion 6.3. 1), so the changes are acceptable without the risk of landscape degradation (score of sub-criterion 6.5. 1). There are three tracks to approach the site (score of sub-criterion 6.4 7.5).

4.7. Mavrolimni (Geosite 24)

Mavrolimni, is a glacial seasonal lake, located on Mount Chelmos at an altitude of 2060 m (37°58′42.43″ N, 22°12′5.15″ E) on bedded limestones of Tripolis zone. To the north of the lake, a moraine ridge was formed due to the movement of glaciers from Psili Korfi or Neraidorachi downstream (Figure 3A). According to Pope et al. [42], the formation of the moraine took place during the Middle Pleistocene. The glacial sediments blocked the drainage network resulting in the formation of the alpine lake.

Figure 3. Representative views of the geosites described in the results. (**A**) Mavrolimni geosite 24, (**B**) Mega Spilaio, geosite 28, (**C**) Ntourntouvana, geosite 35, (**D**) Eroded conglomerates, geosite 38.

Mavrolimni depicts a combination of two types of geological history, (score of sub-criterion 1.1.–2.5). In terms of its representativeness, it is considered low in relation to the geoheritage of the entire study area (score of sub-criterion 1.2. 2.5). The geological features and processes in this geosite consist of evidence for the existence of glaciers during the last glacial periods and the preservation of glacial geomorphs such as the seasonal alpine lake (score of sub-criterion 1.3. 5). In the past, there were other smaller lakes of this type on Mount Chelmos which today are dry or destroyed and thus Mavrolimni is recognized as the only geosite of this kind (score of sub-criterion 1.4.–10). As for the current state of preservation from human activity and natural processes, this geosite remains intact (score of sub-criterion 1.5. 10).

Rare endemic species of flora (score of sub-criterion 2.1. 10), such as *Aquilegia ottonis* subsp. *ottonis, Achillea umbellata, Dianthus tymphresteus, Saxifraga sibthorpii* have been identified around the lake. It is also characterized by high ornithological value. The actual protection status is considered very high throughout the area (score of sub-criterion 2.2. 10).

No important cultural features are associated with this geosite (all cultural criteria are ranked 1). The lake can be seen from the neighboring high peaks (e.g., Neraidorachi) (score of sub-criterion 4.1. 5). It differs to a very high degree from the background due to its aquatic nature in relation to the surrounding rocky environment (score of sub-criterion 4.2. 10).

Visitors do not exceed 5000 per year, as access to the geosite is only possible after hiking on a demanding track (score of sub-criterion 5.1 1) and therefore, it is considered as a local attraction (score of sub-criterion 5.2. 2.5). The official protection of this geosite is international since it belongs to a NATURA 2000 A1 zone (High Peaks of Mount Chelmos-SCI_GR2320002 and SPA_2320013) of Chelmos-Vouraikos National Park (score of sub-criterion 5.3. 1).

There is no use of the site by humans (score of sub-criterion 6.1. 10) and therefore there is no negative impact (score of sub-criterion 6.2. 10). The degree of fragility is considered very high (score of sub-criterion 6.3. 10). No changes are acceptable without the risk of landscape degradation (score of sub-criterion 6.5. 1). The site can be approached only on foot through two different tracks (score of sub-criterion 6.4. 1).

4.8. Mega Spilaio (Geosite 28)

Mega Spilaio geosite is located on the eastern flank of Vouraikos gorge (38°5′25.80″ N, 22°10′28.32″ E) and includes highly elevated peaks (maximum altitude more than 1400 m). The imposing rocks of Mega Spilaio are mainly composed of conglomerates with steep slopes that in some cases form even vertical cliffs with an altitude difference from the riverbed of more than 800 m (Figure 3B). The conglomerates were formed in alluvial fan systems in two phases during the Middle Pliocene to Middle Pleistocene. In a sandy clay horizon, fossil bones of large mammals were found. The presence of alluvial fan sediments in such a high altitude is attributed to the tectonic setting of the wider area, the secondary extensional tectonism of the Corinth rift and the resulting great uplift rates [41].

Hence, Mega Spilaio geosite combines different types of geological history (score of sub-criterion 1.1. 5, score of sub-criterion 1.3. 7.5).

Its representativeness for the geoheritage of the geopark is considered moderate (score of sub-criterion 1.2. 5). Kerpini conglomerates geosite (geosite 5) is also a geosite of this category (score of sub-criterion 1.4. 7.5). This outcrop is moderately deteriorated by human activity (construction of the monastery) and natural processes (cavities from the karstification) (score of sub-criterion 1.5.–5).

Mega Spilaio geosite is a place of very high ecological importance (score of sub-criterion 2.1. 10), because of the existence of the stenotopic endemic chasmophyte species *Silene conglomeratica*, located exclusively in cracks of these specific conglomerates, denoting this way the inseparable relationship between biodiversity and geodiversity. Part of the geosite close to the monastery is protected, while the surrounding outcrops are not (score of sub-criterion 2.2. 5).

The historic monastery of Mega Spilaio was built around 362 AD, in a cavity that had been formed by the erosion of the conglomerates, by two monks and is considered the oldest monastery in Greece. During the 1821 Greek War of Independence, the Monastery was a beacon of resistance against the Otomans (score of sub-criterion 3.2. 10). Every August 15, the monastery celebrates the memory of the Assumption of Virgin Mary (score of sub-criterion 3.1. 2.5).

Due to the large thickness of the sequence, the high and steep slopes and the location of the geosite in higher topographies than the surrounding area, it is easily distinguishable from the regional road Pounta-Kalavryta, from the village of Zachlorou, as well as from the surrounding mountain peaks (score of sub-criterion 4.1. 10). This impressive landscape differs in relation to the background if we consider the canyon of Vouraikos River as the local background (score of sub-criterion 4.2. 7.5).

Visitors exceed 75,000 per year (score of sub-criterion 5.1. 10). Thus, it is considered as an international attraction (score of sub-criterion 5.2. 10). Although it is located close to the NATURA 2000 A3 zone of Vouraikos Gorge, the monastery and the respective outcrops are not included within the protected area. However, the monastery has been declared a protected archaeological site (score of sub-criterion 5.3. 7.5).

The use by humans is considered weak (score of sub-criterion 6.1. 7.5), as the monastery currently does not inflict any further destruction to the geosite (score of sub-criterion 6.2. 1). The degree of fragility of its physical characteristics is considered low regarding the cohesiveness of the conglomerates (score of sub-criterion 6.3. 2.5). Thus, changes are acceptable without the risk of degrading its features (score of sub-criterion 6.5. 10). The geosite is easily accessed through the local Pounta-Kalavryta road, just 10 km before the town of Kalavryta (score of sub-criterion 6.4. 10).

4.9. Ntourntouvana (Geosite 35)

Ntourntouvana (or Pentelia) Peak is located (37°54′51.72″ N, 22°15′11.17″ E) at the southern margins of Mount Chelmos, standing at an altitude of 2109 m (Figure 3C). It develops on Cretaceous neritic limestones of Tripolis zone which are in tectonic contact with thin-bedded white dolomites. Rudists and other bivalve shells can be macroscopically observed in the limestone outcrops. In these intensively karstified limestones one of the largest and most important potholes of the geopark area the "Hole of Feneos" (maximum depth 130 m) is developing. This geosite therefore reflects a large part of the geological history of Mount Chelmos (score of sub-criterion 1.1. 5, score of sub-criterion 1.3. 10).

It is a place with high representativeness for the geoheritage of the geopark (score of sub-criterion 1.2. 7.5). Two more high peaks of Chelmos Mt have been characterized as geosites (Geosite 34: Psili Korfi and Geosite 37: Madero) (score of sub-criterion 1.4. 5). The condition of the geosite, as far as both human activity and natural processes are concerned, is characterized as intact (score of sub-criterion 1.5. 10).

It is a geosite rich in endemic species of mountain-Mediterranean meadows (genus *Nardus*), chasmophytic vegetation and mountain tea (*Sideritis sclandestina* subsp. *peloponnesica* endemic to Peloponnese) (score of sub-criterion 2.1. 5). The existing protection and conservation status are high throughout the Ntourntouvana area (score of sub-criterion 2.2. 10).

This geosite does not display important features of cultural heritage (all cultural criteria are scored 1). Due to its high altitude, this geosite can be observed from many surrounding localities (score of sub-criterion 4.1. 10). Due to the absence of woody plants at Ntourntouvana Peak, the geosite differs from the surrounding area, where forests of black pines and firs grow (score of sub-criterion 4.2. 7.5).

Visitors do not exceed 5000 per year and are limited to mountaineers who hike across its tracks (score of sub-criterion 5.1. 1). Thus, it is considered as a local attraction (score of sub-criterion 5.2. 2.5). Officially, it is included in zone C of the NATURA 2000 area of Chelmos-Vouraikos National Park (score of sub-criterion 5.3. 1).

There is no use of the area by humans (score of both sub-criteria 6.1., 6.2. 10). Due to the rocky substrate, it does not have any degree of fragility (score of sub-criterion 6.3. 1). Therefore, small changes are acceptable without the risk of degrading its physical characteristics (score of sub-criterion 6.5. 10). There are two tracks to approach the site (score of sub-criterion 6.4. 1).

4.10. Eroded Conglomerates (Geosite 38)

East of the monastery of Megalo Spilaio, the Eroded Conglomerates geosite is developing (38°5′0.37″ N, 22°11′4.26″ E). It is composed of different Lower-Middle Pleistocene cohesive conglomerates. They were deposited as fan deposits of larger or smaller rivers [54]. Due to the high-altitude, strong winds dominate the microclimate. They erode the cohesive conglomerates forming surface geomorphs, resulting in an impressive landscape (Figure 3D). The Eroded conglomerates geosite depicts a combination of two types of geological history (score of sub-criterion 1.1. 2.5).

It is not considered representative for the geoheritage of the geopark (score of sub-criterion 1.2. 1), while geodiversity is characterized as low (score of sub-criterion 1.3. 2.5). Along with the Kerpini Conglomerates geosite, they are considered as two geosites depicting similar processes (score of sub-criterion 1.4. 7.5). The continuous erosion of wind and water deforms slowly the geosite. However, this process is considered slow and natural. Moreover, this is what characterizes this geosite. For this reason, it is characterized as weakly deteriorated (score of sub-criterion 1.5. 7.5).

The ornithological value of the site is considered important (score of sub-criterion 2.1. 10), however, the area is not protected (score of sub-criterion 2.2. 1).

No cultural heritage features are related to this specific geosite (all cultural criteria are scored 1). This geosite can be seen from two points on the local road network (score of sub-criterion 4.1. 5). It shows a moderate difference from the rest of the landscape due to the presence of fir trees (score of sub-criterion 4.2. 5). The visitors are limited to those who hike across the Kalavryta to Mega Spilaio georoute (total length 36.5 km). They do not exceed 5000 per year (score of sub-criterion 5.1. 1), and thus it is considered a local attraction (score of sub-criterion 5.2. 2.5). There is no official protection status for this geosite (score of sub-criterion 5.3. 10).

Its use is characterized as moderate due to grazing as well as to the presence of small farming facilities for animal watering (score of sub-criterion 6.1. 5), which alter the aesthetics of the landscape (score of sub-criterion 6.2. 5). In addition, this is a geosite with low fragility (score of sub-criterion 6.3. 2.5) and therefore changes are acceptable (to a moderate degree) without the risk of landscape degradation (score of sub-criterion 6.5. 5). The geosite can be accessed either using the aforementioned track, or from the Waters of Styx-Mega Spilaio georoute. Furthermore, visitors can approach the place by car using a forest road (3.6 km). However, this road is not free to public, and passage is allowed only after a special permit (score of sub-criterion 6.4. 2.5).

4.11. Synthesis of Results

As far as the scientific criteria are concerned, the scores range from 2.1 to 9.5. More specifically, the highest values are presented by Geosite 1 (Niamata) (9.5) and Geosite 23 (Lousoi polje) (8.5). This high score is attributed to the depiction of a large part of the geological history of the study area.

The ecological criteria cover the full scores' range (1 to 10). Geosites 2 (Portes–Triklia), 5 (Kerpini conglomerates), 11 (the Cave of the Lakes) and 24 (Mavrolimni) present the highest score (10). Geosites 20 and 34 (Water of Styx and Psili Korfi, respectively) are classified slightly lower (8.8). These geosites present very high ecological importance in combination to their high protection status.

The rating of the cultural criteria ranges from low to medium (1 to 5.9). More specifically, Geosite 11 (Mega Spilaio) presents the highest score (5.9) because of its very high religious interest. Geosite 8 (Agia Lavra) follows (5.5) with moderate historical and religious interest. Geosite 17 (Doxa lake) with 4.6 is also at the same score level mainly because of its religious and historical significance. Finally, Geosite 33 (Pausanias Vine), with 3.6, is considered a geosite of mainly historical interest.

The aesthetic criteria score shows a wide range as well (1 to 10). Geosites 11 and 17, Cave of the Lakes and Doxa Lake, respectively, present the highest value, (10). Geosite 11 (Cave of the Lakes), although it cannot be observed from other locations being underground, is one of the most popular geosites of the geopark with amazing speleothems. Geosite 17 (Lake Doxa) is highly visible from many viewpoints of the surrounding area and has a very high landscape diversity. Geosites 2 (Portes–Triklia), 7 (Tectonic Graben of Kalavryta), 28 (Mega Spilaio) and 35 (Ntourntouvana) present also a very high score (8.8).

The score of the economic criteria varies from 1 to 10. Geosite 7 (Tectonic Graben of Kalavryta) has been scored with 10. Geosite 28 (Mega Spilaio) also presents a very high score (9.2). The exceptionally high scores of the aforementioned geosites are mainly attributed to their very high visibility.

The score of the criteria for potential use ranges from 3.9 to 9.5. Geosite 13 (Lousoi sinkholes) bear the highest score for this criterion (9.5), followed by geosites 7 and 36 (Tectonic graben of Kalavryta and Chelonospilia).

As suitable geosites for educational activities are considered those geosites with values of V_{edu} exceeding 6 (Figure 4). These are (in descending order) the geosites Portes-Triklia (C 02, V_{edu} = 7.3), the Cave of the Lakes (C 11, V_{edu} = 7.2), Mega Spilaio (C 28, V_{edu} = 6.8), Ntountouvana (C 35, V_{edu} = 6.5), Niamata (C 01, V_{edu} = 6.4), Doxa lake (C 17, V_{edu} = 6.4) and Mavrolimni (C 24, V_{edu} = 6.1). Geosites with moderate educational value ($4 \geq V_{edu} \leq 6$) include the geosites (in descending order) Water of Styx (C 20, V_{edu} = 6), Lousoi polje (C 23, V_{edu} = 5.9), Tsivlos lake (C 19, V_{edu} = 5.9), the Tectonic Graben of Kalavryta (C 07, V_{edu} = 5.6), Psili Korfi (C 34, V_{edu} = 5.5), Kerpini conglomerates (C 05, V_{edu} = 5.1), Water of Styx (C 20, V_{edu} = 5.1), Agia Lavra (C 08, V_{edu} = 4.7), Madero (C 37, V_{edu} = 4.7), Valvousi (C 26, V_{edu} = 4.5), Keramidaki (C 27, V_{edu} = 4.5), Aroanios springs (C 14, V_{edu} = 4.4), Pausanias Vine (C 33, V_{edu} = 4.1), Mati tou Ladona (C 15, V_{edu} = 4) and the Eroded conglomerates (C 38, V_{edu} = 4). Geosites Ntourntouvana, Psili korfi, Water of Styx and Madero remain unused for educational purposes mainly as access to these sites is difficult.

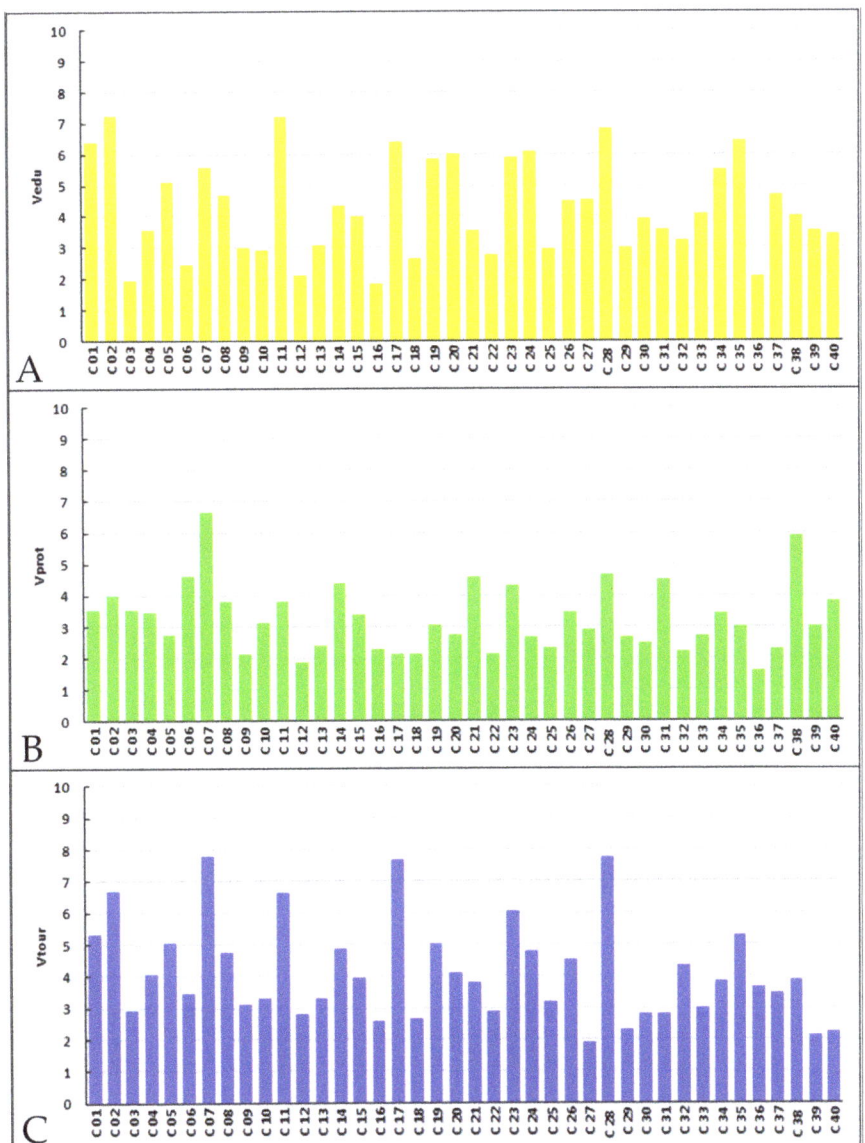

Figure 4. Diagrammatic representation of the results of the calculation of the three indices resulting from the assessment of Chelmos-Vouraikos UGGp geosites, (**A**) V_{edu}, (**B**) V_{prot}, (**C**) V_{tour}.

Regarding the geotouristic value of the geosites index V_{tour} is considered (Figure 4). The highest values are presented by Tectonic Graben of Kalavryta (C 07, V_{tour} = 7.8), Mega Spilaio (C 28, V_{tour} = 7.7), Doxa lake (C 17, V_{tour} = 7.7), the Cave of the Lakes (C 11, V_{tour} = 6.7), Portes–Triklia (C 02, V_{tour} = 6.7) and Lousoi polje (C 23, V_{tour} = 6). Lower values (4 ≥ V_{tour} < 6) are presented by the following geosites (in descending order): Niamata (C 01, V_{tour} = 5.3), Kerpini conglomerates (C 05, V_{tour} = 5.1), Tsivlos lake (C 19, V_{tour} = 5), Aroanios springs (C 14, V_{tour} = 4.9), Agia Lavra (C 08, V_{tour} = 4.8), Mavrolimni (C 24, V_{tour} = 4.8), Valvousi (C 26, V_{tour} = 4.5), Valimi landslides (C 32, V_{tour} = 4.3), Ntourntouvana (C 35,

V_{tour} = 5.3), Waters of Styx (C 20, V_{tour} = 4.1), Trapeza Marine terrace (C 04, V_{tour} = 4.1) and Mati Tou Ladona (C 15, V_{tour} = 4). Ntourntouvana and Water of Styx geosites are not yet touristically exploited due to difficult access.

Geosites that according to the V_{prot} index, show the greatest need for protection are the Tectonic Graben of Kalavryta (C 07, V_{prot} = 6.7) and the Eroded conglomerates (C 38, V_{prot} = 5.9), (Figure 4). The need for protection of the Tectonic Graben of Kalavryta Geosite concerns the very high human activity (farming, industry, etc.) which is spotted in the wider area. The Eroded conglomerates geosite needs protection from overgrazing that has caused alteration in the landscape.

As far as the ecological risk factor (F_{ecol}) is concerned, the highest value is held by the Eroded conglomerates geosite (C 38, F_{ecol} = 10). The F_{ecol} is also particularly high in the Tectonic Graben of Kalavryta (C 07, F_{ecol} = 5). Therefore, there is an urgent need for official protection in these geosites.

5. Discussion

Over the last decades, geoconservation [65] approaches have become very popular, contributing significantly as an important component of nature conservation practices [66]. Deterioration of the environment driven by the constantly increasing human pressure to our planet has made clear the need to record, protect and promote not only biodiversity but geodiversity as well [67]. Preserving geological heritage has thus become a key factor for future legislation and policies that would allow the more effective management of the natural environment through the protection of geosites [68]. To achieve these goals, geosite conservation practices need to be implemented to limit anthropogenic and natural deterioration or destruction [69]. Nevertheless, to plan or take specific geoconservation measures for geosites and particularly in places where geoconservation can be enhanced such as geoparks, geosite assessments need to be implemented first to identify their value, possible threats and the need for protection [69,70].

Except for geoconservation, geoparks in their effort to promote sustainable development and economic benefits for local communities through geotourism and education, have also as main goals the connection of nature with people and the connection of geodiversity with biodiversity, cultural heritage and local communities [68,71,72]. Through geotouristic and educational activities organised by the geoparks, geoscientific knowledge and geoconservation concepts are transferred to the public [72].

To succeed in all these, a more holistic approach for the sustainable management of geoparks is required, that will combine sustainable development activities with effective geoconservation which will promote geoheritage values [71]. The first step, however, before geopark management approaches are developed, is the assessment of a geopark's geosites where scientific, cultural, ecological and economic criteria are considered. Thus, employing a method for the assessment of geosites such as the Fassoulas et al. [19] method, which was particularly developed to meet the needs of UGGps, is considered as a good starting point for the management planning of the Chelmos-Vouraikos UGGp.

The evaluation method of Fassoulas et al. [19] has provided very detailed information regarding the value of each geosite and its needs. It takes into account all the existing parameters so that a geosite can be scored impartially and completely. The large range of the grading system (1 to 10) provides the opportunity for a more detailed assessment of the criteria.

Nevertheless, during the evaluation of the results after the application of this methodology, some minor discrepancies became apparent. Concerning the aesthetic criteria (criteria 4.1. and 4.2.), in the case of caves and generally underground geosites, some improvements are certainly needed. We have to stress here that during the development and testing of the methodology [19], no caves were used under the assessment process and thus no certain specifications are provided for their scoring. In our case, particular caves present low visibility and for this reason they bear a low score on this criterion. However, these geosites are often very popular and aesthetically enhanced and thus this criterion scoring gives a misleading impression. A characteristic example is the Cave of the Lakes geosite. Hence, application of another evaluation methodology which was created for karst systems, by Li et al. [22] was put under consideration. However, this methodology was difficult to be carried out in this case, due to the fact that a lot of different geosites, such as karstic geomorphs, gorges and rivers, should be taken into account as well. Therefore, the above method could have been applied only for these geosites, since most of the geosites of Chelmos-Vouraikos UGGp do not fall into the above category. Thus, this method is not appropriate for the evaluation of all geosites of the geopark but just for a number of them. To overpass the problem with Fassoulas et al. [19] methodology, if criteria 4.1 and 4.2 are treated for caves under their strict meaning (number of view points, or landscape difference), we should consider the underground aesthetic image in a broader sense as it happens with the open air. Thus, the variety of the cave system (number of halls, length, different levels) could count as the number of viewpoints, whereas the landscape difference could, in this case, represent the wealth of speleothems (i.e., Stalagmites, stalactites, gours, curtains, etc.).

A similar problem is encountered in the cultural criteria which are divided into four sub-criteria. A geosite may bear a high score in one sub-criterion because of its exceptional value, however it could be scored low in the rest. As a result, the final score of this criterion is low for this specific geosite even if the value of one of the sub-criteria is exceptionally high. A characteristic example is the Water of Styx geosite which has a very important mythological heritage but has low values of ethics, history and art. It thus presents a final low score on this criterion which undermines the undeniably high cultural value of this geosite. Thus, a more careful inspection and consideration in the broader sense of each of the cultural sub-criteria separately must be taken into account when implementing this methodology, or maybe the introduction of an additional one that could refer to intangible cultural heritage might also be useful.

Finally, the scoring for the number of visitors is easy to be calculated in the case that ticket offices exist (e.g., Cave of the Lakes). There, the number of visitors is recorded with accuracy. However, in geosites where the number of visitors cannot be calculated with absolute values but only approximately, the score of this criterion is questionable. Thus, an additional visitor's estimation system has to be considered under scoring.

The criteria focus on a regional level, which helps to identify the priorities of the geopark. Geosites that have high values in the touristic or educational index should be utilized in corresponding activities. Similarly, the geosites that have a high index for the need of protection are the ones that face higher risks, and it is deemed necessary to implement actions to protect them. Geosites with great touristic and educational value that are already exploited for corresponding activities, it is necessary to be maintained at this level. Conversely, in geosites with high touristic and education importance that are not exploited to date, actions such as opening of new paths or construction of funiculars etc, must be taken, so that they can be approached safely by tourists and students. Characteristic examples are the geosites Water of Styx and Psili Korfi at Chelmos Mt. According to their assessment they are geosites of great interest that certainly need enhancement. For this reason, the Management Body of Chelmos-Vouraikos, after evaluating the results of the assessment, has already planned during the following year a series of interventions for all the geosites found on Chelmos Mt (including Water of Styx and Psili Korfi) to improve accessibility to the geosites and enhance their promotion including activities in collaboration with the local Ski Center, even for people that will not be able to access the actual sites. This is a classic example of how the assessment of geosites can help the managing authorities to prioritize the needs of their geoparks. Moreover, in geosites with high protection-need, more intense protection measures need to be taken. For instance, livestock and respective facilities should be removed, as well as restrictions on crop cultivation within these geosites should be applied.

The immediate next aim of this research is to improve the existing evaluation method. Additionally, an average score for each index or criteria could be calculated for all the geosites of the evaluated geopark. If such a methodology is implemented by different geoparks, a comparison between them would be possible. This comparison would be helpful for the further development of all geoparks, which have as their main objective the conservation of the geological heritage and the promotion of sustainable development. In addition, by using a single rating system by different geoparks, it would be possible to propose response measures for low-scored geosites. Utilization of the above measures would be helpful for other geoparks as well facing similar difficulties.

6. Conclusions

In conclusion, the quantitative assessment of geosites in Chelmos-Vouraikos UGGp revealed the abundant possibilities for educational, scientific and touristic activities that the geosites can offer, such as the understanding of the Corinth rift and the thrusting of the Pindos nappe, the observation of one of the oldest formations of the Peloponnese, the creation of an impressive lake due to a catastrophic landslide as well as many other geological processes. Apart from the geological processes, the flora and fauna as well as the cultural features of the study area are rich. However, due mainly to the extensive livestock breeding in the geopark area, many geosites are endangered with degradation, both in their aesthetics and in their characteristics.

Most geosites with calculated high educational value are already used for corresponding activities from the managing authorities of the geopark. Those with a very high V_{edu} index have great improvement potential for Chelmos-Vouraikos UGGp. Geosites with high touristic value, which so far are not fully exploited, have highlighted the necessity to enhance their touristic development with various actions. Based on the assessment results, the geopark management has already planned actions to develop accessibility and promote the geosites of Chelmos Mt, and particularly the Water of Styx and Psili Korfi that were highly rated. Similarly, geosites with high protection index values (such as the Tectonic Graben of Kalavryta), are considered as places with a great need for protection. Sometimes protection measures such as removing/reducing livestock activities and facilities around geosites are easy to be taken, whereas in other cases protection measures are difficult because of the very high human activity.

Testing the methodology of Fassoulas et al. [19] in the area of Chelmos-Vouraikos made possible the identification of some malfunctions and elaboration problems related to caves, intangible heritage and number of visitors used in the evaluation criteria. We think that these problems can be solved with further refinement and specifications in the criteria description related with the above cases.

The assessment of geosites like the one presented herein for Chelmos-Vouraikos UGGp, performed either with the methodology we have chosen or a similar one, is considered necessary for all geoparks in order to develop effective and productive geoconservation, geoeducation and geotouristic initiatives [19]. It also highlights the importance and the way of utilization of each geosite, always with respect to nature and the environment, as well as identifying those which need further protection.

Author Contributions: Conceptualization, G.I. and P.P.; methodology, C.F. and G.I.; validation, All Authors; formal analysis, All Authors; investigation, All Authors.; data curation, V.G. and E.K.; writing—original draft preparation, V.G., P.P. and G.I.; writing—review and editing, All Authors; visualization, V.G.; supervision, G.I., N.Z., C.F. and A.Z. All authors have read and agreed to the published version of the manuscript.

Funding: Vasilis Golfinopoulos as an MSc student was financially supported by the "Andreas Mentzelopoulos Scholarships for postgraduate studies at the University of Patras", 33720000.

Institutional Review Board Statement: Not applicable.

Acknowledgments: We would like to thank the personnel of Chelmos-Vouraikos UNESCO Global Geopark for all the help they provided during fieldwork and the collection of data. Also we would like to thank Socrates Tsacos and Irena Pappa for their help during the construction of the geological map of the Geopark.

Conflicts of Interest: The authors declare no conflict of interest.

References

1. Zouros, N. Global Geoparks Network and the New UNESCO Global Geoparks Programme. *Bull. Geol. Soc. Greece* **2017**, *50*, 284. [CrossRef]
2. UNESCO. Available online: http://www.unesco.org (accessed on 2 September 2021).
3. Henriques, M.H.; Brilha, J. UNESCO Global Geoparks: A Strategy towards Global Understanding and Sustainability. *Episodes* **2017**, *40*, 349–355. [CrossRef]
4. Keever, P.J.M.; Zouros, N. Geoparks: Celebrating Earth Heritage, Sustaining Local Communities. *Episodes* **2005**, *28*, 274–278. [CrossRef] [PubMed]
5. Henriques, M.H.; dos Reis, R.P.; Brilha, J.; Mota, T. Geoconservation as an Emerging Geoscience. *Geoheritage* **2011**, *3*, 117–128. [CrossRef]
6. UNESCO. Statutes of the International Geoscience and Geoparks Programme (IGGP). Available online: https://unesdoc.unesco.org/ark:/48223/pf0000234539.locale=en (accessed on 28 December 2021).
7. Catana, M.M.; Brilha, J.B. The Role of UNESCO Global Geoparks in Promoting Geosciences Education for Sustainability. *Geoheritage* **2020**, *12*, 1. [CrossRef]
8. Brilha, J.; Gray, M.; Pereira, D.I.; Pereira, P. Geodiversity: An Integrative Review as a Contribution to the Sustainable Management of the Whole of Nature. *Environ. Sci. Policy* **2018**, *86*, 19–28. [CrossRef]
9. UNESCO. Top 10 Focus Areas of UNESCO Global Geoparks. Available online: https://en.unesco.org/global-geoparks/focus#focus (accessed on 28 December 2021).
10. Vlami, V.; Kokkoris, I.P.; Zogaris, S.; Cartalis, C.; Kehayias, G.; Dimopoulos, P. Cultural Landscapes and Attributes of "Culturalness" in Protected Areas: An Exploratory Assessment in Greece. *Sci. Total Environ.* **2017**, *595*, 229–243. [CrossRef]
11. Pereira, P.; Pereira, D.; Alves, M.I.C.; Pereira, P.; Pereira, D.; Cae-Tano Alves, M.I. Geomorphosite Assessment in Montesinho Natural Park. *Geogr. Helv.* **2007**, *62*, 159–168. [CrossRef]
12. Bruschi, V.M.; Cendrero, A.; Albertos, J.A.C. A Statistical Approach to the Validation and Optimisation of Geoheritage Assessment Procedures. *Geoheritage* **2011**, *3*, 131–149. [CrossRef]
13. Zouros, N. Geomorphosite Assessment and Management in Protected Areas of Greece. The Case of the Lesvos Island—Coastal Geomorphosites. *Geogr. Helv.* **2007**, *62*, 169–180. [CrossRef]
14. Zouros, N.; Valiakos, I. Geoparks Management and Assessment. *Bull. Geol. Soc. Greece* **2010**, *43*, 965. [CrossRef]
15. Vujičić, M.D.; Vasiljević, D.A.; Marković, S.B.; Hose, T.A.; Lukić, T.; Hadžić, O.; Janićević, S. Preliminary Geosite Assessment Model (GAM) and Its Application on Fruška Gora Mountain, Potential Geotourism Destination of Serbia. *Acta Geogr. Slov.* **2011**, *51*, 361–376. [CrossRef]
16. Rivas, V.; Rix, K.; Frances, E.; Cendrero, A.; Brunsden, D. GeomoNrphological Indicators for Environmental Impact Assessment: Consumable and Non-Consumable Geomorphological Resources. *Geomorphology* **1997**, *18*, 169–182. [CrossRef]
17. Reynard, E.; Fontana, G.; Kozlik, L.; Scapozza, C. A Method for Assessing "Scientific" and "Additional Values" of Geomorphosites. *Geogr. Helv.* **2007**, *62*, 148–158. [CrossRef]
18. Bruschi, V.M.; Cendrero, A.; Quaternario, I. Geosite Evaluation; Can We Measure Intangible Values? *Il Quat.* **2005**, *18*, 293–306.
19. Fassoulas, C.; Mouriki, D.; Dimitriou-Nikolakis, P.; Iliopoulos, G. Quantitative Assessment of Geotopes as an Effective Tool for Geoheritage Management. *Geoheritage* **2012**, *4*, 177–193. [CrossRef]
20. Brilha, J. Inventory and Quantitative Assessment of Geosites and Geodiversity Sites: A Review. *Geoheritage* **2016**, *8*, 119–134. [CrossRef]
21. Gajek, G.; Zgłobicki, W.; Kołodyńska-Gawrysiak, R. Geoeducational Value of Quarries Located Within the Małopolska Vistula River Gap (E Poland). *Geoheritage* **2019**, *11*, 1335–1351. [CrossRef]
22. Li, Y.; Li, M.; Ding, Z. Study on Methodology of Assessing Synergy between Conservation and Development of Karst Protected Area in the Case of the Diehong Bridge Scenic Area of Jiuxiang Gorge Cave Geopark, Yunnan, China. *Environ. Dev. Sustain.* **2021**, 1–20. [CrossRef]
23. Kubalíková, L. Geomorphosite Assessment for Geotourism Purposes. *Czech J. Tour.* **2013**, *2*, 80–104. [CrossRef]
24. Kubalíková, L.; Kirchner, K. Geosite and Geomorphosite Assessment as a Tool for Geoconservation and Geotourism Purposes: A Case Study from Vizovická Vrchovina Highland (Eastern Part of the Czech Republic). *Geoheritage* **2016**, *8*, 5–14. [CrossRef]
25. Migoń, P.; Pijet-Migoń, E. Viewpoint Geosites—Values, Conservation and Management Issues. *Proc. Geol. Assoc.* **2017**, *128*, 511–522. [CrossRef]
26. Artugyan, L. Geomorphosites Assessment in Karst Terrains: Anina Karst Region (Banat Mountains, Romania). *Geoheritage* **2017**, *9*, 153–162. [CrossRef]
27. Štrba, Ľ.; Rybár, P. Revision of the "Assessment of Attractiveness (Value) of Geotouristic Objects". *Acta Geoturistica* **2015**, *6*, 30–40.
28. Aoulad-Sidi-Mhend, A.; Maaté, A.; Hlila, R.; Martín-Martín, M.; Chakiri, S.; Maaté, S. A Quantitative Approach to Geosites Assessment of the Talassemtane National Park (NW of Morocco). *Estud. Geológicos* **2020**, *76*, 123. [CrossRef]
29. Albani, R.A.; Mansur, K.L.; de Carvalho, I.S.; dos Santos, W.F.S. Quantitative Evaluation of the Geosites and Geodiversity Sites of João Dourado Municipality (Bahia—Brazil). *Geoheritage* **2020**, *12*, 1–15. [CrossRef]

30. Zangmo, G.T.; Kagou, A.D.; Nkouathio, D.G.; Gountié, M.D.; Kamgang, P. The Volcanic Geoheritage of the Mount Bamenda Calderas (Cameroon Line): Assessment for Geotouristic and Geoeducational Purposes. *Geoheritage* **2017**, *9*, 255–278. [CrossRef]
31. Pralong, J.-P. A Method for Assessing Tourist Potential and Use of Geomorphological Sites. *Géomorphologie Relief Processus Environ.* **2005**, *11*, 189–196. [CrossRef]
32. Dercourt, J. *Contribution a l'Étude Géologique d'Un Secteur Du Péloponnèse Septentrional*; University of Paris: Paris, France, 1964.
33. Koukouvelas, I. *Geology of Greece*; Liberal Books: Athens, Greece, 2018; ISBN 9786185012403.
34. Degnan, P.J.; Robertsoj, A.H. *Mesozoic-Early Tertiary Passive Margin Evolution of the Pindos Ocean (NW Peloponnese, Greece)*; Elsevier: Amsterdam, The Netherlands, 1998; Volume 117, pp. 33–70.
35. Dornsiepen, U.; Gerolymatos, E.; Jacobschagen, V. Die Phyllit—Quartzit-Serie Im Fenster von Feneos (Nord-Peloponnes). *IGME Geol. Geophys. Res. Spec. Issue* **1986**, 99–105.
36. Jolivet, L.; Labrousse, L.; Agard, P.; Lacombe, O.; Bailly, V.; Lecomte, E.; Mouthereau, F.; Mehl, C. Rifting and Shallow-Dipping Detachments, Clues from the Corinth Rift and the Aegean. *Tectonophysics* **2010**, *483*, 287–304. [CrossRef]
37. Kydonakis, K.; Kostopoulos, D.; Poujol, M.; Brun, J.P.; Papanikolaou, D.; Paquette, J.L. The Dispersal of the Gondwana Super-Fan System in the Eastern Mediterranean: New Insights from Detrital Zircon Geochronology. *Gondwana Res.* **2014**, *25*, 1230–1241. [CrossRef]
38. Ford, M.; Hemelsdaël, R.; Mancini, M.; Palyvos, N. Rift Migration and Lateral Propagation: Evolution of Normal Faults and Sediment-Routing Systems of the Western Corinth Rift (Greece). In *Geological Society Special Publication*; Geological Society of London: Londok, UK, 2017; Volume 439, pp. 131–168.
39. Ford, M.; Rohais, S.; Williams, E.A.; Bourlange, S.; Jousselin, D.; Backert, N.; Malartre, F. Tectono-Sedimentary Evolution of the Western Corinth Rift (Central Greece). *Basin Res.* **2013**, *25*, 3–25. [CrossRef]
40. Moretti, I.; Sakellariou, D.; Lykousis, V.; Micarelli, L. The Gulf of Corinth: An Active Half Graben? *J. Geodyn.* **2003**, *36*, 323–340. [CrossRef]
41. Ford, M.; Williams, E.A.; Malartre, F.; Popescu, S.-M. Stratigraphic Architecture, Sedimentology and Structure of the Vouraikos Gilbert-Type Fan Delta, Gulf of Corinth, Greece. In *Sedimentary Processes, Environments and Basins*; Blackwell Publishing Ltd.: Oxford, UK, 2007.
42. Pope, R.J.; Hughes, P.D.; Skourtsos, E. Glacial History of Mt Chelmos, Peloponnesus, Greece. In *Geological Society Special Publication*; Geological Society of London: London, UK, 2017; Volume 433, pp. 211–236.
43. Tsakiri, M.; Koumoutsou, E.; Kokkoris, I.P.; Trigas, P.; Iliadou, E.; Tzanoudakis, D.; Dimopoulos, P.; Iatrou, G. National Park and UNESCO Global Geopark of Chelmos-Vouraikos (Greece): Floristic Diversity, Ecosystem Services and Management Implications. *Land* **2021**, *11*, 33. [CrossRef]
44. Dimopoulos, P.; Georgiadis, T. Floristic and Phytogeographical Analysis of Mount Killini (NE Peloponnisos). *Phyton Ann. Rei Bot.* **1992**, *32*, 282–305.
45. Trigas, P.; Tsiftsis, S.; Tsiripidis, I.; Iatrou, G. Distribution Patterns and Conservation Perspectives of the Endemic Flora of Peloponnese (Greece). *Folia Geobot.* **2012**, *47*, 421–439. [CrossRef]
46. Kokkoris, I.; Dimitrellos, G.; Kougioumoutzis, K.; Laliotis, I.; Georgiadis, T.; Tiniakou, A. The Native Flora of Mountain Panachaikon (Peloponnese, Greece): New Records and Diversity. *J. Biol. Res. Thessalon.* **2014**, *21*, 9. [CrossRef] [PubMed]
47. Mikhailenko, A.V.; Ruban, D.A.; Ermolaev, V.A. Accessibility of Geoheritage Sites—A Methodological Proposal. *Heritage* **2021**, *4*, 1080–1091. [CrossRef]
48. Mikhailenko, A.V.; Nazarenko, O.V.; Ruban, D.A.; Zayats, P.P. Aesthetics-Based Classification of Geological Structures in Outcrops for Geotourism Purposes: A Tentative Proposal. *Geologos* **2017**, *23*, 45–52. [CrossRef]
49. Štrba, L.; Kršák, B.; Sidor, C. Some Comments to Geosite Assessment, Visitors, and Geotourism Sustainability. *Sustainability* **2018**, *10*, 2589. [CrossRef]
50. Reynard, E.; Perret, A.; Bussard, J.; Grangier, L.; Martin, S. Integrated Approach for the Inventory and Management of Geomorphological Heritage at the Regional Scale. *Geoheritage* **2016**, *8*, 43–60. [CrossRef]
51. Kokkoris, I.P.; Drakou, E.G.; Maes, J.; Dimopoulos, P. Ecosystem Services Supply in Protected Mountains of Greece: Setting the Baseline for Conservation Management. *Int. J. Biodivers. Sci. Ecosyst. Serv. Manag.* **2018**, *14*, 45–59. [CrossRef]
52. Cheminal, A.; Kokkoris, I.P.; Strid, A.; Dimopoulos, P. Medicinal and Aromatic Lamiaceae Plants in Greece: Linking Diversity and Distribution Patterns with Ecosystem Services. *Forests* **2020**, *11*, 661. [CrossRef]
53. Tan, K.; Iatrou, G. *Endemic Plants of Greece. The Peloponnese*; Gads Forlag: Copenhagen, Denmark, 2001; ISBN 9788712038573.
54. Trikolas, K. Geological Study of The Wider Area of Aegialia And Kalavryta. Ph.D. Thesis, National Technical University of Athens Faculty of Mining Engineering and Metallurgy Section of Geological Sciences, Athens, Greece, 2008. (In Greek). [CrossRef]
55. Trikolas, C.; Alexouli-Livaditi, A. Geological Structure of the Wider Area of Aegialia and Kalavryta (North Peloponnesus). *Bull. Geol. Soc. Greece* **2004**, *36*, 1568. [CrossRef]
56. Jones, W.H.S.; Wycherley, E.; Orrod, H.H. *Pausanias Description of Greece: With an English Translation*; Harvard University Press: Cambridge, MA, USA, 1969.
57. Frazer, J.G. *Apollodorus, The Library, with an English Translation*; Harvard University Press: Cambridge, MA, USA, 1963.
58. Pollio, M.V.; Gwilt, J. The Architecture of Vitruvius, Book VIII. In *The Architecture of Marcus Vitruvius Pollio*; Cambridge University Press: Cambridge, MA, USA, 2015; pp. 227–258.
59. Cave of Lakes. Available online: https://www.kastriacave.gr (accessed on 2 September 2021).

60. Koutsi, R. The Role of Epikarst in the Estimation and Mapping of Karstic Aquifers' Vulnerability, Using the Newly Developed European Method. Ph.D. Thesis, Department of Geology, National and Kapodistrian University of Athens, Athens, Greece, 2007. (In Greek). [CrossRef]
61. Ocasio-Morales, R.G.; Tsopelas, P.; Harrington, T.C. Origin of Ceratocystis Platani on Native Platanus Orientalis in Greece and Its Impact on Natural Forests. *Plant Disease* **2007**, *91*, 901–904. [CrossRef] [PubMed]
62. Tsopelas, P.; Angelopoulos, A. First Report of Canker Stain Disease of Plane Trees, Caused by *Ceratocystis fimbriata* f. Sp. Platani in Greece. *Plant Pathol.* **2004**, *53*, 531. [CrossRef]
63. Zygouri, V.; Koukouvelas, I.K. Landslides and Natural Dams in the Krathis River, North Peloponnese, Greece. *Bull. Eng. Geol. Environ.* **2019**, *78*, 207–222. [CrossRef]
64. Greek Travel Pages. Available online: https://www.gtp.gr (accessed on 2 September 2021).
65. Sharples, C. *Concepts and Principles of Geoconservation*; Tasmanian Parks & Wildlife Service: St Helens, Australia, 2002.
66. Prosser, C.D.; Brown, E.J.; Larwood, J.G.; Bridgland, D.R. Geoconservation for Science and Society—An Agenda for the Future. *Proc. Geol. Assoc.* **2013**, *124*, 561–567. [CrossRef]
67. Gray, M. *Geodiversity: Valuing and Conserving Abiotic Nature*, 2nd ed.; John Wiley & Sons: Chichester, UK, 2013.
68. Gordon, E.J.; Crofts, R.; Díaz-Martínez, E. Geoheritage Conservation and Environmental Policies. In *Geoheritage*; Elsevier: Amsterdam, The Netherlands, 2018; pp. 213–235.
69. Prosser, C.D.; Díaz-Martínez, E.; Larwood, J.G. The Conservation of Geosites. In *Geoheritage*; Elsevier: Amsterdam, The Netherlands, 2018; pp. 193–212.
70. Crofts, R.; Gordon, J.E.; Brilha, J.; Gray, M.; Gunn, J.; Larwood, J.; Santucci, V.; Tormey, D.; Worboys, G.L. *Guidelines for Geoconservation in Protected and Conserved Areas*; Groves, C., Ed.; International Union for Conservation of Nature: Gland, Switzerland, 2020; ISBN 9782831720791.
71. Gordon, J.E. Geoconservation Principles and Protected Area Management. *Int. J. Geoheritage Parks* **2019**, *7*, 199–210. [CrossRef]
72. Farsani, N.T.; Coelho, C.O.A.; Costa, C.M.M.; Amrikazemi, A. Geo-Knowledge Management and Geoconservation via Geoparks and Geotourism. *Geoheritage* **2014**, *6*, 185–192. [CrossRef]

Article

Geological and Mining Heritage as a Driver of Development: The NE Sector of the Linares-La Carolina District (Southeastern Spain)

Rosendo Mendoza [1], Javier Rey [2,*], Julián Martínez [1] and Maria Carmen Hidalgo [2]

[1] Department of Mechanical and Mining Engineering, EPS Linares and CEACTEMA, Technological Scientific Campus, University of Jaén, 23700 Linares, Spain; rmendoza@ujaen.es (R.M.); jmartine@ujaen.es (J.M.)
[2] Department of Geology, EPS Linares and CEACTEMA, Technological Scientific Campus, University of Jaén, 23700 Linares, Spain; chidalgo@ujaen.es
* Correspondence: jrey@ujaen.es

Abstract: Conservation, rehabilitation and post-valuation of the facilities of old mining districts is considered a valid strategy to revitalize these areas. In this study, the northeastern sector of the Linares-La Carolina mining district was analyzed, integrating geological information with mining to assess its value. The characteristics of the three most emblematic veins (consisting of galena, sphalerite, chalcopyrite, pyrite, quartz, ankerite and calcite) were analyzed, namely El Guindo, Federico and El Sinapismo. In this study, each mining exploitation was evaluated according to their geological context. Currently, old mining operations can only be visited from drainage galleries or from some exploration galleries. However, some of the old mining shafts could be adapted for visitation. On the surface, the remains of the most important extraction shafts and part of the associated facilities are still visible. One can also visit old tailings dumps with a high contents of heavy metals associated with ore concentration plants. The contaminating potential of these wastes is being monitored thanks to control piezometers and sensors installed at different depths within the tailings ponds, which assist in controlling evolution in the latter years. Different localities of special interest from geological, mining and mineralogical points of view are indicated. Therefore, the guided tour described in this work is attractive for tourism and educational purposes.

Keywords: geological and mining heritage; underground-overground patrimonial integration; La Carolina; Spain

1. Introduction

In the second half of the 20th century, many of the large mining basins of Spain and the rest of Europe ceased to be profitable and underwent a process of decline and closure. In parallel, these regions endured an intense industrial and economic crisis that led to strong migratory movements. As an end result, these rich mining districts were abandoned, leaving behind completely desolate landscapes [1].

However, the need to respect and preserve this legacy, which was suffering rapid and alarming deterioration, gradually began to be perceived. This geological, mining and industrial legacy is also beginning to be considered a source of wealth since it would allow tourism initiatives to revitalize these territories [1–4]. In Europe, the Le Creusot mining basin, located in French Burgundy, was one of the pioneers in highlighting mining and industrial heritage [5–7]. The examples of Wieliczka in Poland, Lewarde in France, Kerkrade in Holland, Le Grand-Hornu in Belgium, Blaenavon in the United Kingdom and the Zollverein Mine in Germany are also notable [1]. Numerous recent geoheritage, geoconservation and geotourism studies around the world have had positive results in places of geological and mining interest [8–10].

In Spain, at the end of the last century, mining and industrial heritage also began to be perceived as a resource for tourism development. Today, among the many examples, the old gold mines in Las Médulas (León), the copper mines of Riotinto (Huelva), the mercury mines in Almadén (Ciudad Real), the lead and pyrite mines of La Unión (Murcia) and the coal mines of Andorra-Sierra de Arcos in Teruel should be noted [11–14]. The number of visitors has gradually increased, and Industrial and Mining Heritage Tourism seems to be gaining popularity.

Geodiversity in mining heritage is related to natural and anthropogenic processes. Anthropogenic landforms are very important resources for geotourism and geo-educational activities. This secondary geodiversity includes mining landforms as shafts, underground spaces, quarries and old industrial areas, which present great potential for geotourism activities [15,16].

This work focused on the old metallogenic district of Linares-La Carolina (southern Spain, Province of Jaén, Figure 1), which is characterized by the presence of philonian deposits, basically galena (PbS). These mineralizations have historically been intensely extracted, but were abandoned at the end of the 1960s [17]. The richness of these deposits favored the development of important underground mining for centuries. After years of neglect and deterioration, in recent decades, the potential of this enormous mining and industrial heritage location as alternative an alternative tourist site has also been considered. Important steps have already been taken, notably the progressive protection of elements with heritage value through declarations of Goods of Cultural Interest. Awareness-raising activities promoted by local associations such as the Carolinense Cultural Mining Association (ACMICA in Spanish) or the Arrayanes Project or the Society for the Defence of the Geological and Mining Heritage (SEDPEGYM), along with the elaboration of guides to geological-mining itineraries, also highlight the geodiversity of the region [18–21].

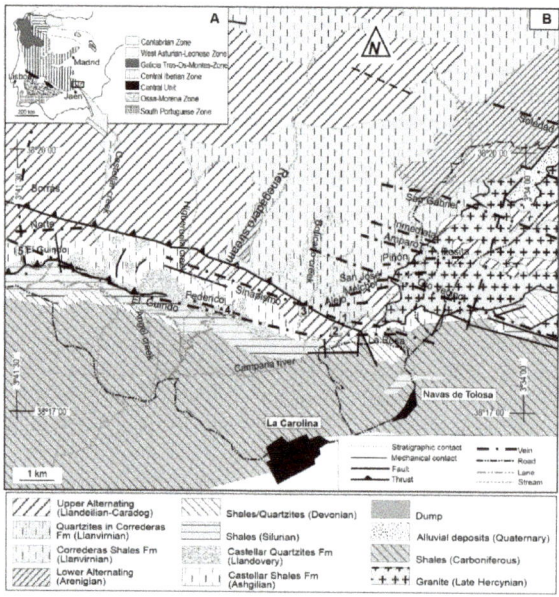

Figure 1. Location map of the Southern Central Iberian Zone [22] (A). Geologic sketch map of the studied region in which the position of the most important veins is indicated [23]. (B) 1, 2, 3, 4, 5: visit areas in the itinerary indicated in the text.

This study aimed to analyze the geological characteristics of the region and associate them with the mining operations of the main veins of the NE sector of the district. The educational and tourism potential of this old mining region is highlighted. Finally, points of special interest are noted to create a guided tour for the visitor.

2. Geological Setting

La Carolina mining district is located on the southeastern slope of Sierra Morena, in the southeastern border of the Hesperian massif [24] in the Central Iberian Zone [25] (Figure 1A). From a regional point of view, two large assemblages of materials can be differentiated, namely, an intensely deformed Paleozoic basement containing mineralizations and a subhorizontal sedimentary cover that fossilizes them.

The Paleozoic basement consists of a succession of metasedimentary rocks (mainly phyllites and quartzites) of Ordovician to Carboniferous age and has been studied previously [23–32] (Figures 1B and 2). A characteristic of this district is the presence of a granitic massif located at the end of the Hercynian orogeny (Castello and Orviz, 1976; Lillo et al., 1998a). The Ordovician series is essentially the host rock of the mineralization, within which a set of units has been defined as a function of the existing quartzite proportion [32], which allows to differentiate, from oldest to youngest, the Armorican Quartzites, the Lower Alternating Fm, the Correderas Shales Fm, the Upper Alternating Fm, and the Castellar Shales Fm (Figures 1B and 2).

The Armorican Quartzites constitute a unit of approximately 450–500 m thick white orthoquartzites, sometimes appearing somewhat pink, with some level of siltstones or mudstones. In these facies, ichnofossils of Skolithos and Cruziana are attributed to the middle of the Arenigian (Tamain, 1972; Ríos, 1977; Lillo et al., 1998b). As can be observed in Figure 2, the presence of an intermediate section of alternations of orthoquartzites and siltstone pelites would allow for the definition of three members [30,32,33].

On the Armorican Quartzites, the "Pochico Strata Fm" appears, with thicknesses of the order of 250 m [30,33]. It is an alternation of quartzites, quartzitic sandstones and more or less sandy pelites. Based on the stratigraphic position and on the ichnofacies present, mostly Cruzian, this lithostratigraphic unit has been attributed to the Arenig [34]. Geological studies of mining companies in the sector have used the term "Lower Alternating" for this unit.

The Correderas Shales Fm, directly on the Armorican Quartzites, range from approximately 400–600 m in thickness. It is a purely slate unit with dark colours. The abundant fauna of trilobites and brachiopods facilitated the dating of the Llanvirnian-Llandeilan [34,35]. The term "Intermediate Unit" has been used in local mining studies to refer to this unit. These lithologies have equivalents for the same age throughout the Central Iberian Zone [23], and they receive different names (Río Shales, Tristani Beds).

The Upper Alternation is deposited on the Correderas Shales Fm. In brief, it is an alternation of pelites and quartzites. It is in this unit that a large portion of the mining of the La Carolina district lies, and it has therefore been studied in great detail. The variable predominance of quartzitic facies allows for the differentiation of different lithostratigraphic units [31,33], namely, Río Quartzite, Botella Shales, Botella Quartzite, Cantera Shales, Mixed Banks, Urbana Limestone (discontinuous) and Chavera Shales (Figure 2). The associations of brachiopods, trilobites or conodonts have allowed for the dating of the Llandeilan, the Caradocian and the Ashgillian [33–36].

On the Ordovician, there is a unit consisting of large quartzite banks with thicknesses varying between 40 and 70 m, defined as Castellar Quartzite [36]. Equivalent materials have also been described in other sectors of the Central Iberian zone [23], where they received local names in Spanish (Cuarcita de Criadero, Cuarcita de Valdelasmanos). In the upper part of this unit, there is an association of graptolites dating from the Silurian (middle Llandovery). Above this unit appear very dark pelitic facies approximately 100 m thick that were defined as the "Graptolitic Shales" [36]. The association of graptolites indicates an upper Llandovery.

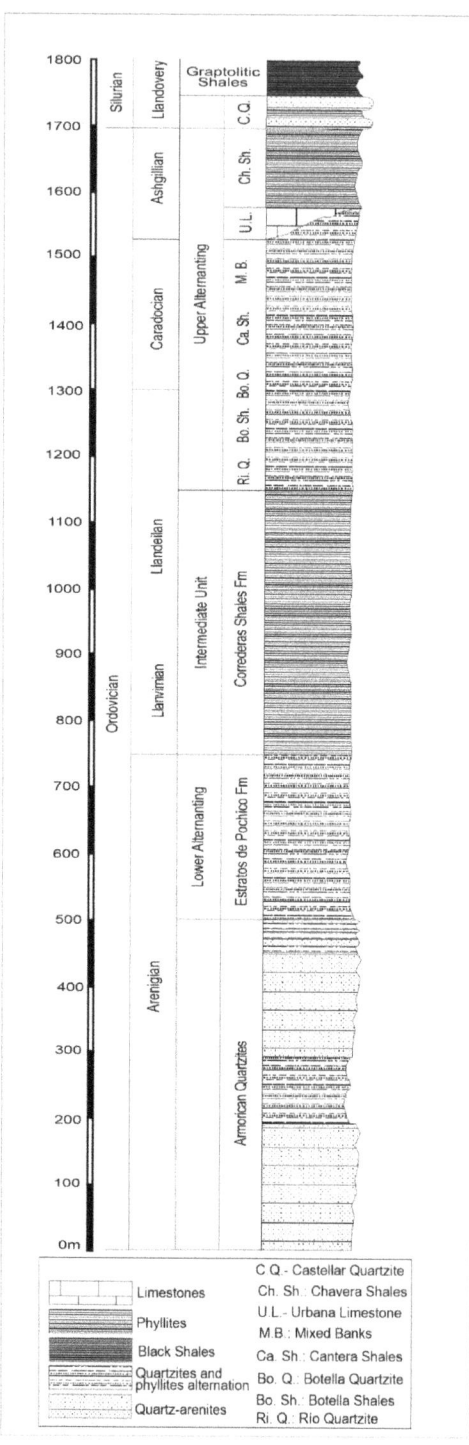

Figure 2. Synthetic stratigraphic column for the Ordovician in the studied region.

3. The Mining District

The philonian rocks in La Carolina are associated with extensional conditions, and the anomalous geothermal gradient originated at the end of the Hercynian orogeny [26,27]. These mineralizations have a hydrothermal origin, whereby the fluid phase and metals were injected through the fractures and discontinuities of the granitic mass and the Paleozoic host rock [26,27]. These veins mainly consist of galena, sphalerite, chalcopyrite, pyrite, quartz, ankerite and calcite and fit in the Paleozoic basement and in the granitic body [29]. In the northeastern sector of the district, they have a predominant N110E/E-W strike (Figure 1B).

In general, the veins have a high dip and extend longitudinally for several kilometers. The mineralizations within the philonian unit are concentrated in lenticular enriched areas. Lithological control is fundamental in mineralization, as the veins were metallized when they were strengthened in quartzites and rapidly became depleted when reaching phyllites (Figures 3 and 4). Mechanical factors were responsible for this process since competent materials (quartzites, sandstones and granite) allow frank and net fracturing, while incompetent materials (phyllites) react to the stress by deforming; thus, creating open spaces was difficult.

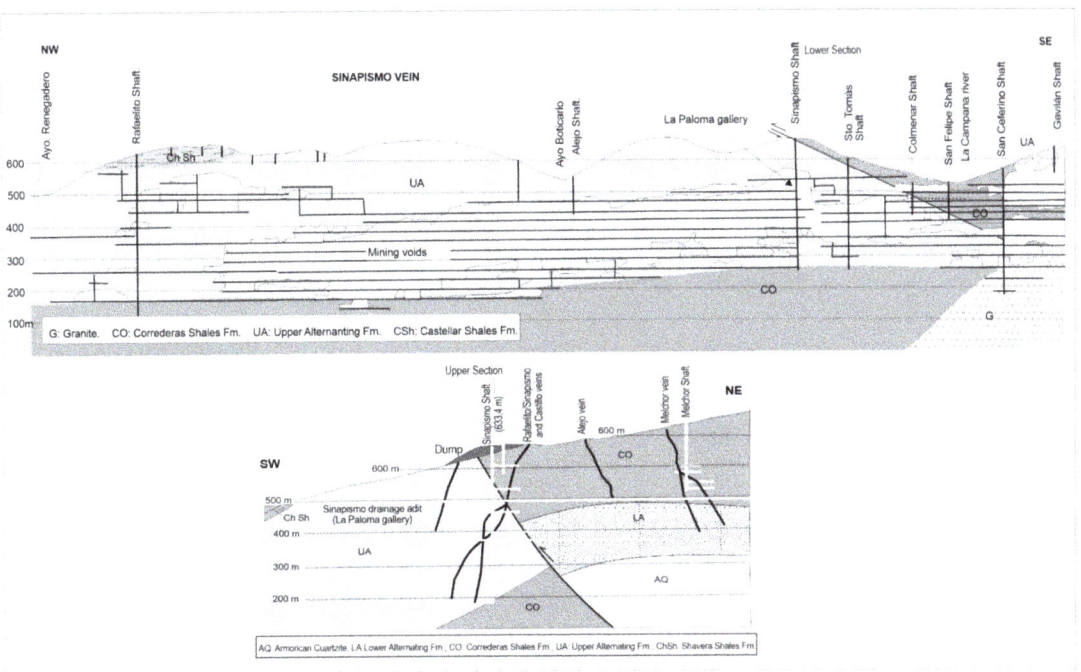

Figure 3. (**Top**) NW-SE geological section coinciding with the trace of the El Sinapismo vein. The La Paloma exploration gallery is indicated (known as the drainage gallery). (**Bottom**) A geological section perpendicular to the previous section is represented, passing through the La Paloma gallery. In both cases, the old mining works are depicted.

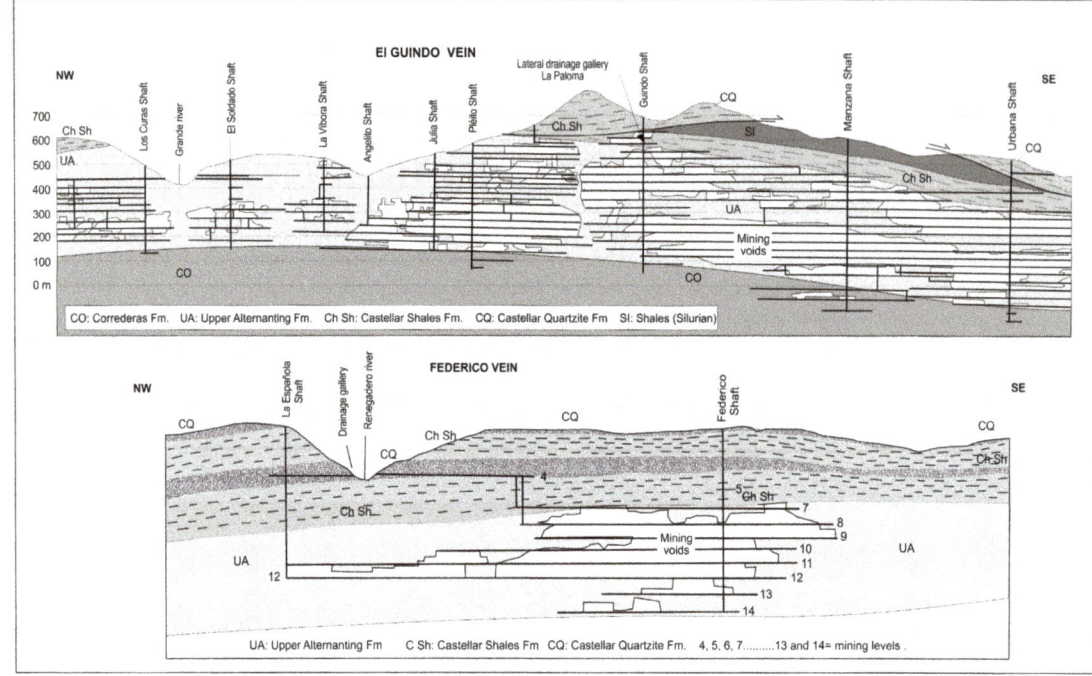

Figure 4. (**Top**) NW-SE geological section coinciding with the trace of the El Guindo vein. (**Bottom**) A NW-SE geological section is represented, coinciding with the trace of the Federico vein. In both cases, the old mining works are depicted.

4. Mineral Exploitation and Concentration Systems in the Mining District

The underground mining method used in this mining district was shrinkage stopping, a common system for philonian deposits with subvertical dips embedded in granites or quartzites [17]. This technique consisted of removing the ore by drilling and blasting by ascending horizontal strips from the bottom up. The blown ore was left in the hole to serve as a working platform and as a temporary support for the gables until finally being evacuated when the chamber excavation was complete [37].

From the stripped materials and after a previous separation operation in the mine, grades between 18% and 20% galena were obtained. Subsequently, the mineral concentration was structured in two stages. First, the material was subjected to a gravimetric process from which a first concentrate of galena was obtained along with a high-grade mixed (composed of host rock and ore minerals) and wastes. The mixed and waste samples were then treated in a flotation process. To release the mineral, flotation required grinding so that the diameter of the particles was below 1 mm (approximately 50–70 µm), resulting in a second concentration of galena and tailings in this process.

The waste from the process was pumped and deposited into tailings dams, which required a dike structure to contain the sludge. The dikes were built with materials from the excavation of mine galleries (blocks and gravel) in the form of closed rings for structures on plains or as a containment dam when using a trough (more common in this sector of the mining district). The regrowth of the dike with the flotation wastes was typically upstream and was carried out by pumping or by guttering (gravity). It should be noted that none of these mining dams were waterproofed in the Linares-La Carolina district, with the tailings being deposited directly on the ground, posing a great environmental risk [38–41]. To obtain an idea of the magnitude of the problem, at least 32 tailings impoundments have been inventoried in the mining district of Linares-La Carolina alone [17].

5. The Main Veins of the Northeastern Sector of the Mining District

The eastern sector of the mining district contains three important veins (Figure 1B), i.e., El Sinapismo, Federico and El Guindo; these veins N110E are subvertical and subparallel.

The El Sinapismo vein is one of the most important veins in the Linares-La Carolina mining district. It was mined for more than 6 km and to depths of 550 m. Stratigraphically, it fits into materials of the Upper Alternating and becomes depleted at depths when penetrating the Correderas Shales Fm (Figure 3). It presents notable bifurcations, and towards the east, it has been mined, which is embedded in the granite. As shown in Figure 3, mining works reached levels below 100 m and were flooded up to 470 m [42] (Figure 3).

In the 1960s, a exploration gallery was constructed perpendicular to the El Sinapismo vein (La Paloma gallery), with the idea of locating new mineralizations [21]. It has a length of 1170 m and cuts a series of secondary veins (Fuente vein, Alejo vein, and Melchor vein, Figure 3). This gallery (also known as La Paloma drainage adit) currently has easy access from the vicinity of the Campana River at an elevation of 495.7 m (Figure 1B). From the observations made inside the adit, it was confirmed that the first 480 m contains an alternation of pelites and quartzites of the Upper Alternating (Figure 3) that dips slightly towards the S-SE. These facies were affected by folding with an inverse flank vergent towards the north. From this contact, the gallery runs through pelitic subhorizontal facies attributed to the Correderas Shales Fm or the upper part of the Lower Alternating (Figure 3). Advancing this gallery from this contact made little sense, since the mineralizations were very impoverished in these pelitic facies; instead, a better use may have been to perform mechanical drilling and verify the existence of mineralizations in the Armorican quartzite at depths suitable for mining.

The El Guindo vein is another one of the most important veins in the mining district, with a run of more than 10 km (Figure 1B). The thickness of the vein, sometimes bifurcated, reached two metres, with quartz and galena fillings. It fit into the Upper Alternating, and metallization disappeared towards the roof in the Shavera Shales and towards the wall in the Correderas Shales Fm (Figure 4). It was mined up to 600 m deep. The three most important shafts of the vein were the N1 of El Guindo, the La Manzana and the Urbana. This last mine shaft can still be visited since it can be accessed from the surface through the gallery of level 1 (Figure 4).

The Federico vein is parallel and very close to El Guindo and could be considered a branch of it (Figure 1B). It has been exploited for a total run of approximately 900 m to a depth of 340 m (Figure 4). The two main shafts were Federico and La Española, linked by level 14. However, it is worth noting the existence of a drainage adit in level 4 that discharges its waters to the Renegadero stream (Figure 4). This point was used for the control and chemical characterization of the waters that flood the vein [42].

6. The Geo-Mining Itinerary

The first stop of the geological mining itinerary is proposed in El Pozo no. 1 of the El Sinapismo vein (Sinapismo shaft) at an elevation of 633 m (1 in Figure 1B). Its elevated position allows for a panoramic view of the mining district and therefore introduces the visitor to the regional geo-mining context (Figure 5a). Here, the mining pit and the remains of the facilities are still visible, offering an introduction to the exploitation systems. Although the old shaft does not intersect the La Paloma exploration gallery, its route passes very close to the gallery (approximately 10 m), so the connection would be relatively simple and would allow for a combined shaft–gallery visit with two exit routes as a safety measure for future guided tours [20,21].

Figure 5. El Sinapismo vein (stops 1, 2 and 3 in the itinerary). (**a**) General photograph in which shaft n° 1 of the El Sinapismo Vein (stop 1) and the La Paloma gallery (stop 2) are positioned. (**b**) Entrance to the gallery from the surface. (**c**) Alternation of quartzites and pelites (Upper Alternating Fm) in the interior of the gallery. (**d**) The gallery cuts small veins filled with quartz, carbonates and some galena. (**e**) Fracturing associated with water percolation and growth of stalactitic structures inside the gallery. (**f**) General view of the Aquisgrana tailings dam (stop 3). The presence of the truck is highlighted to illustrate the size of the dam.

One of the great tourist attractions of the El Sinapismo vein would be the visit to the La Paloma gallery (2 in Figure 1), the entrance of which offers easy access from the surface (Figure 5b). The alternation of pelites and quartzites of the Upper Alternating are visible in the first 480 m (Figure 5c). Some small veins consisting of galena, sphalerite, quartz, ankerite and calcite are also visible (Figure 5d). In addition, recent precipitates that form the stalactitic structures, which possess great beauty, endow the entire complex with enormous tourism potential (Figure 5e).

Additionally, associated with this vein, a visit to the tailings dam of La Aquisgrana is proposed (3 in Figure 1), whose magnitude would attract the attention of any visitor (Figure 5f). Recently, this dam has been the subject of numerous geophysical prospecting studies intended to characterize its internal structure [40]. In addition, rotary drillings have been conducted to take samples for subsequent geochemical analyses, and sensors have been installed in the tailings impoundment to measure some environmental parameters (oxygen, temperature, humidity, electrical conductivity) at different depths to monitor their evolution over time [42,43]. Therefore, this mining dam may be of special interest to students involved in mining or environmental studies.

In the surroundings of Aquisgrana, there are two other points of interest, namely, the old facilities of the flotation-washing plant (at a somewhat lower level than the tailings dam) and the mining interpretation centre managed by ACMICA (in Spanish).

The facilities of the Federico vein are proposed as the fourth stop in this geo-mining itinerary (4 in Figure 1; Figure 6a). This location offers a panoramic view of the trace of the Federico vein, while the facilities of the Federico mine and the La Española mine can be toured, and the medium-sized wastes of the gravimetric processes can be analyzed (Figure 6a). At the back, at a higher elevation, the tailings dam of the flotation processes is still preserved. Recently, this structure has also been characterized by geophysical techniques [41], and work is currently underway for its geochemical characterization (Figure 6b), which would be of interest for a specialized visitor (Figure 6b).

Finally, for this tour, a visit to the most emblematic facilities of the El Guindo vein is proposed (5 in Figure 1). Although separated by a few hundred metres, it is worth approaching the main shaft (El Guindo, Figure 6c,d) and the Urbana shaft. In the latter, the internal workings can be toured with access from the portal of the first-level gallery. From this point, the tailings impoundment of La Manzana, one of the most important dams of the mining district, can be observed; this mining dam has been characterized both geophysically and geochemically in previous studies [38,39].

Figure 6. *Cont.*

Figure 6. (**a**) Federico Mine (stop 4). (**b**) Sampling for the geochemical characterization of the Federico tailings dam. (**c**) Photo from the beginning of the last century in the El Guindo vein (stop 5). (**d**) El Guindo vein today. (**e**) Old facilities in the El Guindo vein and the Manzana tailings dam. (**f**) Mine water sampling from an old shaft in the El Guindo vein.

7. Development Strategy: Discussion

The Linares-La Carolina mining district comprises 1300 mines, with 65 km of main shafts and 786 km of over-seam galleries, being the largest lead producer in the world from 1875–1920. For this reason, it has an extensive mining, geological and cultural heritage that is considered to have a high impact on tourist, educational and scientific projects. In order to attract these projects, official recognition of its geological interest from a highly prestigious institution, such as UNESCO, has been shown to give rise to geotourism [44].

Official recognition would be the most comprehensive strategy to maintain the geo-conservation of these mine sites. However, to reach such a status, the strategy would be to start with this geo-mining itinerary. Although these kinds of revitalization activities required a little economic investment, it should be supported by the local authorities. It has to be noted that currently, tourism development is only carried out by local associations.

Geotourism would be a unique tourist investment in this area, as it offers visitors outdoor activities with a special emphasis on science and education. A recent study reveals that the attractiveness of post-mining areas and the success of tourism development mainly depend on natural landscape, mining heritage sites and architectural features [45]. These mining-related resources are widely available in the study area, and this characteristic is expected to aid in the successful development of geotourism projects. According to previous studies in other former mining districts, mining-related tourist activities are great contributors to the increase in visitors [9,44,45].

On the other hand, it is necessary to disseminate to society the concepts of geoheritage and geotourism, and present the latter as a productive alternative compatible with the local economy. As a current strategy for the enhancement of geological and mining heritage, a combination of GIS and fieldwork evaluation can be used in order to assess and classify the geo-mining sites [46]. This requires the cooperation of the local communities that work in this area with local authorities, tourism companies and the scientific community.

The environmental problems identified in the area would greatly support the involvement of tourism development by the authorities. These anthropogenically influenced landscapes should be managed in parallel with their geoheritage and geotourist values because they are closely linked issues [47,48].

8. Conclusions

Mining-industrial heritage can become a tourist resource capable of acting as a channel for the socioeconomic revitalization of certain territories. This case study aims to enhance the value of the old Linares-La Carolina mining district. For this purpose, an integral use of its geological, mining and metallurgical potential is proposed.

From a geological point of view, there is a thick series of Paleozoic phyllites and quartzites intensely deformed by the Hercynian orogeny and intruded by a granitic batholith. The galena veins are associated with a fracturing phase in the N110E strike. Lithological control was fundamental in the mineralization of these veins since they were metallized when they fitted into quartzites or in the granitic host rock and rapidly depleted when they reached the phyllites. Mechanical factors were responsible for this process since the competent materials (quartzites and granitoids) allow for frank and net fracturing, while the incompetent materials (phyllites) react to stress by deforming, thus hindering the creation of open spaces. All of these factors justify the mining exploitation focusing on materials of the Upper Alternating (predominance of quartzites) and of abandoning the Correderas Shales Fm.

In this study, visits to three of the most important veins have been proposed, namely, El Sinapismo, El Guindo and Federico. The possibility of entering a mining shaft or an exploration gallery is appealing to any visitor. This interest is justified by the number of annual visits to other mining districts already evaluated. In this study, as an added value, it is worth highlighting both the presence of mining and metallurgical facilities and the wastes resultant of all this activity. It is of considerable educational value for disciplines related to mining engineering or the environment.

Author Contributions: Conceptualization and investigation, R.M., J.R., J.M. and M.C.H.; software, J.M. and J.R.; writing and review, R.M., J.R. and M.C.H. All authors have read and agreed to the published version of the manuscript.

Funding: This research was funded by the project FEDER 2020, reference 1380520.

Institutional Review Board Statement: Not applicable.

Informed Consent Statement: Not applicable.

Conflicts of Interest: The authors declare no conflict of interest.

References

1. Richards, G. (Ed.) *Cultural Attracions and European Tourism*; CABI Publishing: Wallingford, UK, 2001; 259p.
2. Cueto, G. El Patrimonio Industrial como motor de desarrollo económico. *Rev. Patrim. Cult. De España* **2010**, *3*, 159–173.
3. Ahmad, S.; Jones, D. Investigating the Mining Heritage Significance for Kinta District, the Industrial Heritage Legacy of Malaysia. *Procedia Soc. Behav. Sci.* **2013**, *105*, 445–457. [CrossRef]
4. Santangelo, N.; Valente, E. Geoheritage and Geotourism Resources. *Resources* **2020**, *9*, 80. [CrossRef]

5. Pillet, F. *Le Patrimoine Industriel Minier du Bassin de Blanzy, Montceau, Le Creusot (Saône-et-Loire)*; Éditions du Patrimoine: Dijon, France, 1999; 47p.
6. Pillet, F. *Le Patrimoine Industriel Métallurgique Autour du Creusot (Saône-et-Loire)*; Itinéraires du Patrimoine; Faton: Paris, France, 2001; 64p.
7. Cañizares, M.C. Patrimonio minero y territorio en la Borboña francesa. El "Museo de la Mina" de Blanzy. *De Re Met.* 2010, 14, 13–22.
8. Gioncada, A.; Pitzalis, E.; Cioni, R.; Fulignati, P.; Lezzerini, M.; Mundula, F.; Funedda, A. The volcanic and mining geoheritage of San Pietro Island (Sulcis, Sardinia, Italy): The potential for geosite valorization. *Geoheritage* 2019, 11, 1567–1581. [CrossRef]
9. Carrión-Mero, P.; Loor-Oporto, O.; Andrade-Ríos, H.; Herrera-Franco, G.; Morante-Carballo, F.; Jaya-Montalvo, M.; Aguilar-Aguilar, M.; Torres-Peña, K.; Berrezueta, E. Quantitative and qualitative assessment of the "El Sexmo" tourist gold mine (Zaruma, Ecuador) as a geosite and mining site. *Resources* 2020, 9, 28. [CrossRef]
10. Briševac, Z.; Maričić, A.; Brkić, V. Croatian Geoheritage Sites with the Best-Case Study Analyses Regarding Former Mining and Petroleum Activities. *Geoheritage* 2021, 13, 95. [CrossRef]
11. Puche Riart, O. Patrimonio minero de España: Aspectos económicos. In *Patrimonio Geológico y Minero: Su Caracterización y Puesta en Valor*; Rábano, I., Mata, J.M., Eds.; IGME: Madrid, Spain, 2006; pp. 15–24.
12. Cañizares, M.C. Almadén: A excepcional mining heritage. *Patrim. L'Industrie/Ind. Patrimony* 2008, 20, 39–46.
13. Cañizares, M.C. Patrimonio, parques mineros y turismo en España. *Cuad. De Tur.* 2011, 27, 133–153.
14. López-García, J.A.; Oyarzun, R.; Sol, A.; Manteca, J.I. Scientific, educational, and environmental considerations regarding mine sites and geoheritage: A perspective from SE Spain. *Geoheritage* 2011, 3, 267–275. [CrossRef]
15. Kubalíková, L. Mining landforms: An integrated approach for assessing the geotourism and geoeducational potential. *Czech J. Tour.* 2017, 6, 131–154. [CrossRef]
16. De Tarso Castro, P.; Nascimento, S.T.; de Paula, S.F. Classification of geo-mining heritage based on anthropogenic geomorphology. *J. Geol. Surv. Braz.* 2021, 4. Available online: https://jgsb.cprm.gov.br/index.php/journal/article/view/127 (accessed on 8 December 2021).
17. Gutiérrez-Guzmán, F. *Minería en Sierra Morena*; Ilustre Colegio de Ingenieros Técnicos de Minas de Linares, Granada, Jaén y Málaga: Linares, Spain, 2007; 586p.
18. Dueñas, J.; Hidalgo, M.C.; Rey, J. Itinerario en el Distrito minero La Carolina (Jaén). *Colección Temas Geológico-Min.* 2000, 31, 457–464.
19. Hidalgo, M.C.; Aguado, R.; López Sánchez-Vizcaíno, V.; Martínez, J.; Rey, J.; de la Torre, M.J. Inventory and valueing of geo-mining heritage in the Linares-La Carolina District (Jaén, Spain). In *Geoevents, Geological Heritage, and the Role of the IGCP*; Marcos, A.L., Ed.; Elsevier: Caravaca de la Cruz, Spain, 2010; Volume 48, pp. 225–226.
20. Galdón, J.M.; Rey, J.; Martínez, J.; Hidalgo, M.C. Application of geophysical prospecting techniques to evaluate geological-mining heritage: The Sinapismo mine (La Carolina, Southern Spain). *Eng. Geol.* 2017, 218, 152–161. [CrossRef]
21. Galdón, J.M. Análisis de Viabilidad Técnica para la Puesta en Valor Turístico del Patrimonio Minero-industrial en el Paraje de la Aquisgrana en La Carolina (Jaén). Ph.D. Thesis, University of Jaén, Jaén, Spain, 2018.
22. Diez Balda, M.A.; Vegas, R.; Lodeiro, F. Structure, Autochthonous Sequences, Part IV Central-Iberian Zone. In *Pre-Mesozoic Geology of Iberia*; Dallmeyer, R.D., Martínez García, E., Eds.; Springer: Berlin, Germany, 1990; pp. 172–188.
23. Ríos, S. Estudio del metalotecto plumbífero del Ordoviense (La Carolina-Santa Elena, Sierra Morena Oriental, Provincia de Jaén). Ph.D. Thesis, University Politécnica, Madrid, Spain, 1977; 271p.
24. Julivert, M.; Fontboté, J.M.; Ribeiro, A.; Conde, L.E. *Mapa Tectónico de la Península Ibérica y Baleares y Memoria Explicativa, Escala 1:1,000,000*; IGME: Madrid, Spain, 1972; 113p.
25. Gutiérrez-Marco, J.C.; De San José, M.A.; Pieren, A.P. Central-Iberian Zone, Autochthonous Sequences. Post-Cambrian Palaeozoic Stratigraphy. In *Pre-Mesozoic Geology of Iberia*; Dallmeyer, R.D., Martínez García, E., Eds.; Springer: Berlin, Germany, 1990; pp. 160–171.
26. Butenweg, P. Geologische Untersuchungen im Ostteil der Sierra Morena nordöstlich von La Carolina (Prov. Jaén, Spaien). Münster. *Forsch. Geol. Paläont.* 1967, 6, 1–126.
27. Azcárate, J.E.; Argüelles, A. Evolución tectónica y estructuras filonianas en el distrito de Linares. In Proceedings of the Congreso Hispano-Luso-Americano de Geología Económica, Madrid, Spain, 19–23 September 1971; Volume 1, pp. 17–32.
28. Castelló, R.; Orviz, F. *Mapa y Memoria Explicativa de la Hoja nº 884 (La Carolina) del Mapa Geológico de España, Escala 1:50,000*; IGME: Madrid, Spain, 1976.
29. Fontboté, J.M. *Mapa Geológico y Memoria explicativa de la Hoja nº 70 (Linares) del Mapa Geológico de España, Escala 1:200,000*; IGME: Madrid, Spain, 1982.
30. Lillo, F.J. Geology and Geochemistry of Linares-La Carolina Pb-ore field (Southeastern border of the Hesperian Massif). Ph.D. Thesis, University of Leeds, Leeds, UK, 1992; 377p.
31. Rey, J.; Hidalgo, M.C. Siliciclastic sedimentation and sequence stratigraphic evolution on a storm-dominated shelf: The Lower Ordovician of the Central Iberian Zone (NE Jaén, Spain). *Sediment. Geol.* 2004, 164, 89–104. [CrossRef]
32. Rey, J.; Hidalgo, M.C.; Martínez-López, J. Upper Ordovician-Lower Silurian Transgressive-Regressive Cycles of the Central Iberian Zone (NE Jaén, Spain). *Geol. J.* 2005, 40, 477–495. [CrossRef]

33. Tamain, G. Recherches géologiques et minières en Sierra Morena orientale (Espagne). Ph.D. Thesis, University of Paris-Sud, Orsay, France, 1972; 648p.
34. Lillo, F.J.; Pieren, A.; Hernández-Samaniego, A.; Ólive, A.; Carreras, F.; Gutiérrez-Marco, J.C.; Sarmiento, G.N.; Fernández, D.C. *Mapa y Memoria Explicativa de la Hoja 862 (Santa Elena) del Mapa Geológico de España, Escala 1:50,000*; IGME: Madrid, Spain, 1998.
35. Lillo, F.J.; López-Sopeña, F.; Pieren, A.; Hernández-Samaniego, A.; Salazar, A.; Gutiérrez-Marco, J.C.; Sarmiento, G.N.; Pardo-Alonso, M.V. *Mapa y Memoria Explicativa de la Hoja 863 (Aldeaquemada) del Mapa Geológico de España, Escala 1:50,000*; ITGE: Madrid, Spain, 1998.
36. Henke, W. Beitrag zur Geologie der Sierra Morena nördlich von La Carolina (Jaén). *Abh. Senckenberg. Naturforsch. Ges.* **1926**, *39*, 205–213.
37. Hoek, E.; Kaiser, P.K.; Bawden, W.F. *Support of Underground Excavations in Hard Rock*; Bolkema Publishers: Rotterdam, The Netherlands, 1995; 225p.
38. Martínez, J.; Rey, J.; Hidalgo, M.C.; Benavente, J. Characterizing abandoned mining dams by geophysical (ERI) and geochemical methods: The Linares-La Carolina District (southern Spain). *Water Air Soil Pollut.* **2012**, *223*, 2955–2968. [CrossRef]
39. Martínez, J.; Rey, J.; Hidalgo, M.C.; Garrido, J.; Rojas, D. Influence of measurement conditions on resolution of electrical resistivity imaging: The example of abandoned mining dams in the La Carolina District (Southern Spain). *Int. J. Miner. Process.* **2014**, *133*, 67–72. [CrossRef]
40. Martínez, J.; Hidalgo, M.C.; Rey, J.; Garrido, J.; Kohfahld, C.; Benavente, J.; Rojas, D. A multidisciplinary characterization of a tailings pond in the Linares-La Carolina mining district, Spain. *J. Geochem. Explor.* **2016**, *162*, 62–71. [CrossRef]
41. Martínez, J.; Mendoza, R.; Rey, J.; Sandoval, S.; Hidalgo, M.C. Characterization of Tailings Dams by Electrical Geophysical Methods (ERT, IP): Federico Mine (La Carolina, Southeastern Spain). *Minerals* **2021**, *11*, 145. [CrossRef]
42. Hidalgo, C.; Rey, J.; Benavente, J.; Martínez, J. Hydrogeochemistry of abandoned Pb sulphide mines: The mining district of La Carolina (southern Spain). *Environ. Earth Sci.* **2010**, *61*, 37–46. [CrossRef]
43. Rojas, D.; Hidalgo, M.C.; Kohfahl, C.; Rey, J.; Martínez, J. Oxidation dynamics and composition of the flotation plant derived tailing impoundment Aquisgrana (Spain). *Water Air Soil Pollut.* **2019**, *230*, 1–16. [CrossRef]
44. García-Sánchez, L.; Canet, C.; Cruz-Pérez, M.Á.; Morelos-Rodríguez, L.; Salgado-Martínez, E.; Corona-Chávez, P. A comparison between local sustainable development strategies based on the geoheritage of two post-mining areas of Central Mexico. *Int. J. Geoheritage Parks 9* **2021**, *9*, 391–404. [CrossRef]
45. Armis, R.; Kanegae, H. The attractiveness of a post-mining city as a tourist destination from the perspective of visitors: A study of Sawahlunto old coal mining town in Indonesia. *Asia-Pac. J. Reg. Sci.* **2020**, *4*, 443–461. [CrossRef]
46. Szepesi, J.; Ésik, Z.; Soós, I.; Németh, B.; Sütő, L.; Novák, T.J.; Harangi, S.; Lukács, R. Identification of Geoheritage Elements in a Cultural Landscape: A Case Study from Tokaj Mts, Hungary. *Geoheritage* **2020**, *12*, 89. [CrossRef]
47. Marescotti, P.; Brancucci, G.; Sasso, G.; Solimano, M.; Marin, V.; Muzio, C.; Salmona, P. Geoheritage values and environmental issues of derelict mines: Examples from the sulfide mines of Gromolo and Petronio valleys (Eastern Liguria, Italy). *Minerals* **2018**, *8*, 229. [CrossRef]
48. AlRayyan, K.; Hamarneh, C.; Sukkar, H.; Ghaith, A.; Abu-Jaber, N. From abandoned mines to a labyrinth of knowledge: A conceptual design for a geoheritage park museum in Jordan. *Geoheritage* **2019**, *11*, 257–270. [CrossRef]

Article

Digital Tools to Serve Geotourism and Sustainable Development at Psiloritis UNESCO Global Geopark in COVID Times and Beyond

Charalampos Fassoulas [1,*], Emmanouel Nikolakakis [2] and Spiridon Staridas [3]

1. Natural History Museum of Crete, University of Crete, 71409 Heraklion, Crete, Greece
2. Unit of Communication & Networking (UCNET), University of Crete, 70013 Heraklion, Crete, Greece; nikolakakis@uoc.gr
3. Staridas Geography, 71201 Heraklion, Crete, Greece; staridasgeography@gmail.com
* Correspondence: fassoulas@nhmc.uoc.gr

Citation: Fassoulas, C.; Nikolakakis, E.; Staridas, S. Digital Tools to Serve Geotourism and Sustainable Development at Psiloritis UNESCO Global Geopark in COVID Times and Beyond. *Geosciences* 2022, 12, 78. https://doi.org/10.3390/geosciences12020078

Academic Editors: Hara Drinia, Panagiotis Voudouris, Assimina Antonarakou and Jesus Martinez-Frias

Received: 30 December 2021
Accepted: 27 January 2022
Published: 7 February 2022

Publisher's Note: MDPI stays neutral with regard to jurisdictional claims in published maps and institutional affiliations.

Copyright: © 2022 by the authors. Licensee MDPI, Basel, Switzerland. This article is an open access article distributed under the terms and conditions of the Creative Commons Attribution (CC BY) license (https://creativecommons.org/licenses/by/4.0/).

Abstract: Digital tools that aid geolocation, geointerpretation and geomodelling are increasingly used in the promotion of geoheritage and geoconservation. UNESCO Global Geoparks (UGGps) are complex regions that require a variety of approaches to advance geoconservation and public awareness, holistic heritage management and sustainable development. UGGps need more diversified and applied digital tools to address these subjects. Additional efforts are made through their commitment to achieving sustainable development goals (SDGs) in the changing and challenging world of the COVID-19 pandemic and the exacerbation of climate change. In this study, we present three new digital applications developed for the Psiloritis UGGp in Southern Greece. These digital tools were developed under the implementation of the "Enhancement Plan" of the geopark via the RURITAGE, a project that supports rural regeneration through conservation, with a focus on local heritage. Digital tools developed in the project include an interactive digital map that demonstrates all properties of local heritage, products and services, two story maps focusing on historic churches and monasteries of the Amari district and on the natural and cultural values of Nida plateau, and a business-listing map with the affiliated geopark enterprises. These digital tools combine multiple applications and methods such as Wordpress webpages, web maps, spherical panoramas, multimedia, site interpretation, geolocation and virtual reality to aid the interpretation of natural and cultural heritage, promote important sites, demonstrate overlaps between nature and human society and support local productivity. Digital tools offer online access to interested parties in any area and are also used for in situ information sites. They are user-friendly, device-adjusted and available for sharing on social media and webpages. The applicability and effectiveness of these digital tools are proven to advance geotourism and the SDGs, in line with the provisions of the "World After roadmap" of UGGps. During the COVID-19 pandemic, the "visibility" of the Psiloritis UGGp was doubled via the use of these digital tools, as they have become popular among the general public.

Keywords: digital tools; story maps; virtual reality; geotourism; sustainable development; RURITAGE; geopark; Psiloritis

1. Introduction

Living in challenging times, i.e., in 2020 and 2021 during the COVID-19 pandemic, requires flexibility, innovation and adaptation, not only for our physical well-being, but for the continuation of our activities, services and products. The shock of the COVID-19 pandemic affected all sectors of social and economic life. The pandemic was particularly devastating to tourism, especially in 2020. Global traveling was banned by 90% of all countries, resulting in a 70–80% decrease in flights and overnight stays, consequently reducing tourism-related income [1].

Among those seriously affected by the effects of the pandemic on tourism are the UNESCO Global Geoparks (UGGps), which depend on geotourism for their sustainable development [2]. By limiting travel, most activities and services in geoparks were frozen, minimizing the income for management structures and inhabitants. The essential connection between geoparks, their stakeholders and their visitors was threatened. To mitigate this problem, geoparks initiated the development of digital initiatives and services to communicate with their inhabitants, promote their territory and support local products and producers [3].

The application of digital technologies in geoparks was initiated before the pandemic. Cayla [4] reported digital technologies used for the management of geoheritage in various geoparks and protected areas, including geolocation and digital mapping, digital imaging and modeling, and hybrid environments (virtual and augmented reality). Laser scanning, digital monitoring and 3D modeling were also implemented in several geoparks and nature parks in France for the study, management, and promotion of karstic systems [5]. The Magma UGGp was possibly the first to integrate virtual reality into the communication and interpretation of geological phenomena under the GEOvisual project in 2015 [6]. In the Sesia Val Grande UGGp, virtual geotrips and digital maps were utilized to enrich geotourism [7]. Special mobile applications have been developed to interpret and communicate geoheritage through the building stones of Torino, Rome and Lausanne [8,9], and to strengthen geoeducation along geotrails [10–12]. Similar applications have been developed in Asia to promote geointerpretation and geoeducation, such as in Mudeung UGGp in Korea [13], and several other Chinese geoparks [14].

The reason for the influx of digital technologies in geointerpretation and promotion is apparent. Digital tools are adaptable, easily modified and updated, open to all, used online and in situ, combine variable resources and means, and are supported by many types of devices (PCs, laptops, tablets, mobile phones, etc.). Digital technologies are user-friendly and openly accepted by younger people, making them suitable for training and education. During the last decade, we have become witnesses of a fast-evolving technology, known as "Accelerating Change" [15], which is pushing technological development to provide products more quickly to a wider demographic range for diversified applications. As more companies invest in a particular field, digital technology changes are accelerated, and target groups increase in size. Virtual reality (VR) is greatly accelerating in development, especially following the commitment of Facebook (also known as Meta since October 2021) to VR and augmented reality (AR) services (https://www.bbc.com/news/technology-58749529, accessed on 28 December 2021).

Digital technology plays a fundamental role in achieving sustainable development goals (SDGs) [16] if it follows the values of equality, harmony with society and the environment and self-determination for our common future [17–19]. Sparviero and Ragnedda [19] defined sustainable digital development as a cluster of values, such as sustainability, that should be applied for the creation and adoption of new technologies to serve a sustainable future. They emphasize that digital sustainability should be achieved by considering the economic, social and environmental goals of sustainable development in conjunction with the rights of individuals to have access to, and benefit from, digital technologies. If individuals, either as consumers or as producers, can benefit from the improvement in digital technology and can be equally considered together with the economy, society and environment when planning for the achievement of SDGs, then a sustainable future is more likely. This concept has been broadly adopted and is being promoted in Central America under various geoconservation and geotourism studies [20]. Williams and McHenry [21] demonstrated that many global geoconservation and geotourism professionals have used digital tools to support them in the decision making, communication of geosite inventory and site map production, and stated that more opportunities would be available if more sophisticated decision-making tools were developed.

The achievement of a sustainable future has been the main goal of the UNESCO Global Geoparks ever since their establishment as a Global Geoparks Network (GGN) in

2004 [22]. This goal is furthered within the "World After roadmap", the Action Plan of UGGps in post COVID-19 times [2]. This proposes UGGps' actions that align with the territorial targets for local sustainable development and SDGs achievement, placing an emphasis on the global climate action. Apart from these future initiatives, it is broadly acknowledged that the UGGps already contribute significantly to the achievement of SDGs [23,24], particularly SDGs 1, 4, 5, 8, 11, 12, 13, and 17. Silva [25] analyzed the contribution of the UGGps to achieving the SDGs and further suggests that they would also contribute to SDGs 6, 7, and 10. All of these actions of the geoparks are described under the four main goals of UGGps that refer to: the conservation and promotion of the natural and cultural environment; raising awareness and training of locals and visitors for sustainability and disaster mitigation; effective management of natural and human resources; and local economic development through geotourism and the promotion of local production [22,26–40].

We present in this study the advantages of new digital applications developed by Psiloritis UGGp. These were developed during the implementation of the RURITAGE project, which focuses on rural regeneration through natural and cultural heritage. We analyze the impact of this technology on geopark promotion and visibility, knowledge communication, support of local economy and tourism, and its contribution to local sustainable development and growth in COVID-19 and post pandemic times.

2. Methods and Materials

2.1. The UNESCO Global Geoparks

UNESCO Global Geoparks are single, unified geographical areas that host sites and landscapes of international geological significance, as well as other natural and cultural sites that are managed in a holistic and sustainable manner [23,38]. Establishment of the UGGps by the International Geosciences and Geoparks Program (IGGP) was approved by all member states of UNESCO in November 2015 [38]. The geopark initiative itself goes further. In 2000, four territories in Europe created the European Geoparks Network (EGN) that was then placed under the umbrella of UNESCO [22,41]. In 2004, with the Madonie declaration, UNESCO assisted the creation of the Global Geoparks Network, which at that time included eighteen European and eight Chinese geoparks. As the geopark initiative was maturing and spreading over the world, the Asia Pacific Geoparks network was created in 2006, followed by the establishment of other regional networks, such as those of Latin America and the Caribbean, and finally those of Africa [34]. At present, 169 territories from forty-four countries are nominated as UGGps and participate in the Global Geoparks Network, a non-profit organization based in Paris [40].

Established as "bottom-up" initiatives, UGGps transmit the local character, highlight individualities, enhance and exploit indigenous and local knowledge for resilience and regeneration, and build the pride of inhabitants and stakeholders [22,40,42]. According to the "Operational Guidelines and Statutes", each UGGp should develop activities focusing on the "Top 10 Focus Areas" [43]: on the natural resources, revealing their importance for modern societies and the need for their sustainable use; on geohazards and climate change to secure local resilience; on the education of inhabitants and visitors to spread knowledge and raise awareness of our geological heritage and its links to natural and cultural heritage; on science and research through the collaboration with academic and research institutions; on culture and celebration of local heritage; on empowerment and equality of women; on indigenous people and their knowledge; on sustainable development through geotourism and local product promotion; and on geoconservation to safeguard the geological value of their territory together with other values, for future generations. All geoparks are evaluated every four years on their achievements and progress in these ten focus areas [23,40].

To develop strategies, methodologies, and tools to achieve SDGs and improve their effectiveness, it is important for geoparks to analyze their resources, operability, and local interactions. UNESCO defines in detail four features that are fundamental for a UGGp [39]: the geological heritage of international significance; effective management structure; proper visibility of the region as an UGGp; and networking at local and international levels. A UGGp (Figure 1) is a web of smaller networks that operate and interact at a local level. These networks may include various assets, including *sites of interest*, such as the geological, biological, historical, cultural, etc., that are further represented in sub-networks: for example, geosites can be subdivided to categories of fossil, geomorphological, petrological, tectonic interest. The network of *infrastructure* that may include info points, trails, museums, views and interpretation sites, is formed by sub-groups that refer to art, archaeology, nature, and folklore. The network of *local heritage* includes tangible and intangible attributes, traditions, ethics, etc. The network of *products* offered in the territory may relate to agricultural, dairy, art, and so on. Networks of *service providers* can include groups of providers for healthcare, recreation, food, and accommodation, and their subgroups. The most crucial network is *human capital*, and this may include indigenous and minority groups, local cooperatives, associations and any human aggregation that interacts with the nature and culture of the area. Each of these networks requires specific management, operation, treatment, promotion, and exploitation that is performed by local authorities, organizations, associations, businesses, and private individuals and their concerns. The role of a geopark in a specific territory depends on its status and its position within a national administration. In general, a geopark has to act as an umbrella for the promotion and coordination of these activities to seek new collaborations between these networks, develop local synergies and new products, and act as a locomotive for local sustainable development.

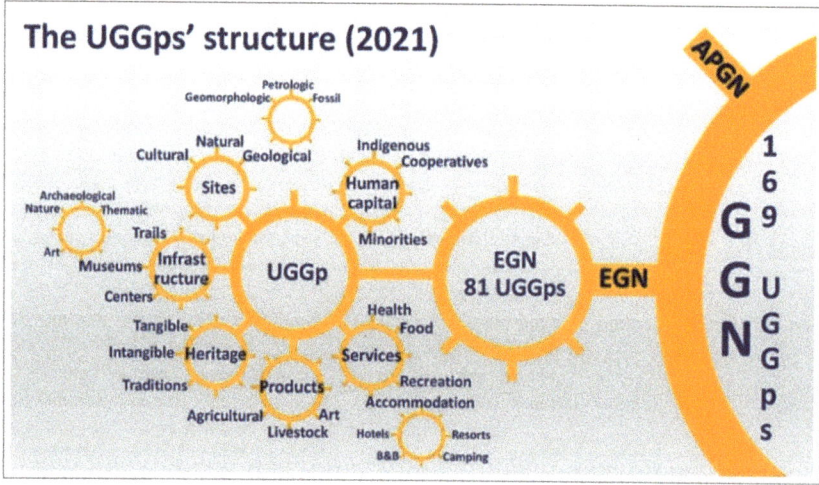

Figure 1. Schematic representation of the UNESCO Global Geoparks' (UGGp) structure and the relative Continental (EGN, APGN) and Global Geoparks Networks (GGN).

On a larger scale (Figure 1) geoparks interact through networking within regional networks, i.e., the European Geoparks Network (EGN), Asia Pacific Geoparks Network (APGN), Latin American and Caribbean Geoparks Network (LACGN), African Geoparks Network (AUGGN) and Global Geoparks Network (GGN), all of which UGGps participate in. Within these networks the geoparks develop synergies, common products and initiatives; magnify their visibility and influence; transfer knowledge and good practices; and organize capacity building activities. To highlight the importance of these interactions, networking is a fundamental consideration in the evaluation processes of UGGps [23].

Another fundamental occupation for geoparks is the development and enhancement of geotourism. Most sustainable development activities in geoparks are achieved through geotourism [23,27,40,44–46]. Geotourism is a form of responsible and sustainable tourism, which is considered as a branch of rural or nature tourism. It addresses a large demographic range of visitors and interests from Earth science lovers to the "general public" travelers [47,48]. Geotourism bonds landscapes, historic cultures, human activities, and local experiences into a new touristic "product" defined by the various participants. Newsome and Dowling [49] and Dowling [50] clearly separate public geotourism from the typical geological field trip, describing it as a sustainable tourism focused on experiencing the geological monuments of our planet in a way that fosters an environmental and cultural understanding, appreciation, and conservation for the benefit of local societies. Large tourism companies such as National Geographic define geotourism as the type of tourism that sustains or enhances the distinctive geographical character of an area, including its environment, heritage, aesthetics, culture, and the well-being of its residents. National Geographic has set thirteen principles for governments and tourism operators [51], and predicts that geotourism is the "future" of touristic travelling. During the International Congress of Geotourism held in 2011 in Arouca UGGp in Portugal under the auspices of UNESCO, geotourism was defined as "tourism which sustains and enhances the identity of a territory, taking into consideration its geology, environment, culture, aesthetics, heritage, and the well-being of its residents", later known as the Arouca Declaration [52].

Stoffelen and Vanneste [53] point out contrasts characterizing geotourism as a geologic- or as a geographic-based specialty. They suggest that geotourism should be reinterpreted as a synergy of landscape science with tourism. These authors [54] consider that a holistic and spatial conceptual framework needs to connect the physical and tangible research of landscapes, with a societal response to the landscape in terms of geotourism. Dowling [50] differentiates between geologic-oriented tourism and a geographic approach to tourism. He concludes that geotourism, as it is performed by geoparks, is both a form and an approach of tourism that is strongly connected to the "geological nature of an areas' sense of place". According to Dowling [50], five key principles characterize geotourism: it is geologically based, sustainable, educative, locally beneficial, and generates touristic satisfaction. A considerable amount of scientific research is being conducted on this topic [46], with the majority of this research focusing on the processes and the methods to assess, interpret, and manage geoheritage, and the minority on geotourism stakeholders, including tourists and local communities, and the context of sustainable development.

To advance geotourism, the UGGps develop infrastructure, tools and the means to enhance and promote the geological, but also the biological, environmental and cultural heritage of their territory. They promote educational products and initiatives to train inhabitants and visitors on the necessity to conserve and manage their region in a sustainable way; implement actions to improve their visibility and promote their territory within their country and abroad; and develop initiatives to support local production and services by creating new experiences and tourism services [23,40].

2.2. The Psiloritis UGGp

Psiloritis UNESCO Global Geopark (www.psiloritisgeopark.gr) is located on the island of Crete in Southern Greece, hosting the highest mountain on the island, Psiloritis (up to 2456 m), (Figure 2). The rugged terrain and the rural environment are among the most popular touristic destinations of Crete. Psiloritis UGGp covers an area of 1270 km^2 that includes the territories of 8 municipalities, 96 small towns and villages and 42,000 inhabitants. Psiloritis became one of the first members of EGN in 2001, and in 2015, it entered the IGGP [55].

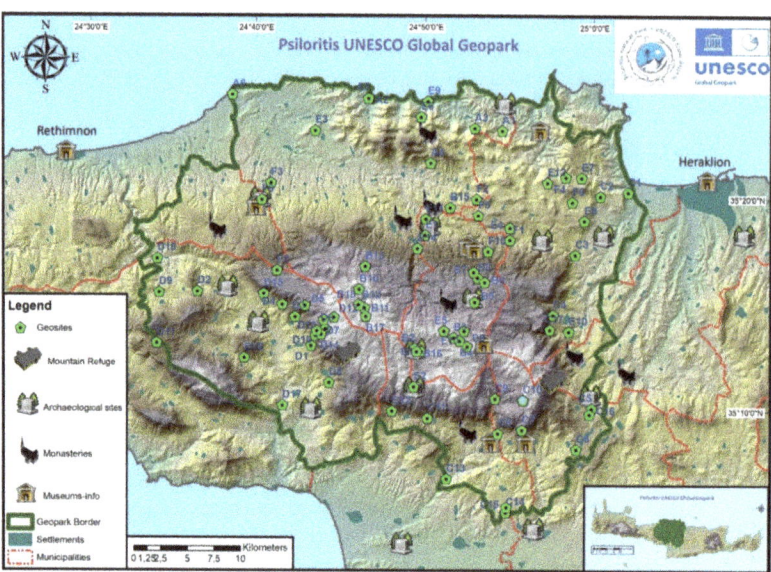

Figure 2. Main morphological, administrative and geoheritage elements of Psiloritis UGGp, associated with the most important historical, cultural and tourist infrastructures. In the embedded map the island of Crete. Labels of geosites refer to the geoparks geosite list (www.psiloritisgeopark.gr).

Psiloritis UGGp is an excellent destination for understanding mountain-building processes and the development of the Hellenic active tectonic arc and subduction zone. It has an exceptionally diverse geological heritage that includes seven tectonic nappes, more than 150 rock types, prominent tectonic features, more than 2000 caves and a great variety of landscape types [56]. So far, eighty-one geosites were identified (Figure 2), and most of them were assessed qualitatively and quantitatively in such a way to establish their global importance, economic value, capacity for geotourism and education, as well as their vulnerability with respect to human and natural pressures [57,58]. Fifty-six percent of the geopark's territory is included in the Natura 2000 network due to the inherent importance of its ecologic–environmental settings. In the geopark, very rare and even endemic species can be found within its unique ecosystems. The geopark territory hosts a large number of archaeological, historical and cultural sites, demonstrating the continuous presence of humans and their civilization in this island setting. The intangible heritage of Psiloritis includes its local foods, livestock products, weaving, architecture, music, and dances. The main economic activities within the geopark are livestock breeding, farming and tourism. However, the geopark comprises a remote, rural and less-developed area compared to the nearby coastal touristic destinations, such as the cities of Heraklion and Rethimnon (Figure 2).

The geopark is managed by a nonprofit enterprise established by the local authorities, in which all local organizations, associations and stakeholders participate. The geopark's info center is based at its headquarters in Anogia village. A network of thematic museums, geopark exhibitions and info points was developed in collaboration with local stakeholders. Educational activities are carried out either by geopark's staff or by its partners, i.e., the Anogia Environmental Education Center and the Natural History Museum of Crete. Since 2007, the geopark sustains a local "Quality Agreement" awarding privileges to special products or services labelling, which fulfill certain quality and sustainability criteria [59]. This network of stakeholders and supporters includes "show" caves, museums and exhibitions, local producers, artists, tourism service providers, and individuals located or intervening within the geopark territory. The visibility of the geopark is displayed within the area by any possible means, such as welcoming road signs, geosite interpretation panels, QR codes, geotrail interpretation panels, printed and digital tools (websites and social network sites). Most of these infrastructures and tools were developed through networking with other geoparks or partners, and with the support of European research or development funds (LEADER, INTERREG, HORIZON 2020).

2.3. RURITAGE Project

Psiloritis UGGp, through its main partner, the Natural History Museum of the University of Crete, is participating in a HORIZON 2020 research project entitled RURITAGE, focusing on the rural regeneration through local heritage [60]. RURITAGE consists of a large consortium with thirty-eight partners from fourteen EU countries and four more from the rest of the world, and is coordinated by the University of Bologna in Italy (www.ruritage.eu). The project focused on six systemic innovation areas (SIAs), namely pilgrimage, local food, human migration, art and festivals, resilience, and landscape. These factors serve as frameworks to identify unique heritage potential within the rural communities. The RURITAGE project adapts an innovative approach to research and implementation using some partners as "Role Models" and other as "Replicators". "Role Models" are areas that can demonstrate and transfer good practices and achievements on the systemic innovation areas, while "Replicators" are territories that benefit from the knowledge of "Role Models" and the RURITAGE tools. Some associates of the RURITAGE participants are "Knowledge Facilitators"; these are research and innovation institutions that develop tools and methodologies to facilitate the transfer of knowledge and outcomes among the partnership.

Six UGGps participate in RURITAGE either as "Role Models" or "Replicators", and Psiloritis UGGp is a "Role Model" on resilience, based on its experience in raising awareness and mitigate disaster risk through education and training. Each territory developed its own "Enhancement plan", where was more comprehensive and ambitious for the "Replicators", but simple and generic for "Role Models". "Role Model" geoparks have the opportunity to develop activities in other SIAs that are crucial for their territories. Psiloritis' "Enhancement plan" aims to enhance and promote the local natural and cultural heritage, as well as its excellent quality products and services, and transform them into a strong and recognizable development tool to increase local pride and improve the well-being of residents and visitors. Psiloritis' plan is based on SIAs of landscape, local food, pilgrimages, and art and festivals. This is analyzed in three axes focusing on: the use of modern technologies (virtual maps, tours, and VR) to enhance and promote the natural and cultural heritage of Psiloritis; the establishment of local participative processes to manage and conserve local landscape and heritage; and organizing general public/stakeholder events to encourage local identity and pride.

The digital applications we describe in the next section were developed by the Natural History Museum of Crete of the University of Crete in collaboration with external experts.

3. Results

Psiloritis UGGp received development support under a former INTERREG project entitled "GEO-IN" (www.geoin.eu). This provided several digital "apps" for geotouristic and training activities, such as: an interactive, web-based map; 360° spherical panoramas; a storytelling map for the geopark; and two educational "apps" for mobile devices (treasure hunt games titled "e-Geodiscover") that focus on the natural and cultural heritage along two geotrails [12,61]. During the implementation of the "Enhancement Plan" of the RURITAGE project Psiloritis UGGp further elaborated and expanded the interactive map and developed new "apps", such as story maps, business listings and virtual tours. The specific features of these tools are described in the following sub-sections.

We initiated the development of the new digital "apps" to aid axes 1 and 2 of the "Enhancement Plan" and to enrich the geotouristic product and visitation of the geopark. The new applications are continuously promoted by the geopark through its social media and webpages. The Psiloritis geopark launched the exploitation of these "apps" in the spring and autumn of 2020 during the first global lockdown due to COVID-19. During that period, the geopark participated in a campaign organized by the Hellenic Geoparks Forum titled "Experience with safety natural and cultural monuments", by promoting its supporters, affiliated partners, and local goods under the following motto: "The geopark supports its supporters". During 2021, when Psiloritis geopark celebrated its 20th anniversary, a new campaign was launched on the social media (Facebook and Instagram) using posters and images produced under the RURITAGE project to highlight and emphasize the natural and cultural wealth of Psiloritis.

3.1. Interactive Geopark Map

The interactive digital map of Psiloritis UGGp can be regarded as an online book with interactive chapters, presenting a strong spatial reference (https://tours.nhmc.uoc.gr/geo/psiloritis/, accessed on 28 December 2021). It was initially developed as a web map by the Natural History Museum of Crete, together with a similar map for Sitia UGGp under GEO-IN project [61], and was further expanded and enriched under the current RURITAGE project to benefit the sustainable development of Psiloritis. It features two interconnected areas, the map area itself, and the side panel area (Figure 3). On the side panel, the menu (chapters and subchapters) is presented in "accordion" mode. By clicking on any topic, it expands the chapter relevant to that topic, and at the same time spatial data are added on the map. The map area covers the largest part of the screen and provides tools for zooming in/out, viewing in "full screen", returning to the home page or geolocating a user (if the user is in a field with a mobile device) (Figure 3a). The map has eighteen zoom levels, from the global scale (global map) to the local scale up to 1:2500, and always displays a scale bar. As the reader scrolls and reads or clicks on information in the side panel, the map pans and zooms, facilitating the display of the spatial context of the information. Additionally, in reverse, when clicking a feature on the map, it scrolls automatically through the side panel until the relevant information reaches the top of the page (Figure 3b).

The map has been developed with the Leaflet.js API that uses "feature" and "tile layers" from the ArcGIS Online account of the Natural History Museum of Crete. Since it is data-centered, all data were collected in a uniform geodatabase (Figure 4), and analyzed in ArcGIS Pro. The geodatabase was initially developed for the geoheritage assessment and geoconservation [57,58], and later on, during the GEOIN project implementation, was updated to incorporate geointerpretation and geotourism. At present, the geodatabase hosts information and documentation of geoheritage and all natural, cultural and economic values of the geopark; it is best regarded as an information depository. The geodatabase includes previously established information on the Psiloritis geopark, and new information digitized during the GEO-IN or RURITAGE projects. The data are separated into divisions: some are static (such as environmental and geological polygons) and some are dynamic (such as geosites, local businesses, cultural sites, etc.). Static data were uploaded to ArcGIS Online as "tile layers" to save storage and decrease loading time. Dynamic data were uploaded as "feature layers", further edited in either ArcGIS Online or ArcGIS Pro. This offers a huge advantage as any new entry or correction can happen in real time (on both the server and user side). The maintenance team can easily perform updates, thus keeping the map alive!

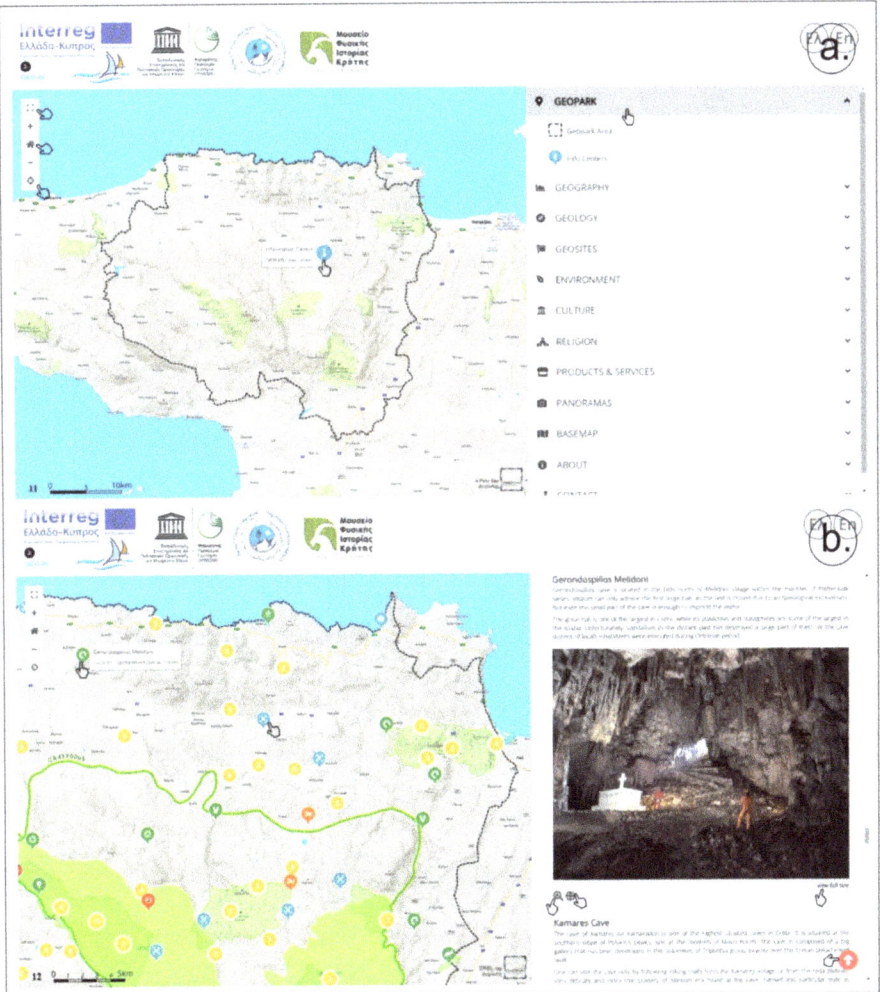

Figure 3. Screenshots from the geopark interactive map. (**a**) The app layout, that is separated into two parts: the map and the data panel that expose the main chapters of information. In the map area, specific features can be clicked showing a popup window with information, zoom in and out buttons, as well as geolocation and home buttons that can be found on the top left, while at the bottom left, the zoom level appears; (**b**) Overlay of different areas of the map, such as the geosites, protected areas (Natura 2000 and Wild Nature Reserves) and spherical panoramas, each presented with a different symbol. By clicking an item on the map, a popup window label appears, while a description and an image (in most cases) is exposed on the data panel. There, buttons enable to zoom in the map or to view in google maps or full screen.

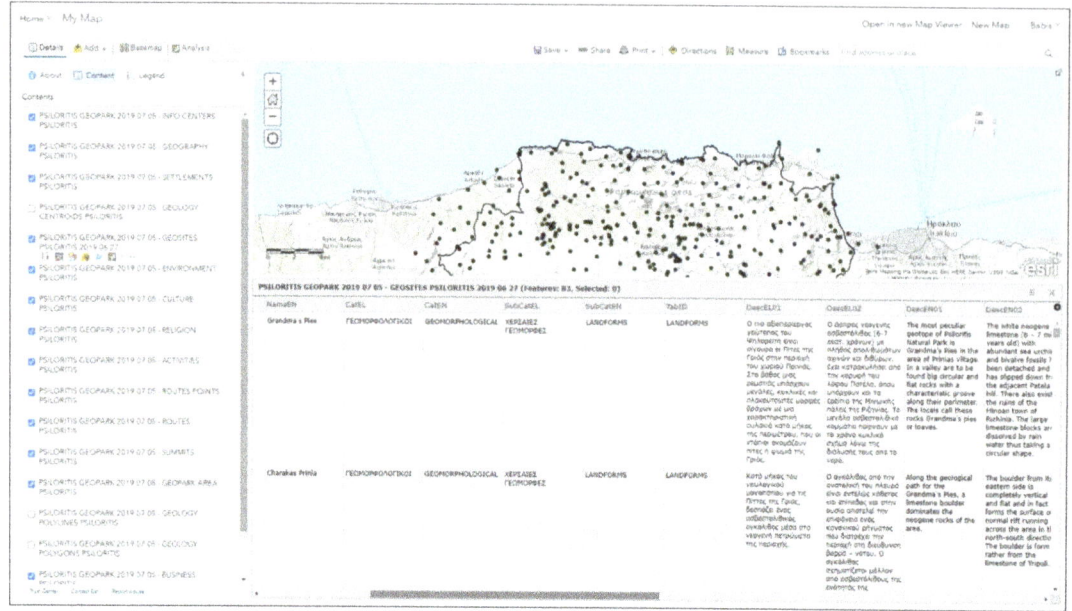

Figure 4. Screenshot of the geodatabase that hosts all information used for the interactive map and other applications.

We emphasize the responsive design of this map application: it is compatible to the user's device, either a PC, laptop, mobile, etc., and adapts to screen dimensions and orientation. The map includes a geolocation function on portable devices so that it can be used as a navigational tool when used in outdoor activities. While such navigational tools are not a new technology, we introduced the facility to click a geolocation button and immediately identify nearby sites of interest, discover other aspects of local heritage, find accommodation and food options, and, in real time, re-design the visit in the geopark. The map at present is bilingual (Greek and English). For communication purposes, the geopark developed three static information units, with touchscreens and personal computer monitors that permit visitors to museums or info centers to navigate through the app (Figure 5). At present, such units are located in our info center at Anogia, the exhibition of the Natural History Museum of Crete at Heraklion, and at the info kiosk of Meronas village at Amari.

The topics that are presented on the map were carefully selected to display the natural and cultural values of Psiloritis, to promote territories within the geopark, and to enhance geotouristic aims. They include information on the local heritage, i.e., geography and geomorphology, the geological features and geosites, environmental and protected areas, and sites of cultural and religious significance. Corrections were made in available geological polygons and a whole new classification system for the geosites was established following the assessment of the geopark [58] and incorporating an international nomenclature [62]. Geosites are classified via geomorphological, petrological, tectonic, hydrogeological, paleontological, geocultural parameters. All heritage features are noted on the map and interpreted on the side panel using text or/and images that can be spotted on the map or shared via Google Maps. Various information levels can be combined and shown on the map panel.

Figure 5. One of the units presenting the interactive geopark's map, associated with a guiding poster, located at the geopark's info center in Anogia.

A great asset of this digital application is a special menu featuring aerial- and ground-based 360° spherical panoramas that have been captured by drones and other sophisticated devices. Under the RURITAGE project, more than 130 panoramas were added. The location of each "capture" is on the map, and the user can watch a virtual tour by clicking and opening the information window. The map user has the ability to "jump" from one panorama to another and create his/her own virtual trip through the popup window or view it on a new window. At present, more than 400 panoramas have been produced for the geopark, and all can be shared free-of-charge by anyone via the relevant social media or websites.

A new feature of this map is its display of local enterprises participating in the "Quality Agreement" initiative (Figure 3b). These are listed under the chapter "Products and Services" that includes five categories: Accommodation, Local Cuisine, Local Products, Local art, and Alternative tourism. Twenty-eight producers, artists and service providers are promoted at this time. By clicking on the "Products & Services" chapter, all of them are presented on the map. Contact information for each partner is provided by clicking on the map or at the side area, as well as their equivalent entry on other business platforms, such as Google Maps. All data layers can be overlain and the user him/herself can combine various information on the map (Figure 3b). Using the geolocation tool and other base maps, the user can identify nearby heritage sites, villages, walking trails or any other information provided by terrain or satellite maps. This tool can be used for planning trips.

3.2. Story Maps

The idea behind story maps derives from the respective ESRI ArcGIS web "apps", where spatial information depicted in one or more maps plays a significant role in how the information is presented to the user [63]. In the previous GEOIN project, web tools and templates provided by ESRI were used to build a "classic" story map [64], and create an individualized storytelling map for Psiloritis UGGp (https://tours.nhmc.uoc.gr/geoparks/map/idi/en/index.html, accessed on 28 December 2021). However, we found limitations when the map was based only on these "classic" story map features. Creating multiple story maps under the same domain, presenting them in the same thematic website, and translating them into different languages were accomplished by an alternative solution within the Wordpress websites. We then developed a Wordpress website (https://storymaps.nhmc.uoc.gr/, accessed on 28 December 2021) interlinked with the geopark's website that hosted new thematic story maps and a business listing. We present thematic story maps that depict the historic churches of the Amari area and the geosites of Nida plateau. We also list the "Affiliated Businesses" who are participating in the "Quality agreement" of the geopark as an independent listing directory rather than a story map. The business listing directories are webpages providing information, that lists businesses within niche-based categories.

Our thematic webpages are an interconnection and interplay of three main standalone applications: the Wordpress webpage; the web maps; and the virtual tours (Figure 6). Several plugins of Wordpress applications host most of the textual information, photos and videos including galleries, carousels, and slideshows. The website itself connects the texts and the visual data, offering a unified environment for the user's experience. The other two components (web maps and virtual tours) are embedded into it. The web maps were developed in ESRI ArcGIS JavaScript API 4.13 and the spatial data used in the maps are hosted in the Natural History Museum of the University of Crete ArcGIS Online Portal. These maps function as standalone web "apps" for the purposes of panning and zooming in and out. The virtual tours are HTML5 webpages that can be viewed as standalone web "apps" in browsers such as Chrome, Safari, Firefox, Edge, etc. The virtual tours are comprised of spherical photo panoramas that are shots from a specific location in the geopark. Most of these panoramas are aerial and allow a better view of the surrounding area. Virtual tours facilitate standalone embedded maps that include the locations of the panoramas and hotspots, and allow menu navigation from one panorama to another. Virtual tours provide supplementary visual information and interpretation by allowing the users to have control of what part of the photo they want to inspect. These virtual tours also have WebXR (https://www.w3.org/TR/webxr/, accessed on 28 December 2021) capabilities that allow the supporting browsers to be viewed in a Virtual Reality (VR) mode.

Figure 6. Screenshots from the story maps of Psiloritis. (**a**) The historic Churches of Amari area. Each story map is composed of a Wordpress webpage (red frame), an embedded web map (blue frame) and a popup window to present, when clicking an item on the map, either an image or a panorama. Buttons on the map enable zoom in and out (**top left**), or display legends (at **bottom left**), while in the popup window, similar buttons zoom in on the map, enable viewing in a new table on the main body of the webpage, or interlink connecting text with the map; (**b**) The front page of Nida plateau story map provides, as with all maps, additional visual information such as videos and photo galleries; (**c**) The "Affiliated businesses" webpage hosts additional features to permit the promotion of enterprise products and services (such as photo gallery or videos), Google Maps with the correct location, as well contact information and communication form.

Links within the Wordpress webpage enable an interaction with the embedded web map and can cause it to focus on respective locations, while an information window (popup) appears that includes the title of the site (church, location, enterprise), and either a photo (in the Churches of Amari) or a panorama from the virtual tour (in the Nida plateau and some

Churches at Amari). All three of these components present information by corresponding to the size of the device that is used to view the website. The Wordpress webpage approach offers the opportunity to add more story maps and integrate more information within them. Each webpage suggests trips that can be planned through integrating the interactive map.

The story map webpage for the *Churches of Amari* hosts information and images of thirty historic Christian churches, monasteries, and chapels dating back to the 7th century A.D. Amari is located in the southwestern part of the geopark (Figure 6a). It is an area of rich natural and cultural heritage popular with alternative tourists seeking birdwatching, botanical activities, hiking and climbing within the gorges of Platania, Patsos and Fourfouras, and trekking to the peak of Psiloritis.

The Nida plateau is one of the most important areas of the geopark, hosting a remarkable bio- and geodiversity, as well as valuable cultural assets. In the broader area of the plateau, five important geosites occur: the Idaion Andro and Kamares archaeological caves, with the first one being the most important sacred place from Neolithic to Roman times in Crete; the Cretan detachment fault; the Idaion active fault; and the Nida plateau itself. Several endemic species of local flora live in this area, as well as endangered and endemic fauna. The intangible heritage of the "Mitata" includes the stone buildings used by the shepherds. The customs and ethics of these shepherds are still extant, and their centuries of grazing activities dominate the plateau landscape. This heritage is shown on the *Nida plateau* story map webpage with a narration that focuses on a virtual tour along the plateau, also including the geomorphic sinkhole topography and the peaks of Mavri Korifi within an embedded interpretative map and a photo gallery (Figure 6b).

The third, the business listing webpage, hosts the *Affiliated Businesses* of the geopark (Figure 6c). Contact details and information regarding products and services enabled the allocation of businesses into six categories: "Local cuisine" with five tavernas and local restaurants; "Alternative tourism" with eight travel agencies or tourism service providers offering accommodation and eco-touristic activities; "Accommodation" with ten local resorts, hotels, villas or studios; "Local products" with nine producers of local goods (such as honey, dairy, cookies, wine, or olive products); and "Local art" with three local artists and their pottery, wood-carving or glass workshops. The enterprises are presented through an embedded map but also through individual webpages that include descriptions and contact information for the enterprise, an embedded map, direct contact tools and photo or/and video gallery.

All three of these webpages are linked with the interactive map of the geopark, increasing the accessibility of the knowledge base of natural and cultural heritage, and the products and experiences offered. This link is found under the "prompt" that leads to planning a trip through the geopark of Psiloritis.

3.3. The Virtual Reality Tour

We have created a Virtual Reality (VR) tour for the Nida plateau area that is a "standalone" product (https://tours.nhmc.uoc.gr/geoparks/nida/, accessed on 28 December 2021). Since this uses locally stored files, there are no internet bandwidth limitations, and so it can reproduce super-high-definition (SHD) 360° panoramic videos. This VR tour was designed to further exploit these 360° panoramas and enrich the user's experience via WebXR capabilities that allow web browsers of mobile devices and special VR equipment to view the panorama in a virtual reality. The tour is composed of nineteen SHD aerial and ground spherical panoramic videos that cover the whole plateau and its surrounding areas, offering shots even from inside the Idaion Andro cave, as well as embedded maps, videos, images, and text information (Figure 7). Narrations including interpretations of the Cretan detachment fault and the Idaion fault geosites [65,66]; videos and panoramas on the social life of shepherds, and a large number of images are offered to the users.

Figure 7. The web browser appearance of the virtual reality tour in Nida area. The top menu, among other options, enables the VR experience on mobile devices. At the welcome page, the nearby spherical panoramas appear as circles by turning the screen view, which is shown on an embedded map (at web browser mode), whereas additional features such as an introductory video, other videos describing sites and activities, and images are shown with different symbols. In the bottom left, the connection of the panorama to Facebook or Twitter accounts can be selected.

Devices that support WebXR (VR mode) include most modern mobile phones and VR headsets. In the VR mode of mobile devices, the screen splits into two parts, one for each eye, and the WebXR-enabled device controls the area that is visible to the user by detecting motion and the angle of view. As the device turns (e.g., with the head movement) the angle of view of the virtual tour changes accordingly. In order to view the Virtual Reality tour properly on a mobile phone that supports WebXR, an inexpensive VR Cardboard headset is required. However, using this headset limits VR capabilities, and the interface is stripped of all the virtual tour components, except the hotspots that change the location (spherical photographic panoramas) of the virtual tour. By using more complex commercial VR headsets, the virtual tour can be either viewed as a normal website, or in its WebXR/VR mode. This provides all the benefits of VR technology such as 3D view, spotting and enabling videos, images, narrations, site interpretations and jumping from one panorama to another.

Two commercial headsets were purchased to be used by the visitors to the geopark's info center at Anogia, and the Natural History Museum of Crete at Heraklion.

3.4. Applications' Use during COVID-19 Times

During the COVID-19 pandemic these digital "apps" were applied in the Psiloritis geopark through several opportunities and campaigns. Soon after the first global lockdown in spring 2020, the geopark joined the campaign of the European Geoparks Network [3], promoting geoparks as "Territories of Resilience", and supporting the Hellenic Geoparks Forum initiative "Experience with safety natural and cultural monuments" [67]. From May to September 2020, Psiloritis started the promotion of affiliated partners through social media, dedicating one post to each partner. This campaign was enthusiastically accepted by the geopark's followers and stakeholders, as well as by the affiliated enterprises that shared posts in large numbers. Statistics (Figure 8a,b) indicate that posts related to enterprise promotion (depicted as characteristic peaks on the relative graph) received many more views than the ordinary posts of the geopark. This campaign added more visitors to the official webpage of geopark (through the followers of the enterprises) and strengthened the connections of enterprises with the geopark during the economically challenging times of the pandemic. With satisfying parts of the "Enhancement Plan" and the celebration of the 20th anniversary of Psiloritis geopark, a new social media campaign was launched in February 2021 that lasted through October of the same year. The statistics of posts (Figure 8c) show a considerable increase in views as an overall increase of up to 90%. Specific peaks can be identified with the posts promoting local enterprises, specific landscapes and geosites or the introduction of the new "apps" (Figure 8d). These statistics demonstrate (Figure 8d) that the geopark's audience is within the "economically active" ages: mainly 25–34, 35–44 (the highest demographic) and 45–54 groups. Most of this audience is women (56.1% compared to 43.9% men).

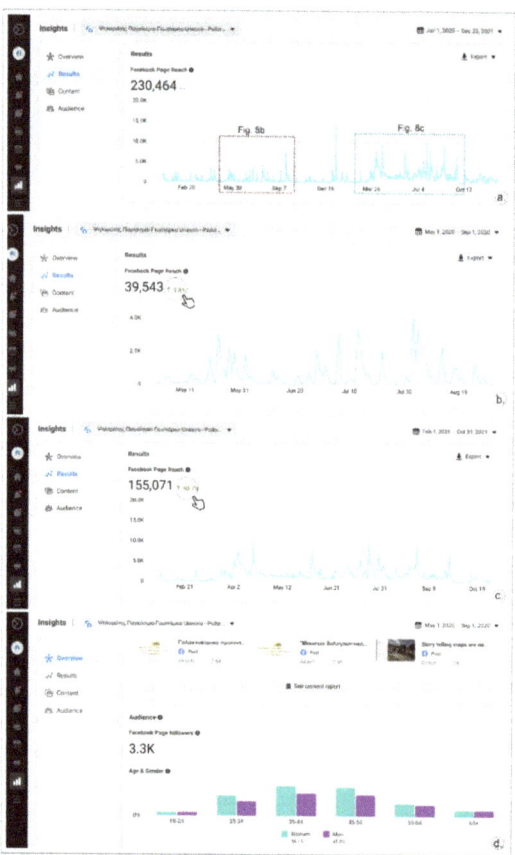

Figure 8. Facebook statistics of Psiloritis UGGp page for the last two years (2020–2021). (a) The total post views over the last two years; in boxes, the particular campaign periods of Figure 8b,c, are marked; (b) Post views during the geopark's campaign to promote its local enterprises and products during the first lockdown (May to September 2020), with the peaks representing followers' responses in the promoted enterprises; (c) The post views for the period from February to October 2021 when the 20th anniversary campaign was launched; (d) The audience gender allocation for the reference period.

It is still too early to fully assess the contribution of these digital tools within the sustainable development of the geopark and the local economy, but we can show that these digital "apps" increased the visibility of our territory, enhanced the understanding and appreciation of our natural and cultural heritage, enriched the products and experiences that our visitors (potential and actual) and inhabitants can enjoy, and supported our local enterprises and products, strengthening their bonds with the geopark. We hope that through continuing the promotion and sharing of these digital "apps", the completion of the "Enhancement Plan" and following the return to "normal", post pandemic life, we will be able to process the direct impact on geotourism and local income that these "apps" provide.

4. Discussion and Conclusions

Digital tools such as web-based maps, 360° panoramas, storytelling maps, and virtual reality environments are continuously evolving. Regarding their use in geoparks, these tools promote many activities related to geoheritage mapping, geolocation, geovisualization, geointerpretation and geotourism [4]. During the past decade, unmanned aerial vehicles (drones), multimedia, and VR technologies have facilitated the promotion and interpretation of geodiversity [68]. These technological "apps" utilize the knowledge base of the existing infrastructure of geoparks derived from geosite inventories, geologic interpretation, trails, and info centers. They are accessible in situ and indoor as web-accessible localities, and are available during and out of tourism seasons. Ólafsdóttir and Tverijonaite [46] point out that the majority of these tools, as described in the literature, do not always cover the needs of geotourism stakeholders within the context of sustainable development. To date, most uses of these tools do not consider their beneficial impact on individuals, either as consumers (visitors) or producers (locals), within the promotion of digital sustainability [19], which is one of the main targets of the UNESCO Global Geoparks. Psiloritis Geopark has exploited and expanded the capabilities of these digital tools: we developed new and innovative applications to support our stakeholders and to offer holistic, inclusive and integrated experiences to our on-site and "web" visitors.

A UNESCO Global Geopark is a complex body with diverse components (local heritage and stakeholders) and multiple tasks to complete. These components act analogously to the gears of a well-set engine to aid the geopark in reaching its targets [29]. Psiloritis is composed of interworking features with a great and imposing geodiversity, profound living environment, well-known and famous culture and human history, as well as scattered human activities. Located in one of the most touristic islands of Greece and the Mediterranean it faces the goal of the responsible development of geotouristic activities in an economically challenged region. A natural environment that is healthy and left intact by tourism and human pressures, a long-lasting history and culture, and the superb quality of agricultural and livestock products are considered strong merits of the territory [59]. The geopark has undertaken the responsibility to coordinate local groups, develop actions and synergies, promote local products and services, and build geotouristic initiatives. Based on these activities it has successfully been evaluated by the GGN and UNESCO several times during the twenty years of its existence.

The challenges imposed by the COVID-19 pandemic and the climate crisis require the flexibility and adaptation of geoparks to the new social, environmental and economic conditions. As stated at the "World After roadmap" [2], it is crucial for UGGps to focus on services in addition to geointerpretation and geotourism. In this way, geoparks can cope with the needs of citizens and visitors, improve the resilience of the society and mitigate various risks. Education and raising awareness can aid the general public's understanding of our fragile planet. The creation of new digital tools can help to meet these needs, even in cases of the recent pandemic, and should be the goals of UGGps according to the "Top 10 Focus Areas" [2,43]. It is apparent that the new digital applications developed by the Psiloritis geopark contribute significantly to achieving the goals of the "Top 10 Focus Areas": digital tools promote geological and broader natural landscapes of Psiloritis, encompass new scientific methodologies and technological achievements, inform and educate visitors and inhabitants on the values of the territory, contribute to sustainable development by enforcing geo and sustainable tourism, and are inclusive and integrated. They help to raise awareness of the natural risks and the need for geoconservation. Digital applications are also compatible with the provisions of the "World After roadmap" with respect to the need to raise awareness of natural risks and adapt to our challenging future.

The new applications come as a result of the implementation of the "Enhancement Plan" for Psiloritis geopark developed under the RURITAGE project. Although Psiloritis serves as a "Role Model" within RURITAGE, project implementation was impeded by the pandemic conditions and provided Psiloritis with the resources needed to develop its own regeneration and "Enhancement Plan". These applications were developed under the first axis of the plan, which includes the "use of modern technologies (virtual maps, tours and VR) to enhance and promote the natural and cultural heritage of Psiloritis". Less directly, they support the needs of the third axis to "strengthen local identity and pride based on the natural and cultural values of the area". The interactive map through its chapters and subchapters, promotes the geological, natural, and cultural heritage of the geopark. The story maps for the Amari churches and the Nida plateau, and the VR tour at the Nida area connect the natural landscapes with the cultural environment and human resources. This combination is a new touristic product easily accessed and shared. The new applications support local products, artefacts and services via an interactive map and the webpage for the "Affiliated Enterprises" of the geopark, thus contributing to the promotion of the area and the local economic development. We can now claim that these products are in agreement with the provision of the "Enhancement Plan" to promote local heritage and goods and transform them into a strong and recognizable development tool that will raise local pride and improve the well-being of residents and visitors. The new products fit the basic tenet that a geopark should represent territories "built by people and for people" [2].

As UNESCO has denoted, geoparks contribute to several SDGs and most profoundly to SDGs 1, 4, 5, 8, 11, 12, 13 and 17. Following SDGs indicators [69], our new applications contribute directly or indirectly to the achievement of these goals. For SDG 1, the applications communicate information on geohazards existing in geopark, thus improving the knowledge level and resilience of the local populations. These digital promotions aid in the designation of local products that can help in critical situations such as pandemics. All applications can be used for lifelong learning, training, and education, and are offered to all free-of-charge, even for disabled people and those unable to travel; thus, they are inclusive, integrated, and equal according to SDG 4. These easily accessible applications grant access to knowledge and information, promote the collaborations of women through "coops" and similar initiatives in the territory, thus promoting the gender equality and female empowerment foreseen under SDG 5. These applications provide powerful, inclusive and innovative geotouristic tools that support, through sustainable tourism development, small- and medium-scale local businesses, their products and their services for the benefit of visitors and inhabitants. Therefore, their support of SDG 8 is very profound. In addition, by providing access to all geological, natural, and cultural values of the territory of Psiloritis, and highlighting the special relations between urban and rural areas, the "apps" contribute to making cities and settlements more safe, resilient, and sustainable, as SDG 11 addresses. The webpage for the "Affiliated Enterprises" and the interactive map promote local products and their consumption in a sustainable manner, as SDG 12 foresees. Similarly, these tools contribute to SDG 13 in the fight against the climate crisis by minimizing long-distance trading, product transportation (considering the islandic nature of Crete), and supporting traditional cultivation and livestock farming practices.

As these digital applications were developed only for Psiloritis UGGp, their impact on the achievement of SDG 17 may be regarded as insignificant. However, the "apps" are outcomes of an international project consortium and they will be shared to more than thirty-eight partners. These tools in our opinion, aid the initiation of a global partnership in support of sustainable development. In addition to the SDGs denoted by UNESCO for the UGGps, these "apps" follow the guidelines recommended by SDG 9 by promoting the development of local infrastructures and networks that are resilient, reliable, and sustainable (as they are easily updated, enriched and maintained), and simultaneously support local well-being and equitable access for all. These applications promote the use of scientific research and upgrade the technological capabilities of small- and medium-scale local enterprises of Psiloritis, facilitating free access to information and communication, as described under SDG 9.5. Finally, these "apps" aid in the promotion, interpretation, and recognition of the value of terrestrial ecosystems and landscapes, and increase the local appreciation and pride. In addition, they make conservation efforts more effective and easier to facilitate, thus helping local managers to achieve the sustainable management of ecosystems, reverse land degradation and halt biodiversity loss, as foreseen under SDG 15.

Based on these facts, we can assert that the new "apps" of Psiloritis UGGp reduce the existing gap [46] between the scientific achievements in the topics of geomapping, geolocation and geomodeling with the real needs of local people and tourists, allowing them to experience the sense of landscape and cultural ecosystem services via modern technology and devices [47]. Geotourism can now be considered a powerful tool for local managers to support geologic and nature conservation; visitors and inhabitants become aware of their surrounding environment, appreciate its value and the need for its maintenance, and can participate in landscape and heritage management without conflicting with the needs of local development.

In summary, the new digital applications that were developed by Psiloritis UGGp under the implementation of the RURITAGE "Enhancement Plan" enrich the existing interactive geopark map, add new story maps of the geopark and a business listing of "Affiliated Enterprises", and provide a virtual reality tour of the Nida area. These tools for the geopark are based on modern technologies for geolocation, geovisualization and geointerpretation, and promote the natural and cultural heritage of Psiloritis and their interconnections and interrelations with human society. The tools support thematic tourism on the topics of religious sites and geology, as well as regional cuisine and local products. Webmaps, 360° spherical panoramas and virtual tours are incorporated into the interactive map and the Wordpress webpages, and include videos, narratives, images and site interpretation. The Nida website offers WEBXR features to support its virtual reality tour.

As with most digital applications, our maps are accessible free-of-charge by all web browsers. They provide access to information and data for all those who are interested and are especially helpful to those who are unable to visit the geopark, including people with special mobility needs. These "apps" are device-customized to provide an optimum performance in all desktop and mobile devices and are bilingual. To be consistent with the principles of digital sustainable development, the "apps" link all natural, cultural and human values in the Psiloritis territory for geo and Earth scientists, as well as the general public, visitors and inhabitants. These new diversified digital products combine older applications and geopark experiences with new concepts that connect the landscape's spirit with the thousands of years of human culture in the area.

These new tools are in compliance with the "Top 10 Focus Areas" of the UGGps, and meet most provisions of the "World After roadmap", as suggested by the GGN. Directly or indirectly, these digital tools support the achievement of the eight SDGs denoted by UNESCO for UGGps, as well as SDGs 9 and 15. The promotion of these tools through the geopark's webpage and social media during the two years of the COVID-19 pandemic demonstrated their great acceptance by the public and their significant contribution in raising the residents' and visitor's awareness of the local heritage. The regional and global visibility of the geopark was maintained and advanced further. Through geotourism, these new digital tools are expected to contribute significantly to the post pandemic regeneration and local sustainable development of Psiloritis UGGp.

Author Contributions: Conceptualization, C.F.; methodology, C.F., E.N. and S.S.; interactive map development, S.S. and C.F.; digitization, E.N.; app development, S.S., E.N. and C.F.; writing—original draft preparation, C.F.; writing—review and editing, C.F., E.N. and S.S.; supervision, C.F. All authors have read and agreed to the published version of the manuscript.

Funding: C.F. and E.N. have received funding from the European Union's Horizon 2020 Research and Innovation Programme, RURITAGE project, grant agreement no. 776465. The contents reflect only the authors' views, and the European Union is not liable for any use that may be made of the information contained therein.

Institutional Review Board Statement: Not applicable.

Informed Consent Statement: Not applicable.

Data Availability Statement: No new data were created or analyzed in this study. Data sharing is not applicable to this article.

Acknowledgments: We wish to thank AKOMM PSILORITIS SA and its director, Dimitrios Pattakos, who developed, undertook, and coordinated several initiatives for Psiloritis UGGp, such as the "Local Quality Agreement" and GEO-IN project. We are grateful to Anne Ewing Rassios for correcting the English text. Three anonymous reviewers are also thanked for their critical and helpful comments.

Conflicts of Interest: The authors declare no conflict of interest.

References

1. Yang, Y.; Altschuler, B.; Liang, Z.; Li, X. (Robert) Monitoring the global COVID-19 impact on tourism: The COVID-19 tourism index. *Ann. Tour. Res.* **2021**, *90*, 103120. [CrossRef] [PubMed]
2. Martini, B.G.; Zouros, N.; Zhang, J.; Jin, X.; Komoo, I.; Border, M.; Watanabe, M.; Frey, M.L.; Rangnes, K.; Van, T.T.; et al. UNESCO Global Geoparks in the "World after": A multiple-goals roadmap proposal for future discussion. *Episodes* **2021**. [CrossRef] [PubMed]
3. EGN. European Geoparks Network, EGN Magazine 2021, 18. Available online: http://www.europeangeoparks.org/?page_id=395 (accessed on 28 December 2021).
4. Cayla, N. An Overview of New Technologies Applied to the Management of Geoheritage. *Geoheritage* **2014**, *6*, 91–102. [CrossRef]
5. Hoblea, F.; Delannoy, J.-J.; Jaillet, S.; Ployon, E.; Sadier, B. Digital Tools for Managing and Promoting Karst Geosites in Southeast France. *Geoheritage* **2014**, *6*, 113–127. [CrossRef]
6. Gentilini, S.; Thjømøe, P.; Birkenes, M. Responsible use of natural and cultural heritage. In Proceedings of the 13th International Geoparks Congress, Rokua Geopark, Finland, 3–6 September 2015. Available online: http://www.europeangeoparks.org/\protect\unhbox\voidb@x\hbox{wp-content}/uploads/2012/02/Book-of-Abstracts-EGN-conference-2015.pdf (accessed on 28 December 2021).
7. Perotti, L.; Bollati, I.M.; Viani, C.; Zanoletti, E.; Caironi, V.; Pelfini, M.; Giardino, M. Fieldtrips and Virtual Tours as Geotourism Resources: Examples from the Sesia Val Grande UNESCO Global Geopark (NW Italy). *Resources* **2020**, *9*, 63. [CrossRef]
8. Gambino, F.; Borghi, A.; D'Atri, A.; Gallo, L.M.; Ghiraldi, L.; Giardino, M.; Martire, L.; Palomba, M.; Perotti, L.; Macadam, J. TOURinSTONES: A Free Mobile Application for Promoting Geological Heritage in the City of Torino (NW Italy). *Geoheritage* **2019**, *11*, 3–17. [CrossRef]
9. Pica, A.; Reynard, E.; Grangier, L.; Kaise, K.; Ghiraldi, L.; Perotti, L.; Del Monte, M. GeoGuides, Urban Geotourism Offer Powered by Mobile Application Technology. *Geoheritage* **2018**, *10*, 311–326. [CrossRef]
10. Aldighieri, B.; Testa, B.; Bertini, A. 3D Exploration of the San Lucano Valley: Virtual Geo-routes for Everyone Who Would Like to Understand the Landscape of the Dolomites. *Geoheritage* **2016**, *8*, 77–90. [CrossRef]
11. Alfonso, J.L.M.; Piedrabuena, M.; Ángel, P.; Bergua, S.B.; Arenas, D.H. Geotourism Itineraries and Augmented Reality in the Geomorphosites of the Arribes del Duero Natural Park (Zamora Sector, Spain). *Geoheritage* **2021**, *13*, 16. [CrossRef]
12. Fassoulas, C.; Kefalogianni, Z.; Stathi, I.; Staridas, S. Interpreting cultural assets through traditional and innovative educational tools: The case of Mygia trail at Psiloritis Geopark. In Proceedings of the 15th Conference of European Geoparks, Sierra Norte Sevilla, Spain, 25–27 September 2019; p. 55.
13. Kim, H.-S.; Lim, C. Developing a geologic 3D panoramic virtual geological field trip for Mudeung UNESCO global geopark, South Korea. *Episodes* **2019**, *42*, 235–244. [CrossRef]
14. Li, Q.; Tian, M.; Li, X.; Shi, Y.; Zhou, X. Toward smartphone applications for geoparks information and interpretation systems in China. *Open Geosci.* **2015**, *7*, 663–677. [CrossRef]
15. Kurzweil, R. The Law of Accelerating Returns. In *Alan Turing: Life and Legacy of a Great Thinker*; Springer International Publishing: Berlin/Heidelberg, Germany, 2004; pp. 381–416. [CrossRef]
16. UN. United Nations, Transforming our World: The 2030 Agenda for Sustainable Development. 2015. Available online: https://en.unesco.org/global-geoparks/focus#sdg (accessed on 28 December 2021).
17. George, G.; Merrill, R.K.; Schillebeeckx, S.J.D. Digital Sustainability and Entrepreneurship: How Digital Innovations Are Helping Tackle Climate Change and Sustainable Development. *Entrep. Theory Pract.* **2020**, *45*, 999–1027. [CrossRef]
18. Pulsiri, N.; Vatananan-Thesenvitz, R.; Tantipisitkul, K.; Aung, T.H.; Schaller, A.-A.; Schaller, A.-M.; Methananthakul, K.; Shannon, R. Achieving Sustainable Development Goals for people with disabilities through digital technologies. In Proceedings of the Achieving Sustainable Development Goals for People with Disabilities through Digital Technologies, Portland International Conference on Management of Engineering and Technology (PICMET), Portland, OR, USA, 25–29 August 2019; pp. 1–10. [CrossRef]
19. Sparviero, S.; Ragnedda, M. Towards digital sustainability: The long journey to the sustainable development goals 2030. *Digit. Policy Regul. Gov.* **2021**, *23*, 216–228. [CrossRef]
20. Quesada-Román, A.; Torres-Bernhard, L.; Ruiz-Álvarez, M.A.; Rodríguez-Maradiaga, M.; Velázquez-Espinoza, G.; Espinosa-Vega, C.; Toral, J.; Rodríguez-Bolaños, H. Geodiversity, Geoconservation, and Geotourism in Central America. *Land* **2021**, *11*, 48. [CrossRef]
21. Williams, M.; McHenry, M. The increasing need for Geographical Information Technology (GIT) tools in Geoconservation and Geotourism. *Geoconserv. Res.* **2020**, *3*, 17–32. [CrossRef]
22. McKeever, P.J.; Zouros, N. Geoparks: Celebrating Earth heritage, sustaining local communities. *Episodes* **2005**, *28*, 274–278.

23. UNESCO. UNESCO Global Geoparks, Celebrating Earth Heritage, Sustaining local Communities. 2016. Available online: https://unesdoc.unesco.org/ark:/48223/pf0000243650 (accessed on 28 December 2021).
24. UNESCO. UNESCO Global Geoparks & Sustainable Development. Geoparks Fundamental Features, Our Commitment to the Sustainable Development Goals. 2021. Available online: https://en.unesco.org/global-geoparks/focus#sdg (accessed on 28 December 2021).
25. Silva, E.M.R. The Contribution of the European UNESCO Global Geoparks for the 2030 Agenda for Sustainable Development—A Study Based On Several Data Sources. Ph.D. Thesis, Universitade Nova, Lisbon, Portugal, 2021. Available online: http://hdl.handle.net/10362/114994 (accessed on 28 December 2021).
26. Catana, M.M.; Brilha, J.B. The Role of UNESCO Global Geoparks in Promoting Geosciences Education for Sustainability. *Geoheritage* **2020**, *12*, 1. [CrossRef]
27. Duarte, A.; Braga, V.; Marques, C.; Sá, A.A. Geotourism and Territorial Development: A Systematic Literature Review and Research Agenda. *Geoheritage* **2020**, *12*, 65. [CrossRef]
28. Fassoulas, C.; Watanabe, M.; Pavlova, I.; Amorfini, A.; Dellarole, E.; Dierickx, F. UNESCO Global Geoparks: Living Laboratories to Mitigate Natural Induced Disasters and Strengthen Communities' Resilience. In *Natural Hazards and Disaster Risk Reduction, Geographies of the Anthropocene Series Book*; Antronico, L., Marincioni, F., Eds.; Il Sileno Edizioni: Rende, Italy, 2018; pp. 175–197.
29. Frey, M.-L. Geotourism—Examining Tools for Sustainable Development. *Geoscience* **2021**, *11*, 30. [CrossRef]
30. Justice, S.C. UNESCO Global Geoparks, Geotourism and Communication of the Earth Sciences: A Case Study in the Chablais UNESCO Global Geopark, France. *Geoscience* **2018**, *8*, 149. [CrossRef]
31. Pásková, M. Can Indigenous Knowledge Contribute to the Sustainability Management of the Aspiring Rio Coco Geopark, Nicaragua? *Geosciences* **2018**, *8*, 277. [CrossRef]
32. Quesada-Román, A.; Pérez-Umaña, D. State of the Art of Geodiversity, Geoconservation, and Geotourism in Costa Rica. *Geoscience* **2020**, *10*, 211. [CrossRef]
33. Quesada-Román, A.; Pérez-Umaña, D. Tropical Paleoglacial Geoheritage Inventory for Geotourism Management of Chirripó National Park, Costa Rica. *Geoheritage* **2020**, *12*, 58. [CrossRef]
34. Rosado-González, E.M.; Sá, A.A.; Palacio-Prieto, J.L. UNESCO Global Geoparks in Latin America and the Caribbean, and Their Contribution to Agenda 2030 Sustainable Development Goals. *Geoheritage* **2020**, *12*, 36. [CrossRef]
35. Pérez-Umaña, D.; Quesada-Román, A.; Tefogoum, G.Z. Geomorphological heritage inventory of Irazú Volcano, Costa Rica. *Int. J. Geoheritage Park.* **2020**, *8*, 31–47. [CrossRef]
36. Silva, E.; Sá, A.A. Educational challenges in the Portuguese UNESCO Global Geoparks: Contributing for the implementation of the SDG 4. *Int. J. Geoheritage Park.* **2018**, *6*, 95–106. [CrossRef]
37. Tefogoum, G.Z.; Román, A.Q.; Umaña, D.P. Geomorphosites inventory in the Eboga Volcano (Cameroon): Contribution for geotourism promotion. *Géomorphologie Relief Processus Environ.* **2020**, *26*, 19–33. [CrossRef]
38. UNESCO. Statutes of the International Geoscience and Geoparks Programme (IGGP). 2015. Available online: https://unesdoc.unesco.org/ark:/48223/pf0000234539.locale=en (accessed on 28 December 2021).
39. UNESCO. Geoparks Fundamental Features. 2021. Available online: https://en.unesco.org/global-geoparks/focus#fundamental (accessed on 28 December 2021).
40. Zouros, N. Global Geoparks Network and the New Unesco Global Geoparks Programme. *Bull. Geol. Soc. Greece* **2017**, *50*, 284–292. [CrossRef]
41. Zouros, N.; Rangnes, K. The European Geoparks Network: Operation and Procedures. *Schriftenr. Dt. Ges. Geowiss.* **2016**, *88*, 31–36. [CrossRef]
42. Martini, G.; Zouros, N. Geoparks, a vision of the future. *Geosciences* **2008**, *7*, 182–189. [CrossRef]
43. UNESCO. Top 10 Focus Areas of UNESCO Global Geoparks. 2021. Available online: https://en.unesco.org/global-geoparks/focus#focus (accessed on 28 December 2021).
44. Farsani, N.T.; Coelho, C.; Costa, C. Geotourism and geoparks as novel strategies for socio-economic development in rural areas. *Int. J. Tour. Res.* **2011**, *13*, 68–81. [CrossRef]
45. Herrera-Franco, G.; Montalván-Burbano, N.; Carrión-Mero, P.; Apolo-Masache, B.; Jaya-Montalvo, M. Research Trends in Geotourism: A Bibliometric Analysis Using the Scopus Database. *Geoscience* **2020**, *10*, 379. [CrossRef]
46. Ólafsdóttir, R.; Tverijonaite, E. Geotourism: A Systematic Literature Review. *Geoscience* **2018**, *8*, 234. [CrossRef]
47. Gordon, J.E. Geoheritage, Geotourism and the Cultural Landscape: Enhancing the Visitor Experience and Promoting Geoconservation. *Geosciences* **2018**, *8*, 136. [CrossRef]
48. Hose, T.A. 3G's for Modern Geotourism. *Geoheritage* **2012**, *4*, 7–24. [CrossRef]
49. Newsome, D.; Dowling, R.K. Setting an agenda for geotourism. *Geotour. Tour. Geol. Landsc.* **2010**, 1–12. [CrossRef]
50. Dowling, R.K. Global Geotourism–An Emerging Form of Sustainable Tourism. *Czech J. Tour.* **2013**, *2*, 59–79. [CrossRef]
51. National Geographic. Geotourism Principles. 2021. Available online: https://www.nationalgeographic.com/travel/article/geotourism-principles-1 (accessed on 28 December 2021).
52. EGN. Arouca Declaration. In Proceedings of the International Congress of Geotourism, Arouca, Portugal, 9–13 November 2011. Available online: http://aroucageopark.pt/documents/78/Declaration_Arouca_EN.pdf (accessed on 28 December 2021).
53. Stoffelen, A.; Vanneste, D. An integrative geotourism approach: Bridging conflicts in tourism landscape research. *Tour. Geogr.* **2015**, *17*, 544–560. [CrossRef]

54. Vanneste, D.; Stoffelen, A. *Integrating Natural and Cultural Heritage Assets for Tourism: A Critical Reflection on Bridging Concepts for Future Research*; Edward Elgar Publishing: Groningen, The Netherlands, 2020; pp. 49–62.
55. Fassoulas, C. Psiloritis Geopark: Protection of Geological Heritage Through Development. In *Natural and Cultural Landscapes: The Geological Foundation*; Parkes, M.A., Ed.; Royal Irish Academy: Dublin, Ireland, 2004; pp. 291–295.
56. Fassoulas, C.; Zouros, N. Evaluating The Influence Of Greek Geoparks To The Local Communities. *Bull. Geol. Soc. Greece* **2017**, *43*, 896. [CrossRef]
57. Fassoulas, C.; Paragamian, K.; Iliopoulos, G. Identification and assessment of Cretan geotopes. *Bull. Geol. Soc. Greece* **2007**, *40*, 1780–1795. [CrossRef]
58. Fassoulas, C.; Mouriki, D.; Dimitriou-Nikolakis, P.; Iliopoulos, G. Quantitative Assessment of Geotopes as an Effective Tool for Geoheritage Management. *Geoheritage* **2012**, *4*, 177–193. [CrossRef]
59. Skoula, Z.; Fassoulas, C. Building participative processes and increasing the economic value of goods in Psiloritis Natural Park. In Proceedings of the 2nd UNESCO International Conference on Geoparks, Belfast, Ireland, 17–21 September 2006; p. 111.
60. de Luca, C.; López-Murcia, J.; Conticelli, E.; Santangelo, A.; Perello, M.; Tondelli, S. Participatory Process for Regenerating Rural Areas through Heritage-Led Plans: The RURITAGE Community-Based Methodology. *Sustainanility* **2021**, *13*, 5212. [CrossRef]
61. Fassoulas, C.; Staridas, S.; Nikolakakis, E.; Perakis, E. New digital applications to promote the geological heritage of Cretan UNESCO Geoparks under GEOIN project. In Proceedings of the 15th International Congress of the Geological Society of Greece, Athens, Greece, 22–24 May 2019; pp. 259–260.
62. Gray, M. *Geodiversity: Valuing and Conserving Abiotic Nature*; J. Wiley & Sons, Ltd.: New York, NY, USA, 2004; p. 434.
63. ArcGIS. ArcGIS StoryMaps, Storytelling that Resonates. 2021. Available online: https://www.esri.com/en-us/arcgis/products/arcgis-storymaps/overview (accessed on 28 December 2021).
64. ArcGIS. Classic Story Maps. 2021. Available online: https://storymaps-classic.arcgis.com/en/ (accessed on 28 December 2021).
65. Fassoulas, C.; Kilas, A.; Mountrakis, D. Post-nappe stacking extension and exhumation of the HP/LT rocks in the island of Crete, Greece. *Tectonics* **1994**, *13*, 121–132. [CrossRef]
66. Nicol, A.; Mouslopoulou, V.; Begg, J.; Oncken, O. Displacement accumulation and sampling of paleoearthquakes on active normal faults of Crete in the eastern Mediterranean. *Geochem. Geophys. Geosyst.* **2020**, *21*, 009265. [CrossRef]
67. Fassoulas, C. Facing the Consequences of COVID-19 in Psiloritis UGGp. *EGN Mag.* **2021**, *18*, 14. Available online: http://www.europeangeoparks.org/?page_id=395 (accessed on 28 December 2021).
68. Santos, I.D.O.; Henriques, R.; Mariano, G.; Pereira, D.I. Methodologies to Represent and Promote the Geoheritage Using Unmanned Aerial Vehicles, Multimedia Technologies, and Augmented Reality. *Geoheritage* **2018**, *10*, 143–155. [CrossRef]
69. UN. United Nations, Global Indicator Framework for the Sustainable Development Goals and Targets of the 2030 Agenda for Sustainable Development. 2017. Available online: https://unstats.un.org/sdgs/indicators/indicators-list/ (accessed on 28 December 2021).

Article

Comparative Analysis of Two Assessment Methods for the Geoeducational Values of Geosites: A Case Study from the Volcanic Island of Nisyros, SE Aegean Sea, Greece

George Zafeiropoulos * and Hara Drinia

Department of Geology and Geoenvironment, National and Kapodistrian University of Athens, Panepis-timiopolis, Athens 15784, Greece; cntrinia@geol.uoa.gr
* Correspondence: georzafeir@geol.uoa.gr

Abstract: In this study, the geoeducational value of five geosites, located in the aspiring geopark of the volcanic island of Nisyros, SE Aegean Sea, was assessed by means of two methods: the G-P method of Brilha (2016) and the M-GAM method. The first method takes into account 12 criteria belonging to the educational potential. The M-GAM method, on the other hand, takes into account the opinions of visitors who, as non-experts, express a different point of view that is rarely calculated or evaluated in different geosite assessment methods. For the better and more objective comparison of the two methods of evaluation of the educational potential of the study areas, the results were converted to a percentage scale (%). The first G-P method clearly highlights the high geological value of the studied geosites, which have a relatively high score and can be used for geotourism and geoeducation. The second method, on the other hand, yields a moderate score in areas with objectively high geological value. This is clearly evident, as this method considers the opinions of visitors who lack the necessary cognitive geological background, thereby underestimating the significance and potential of certain geological features due to lack of formal training.

Keywords: geoeducation; geoheritage; geosite quantitative assessment; Nisyros Island

1. Introduction

In recent years, an effort has been made by the geoscientific community to record, evaluate, and highlight sites of high geological interest. A key role in this initiative was played by the Convention on the Protection of the World Cultural and Natural Heritage that took place in Paris (France, 1972) and, some years later, by the International Declaration on the Rights of the Memory of the Earth, held in Digne (France, 1991) [1]. These two events were the precursors to the creation of a European initiative for the protection of geoheritage and geodiversity. Indeed, the European Geoparks Network (EGN) was established in 2000 with its goal being the systematic dissemination of procedures that would ensure the protection and conservation of geodiversity [2]. A few years later, in 2004, the Global Geoparks Network was established to enhance the value of geological heritage to both geoscientists and the public [3], as well as to promote sustainable development in areas hosting such geoparks. The above initiatives have led to the emergence and documentation of the new geoscientific concepts of geodiversity and geoconservation. According to Zwolinski, [4] geodiversity is defined as the variety in the earth's materials and the forms and the processes that compose and shape the Earth. Sharples [5] states that geoheritage incorporates the protection of dynamic geological processes and geodiversity. Thus, a new need arises for the protection and conservation of areas of high geological value. This need has led to the concept of geoconservation, which denotes all actions taken to preserve and enhance geological and geomorphological features, processes, sites, and specimens [6,7].

In addition, a set of innovative activities is beginning to develop, closely linked to geoeducation and geotourism, based on the principles of sustainability and rational environmental management [8,9] and the conservation of the geoenvironment, geoheritage, and geodiversity of a geologically important area. Geotourism is gaining popularity as an alternative form of tourism that focuses on criteria of social, cultural, environmental, and economic sustainability, in order to achieve benefits not only for society and its citizens, but also for the environment itself.

Quantitative assessment of geoheritage is now considered essential for the development of geotourism and geoeducational activities. In this respect, a number of inventory and assessment methodologies have been developed to protect and promote geoheritage and to document its geoeducational and geotouristic value (e.g., [10–21]). These assessment methods were developed to evaluate the scientific, educational, touristic, and other values of geosites to determine which types of geosites are the most valuable and can be used as tourist attractions or for geoeducational activities. These methods differ from each other, mainly in the criteria they adopt, which in most cases are dependent on the perspective of each researcher. This results in the quantification of the criteria not being carried out in a purely objective way, as subjectivity enters, causing distortions in the final results [13,22–33]. For this reason, there is a risk of misinterpretation and misjudgment when the evaluation is not carried out based on well-documented and objective criteria [34,35]. On the other hand, there are assessment methods that employ mathematical approaches and models to provide a more quantitative and multidisciplinary perspective of areas of high geoscientific interest. In fact, some models consider not only the scientific value that may arise but also the geoeducational perspective on and potential of these areas [13,22,34–40].

The main purpose of this article is to assess geosites using two quantitative assessment methodologies that approach the geoeducational value of a geosite in different ways. The first method, used by Brilha [17], is considered a general-purpose method (G-P method) designed to assess any type of geosite, considering a wide spectrum of criteria. This method emphasizes four parameters: the scientific value (SV) determined by the study area, the potential educational use (PEU) offered by the potential geosite, the potential tourist use (PTU), and the risk of degradation (DR) of the area. It is one of the most popular and applied inventory methods.

The M-GAM (Modified Geosite Assessment Model) developed by Tomić and Božić [36] is a combination of the GAM model created by Vujičić et al. [41] and the importance factor (Im) first introduced by Tomić [42] in his research. In this research project, the viewpoints of visitors were considered during the evaluation process. A survey was used to gather information. Along with the assessment criteria from Vujičić et al. [41], a new element called the importance factor (Im) was added to the evaluation process. This factor enabled visitors and tourists to express their thoughts on the significance of each subindicator in the assessment model. The advantage of this evaluation model is that it incorporates the perspectives of both experts and visitors [36,37]. This is the first time that this method has been applied for the evaluation of geosites in Greece.

The ultimate goal is to compare the results of the two methodologies and to decide which method is most appropriate for determining the educational value of a geosite. The island of Nisyros was selected as a case study.

The volcanic island of Nisyros is an aspiring geopark, located in the SE Aegean Sea, in the Dodecanese island complex. It is distinguished from the neighboring islands by significant geodiversity related to its long and complicated geological history. Particularly interesting is that in this small region, there are volcanic rocks representing five episodes of volcanism. Therefore, Nisyros is widely recognized as a geological museum, attracting many geologists as well as alternative tourists who are awestruck by its wild natural beauty. Its rich volcanic history, steaming hydrothermal craters, intense smell of sulfur and fumarolic gases, and hot springs, together with the island's rich human history, astound visitors [43].

2. Materials and Methods

2.1. The Study Area

Nisyros Island is the youngest volcano of the South Aegean Active Volcanic Arc, which resulted from the subduction of the Eastern Mediterranean lithosphere beneath the active Hellenic margin of the European plate. It is part of the Kos–Yali–Nisyros Volcanic Field, which is located on the easternmost edge of the South Aegean volcanic island arc. (Figures 1 and 2) [43].

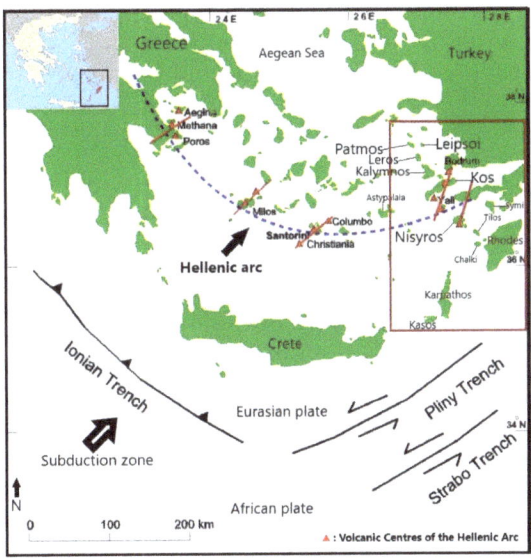

Figure 1. Location of the Dodecanese complex in the southeastern branch of the South Aegean Active Volcanic Arc, at the convergence limits of the two lithospheric plates, the Eurasian and the African.

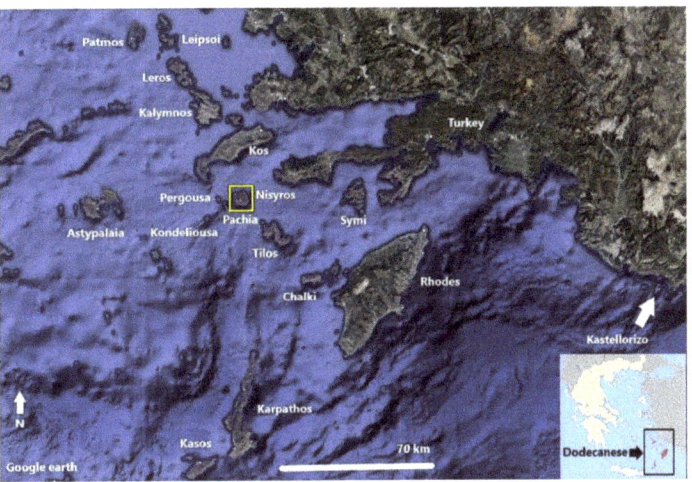

Figure 2. Satellite photo of the Dodecanese island complex, SE Greece, indicating the location of Nisyros Island; Inlet: Sketch map of Greece indicating the location of Dodecanese island complex.

Nisyros has a stratovolcano-like structure formed during the Late Pleistocene–Holocene period within an ENE–WSW-trending neotectonic graben [44,45]. Its stratigraphy is characterized by intercalations of andesitic lavas with andesitic pyroclastic deposits bound by feeder systems of sills and dykes of similar composition that can be seen in the cores of geothermal energy test drillholes (Figure 3). The exposed stratigraphy begins with pillowed basaltic andesite and pillow breccia and progresses to more felsic volcanism, culminating in rhyodacitic post-caldera domes [44,46].

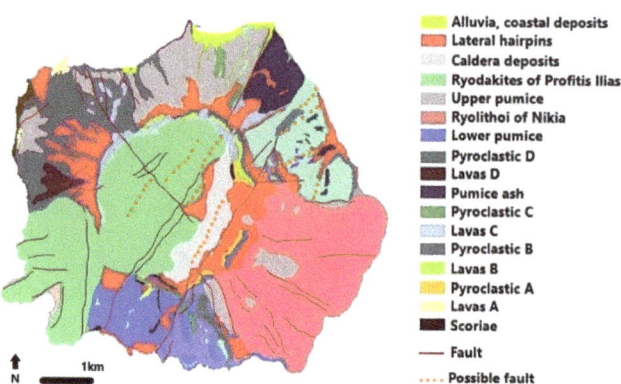

Figure 3. Geological map of Nisyros [44], modified.

The recent form of the caldera is a well-defined circular topographic feature with a diameter of 4 km, with a presently flat caldera floor intercepted by phreatic craters with cliffs of up to 300 m [43]. The top of the Profitis Ilias (Figure 4) post-caldera dome is the highest point (698 m) [44,47–50]. The exposed stratigraphy has an age of −160,000 years [49], and the most recent phreatic eruption occurred in 1867 A.D. [51].

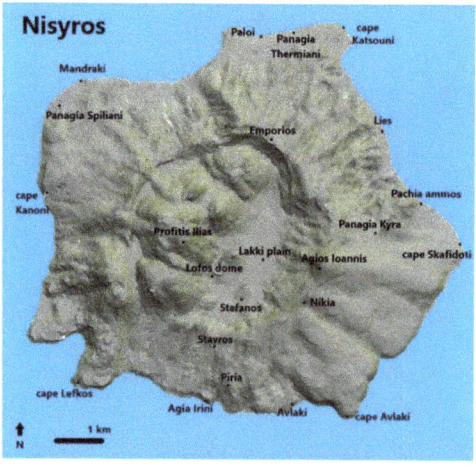

Figure 4. 3D representation of the surface of Nisyros Island, indicating the location of places referred in the text [52], modified.

The caldera hosts a well-known hydrothermal-fumarolic field, whose hydrothermal activity is expressed on the surface by a network of intersecting hydrothermal craters, located in the southern part of the Lakki plain (Figure 4). These craters are governed by diffuse degassing structures. [44,53–57]. The most well-known hydrothermal craters are: Stefanos, Phlegethon, Megalos Polyvotis, and Mikros Polyvotis (Figure 5). [44,53]. The volcano's last crater, "Mikros Polyvotis", was formed in 1887, following the volcano's last—so far—hydrothermal explosion.

Figure 5. Satellite photo of the craters [58], modified.

The Nisyros caldera is also particularly noteworthy for the presence of epithermal gold in the Lofos dome area (82 ppb) and the adjacent Profitis Ilias area (2500 ppb), indicating the presence of significant hydrothermal circulation phenomena [57].

Various authors have studied the evolution of Nisyros Volcano over the last 160,000 years, as well as the succession of calc-alkaline lavas and pyroclastic rocks. The first geographic and geological studies on Nisyros Island were conducted by the Italian geologists Martelli [51] and Desio [59]. Detailed geological studies began in the late 1960s [60] and were carried on by Di Paola [49] and Papanikolaou et al. [44]. According to them, the island's volcanic history is divided into five stages:

(1) The lower volcanic rocks visible on the northern coast near Mandraki were built up by an underwater volcano with erupting basaltic and andesitic pillow-lavas;
(2) For more than 100 ka, a 500–700 m high stratovolcano grew on top of these partly submarine lavas;
(3) Two major rhyodacitic plinian eruptions covered the entire island with pyroclastic flows and pumice falls after several eruptive phases of gas and steam explosions;
(4) At 20 ka BP, a major central, vertical collapse of the volcano left a large caldera; and
(5) The western part of the caldera depression was filled with a series of rhyodacitic domes during prehistoric times, the highest of which, Profitis Ilias, rises 698 m a.s.l.

For at least 25 ka, no volcanic activity is known to have occurred on the island following the formation of the domes; the only reported historical explosions are associated with the formation of several phreatic craters inside the caldera, such as Alexandros, Polyvotis, Stephanos, Phlegethon, and Achelous, which are still emitting fumaroles. The most recent hydrothermal eruptions in 1871–1873 and 1887 AD were accompanied by violent earthquakes, gas detonations, steam blasts, and mudflows [57].

The major distinction in Nisyros' volcanic history is the first period of stratovolcano formation, which ended with a major eruption (Nikia rhyolites) and caldera formation, followed by the second period of volcanic dome formation, which disrupted the former caldera rim (now observed at about 300 m of elevation) and formed the highest actual mountain of Prophitis Ilias (698 m). Recent volcanic formations in the submarine area around Nisyros have created the volcanic centers of Pergousa, Yali, Strongyli, Pachia, and Kondeliousa (Figure 6) [61,62].

Figure 6. Nisyros and the other smaller island around [63], modified.

After several years of inactivity, an intense seismic activity began at the end of 1995 and lasted until 1998, with the largest event recorded on 27 August 1997, with a Ms of 5.3. [64]. This activity resulted in significant variations in fumarole geochemical parameters and progressive uplift and E–W extension of the island's central parts, as well as a possible magma input at greater crustal depth. [63–65]. This gradual uplift resulted in a large N–S trending fracture known as the "Lakki rupture" in the caldera's Lakki plain in early December 2001 [65].

Due to the volcanic activity, there is a significant risk in the wider area for both residents and visitors. As a result, the Volcanological Observatory of Nisyros was established with a suitable and equipped network for monitoring the volcano's physicochemical parameters. In this manner, a valid prediction of the volcano's reactivation can be made in order to take immediate protective measures.

2.2. Geosites

Nisyros Island hosts numerous interesting geosites, ranging from the enormous volcanic craters of the caldera and the thermal springs to the volcanic islets that surround it. The island's landscape is home to a diverse range of magnificent volcanic landforms shaped by natural processes, where visitors can experience the immense power of volcanoes. All the geosites provide accessibility to enjoyable recreation activities. In addition to the fumarolic activity found in the well craters, gas escape is also observed along the active tectonic zones that intersect the island. Thermal springs with temperatures ranging from 27° to 43 °C are situated near the coast. Isotopic analysis of Nisyros thermal water samples revealed a mixture of seawater, magmatic water, and geothermal steam, as well as the possibility of groundwater and/or meteoric water involvement.

In this study, five geosites were selected (Figure 7). Each geosite has been labeled with an ordinal number and the letters "GS".

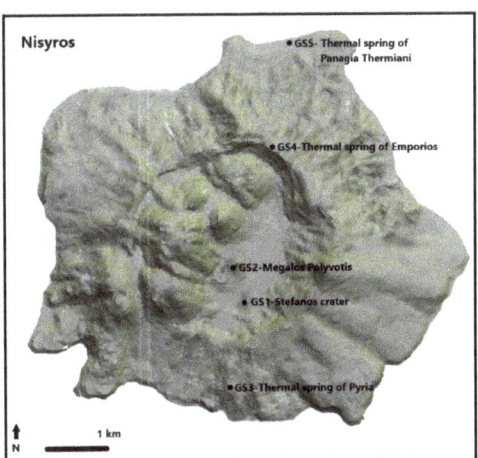

Figure 7. 3D representation of Nisyros Island, indicating the location of the studied geosites [57], modified.

GS1—Stefanos Crater (Figure 8a,b): With dimensions of 260 × 350 m, the elliptical Stefanos Crater is one of the world's largest phreatic craters. Its shape appears to be the result of two main NE-trending active faults characterized by an alignment of fumarolic vents. The crater has a maximum depth of 27 m. The age of the crater's formation is unclear. On the eastern walls of Stefanos, seven stratigraphic layers have been recognized: talus of magmatic lithics, epiclastics and fine argillitic layers, fine-grained lacustrine deposits, solid deposits of explosive compounds generated by the Kaminakia craters, deposits from Stefanos' explosion, and a thin coating of explosive products from Polyvotis are among them. The surrounding area of the crater of Stefanos is characterized by intense and spectacular activity, which is due to the release of gases with a temperature of 100 °C. The gases consist of water vapor and carbon dioxide. Hydrogen sulfide, nitrogen, and methane are released at a smaller rate of about 0.5% [57,66–68]. It is worth noting that amorphous sulfur crystals are deposited at the mouths of the holes from the gas outlet, while the liquefied water vapor irrigates the surrounding soil with dilute sulfuric acid, due to the dissolution of hydrogen sulfide in the steam. Stefanos crater favors the concentration of gases due to its elliptical shape (260 × 190 m). Therefore, it has been observed that in periods of volcanic activity and intense hydrothermal vents, large amounts of gases are released and seismic events occur, which in turn cause landslides.

The characteristic strong and not-so-pleasant smell that exists in the surrounding area is due to the existence of hydrogen sulfide, which even in infinitesimal concentrations is felt by every visitor.

GS2—Megalos Polyvotis (Figure 8a): Megalos Polyvotis was formed because of the Lofos area's first and most powerful hydrothermal explosive cycle. It is an elliptical (180 × 350 m) crater with 3–5 m thick ejecta that is partly covered by products originating from Flegethron or Alexandros crater (Figure 8c), which is a large elliptical-shaped crater that occupies the southeastern part of the area, intersecting with Megalos Polyvotis, from a later event. The crater's material is made up of altered lava fragments and rhyodacitic blocks in a clayey-to-sandy matrix. The lava blocks are surrounded by brown-reddish oxide coatings and cut by anhydrite veins. Its stratigraphy is like that of Stefanos, with lacustrine sediments and unconsolidated clay material at the bottom and chaotic ejecta of earlier magmatic products at the top. After heavy rainstorms, the western sector of the crater floor usually turns into a lake, and it is composed of 1.5 m-thick yellow- and purple-colored varved clayey layers, indicating the past presence of a lake.

GS 1 and GS 2 strongly differ in their morphological characteristics. The crater of Stefanos (GS1) creates a negative relief, whereas the crater of Polyvotis (GS2) creates a positive one. The positive relief of Polyvotis is probably due to the existence of ridges from the adjacent post-caldera structures, which are located around the perimeter of this crater along with the materials from the well explosions. The deposits of Stefanos and Polyvotis craters consist mainly of clay materials that inhibit the deeper penetration of water, resulting in the retention of rainwater, leading to an increase in soil moisture. The simultaneous vapor activity results in the formation of small craters of hot mud.

Figure 8. Geosites of Nisyros Island: (**a**) the Stefanos and Polyvotis craters, (**b**) the Stefanos crater, (**c**) the Alexandros crater, (**d**) the Piria hot spring, (**e**) natural hot spring of Emporios, and (**f**) Panagia Thermiani thermal spring.

GS 3—thermal spring of Pyria (southern part of the island, Figure 8d): In 1841, during a visit to the island by the German archaeologist Ludwig Ross from 9–11 August, he described and depicted several areas with hydrothermal activity. In particular, he states that in Pyria (Arodafnes area), a great heat is exuded, probably due to the strongly cracked southern slopes of the caldera of the island. In fact, these cracked areas appear along the rupture zone that runs in a SW–NE direction [43,44]. In its original form, the thermal spring included a stone building complex with 5–6 chambers. Today, only the chamber where natural steam at a temperature of 40–45 °C is released is intact and in good condition. The source is still used by locals and visitors for the same purpose.

GS4—thermal spring of Emporios (Figure 8e): This is located at the entrance of the settlement of Emporios (northeast of the island) and exhibits temperatures between 36 and 40 °C. This phenomenon is due to the cracked zone (NE) that exists in contrast to other systems [44]. In addition, it shows a divergent direction compared to the other fault systems (south, west, and east sides of the island), resulting in the transport and circulation of hydrothermal fluids in several cases [43]. In fact, the length of the fault throw created by the faults in these areas in places reaches up to 100 m; therefore, it is perceived that this location is inextricably linked to geothermal interest [44]. As a result, in this small chamber, the diffusion of heat is perceived as well as the sulfur that is perceived (yellowish appearance on the surface of the rocks) on the inner surface of the space.

GS5—thermal spring of Panagia Thermiani: This water spring is located in the northern part of the island (near the settlement of Paloi), next to the small church of Panagia Thermiani (Figure 8f). Thermiani is one of the Nisyros thermal springs that have been evidenced since antiquity. Roman baths have been found there, at least ruins and inscriptions proving their past glory. In 1889, a doctor named Pantelis Pantelidis operated a medical unit in which he exploited the water that gushed from this source. The high salinity of this spring may be due to the underground circulation of water in pyroclastic formations, as well as to geological structures that reveal fault zones. The spring is affected by the infiltration of seawater; therefore, there is an admixture of several minerals and various meteoric components, since a perpetual interaction is observed between the sources and the specific geological formations. The source is rich in SO_4^{2-}, Mg^{2+}, Cl^-, and HCO^3. Water here is a mere 33 degrees, but it is equally refreshing and healing.

This spring is due to the geothermal activity of the island as well as to the northern ruptured zone that exists [68–72].

2.3. Methodology

In this study, two methods of geosite assessment were applied. The first method is the general-purpose method of Brilha [17] (G-P method). This method was chosen due to the broad criteria that allow the evaluation of any type of geosite and have been applied in several studies. It provides a quantification proposal consisting of four factors: scientific value (SV), potential educational use (PEU), potential touristic use (PTU), and degradation risk (DR). The quantitative evaluation of the scientific value (SV) of a geosite includes the following seven criteria: representativeness, key locality, scientific knowledge, integrity, geological diversity, rarity, and use limitations. The first term refers to a geosite's potential to highlight a geological process or a variety of features. The term "key locality" corresponds to the geosite's significance as a reference point for various geological features. If there are national or international publications, "scientific knowledge" plays an important role. "Integrity" denotes the geosite's conservation status. The term "geological diversity" corresponds to the quantity of different geological elements found in each geosite. The term "rarity" is used to explore whether there are geosites with similar characteristics in the same area. Finally, the term "use limitations" refers to the obstacles and limitations that may make research and study of the geosite difficult.

Regarding the quantitative assessment of the educational potential use (PEU), twelve criteria are estimated: vulnerability of a geosite, its accessibility, use limitations, safety, logistics, density of population, association with other values, scenery, uniqueness, observation conditions, and didactic potential. The vulnerability is centered on the presence of geological elements that can be affected by visitors. Beyond that, this method investigates the accessibility of the geosite, including the use limitations if there are any. Furthermore, the status of safety and the potential facilities that could be provided are investigated. In addition, the population density and the association of the geosite with other values such as cultural, aesthetic, and so on are determined. The method is then based on the scenery of the area and its uniqueness. Furthermore, the observation conditions and didactic potential are investigated.

For the quantitative assessment of the geotouristic potential (PTU), 13 criteria are used, the first ten of which are similar to those used for educational purposes, and the remaining three of which take into account the interpretative potential, the economic level of the people who live in the area, and the proximity of recreational areas.

Each of the criteria was assigned a score from 1 to 4, with 1 indicating a low possibility of use and 4 indicating a high possibility of use for SV, PEU, and PTU.

The degradation risk (DR) refers to the possibility of a geosite being damaged or destroyed, i.e., losing any of the characteristics that make it valuable as a geosite [29,30,73,74]. Finally, five criteria for evaluating and quantifying degradation risk (DR) are considered: deterioration of geological elements, proximity to areas/activities with potential to cause degradation, legal protection, accessibility, and population density.

Weighing criteria are used to complete the geosite quantification process. Weights were applied to each of the quantification criteria based on their importance in order to examine the potential for scientific, educational, and tourist use. Weights are also assigned to each of the criteria based on their importance in assessing the degradation risk of geosites. The weights used in this study are according to Brilha [17] and are shown in Table 1.

According to G-P method categorization, the geosites can be classified as having a low, medium, or high degradation risk based on the criteria.

Regarding the second method, M-GAM (Modified Geosite Assessment Model), this originated from the differentiation of the GAM method introduced by Vujičić et al. [41]. The method is based on previous geosite assessment methods modified by Tomić and Božić [36] and applied by several scientists [75–85]. The innovation of this method is that it considers the opinion not only of experts but also of visitors to a specific geological area. As a result, M-GAM is used to assess the scientific, educational, and tourist value of a geosite while also assessing the visitors' point of view, regardless of their geological knowledge background. However, it should be mentioned that in the present method, the evaluation is done with a mathematical model, so that there is an equal estimation of both methods, without affecting each other.

In this method, two main components are considered, the main values (MV) and the additional values (AV). The main values are divided into 12 criteria and the additional values into 15. The main values consider the abiotic characteristics of a geosite, while the additional values mostly concern the human activities carried out in relation to each geosite.

The following three aspects are included in the main values: scientific/educational value (VSE), scenic/aesthetic value (VSA), and protection (VPr). The following characteristics are included in the VSE field: rarity, representativeness, scientific issue knowledge, and interpretation level. The following aspects are included in the VSA field: viewpoints, surface, surrounding landscape and nature, and environmental fit, meaning contrast to the nature, contrast of colors, appearance of shapes, etc., of site. Finally, the following factors are involved in the VPr field: current condition, protection level, vulnerability, and appropriate number of visitors. The main values are estimated using the following equation: $MV = VSE + VSA + VPr$.

The additional values comprise two parameters: functional values (VFn) and touristic values (VTr). The estimation of the additional values is computed from the equation $AV = VFn + VTr$.

Table 1. Calculation of the final scores in the G-P method.

Factors	Criteria	
Scientific Value (SV) = SUM of the criteria	Representativeness	30 × score
	Key locality	20 × score
	Scientific knowledge	5 × score
	Integrity	15 × score
	Geological diversity	5 × score
	Rarity	15 × score
	Use limitations	10 × score
Educational Potential Use (PEU) = SUM of the criteria	Vulnerability	10 × score
	Accessibility	10 × score
	Use limitations	5 × score
	Safety	10 × score
	Logistics	5 × score
	Density of population	5 × score
	Association with other values	5 × score
	Scenery	5 × score
	Uniqueness	5 × score
	Observation conditions	10 × score
	Didactic potential	20 × score
	Geological diversity	10 × score
Tourism Potential Use (PTU) = SUM of the criteria	>Vulnerability	10 × score
	Accessibility	10 × score
	Use limitations	5 × score
	Safety	10 × score
	Logistics	5 × score
	Density of population	5 × score
	Association with other values	5 × score
	Scenery	15 × score
	Uniqueness	10 × score
	Observation conditions	5 × score
	Interpretative potential	10 × score
	Economic level	5 × score
	Proximity of recreational areas	5 × score
Degradation Risk (DR) = SUM of the criteria	Deterioration of geological features	35 × score
	Proximity to areas/activities with potential to cause degradation	20 × score
	Legal protection	20 × score
	Accessibility	15 × score
	Density of population	10 × score

As previously stated, this method takes into account the visitors' viewpoint. This is incorporated in the importance factor (Im), which is independently assessed by the visitor for each field (rarity, representativeness, etc.) and multiplied by each corresponding subindicator (which is given by experts). In our study, we relied on Tomić and Božić's [36] research for the importance factor, which has values including 0.00 (not at all important), 0.25 (not quite important), 0.50 (neither insignificant nor important), 0.75 (a little important), and 1.00 (quite great importance). The importance factor is defined by the following formula:

$$Im = \frac{\sum_{k=1}^{K} Ivk}{K}$$

where Ivk is the evaluation/score of each visitor for each field (field-subindicator), and K is the final number of visitors.

Following in the footsteps of Antić, Tomić, and Marković [37], who used the data and results of Božić and Tomić [86] in a geosite assessment survey, we used the importance factor values from the same publication of Božić and Tomić [86].

The final results coming out from this method are depicted in a diagram, where the 12 categories studied in the main values are placed on the vertical axis, while the 15 categories of additional values are placed on the horizontal axis. Furthermore, this diagram is divided into 9 fields, which are as follows: starting from the beginning of the axes and referring to the x-axis in the fields Z11, Z21, and Z31; moving up to the y-axis in the fields Z12, Z22, Z23; and finally, the fields Z13, Z23, and Z33. Each study area's final score is represented by the diagram.

For the better and more objective comparison of the two methods of evaluation of the educational potential of the study areas, the results were reduced to a percentage scale (%). This process resulted in a disparity between the results of the two evaluation methods. In the context of our research, this reduction was made only for the educational value, since the objective of our study is the evaluation of the geoeducational potential of the geological heritage. More specifically, the maximum possible value by which a study area can be evaluated in the G-P method is 400. As a result, all scores are reduced to a scale with a maximum value of 100. Similarly, the M-GAM method, which can have a maximum effect value in the Z33 region, follows the same approach. When the horizontal axis (Main Values) is marked with the value 12 and the vertical axis (Additional Values) is marked with the value 15, the maximum possible distance from the beginning of the axes is achieved. The following formula is used to calculate the distance of a point from the beginning of the axes to the plane: $d = \sqrt{MV^2 + AV^2}$ (Figure 9).

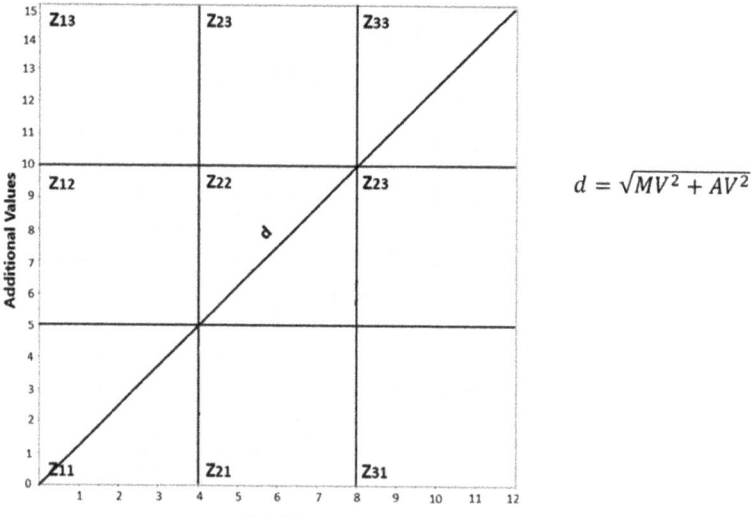

Figure 9. Maximum potential distance from the origin 0 (0,0).

In our study, there is a maximum possible value for MV = 12 and AV = 15:

$$d_{max} = \sqrt{12^2 + 15^2} \cong 19.2$$

Finally, the results are converted to a percentage. As a result, size comparisons on a single scale are possible.

The two aforementioned methods were chosen for two primary reasons. The G-P method is a quantitative evaluation purely for educational purposes, with a set of 12 criteria, each with its own corresponding weight in the final evaluation. As a result, this method approaches a mathematical result with greater objectivity in order to ultimately determine whether the geosite is of educational interest. Second, the M-GAM method is groundbreaking in that it incorporates public opinion into mathematical calculations, allowing the scientific community to discern citizens' perspectives as well as perspectives that may not have been considered previously.

3. Results

The results of the quantitative assessment of educational use value are described below.

3.1. The General-Purpose Model—G-P Method

The results of this method are shown in Tables 2–6.

Table 2. Quantitative evaluation of geosites for scientific value (SV).

Scientific Criteria	Weight	Craters		Natural Sauna Points		Thermal Spring
		Stefanos	Polyvotis	Emporios	Piria	Panagia Thermiani
Representativeness	30	3	3	2	2	2
Key locality	20	2	2	2	1	2
Scientific knowledge	5	4	4	2	2	2
Integrity	15	4	4	4	4	4
Geological diversity	5	4	4	2	2	2
Rarity	15	3	3	2	2	2
Use limitations	10	4	4	4	4	4
Total score		315 High	315 High	250 Moderate	230 Moderate	250 Moderate

<200 Low, 201–300 Moderate, >301 High [87].

Table 3. Quantitative evaluation of geosites for educational potential use (PEU).

Educational Criteria	Weight	Craters		Natural Sauna Points		Thermal Spring
		Stefanos	Polyvotis	Emporios	Piria	Panagia Thermiani
Vulnerability	10	4	4	4	4	4
Accessibility	10	4	4	4	1	4
Use limitations	5	4	4	4	4	4
Safety	10	3	3	3	3	3
Logistics	5	4	4	4	4	4
Density of population	5	3	3	3	3	3
Association with other values	5	4	4	3	3	4
Scenery	5	4	4	3	3	3
Uniqueness	5	4	3	4	4	4
Observation conditions	10	4	3	4	4	4
Didactic potential	20	4	4	2	2	2
Geological diversity	10	4	4	2	2	2
Total score		385 High	370 High	315 High	285 Moderate	320 High

<200 Low, 201–300 Moderate, >301 High [87].

The application of the G-P method shows that no area has a low scientific value, and in fact, two of them have a fairly high score (>301). Studying the educational potential of the areas, it is found that all sites present high scores, except for the geosite of Piria, which presents a moderate score (201–300). In terms of potential tourist use (PTU), it is observed that all geosites present high scores, which indicates the strong geotouristic dynamic of these places. Finally, checking the risk of degradation, it is found that all geosites have a low score, except for Emporios, which shows a moderate score (201–300).

Table 4. Quantitative evaluation of geosites for tourism potential use (PTU).

Touristic Criteria	Weight	Craters		Natural Sauna Points		Thermal Spring
		Stefanos	Polyvotis	Emporios	Piria	Panagia Thermiani
Vulnerability	10	4	4	4	4	4
Accessibility	10	4	4	4	1	4
Use limitations	5	4	4	4	4	4
Safety	10	3	3	3	3	3
Logistics	5	4	4	4	4	4
Density of population	5	3	3	3	3	3
Association with other values	5	4	4	3	3	4
Scenery	15	4	4	3	3	3
Uniqueness	10	4	3	4	4	4
Observation conditions	5	4	3	4	4	4
Interpretative potential	10	4	4	4	4	3
Economic level	5	2	2	2	2	2
Proximity of recreational areas	5	4	4	4	3	4
Total score		375	360	355	325	350
		High	High	High	High	High

<200 Low, 201–300 Moderate, >301 High [87].

Table 5. Degradation risk evaluation of geosites.

Scientific Criteria	Weight	Craters		Natural Sauna Points		Thermal Spring
		Stefanos	Polyvotis	Emporios	Piria	Panagia Thermiani
Deterioration of geological features	35	1	1	2	1	1
Proximity to areas/activities with potential to cause degradation	20	1	1	4	1	4
Legal protection	20	2	2	2	4	2
Accessibility	15	4	3	4	1	4
Density of population	10	3	3	3	3	3
Total score		185	170	280	180	245
		Low	Low	Moderate	Low	Moderate

<200 Low, 201–300 Moderate, >301 High [17,87].

Table 6. Final scores for G-P method.

Values	Craters		Natural Sauna Points		Thermal Spring
	Stefanos	Polyvotis	Emporios	Piria	Panagia Thermiani
Scientific value	315 High	315 High	250 Moderate	230 Moderate	250 Moderate
Educational value	385 High	370 High	315 High	285 Moderate	320 High
Tourism value	375 High	360 High	355 High	325 High	350 High

3.2. The M-GAM Method

The application of the M-GAM quantitative evaluation method (Tables 7 and 8) reveals a corresponding picture of the results. In more detail, the geosite of Piria shows the lowest overall scores, whereas the craters of Stefanos and Polyvotis have the highest scores. These two geosites highlight important geoeducational elements, particularly magmatogenesis and the formation of geothermal fields. Remarkable scores also appear in the thermal water springs of Panagia Thermiani (GS5), Emporio (GS3), and Piria (GS4), where heat diffusion from the inside of the earth occurs due to the intense fault zones. Therefore, it can be said that the two methods show a fairly remarkable correlation. Moreover, this method may not take into account many parameters, but it includes, in addition to the opinion of experts, the opinion of visitors and the public.

Table 7. Calculation by M-GAM method of geosites [36,86].

	M-GMAM Method										
	Values Given by Experts						Total Value (with Im Factor)				
	Craters		Natural Sauna Points		Thermal Spring		Craters		Natural Sauna Points		Thermal Spring
	Stefanos	Polyvotis	Emporios	Piria	Panagia Thermiani	Im	Stefanos	Polyvotis	Emporios	Piria	Panagia Thermiani
Main Values (MV)											
Scientific/educational value (VSE)											
1. Rarity	0.75	0.75	0.5	0.5	0.25	0.89	0.6675	0.6675	0.445	0.445	0.2225
2. Representativeness	0.75	0.75	0.5	0.5	0.25	0.79	0.5925	0.5925	0.395	0.395	0.1975
3. Knowledge of scientific issues	1.00	1.00	0.5	0.5	0.5	0.45	0.45	0.45	0.225	0.225	0.225
4. Level of interpretation	0.50	0.50	0.25	0.25	0.25	0.85	0.425	0.425	0.2125	0.2125	0.2125
Scenic/aesthetic (VSA)											
5. Viewpoints	0.75	0.75	0.5	0.25	0.50	0.79	0.5925	0.5925	0.395	0.1975	0.395
6. Surface	1.00	1.00	0.5	0.50	0.50	0.54	0.54	0.54	0.27	0.27	0.27
7. Surrounding landscape and nature	1.00	1.00	0.75	0.75	0.50	0.95	0.95	0.95	0.7125	0.7125	0.475
8. Environmental fitting of sites	1.00	1.00	0.5	0.50	0.50	0.68	0.68	0.68	0.34	0.34	0.34
Protection (VPr)											
9. Current condition	1.00	1.00	0.50	0.50	0.50	0.83	0.83	0.83	0.415	0.415	0.415
10. Protection level	0.75	0.75	0.25	0.25	0.25	0.76	0.57	0.57	0.19	0.19	0.19
11. Vulnerability	1.00	1.00	0.50	0.50	0.50	0.58	0.58	0.58	0.29	0.29	0.29
12. Suitable number of visitors	1.00	1.00	1.00	1.00	1.00	0.42	0.42	0.42	0.42	0.42	0.42
Additional values (AV)											
Functional values (VFn)											
13. Accessibility	1.00	1.00	1.00	0.25	1.00	0.75	0.75	0.75	0.75	0.1875	0.75
14. Additional natural values	0.75	0.75	0.75	0.50	0.50	0.71	0.5325	0.5325	0.5325	0.355	0.355
15. Additional anthropogenic values	0.75	0.75	0.50	0.25	0.50	0.70	0.525	0.525	0.35	0.175	0.35
16. Vicinity of emissive centers	0.00	0.00	0.25	0.00	0.00	0.48	0.00	0.00	0.12	0.00	0.00
17. Vicinity of important road network	0.25	0.25	0.25	0.00	0.25	0.62	0.155	0.155	0.155	0.00	0.155
18. Additional functional values	0.50	0.50	0.50	0.00	0.50	0.59	0.295	0.295	0.295	0.00	0.295
Touristic values (VTr)											
19. Promotion	0.50	0.50	0.25	0.25	0.25	0.85	0.425	0.425	0.2125	0.2125	0.2125
20. Organized visits	1.00	1.00	1.00	0.00	0.75	0.56	0.56	0.56	0.56	0.00	0.42
21. Vicinity of visitors' centers	0.5	0.50	0.50	0.50	0.50	0.87	0.435	0.435	0.435	0.435	0.435
22. Interpretative panels	0.25	0.25	0.25	0.00	0.00	0.81	0.2025	0.2025	0.2025	0.00	0.00
23. Number of visitors	0.75	0.25	0.25	0.25	0.25	0.43	0.3225	0.1075	0.1075	0.1075	0.1075
24. Tourism infrastructure	0.75	0.75	0.50	0.00	0.25	0.73	0.5475	0.5475	0.365	0.00	0.1825
25. Tour guide service	0.25	0.25	0.00	0.00	0.00	0.87	0.2175	0.2175	0.00	0.00	0.00
26. Hostelry service	0.5	0.50	0.50	0.50	0.50	0.73	0.365	0.365	0.365	0.365	0.365
27. Restaurant service	1.00	1.00	1.00	0.50	0.75	0.78	0.78	0.78	0.78	0.39	0.585

Table 8. Final results of the application of the M-GAM method.

Geosite	Main Values		Additional Values		
	VSE + VSA + VPr	SUM	VFn + VTr	SUM	Field Area
Stefanos crater	2.135 + 2.7625 + 2.4	7.2975	2.2575 + 3.855	6.1125	Z22
Polyvotis crater	2.135 + 2.7625 + 2.4	7.2975	2.2575 + 3.64	5.8975	Z22
Emporios (natural sauna)	1.2775 + 1.7175 + 1.315	4.31	2.2025 + 3.0275	5.23	Z22
Piria (natural sauna)	1.2775 + 1.52 + 1.315	4.1125	0.7175 + 1.51	2.2275	Z21
Panagia Thermiani (thermal spring)	0.8575 + 1.48 + 1.315	3.6525	1.905 + 2.3075	4.2125	Z11

In more detail, the M-GAM method shows the highest scores for the two craters of Stefanos (GS1) and Polyvotis (GS2), respectively. Also, these two sites are geosites that highlight important geoeducational elements, especially the processes of magmatogenesis, creation, and deposition of geological formation, as well as hydrothermal vapors. In addition, with the second method, remarkable scores appear in three other sites: the thermal water springs of Panagia Thermiani (GS5), Emporio (GS3) and Piria (GS4), where the diffusion of heat from the inside of the earth occurs due to the intense fault zones. Therefore, from a geoeducational point of view, the public can be informed about the importance of fault zones and how more superficial geological formations can be affected. Therefore, students and visitors can understand in practice how the earth works internally and what effects it has. In addition, through the geosite of Panagia Thermiani, the meaning and significance of the hydrological cycle of water can be even better understood, since meteoric water and groundwater are united and influenced by endogenous forces, resulting in being warmer.

The diagram below (Figure 10) depicts the method's final results (Table 8). In terms of results, field Z22 contains three of the five areas investigated. In terms of geological peculiarities, these three areas share values and characteristics. Fields Z11 and Z21 each host one region with the lowest score when compared to the other three.

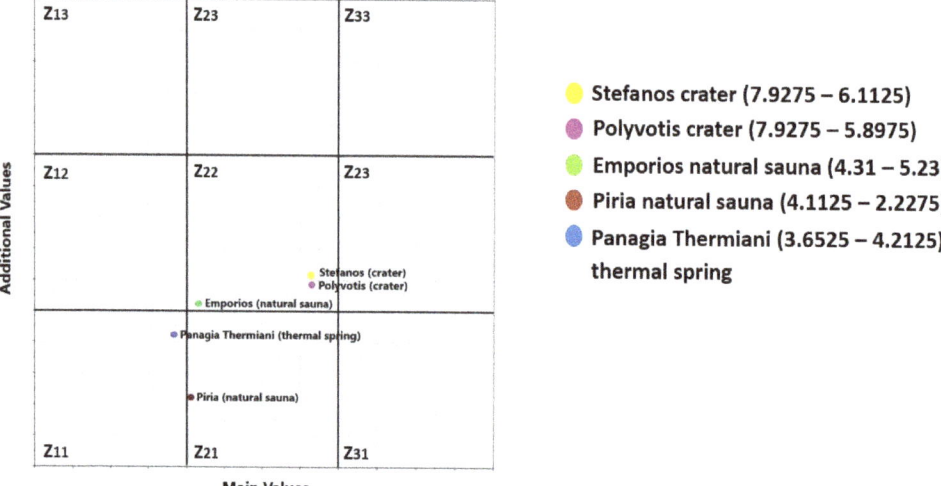

Figure 10. The display of results from the M-GAM method.

4. Discussion

Comparative Analysis

"Telling the geological story of a protected area is the equivalent of telling people about a slice of Earth's history," writes Tormey [88]. In this sense, a site with significant geological and geomorphological features and processes must have educational value in order to help increase knowledge about dynamic phenomena that occur on the earth's surface on a continuous basis. The educational value of a specific geoheritage site will also aid in the conservation of natural resources, which is one of the key principles of Geotourism. The evaluation of educational value typically assesses the representativeness of the features or processes, exemplarity, and educational usage of the specific geoheritage site.

The goal of this study was to assess the geoeducational potential of five habitats on the volcanic island of Nisyros' aspiring geopark by employing two different methods, each with a different philosophy, comparing the results, and exploring the limitations that may be contained in each. The general-purpose Model (G-P method) was chosen because it examines a wide range of fields that are clearly graded and evaluated in terms of geoeducational potential. The M-GAM method, on the other hand, takes into account the opinions of visitors who, as non-experts, express a different point of view that is rarely calculated or evaluated in different geosite assessment methods.

The results were reduced to a percentage scale (%) for a better and more objective comparison of the two methods of evaluating the educational potential of the study areas.

Table 9 displays the final results obtained from applying the methods, based on the grading scale for the geoeducational value of the respective geosites studied. In addition, the final score is recorded after the results have been converted and reduced to a common scale.

Table 9. Final results after the application of the two methods and after reduction to a common scale.

Geosites	G-P Method—Final Score of Education Use	M-GAM Method—Final Score of Education Use		Conversion to a Percentage for G-P Method	Conversion to a Percentage for M-GAM Method
		Main Value	Additional Value		
GS1-Stefanos crater	385	7.2975	6.1125	96.25	49.53
GS2-Megalos Polyvotis crater	370	7.2975	5.8975	92.5	48.85
GS3-Thermal spring of Pyria	315	4.31	5.23	78.75	35.26
GS4-Thermal spring of Emporios	285	4.1125	2.2275	71.25	24.32
GS5-Thermal spring of Panagia Thermiani	320	3.6525	4.2125	80.00	29.01

According to the results shown in the table above, there is great difference in the values obtained by the application of the two evaluation methods. The first G-P method clearly highlights the high geological value of the studied geosites, which have a relatively high score and can be used for geotourism and geoeducation. Because it considers more fields and criteria, the G-P method more accurately testifies to and captures the dynamics of an area with strong geological features. This is accomplished by providing a more positive assessment and emphasizing both geographical and geoenvironmental value, both of which can be demonstrated to the general public through geoeducational activities.

The second method, on the other hand, yields a moderate score in areas with objectively high geological value. This is evidently becaause this method considers the opinions of visitors who do not have the necessary cognitive geological background, thus underestimating the importance and potential of certain geological features due to lack of formal training.

As a result, the comparison of the two methods reveals that the G-P method necessitates more parameters (12) to document the geoeducational value of the sites and approaches the geoeducational dimension of places of high geological interest in a broader and deeper manner. The second method, on the other hand, may not take into account as many parameters, but it does include, in addition to the opinion of experts, the opinion of visitors and the general public.

However, both assessment methods used in the current study clearly demonstrated that the geoeducational potential of the selected geosites can be developed. The G-P method clearly highlights the geoeducational potential of each geosite, because it includes didactic potential as a separate field in its analysis and gives it a special emphasis. It reveals that utilization is possible even at the highest educational level. As a result, the magnitude of the educational value and perspective became more understood, encouraging geoeducational activities aimed at primary, secondary, and higher education.

On the other hand, the low scores for the geoeducational potential of these geosites obtained using the M-GAM method are due to the inclusion of visitor feedback. The general public, who lacks sufficient geological knowledge, cannot comprehend the educational perspective and potential of a region with strong geological features. The lack of geoeducational knowledge in culture becomes apparent, and the need for geoeducational activities to be implemented is identified. This will result in the dissemination and promotion of the geoheritage value of areas of high geological importance.

5. Conclusions

The island of Nisyros is undoubtedly a living geological laboratory, in which there are huge possibilities and prospects for the development of mainly geoeducational activities. In fact, the various geoeducational programs that can be implemented in the field will be able to contribute to the dissemination and promotion of geoeducation at various educational levels. In this way, issues such as geoheritage, geoethics, and geoconservation will become more understandable to each visitor and clear to the local community.

The two different methods implemented in this study could improve further evaluation of the geoeducational potential of five geosites located around the caldera of the volcanic island of Nisyros.

The first method, used by Brilha [17], is considered a general-purpose method (G-P method) designed to assess any type of geosite, considering a wide spectrum of criteria. The inclusion of 12 criteria for assessing the geoeducational value of studied geosites leads to more objective results. On the other hand, the M-GAM method incorporates the perspectives of the public, which in the present study clearly illustrates the lack of geoenvironmental awareness and knowledge.

The two methods provide different perspectives on the geoenvironmental value of a given geosite.

The first method addresses the geoeducational dimension of places of high geological interest in a broader and more in-depth way. The second method, while not taking into account as many parameters, does include the opinion of visitors and the general public.

For the most comprehensive evaluation of the geoeducational perspective on a geosite, the combination of the two aforementioned methods is considered necessary.

Author Contributions: Conceptualization, G.Z. and H.D.; methodology, G.Z. and H.D.; formal analysis, G.Z.; investigation, G.Z.; resources, G.Z.; data curation, G.Z.; writing—original draft preparation, G.Z.; writing—review and editing, H.D.; supervision, H.D. All authors have read and agreed to the published version of the manuscript.

Funding: This research received no external funding.

Informed Consent Statement: Not applicable.

Data Availability Statement: The data presented in this study are available on request from the corresponding author.

Acknowledgments: The authors gratefully thank the journal editor and the three reviewers for their thorough consideration of this paper.

Conflicts of Interest: The authors declare no conflict of interest.

References

1. Martini, G. (Ed.) Actes du premier symposium international sur la protection au patrimonie geologique. In *Memoires de la Societe 656 geologique de France, Proceedings of the First Symposium on Earth Heritage Conservation, 11–16 June 1991*; numero special 165; Société géologique de France: Digne, France, 1993; 276p.
2. Zouros, N. The European Geoparks Network. *Episodes* **2004**, *27*, 165–171.
3. Zouros, N.; Valiakos, I. Geoparks Management and Assessment. *Bull. Geol. Soc. Greece* **2017**, *43*, 965–977. [CrossRef]
4. Zwolinski, Z. Geodiversity. In *Encyclopedia of Geomorphology*; Goudie, A.S., Ed.; Routledge: Oxfordshire, UK, 2004; Volume 1, pp. 417–418.
5. Sharples, C. Geoconservation in forest management—Principles and practice. *Tasforests* **1995**, *7*, 37–50.
6. Burek, C.V.; Potter, J. *Local Geodiversity Action Plans-Sharing Good Practice Workshop, Peterborough, 3 December 2003*; English Nature: Peterborough, UK, 2004.
7. Burek, C.V.; Potter, J. *Local Geodiversity Action Plans-Setting to Context for Geological Conservation*; English Nature: Peterborough, UK, 2006.

8. Herrera-Franco, G.; Montalván-Burbano, N.; Carrión-Mero, P.; Apolo-Masache, B.; Jaya-Montalvo, M. Research trends in geotourism: A bibliometric analysis using the scopus database. *Geosciences* **2020**, *10*, 379. [CrossRef]
9. Ólafsdóttir, R.; Tverijonaite, E. Geotourism: A systematic literature review. *Geosciences* **2018**, *8*, 234. [CrossRef]
10. Panizza, M. Geomorphosites: Concepts, methods and examples of geomorphological survey. *Chin. Sci. Bull.* **2001**, *46*, 4–5. [CrossRef]
11. Trueba, J.J.G.; Cañadas, E.S. La valoración del patrimonio geomorfológico en espacios naturales protegidos. Su aplicación al parque nacional de los picos de Europa. *Bull. Assoc. Span. Geogr.* **2008**, *47*, 175–194, Published online 2008.
12. De Wever, P.; Baudin, F.; Pereira, D.; Cornee, A.; Egoroff, G.; Page, K. The Importance of Geosites and Heritage Stones in Cities—A Review. *Geoheritage* **2017**, *9*, 561–575. [CrossRef]
13. Reynard, E.; Fontana, G.; Kozlik, L.; Scapozza, C. A method for assessing the scientific and additional values of geomorphosites. *Geogr. Helv.* **2007**, *62*, 148–158. [CrossRef]
14. Ruban, D.A. Quantification of geodiversity and its loss. *P. Geologist. Assoc.* **2010**, *121*, 326–333. [CrossRef]
15. Ruban, D.A. Geotourism—A geographical review of the literature. *Tour. Manag. Perspect.* **2015**, *15*, 1–15. [CrossRef]
16. Skentos, A. Geotopes of Greece. Master's Thesis, University of Athens, Athens, Greece, 2012.
17. Brilha, J.B. Inventory and Quantitative Assessment of Geosites and Geodiversity Sites: A Review. *Geoheritage* **2016**, *8*, 119–134. [CrossRef]
18. Henriques, M.H.; Brilha, J. UNESCO Global Geoparks: A strategy towards global understanding and sustainability. *Episodes* **2017**, *40*, 349–355. [CrossRef]
19. Bruschi, V.M.; Coratza, P. Geoheritage and Environmental Impact Assessment (EIA). In *Geoheritage: Assessment, Protection and Management*; Reynard, E., Brilha, J., Eds.; Elsevier: Amsterdam, The Netherlands, 2018; pp. 251–264.
20. Zwoliński, Z.; Najwer, A.; Giardino, M. Methods for Assessing Geodiversity. In *Geoheritage: Assessment, Protection, and Management*; Reynard, E., Brilha, J., Eds.; Elsevier: Amsterdam, The Netherlands, 2018; pp. 27–52.
21. Drinia, H.; Tsipra, T.; Panagiaris, G.; Patsoules, M.; Papantoniou, C.; Magganas, A. Geological heritage of Syros Island, Cyclades complex, Greece: An assessment and geotourism perspectives. *Geosciences* **2021**, *11*, 138. [CrossRef]
22. Fassoulas, C.; Mouriki, D.; Dimitriou-Nikolakis, P.; Iliopoulos, G. Quantitative Assessment of Geotopes as an Effective Tool for Geoheritage Management. *Geoheritage* **2012**, *4*, 177–193. [CrossRef]
23. Cendrero, A. El patrimonio geológico. Ideas para su protección, conservación y utilización. In *El patrimonio geológico. Bases para su valoración, protección, conservación y utilización*; Serie Monografías del Ministerio de Obras Públicas, Transportes y Medio Ambiente; Ministerio de Obras Públicas, Transportes y Medio Ambiente: Madrid, Spain, 1996; pp. 17–27.
24. Cendrero, A. Propuestas sobre criterios para la clasificación y catalogación del patrimonio geológico. In *El patrimonio geológico. Bases para su valoración, protección, conservación y utilización*; Serie Monografías del Ministerio de Obras Públicas, Transportes yMedio Ambiente; Ministerio de Obras Públicas, Transportes y Medio Ambiente: Madrid, Spain, 1996; pp. 29–38.
25. Coratza, P.; Giusti, C. Methodological proposal for the assessment of the scientific quality of geomorphosites. *Il Quat.* **2005**, *18*, 307–313.
26. Pralong, J.P.; Reynard, E. A proposal for the classification of geomorphological sites depending on their tourist value. *Quaternario* **2005**, *18*, 315–321.
27. Pereira, P.; Pereira, D.; Alves, M.I.C. Geomorphosite assessment in Montesinho Natural Park (Portugal). *Geogr. Helv.* **2007**, *62*, 159–168. [CrossRef]
28. Bruschi, V.M.; Cendrero, A. Direct and parametric methods for the assessment of geosites and geomorphosites. In *Geomorphosites*; Reynard, E., Coratza, P., Regolini-Bissig, G., Eds.; Verlag Dr. Friedrich Pfeil: München, Germany, 2009; pp. 73–88.
29. Reynard, E. The assessment of geomorphosites. In *Geomorphosites*; Reynard, E., Coratza, P., Regolini Bissig, G., Eds.; Verlag Dr. Friedrich Pfeil: Munchen, Germany, 2009; pp. 63–71.
30. Pereira, P.; Pereira, D. Methodological guidelines for geomorphosite assessment. Géomorphologie: Relief, processus. *Environnement* **2010**, *16*, 215–222.
31. Bruschi, V.M.; Cendrero, A.; Albertos, J.A.C. A statistical approach to the validation and optimisation of geoheritage assessment procedures. *Geoheritage* **2011**, *3*, 131–149. [CrossRef]
32. Pereira, P.; Pereira, D.I. Assessment of geosites tourism value in geoparks: The example of Arouca Geopark (Portugal). In Proceedings of the 11th European Geoparks Conference, Arouca, Portugal, 19–21 September 2012; pp. 231–232.
33. Bollati, I.; Smiraglia, C.; Pelfini, M. Assessment and selection of geomorphosites and trails in the Miage Glacier area (Western Italian Alps). *Environ. Manag.* **2013**, *51*, 951–967. [CrossRef] [PubMed]
34. Gray, M. *Geodiversity: Valuing and Conserving Abiotic Nature*; Wiley: Chichester, UK, 2004; p. 448.
35. Brilha, J.; Gray, M.; Pereira, D.I.; Pereira, P. Geodiversity: An integrative review as a contribution to the sustainable management of the whole of nature. *Environ. Sci. Policy* **2018**, *86*, 19–28. [CrossRef]
36. Tomić, N.; Božić, S. A modified geosite assessment model (MGAM) and its application on the Lazar Canyon area (Serbia). *Int. J. Environ. Res.* **2014**, *8*, 1041–1052.
37. Antić, A.; Tomić, N.; Marković, S.B. Karst geoheritage and geotourism potential in the Pek River lower basin (eastern Serbia). *Geogr. Pannoni.* **2019**, *23*, 32–46. [CrossRef]
38. Gray, M. *Geodiversity: Valuing and Conserving Abiotic Nature*, 2nd ed.; Wiley-Blackwell: Hoboken, NJ, USA, 2013; pp. 3–14.

39. Megía, M.V.; Alarcón, J.C.B.; Romero, J.S.G.; Muñoz, A.B.P. El inventario andaluz de georrecursos culturales: Criterios de valoración. *Rev. Soc. Española Def. Patrim. Geológico Min.* **2004**, *3*, 9–22.
40. Kozlowski, S. Geodiversity. The concept and scope of geodiversity. *Prz. Geol.* **2004**, *52*, 833–837.
41. Vujičić, M.D.; Vasiljević, Đ.A.; Marković, S.B.; Hose, T.A.; Lukić, T.; Hadžić, O.; Janićević, S. Preliminary geosite assessment model (GAM) and its application on Fruška Gora Mountain, potential geotourism destination of Serbia. *Acta Geogr. Slov.* **2011**, *51*, 361–377. [CrossRef]
42. Tomić, N. The potential of Lazar Canyon (Serbia) as a geotourism destination: Inventory and evaluation. *Geogr. Pannonica* **2011**, *15*, 103–112. [CrossRef]
43. Dietrich, V.J.; Lagios, E. (Eds.) *Nisyros Volcano, Active Volcanoes of the World*; Springer: Berlin, Germany, 2018; pp. 13–55. ISBN 978-3-319-55458-7.
44. Papanikolaou, D.; Lekkas, E.; Sakellariou, D. Geological structure and evolution of the Nisyros volcano. *Bull. Geol. Soc. Greece* **1991**, *25*, 405–419.
45. Tibaldi, A.; Pasquarè, F.A.; Papanikolaou, D.; Nomikou, P. Tectonics of Nisyros Island, Greece, by field and offshore data, and analogue modeling. *J. Struct. Geol.* **2008**, *30*, 1489–1506. [CrossRef]
46. Francalanci, L.; Vougioukalakis, G.E.; Perini, G.; Manetti, P. A West-East Traverse along the magmatism of the south Aegean volcanic arc in the light of volcanological, chemical and isotope data. *Dev. Volcanol.* **2005**, *7*, 65–111.
47. Khaleghi, M.; Ranjbar, H.; Abedini, A.; Calagari, A.A. Synergetic use of the Sentinel-2, ASTER, and Landsat-8 data for hydrothermal alteration and iron oxide minerals mapping in a mine scale. *Acta Geodyn. Geromater.* **2020**, *17*, 311–329. [CrossRef]
48. Rajan Girija, R.; Mayappan, S. Mapping of mineral resources and lithological units: A review of remote sensing techniques. *Int. J. Image Data Fusion.* **2019**, *10*, 79–106. [CrossRef]
49. Di Paola, G.M. Volcanology and petrology of Nisyros Island (Dodecanese, Greece). *Bull. Volcanol.* **1974**, *38*, 944–987. [CrossRef]
50. Hunziker, J.C.; Marini, L. (Eds.) *The geology, Geochemistry and Evolution of Nisyros Volcano (Greece): Implications for the Volcanic Hazards*; Section des sciences de la Terre, Université de Lausanne: Lausanne, Switzerland, 2005; Volume 44.
51. Martelli, A. *Il gruppo eruttivo di Nisiro nel Mare Egeo, 1917, Memorie della Societa Italiano della Scienze detta dei XL Serie 3a T. XX*; Accademia dei Lincei: Rome, Italy, 1917.
52. Marini, L.; Fiebig, J. Fluid geochemistry of the magmatic-hydrothermal system of Nisyros (Greece). In *The Geology, Geochemistry and Evolution of Nisyros Volcano*; Mémoire de Géologie: Lausanne, Switzerland, 2005.
53. Ambrosio, M.; Doveri, M.; Fagioli, M.T.; Marini, L.; Principe, C.; Raco, B. Water–rock interaction in the magmatic-hydrothermal system of Nisyros Island (Greece). *J. Volcanol. Geother. Res.* **2010**, *192*, 57–68. [CrossRef]
54. Gorceix, M.H. Sur l'état du volcan de Nisyros au mois de mars 1873. *C. R. Seances Acad. Sci. Paris* **1873**, *77*, 597–601.
55. Gorceix, M.H. Sur l'éruption boueuse de Nisyros. *C. R. Seances Acad. Sci.* **1873**, *77*, 1474–1477.
56. Gorceix, M.H. Etude des fumerolles de Nisyros et de quelques-uns des produits des éruptions dont cette ile a été le siège en 1872 et 1873. *Ann. Chim. Phys. Paris* **1874**, 333–354.
57. Marini, L.; Principe, C.; Chiodini, G.; Cioni, R.; Fytikas, M.; Marinelli, G. Hydrothermal eruptions of Nisyros (Dodecanese, Greece). Past events and present hazard. *J. Volcanol. Geother. Res.* **1993**, *56*, 71–94. [CrossRef]
58. Vassilopoulou, S.; Hurni, L. The use of digital elevation models in emergency and socio-economic planning: A case study at Kos-Yali-Nisyros-Tilos islands, Greece. In Proceedings of the 20th International Cartographic Conference, Beijing, China, 6–10 August 2001; Chinese Society of Geodesy, Photogrammetry and Cartography: Beijing, China; pp. 3424–3431.
59. Desio, A. Le isole italiane dell'Egeo. *Mem. Carta Geol. D'Ital.* **1931**, *24*, 534.
60. Davis, E.N. Zur geolofie und Petrologie der Inseln Nisyros und Jail (Dodekanes). *Prakt. Acad. Athens* **1967**, *42*, 235–252.
61. Piper, D.J.W.; Pe-Piper, G.; Anastasakis, G.; Reith, W. The volcanic history of Pyrgousa—Volcanism before the eruption of the Kos Plateau Tuff. *Bull. Volcanol.* **2019**, *81*, 32. [CrossRef]
62. Papanikolaou, D.; Nomikou, P. Tectonic structure and volcanic centres at the eastern edge of the Aegean volcanic arc around Nisyros Island. *Bull. Geol. Soc. Greece* **2001**, *34*, 289–296. [CrossRef]
63. Lagios, E.; Sakkas, V.; Parcharidis, I.; Dietrich, V. Ground Deformation of Nisyros Volcano (Greece) for the period 1995–2002: Results from DInSAR and DGPS observations. *Bull. Volcanol.* **2005**, *68*, 201–214. [CrossRef]
64. Papadopoulos, G.A.; Sachpazi, M.; Panopoulou, G.; Stavrakakis, G. The volcanoseismic crisis of 1996–1997 in Nisyros, SE Aegean Sea, Greece. *Terra Nova* **1998**, *10*, 151–154. [CrossRef]
65. GEOWARN-IST 12310. Geological Map of Greece, 1:10,000. Geo-SpatialWarning Sys-tems Nisyros Volcano (Greece): An Emergency Case Study. Information Society Tech-nologies Programme. Available online: www.geowarn.ethz.ch (accessed on 13 January 2022).
66. Sykioti, O.; Kontoes, C.; Elias, P.; Briole, P.; Sachpazi, M.; Paradissis, D.; Kotsis, I. Ground deformation at Nisyros volcano (Greece) detected by ERS-2 SAR differential interferometry. *Int. J. Remote Sens.* **2003**, *24*, 183–188. [CrossRef]
67. Venturi, S.; Tassi, F.; Vaselli, O.; Vougioukalakis, G.E.; Rashed, H.; Kanellopoulos, C.; Caponi, C.; Capecchiacci, F.; Cabassi, J.; Ricci, A.; et al. Active hydrothermal fluids circulation triggering small-scale collapse events: The case of the 2001–2002 fissure in the Lakki Plain (Nisyros Island, Aegean Sea, Greece). *Nat. Hazards* **2018**, *93*, 601–626. [CrossRef]
68. Chiodini, G.; Cioni, R.; Marini, L. Reactions governing the chemistry of crater fumaroles from Vulcano Island, Italy, and implications for volcanic surveillance. *Appl. Geochem.* **1993**, *8*, 357–371. [CrossRef]

69. Kavouridis, T.; Kuris, D.; Leonis, C.; Liberopoulou, V.; Leontiadis, J.; Panichi, C.; La Ruffa, G.; Caprai, A. Isotope and chemical studies for a geothermal assessment of the island of Nisyros (Greece). *Geothermics* **1999**, *28*, 219–239. [CrossRef]
70. Chiodini, W.; Brombach, T.; Caliro, S.; Cardellini, C.; Marini, L.; Dietrich, V. Geochemical indicators of possible ongoing volcanic unrest at Nisyros Island (Greece). *Geophys. Res. Lett.* **2002**, *29*, 1759. [CrossRef]
71. Brombach, T.; Caliro, S.; Chiodini, G.; Fiebig, J.; Hunziker, J.C.; Raco, B. Geochemical evidence for mixing of magmatic fluids with seawater, Nisyros hydrothermal system, Greece. *Bull. Volcanol.* **2003**, *65*, 505–516. [CrossRef]
72. Dotsika, E.; Poutoukis, D.; Michelot, J.; Raco, B. Natural tracers for identifying the origin of the thermal fluids emerging along the Aegean Volcanic arc (Greece): Evidence of Arc-Type Magmatic Water (ATMW) participation. *J. Volcanol. Geotherm. Res.* **2009**, *179*, 19–32. [CrossRef]
73. Carcavilla Urquí, L.; López Martínez, J.; Durán Valsero, J.J. *Patrimonio Geológico y Geodiversidad: Investigación, Conservación, Gestión y Relación Cuadernos*; Instituto Geológico y Minero de España (IGME): Madrid, Spain, 2007; ISBN 9788478407101.
74. Fuertes-Gutiérrez, I.; Fernández-Martínez, E. Mapping geosites for geoheritage management: A methodological proposal for the Regional Park of Picos de Europa (León, Spain). *Environ. Manag.* **2012**, *50*, 789–806. [CrossRef] [PubMed]
75. Vuković, S.; Antić, A. Speleological approach for geotourism development in Zlatibor county (west Serbia). *Turizam* **2019**, *23*, 53–68. [CrossRef]
76. Tomić, N.; Antić, A.; Marković, S.B.; Đorđević, T.; Zorn, M.; Breg Valjavec, M. Exploring the potential for speleotourism development in eastern Serbia. *Geoheritage* **2019**, *11*, 359–369. [CrossRef]
77. Tičar, J.; Tomić, N.; Breg Valjavec, M.; Zorn, M.; Marković, S.B.; Gavrilov, M.B. Speleotourism in Slovenia: Balancing between mass tourism and geoheritage protection. *Open Geosci.* **2018**, *10*, 344–357. [CrossRef]
78. Antić, A.; Tomić, N. Assessing the speleotourism potential together with archaeological and palaeontological heritage in Risovača Cave (Central Serbia). *Acta Geoturistica* **2019**, *10*, 1–11.
79. Miljković, Đ.; Božić, S.; Miljković, L.; Marković, S.B.; Lukić, T.; Jovanović, M.; Bjelajac, D.; Vasiljević, Đ.A.; Vujičić, M.D.; Ristanović, B. Geosite assessment using three different methods; a comparative study of the Krupaja and the Žagubica Springs—Hydrological Heritage of Serbia. *Open Geosci.* **2018**, *10*, 192–208. [CrossRef]
80. Pál, M.; Albert, G. Comparison of geotourism assessment models: And experiment in Bakony–Balaton UNSECO Global Geopark, Hungary. *Acta Geoturistica* **2018**, *9*, 1–13. [CrossRef]
81. Jonić, V. Comparative analysis Devil's town and Bryce canyon geosites by applying the modified geosite assessment model (M-GAM). *Researches Review the Department Geography. Tour. Hotel Manag.* **2018**, *47*, 113–125. [CrossRef]
82. Antić, A.; Tomić, N. Geoheritage and geotourism potential of the Homolje area (eastern Serbia). *Acta Geoturistica* **2017**, *8*, 67–78. [CrossRef]
83. Vukoičić, D.; Milosavljević, S.; Valjarević, A.; Nikolić, M.; Srećković-Batoćanin, D. The evaluation of geosites in the territory of National park 'Kopaonik' (Serbia). *Open Geosci.* **2018**, *10*, 618–633. [CrossRef]
84. Tomić, N.; Marković, S.B.; Korać, M.; Mrđić, N.; Hose, T.A.; Vasiljević, D.A.; Jovičić, M.; Gavrilov, M.B. Exposing mammoths: From loess research discovery to public palaeontological park. *Quat. Int.* **2015**, *372*, 142–150. [CrossRef]
85. Tomić, N.; Marković, S.B.; Antić, A.; Tešić, D. Exploring the potential for geotourism development in the Danube Region of Serbia. *Int. J. Geoheritage Park.* **2020**, *8*, 123–139. [CrossRef]
86. Božić, S.; Tomić, N. Canyons and gorges as potential geotourism destinations in Serbia: Comparative analysis from two perspectives—General geotourists' and pure geotourists'. *Open Geosci.* **2015**, *7*, 531–546. [CrossRef]
87. Lima, F.F. *Proposta metodológica para inventariação do patrimônio geológico brasileiro*. Dissertação (Mestrado em Patrimônio Geológico e Conservação); Universidade do Minho: Braga, Portugal, 2008.
88. Tormey, D. New approaches to communication and education through geoheritage. *Int. J. Geoheritage Park.* **2019**, *7*, 192–198. [CrossRef]

Article

A Step towards a Sustainable Tourism in Apennine Mountain Areas: A Proposal of Geoitinerary across the Matese Mountains (Central-Southern Italy)

Francesca Filocamo [1],*, Carmen Maria Rosskopf [1], Vincenzo Amato [1] and Massimo Cesarano [2]

[1] Department of Biosciences and Territory, University of Molise, Contrada Fonte Lappone, 86090 Pesche, IS, Italy; rosskopf@unimol.it (C.M.R.); vincenzo.amato@unimol.it (V.A.)
[2] Institute of Environmental Geology and Geoengineering (IGAG), National Research Council (CNR), Via Salaria km 29,300, 00015 Montelibretti, RM, Italy; massimo.cesarano@igag.cnr.it
* Correspondence: francesca.filocamo@gmail.com; Tel.: +39-0874-404168

Abstract: The Apennine mountain areas suffer progressive abandonment and marginality, although being characterized by an extraordinary richness in natural and cultural resources, and landscapes of great beauty. Therefore, their natural heritage, and especially their geoheritage, tranformed into geotourism initiatives, can represent an essential resource to support local economy andsustainable development. The present study illustrates the case of Matese Mountains (Southern Apennines), particularly rich in protected areas, including the Matese National Park currently taking off, which is characterized by a rich geoheritage, based on 59 geosites. Among these geosites, examining the specially built geosite GIS database, 16 geosites were selected to construct a geoitinerary crossing the Matese Mountains. The geoitinerary was delineated to optimally represent the major geomorphological and geological (especially geohistorical) features of the Matese area. The selected geosites were associated to a new procedure to assess their Scientific Value (SV) and Potential Tourism Use (PTU), and to confirm their suitability for the purpose. To illustrate the geoitinerary, a geoitinerary map, and illustration material such as descriptive cards were produced. As an overall result, the proposed geoitinerary represents a valuable contribution for the geotourism promotion of the Matese Mountains on which to base future studies and initiatives in this perspective.

Keywords: inner areas; natural resources; geosites; geoheritage; geotourism; Matese National Park; Southern Apennines

1. Introduction

The Mediterranean mountain ranges (Alps, Apennines, Pyrenees, Atlas, etc.) are areas of priority interest due to their natural resources, especially for their richness in fauna and flora and relatively high biodiversity. Likewise, these mountainous areas are also of high geological interest as they testify important steps of the geological history of the Earth and host highly diversified landscapes, which result from the prolonged interplay of endogenic and exogenic processes under variable climate conditions. This makes the need for natural resource protection a priority for Mediterranean mountain areas, implicating a substantial maintenance of their environmental features and values.

At the same time, Mediterranean mountain areas are largely part of the so-called inner areas, i.e., rural areas that experience marginalization due to their geographical and socio-economic conditions ([1,2] and references therein), and are significantly affected by demographic decline and population ageing, as well as landscape degradation caused primarily by agricultural abandonment ([3] and references therein).

The set of these characteristics, together with the significant geographical space that mountain areas occupy, make them become priority objects of sustainable development policies (e.g., [4]).

In Europe, among the several initiatives aimed at reversing depopulation and marginalization of peripheral areas, one worthy of mentioning is the National Strategy for Inner Areas (NSIA), launched in 2012 by the Italian Government, that counts among its main themes the defense and usage of cultural heritage [5]. This theme implicates a contrast between interventions aimed at nature protection, especially the institution and management of protected areas (from special protection areas up to national parks), widespread at the European scale, and others that focus on the exploitation of the cultural heritage and the socio-economic development of the territory.

Obviously, the Mediterranean mountain areas, thanks to their landscapes of exceptional aesthetic quality [6] and richness in natural and cultural resources, are important potential destinations of tourism activities that, however, have to "unite under the same umbrella" the need of environment preservation with the desired socio-economic development. It is also for this reason that concepts such as mountain tourism [6,7], sustainable tourism, eco-tourism and slow travel ([8,9] and references therein), as well as geotourism [9–12], are progressively developing and gaining increasing importance.

In our specific case, research focuses on geotourism, which is intended as a geology-based tourism [12] and, according to [9], as tourism which focuses on an area's geology and landscape as the basis of fostering sustainable tourism development. Geotourism is one of the newest concepts in tourism studies today. It has grown rapidly over the past few decades and the potential for geotourism development is largely going to be explored in European and Mediterranean countries (e.g., [13–16] and references therein). Particularly, geotourism is among the novel strategies used for socio-economic development in rural areas [17,18], and has been demonstrated to have positive economic effects in several contexts and especially in Geopark areas, both at the European and global scale (e.g., [17,19–21] and references therein).

Geotourism activities essentially concern the knowledge and exploitation of the geological heritage of a territory. They can represent a valid alternative or integration to other more or less traditional tourism activities in mountain areas, especially for summer seasons, even more as being able to respond to the need to promote scientific research and environmental education. Numerous are the activities that can be realized in several contexts, to promote geological heritage and related geotourism purposes, e.g., [22–25]. Among these, geological itineraries are a powerful tool for the dissemination of geosciences and geotourism development [26–29]. In fact, a consistent part of the recent and rich literature on geotourism concerns the proposal, design and/or illustration of geoitineraries [30–40], especially in Parks and/or Geoparks (i.e., [41–49]).

Among the major mountain areas in Italy, the Matese Mountains (Figure 1) well meet several of the previously mentioned characteristics of Mediterranean mountain areas. However, as regards the current exploitation of its geological heritage, there are only a few initiatives aimed at the promotion of geotourism. Among them, worth mentioning are the geoitineraries proposed respectively for the Molise sector [50,51] and for the Campanian sector [52,53], the two sectors into which the Matese mountains are subdivided from the administrative point of view (see below).

To contribute to the enhancement of the geological heritage of the Matese Mountains, in order to take a step towards the development of sustainable tourism, we have developed a proposal for a geoitinerary that crosses the entire Matese Mountains.

The geosite selection and the definition of the itinerary were based on several criteria (see below), to respond to the following essential requirements: (i) to best enhance the overall geological heritage of the Matese area, (ii) to illustrate in the most complete and optimal way the main steps of its geological history and landscape evolution, and (iii) to overcome the administrative and physiographic boundaries between the Campanian and Molise sectors of the Matese area.

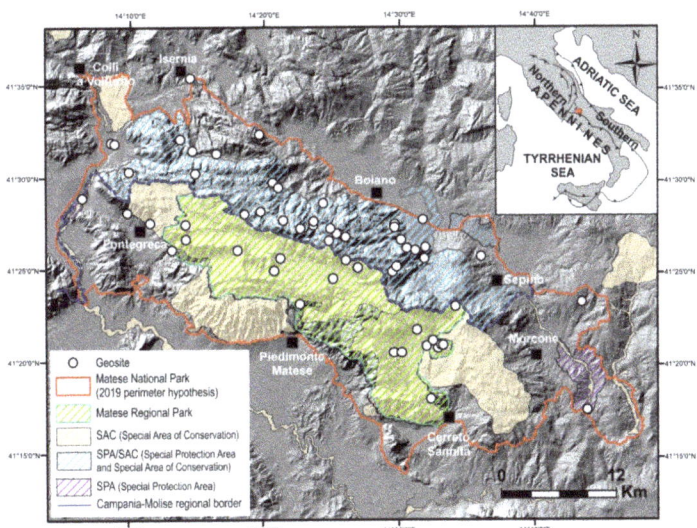

Figure 1. The Matese Mountains. Major protected natural areas and the 2019 perimeter hypothesis of the future Matese National Park are shown.

2. Study Area

The Matese Mountains formed during the Apenninic orogenesis that started following the closure of the Tethys Sea, with the deformation, piling up and uplift of thousands of meters thick marine sedimentary successions that had mainly deposited in carbonate platform environments (e.g., [54]). This group of mountains is placed in the junction zone between the southern and the northern Apennine arcs (Figure 1), and has been object of numerous geological studies since the end of the 1700 [55]. The major geological topics dealt with in the literature, for this sector, concern the tectonic evolution and deformation styles that have characterized it (e.g., [56–59]), the stratigraphy of the Mesozoic-Cenozoic successions (e.g., [60] and references therein), as well as the related palaeogeographic, paleoenvironmental and paleontological aspects (e.g., [54] and references therein, [61,62]).

The Matese Mountains are prevailingly composed of shallow water limestones and dolostones of Triassic to Miocene age (Figure 2), referring to carbonate platform domains. Towards their southeast, a major N-S tectonic feature puts these successions in contact with a tectonic unit composed of varicoloured clays, limestones, marls and arenites belonging to the basinal Sannio Units, and of sandstones and conglomerates of the San Bartolomeo Flysch (Figure 2). Furthermore, sandstones, clays and conglomerates belonging to the Miocene Molise Flysch are locally present both along the borders of the carbonate massif and inside it, within some major tectonic depressions (Figure 2). Rocks of Pliocene age are instead totally lacking in the Matese area. Finally, Quaternary deposits, which are mainly of alluvial and volcanic origin (Figure 2), crop out widespread in the basins and river plains surrounding the Matese Mountains, as well as in most of its major intramountainous tectonic depressions.

From the Late Pliocene onwards, the activity of extensional, mainly NW-SE to W-E and NE-SW oriented faults, caused the progressive tectonic fragmentation of the Matese Mountains. Clear evidence of this is found in the major intramountainous depressions (such as those that host the Matese and Gallo lakes) and the staircases of normal faults, responsible for the progressive relative tectonic lowering, especially towards NE and SW [63–67], of the external sectors of the massif, as well as of the basins around it (Figures 2 and 3).

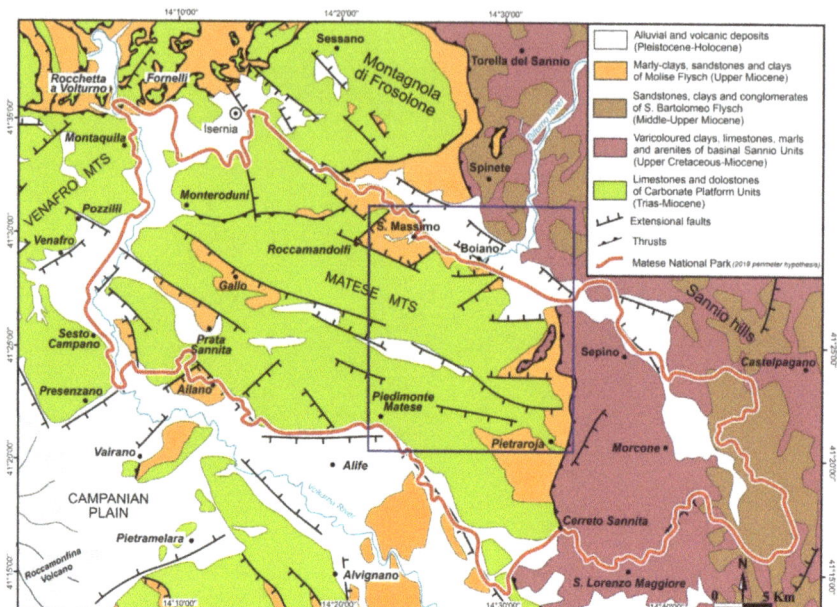

Figure 2. Schematic geologic map of the Matese area and surroundings. The blue frame limits the area selected for the geoitinerary proposal.

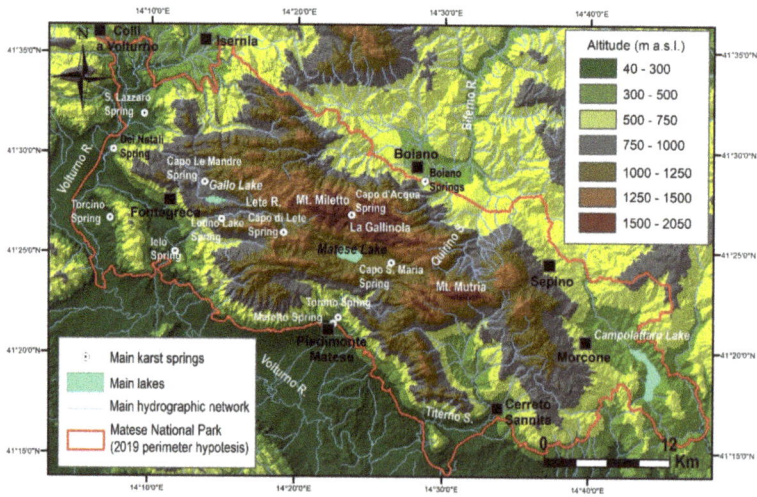

Figure 3. Main altitudinal-hydrographic features of the study area.

Extensional tectonics has significantly controlled the orographic-hydrographic setting of the Matese Mountains. The latter, in fact, are characterized by a central wide mountain to high mountain plateau-like sector with altitudes of above 1000 m and up to 2050 m a.s.l. (Figure 3), which is limited towards the surrounding hills and plains by huge, up to several hundreds of meters high carbonate fault slopes and structural-controlled slopes (Figures 2 and 3). The surface water drainage within the massif appears deeply controlled by fault and thrust alignments. Furthermore, major karst springs (see for instance the Boiano and the Torano-Maretto springs that emerge respectively along the central northern and southern edges of the Massif, Figure 3), are located along the contact between the carbonate karstified system and the surrounding low-permeability to impermeable rocks.

From a morphodynamic point of view, the Matese Massif typically represents the Mediterranean and, especially, the Apennine mountain landscape [68], whose landforms have evolved under the long lasting influence and interplay of tectonics and climate. In particular, with a maximum height of 2050 m (Mount Miletto), the Matese massif is one of the few mountainous areas in central-southern Italy that hosts significant evidence of Middle to Late Pleistocene glaciations, particularly relicts of glacial landforms (cirques and troughs) and remnants of moraines [69,70]. Apart from these important paleoclimatic relicts, the Matese Mountains are particularly rich in karst landforms due to their carbonate nature, and are furthermore characterized above all by periglacial, tectonic-structural and fluvial landforms [68,70].

From an administrative and geographical point of view, the Matese area is divided into a southern and northern sector, falling respectively in the Campanian and Molise region (Figure 1). This circumstance has surely contributed to a certain fragmentation of the mountain territory, and poor road network especially in the central, mountain to high-mountain sector. In fact, villages are located mostly in the external sectors, and are normally reachable through roads starting from the plain areas surrounding the massif, while only a few roads, today partly inaccessible, cross the massif.

The administrative and territorial fragmentation has also played an important role in the environmental valorization and promotion of the Matese Mountains that until now have been managed in an uncoordinated way by the Campania and Molise regions, each restricted exclusively to their own territorial competences. In this regard, the establishment, in 1993, of the Matese Regional Park (Figure 1), which is entirely located in Campanian territory, represented a fundamental, albeit "partial" step towards the conservation and sustainable fruition of the rich natural heritage of the Matese Mountains. Conversely, the Molise Regional Park did never become a reality, and it took nearly three decades to tackle concretely the Matese National Park project. The latter is finally being set up in recent years, even if with some problems related to its perimetration are still under discussion (see perimeter hypothesis 2019 in Figure 1).

A substantial part of the rich natural heritage of the Matese area is linked to the wide extension of the protected areas, consisting in the Matese Regional Park area and 13 partially overlapping Special Protection Areas (SPAs) and Special Areas of Conservation (SACs) (Figure 1). To this are added the great beauty and diversity of its mountain landscape together with its elevated geological heritage. The Molise portion of the Matese area, in fact, is characterized by the highest density of geosites at the regional scale (macro-area Matese-Boiano-Sepino basins, [68]), while a total of 59 geosites are found within the hypothesized perimeter of the Matese National Park (see Figure 1 and below for further details).

Both the biological and geological resources are surely precious and essential for the green growth and sustainable development of the Matese area and, especially, for the success of the future Matese National Park [71], making it a good candidate for future promotion as a Geopark. These resources could contribute, furthermore, to contrast among others the trend to consistent depopulation, also coupled with a net increase of the old-age index, which has affected most of the 37 municipalities of the Matese area documented for the period 1971–2011 [72].

3. Materials and Methods
3.1. Selection and Evaluation of the Geosites

To define the geoitinerary and select the most suitable geosites for it, we followed the workflow illustrated in Figure 4. All data available for the geosites falling in the hypothesized perimeter of the Matese National Park (Figure 1) were examined. Among various data sources, such as literature, geosite inventories and the data archive of the authors, the main sources of data are represented by the following official geosite inventories/projects (Figure 4): the Geosite Inventory of Molise region [68,73,74], the Italian

Geosites Inventory of ISPRA [75], and the "Census of geosites and cartography of the geological-environmental itineraries of Campania" project (geosites included in the Geosite Map of Campania [76,77]. Based on the data collected for all geosites, we created a relative database in a GIS environment (Database of the Matese National Park geosites, Figure 4).

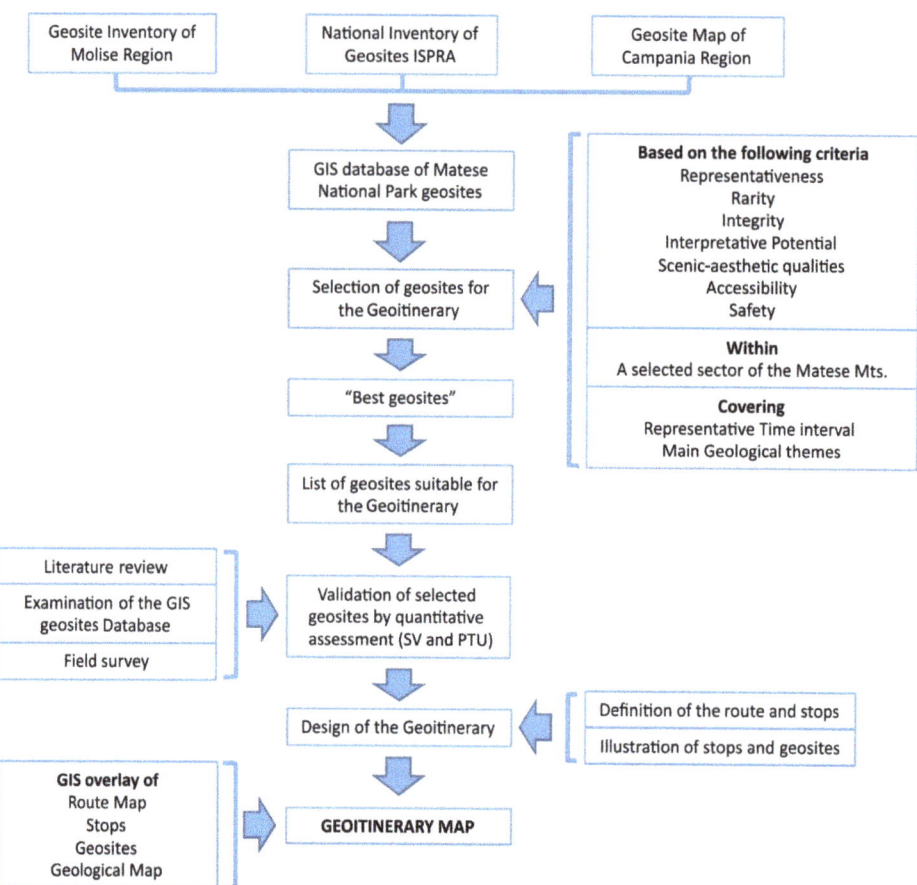

Figure 4. Flow chart showing the criteria and steps used for the design of the geoitinerary, the selection of geosites and the realization of the Geoitinerary Map.

For the Molise geosites, inventory cards containing their description and other information, such as the ages of rock formations and genetic processes involved, the main scientific interests and the relevance of the geosites, are available [74].

For the Campanian geosites included in the Geosite Map of Campania, such cards are not available. Therefore, first information on the Campanian geosites included in our database was mainly extracted from the cards available on the ISPRA website [75] and from literature [68,78,79].

The compilation of the Matese National Park Geosite Database allowed for storing information provided by the consulted data sources, concerning in particular the geological periods covered by the geosites, their primary scientific interests and the main geological themes they deal with.

Compatibly with this information and the overall distribution of geosites, we have individuated an area particularly representative as regards the geological and geomorphological aspects of greatest importance of the Matese massif, where to individuate the "best geosites" for the geoitinerary.

Regarding the selection of the "best geosites", we used an integrated approach, based on several criteria. Starting conditions were that the proposed itinerary could best describe major steps of the geological and geomorphological evolution of the Matese massif involving geosites that cover significant time intervals in this sense and can excellently illustrate the main geological themes identified.

To ascertain the scientific value of these geosites, we took into account the criteria Representativeness, Rarity and Integrity (present conservation status) which are widely used worldwide (e.g., [80–89]). Furthermore, we considered the selection criteria Safety, Accessibility, Scenic-aesthetic qualities and Interpretative potential, which are essential for assessing sites suitable for geotourism use (e.g., [77,89]). Particularly, the scenic-aesthetic qualities and the interpretative potential of geosites are important features to approach a wide audience and to disseminate geological information to non-geologists. Therefore, in order to avoid duplications, in cases of two or more geosites illustrating similar geological features or geomorphological processes, having all other values approximately equal, we selected the one with the highest interpretative potential and scenic-aesthetic appeal.

In addition, the necessary movements from site to site were considered and the connection between sites through a route was ensured. Based on this comparative analysis, we selected 16 geosites for the itinerary.

However, such analyses and related geosite evaluations can be considered not sufficient or may imply errors, as the used information derives from different inventories, which are based on different evaluation methods, essentially qualitative for the Campanian geosites [90] but quantitative instead for the Molise geosites [68].

Therefore, to have an evaluation be valid equally for all selected geosites, allowing the possibility of a real comparison between geosites, we have subjected them to a new quantitative assessment procedure. According to Mucivuna et al. 2022 [91], to evaluate the scientific value of geosites, in the absence of specific features to be evaluated (as, for example, in the case of urban or underwater sites), instead of creating new methods, the use of existing, validated methods is preferable as a priorityand, specifically, that of general-purpose quantitative methods that can be applied well to both geosites and geomorphosites. Therefore, considering the features of the Matese geosites, among the many methods developed (e.g., Mucivuna et al., 2019 [92], and reference therein), we have chosen to use the one proposed by Brilha 2016 [86]. This method is a widely used general-purpose method that allows for deriving the scientific value (SV) and Potential Tourist Use (PTU) of geosites, providing a maximum achievable value of 400 for each index.

We based the assessment of SV and PTU on the data provided by our GIS geosite database coupled with a detailed literature review and, where appropriate, new field surveys. Regarding in particular the parameters/criteria used for the PTU assessment, we integrated these data with those extractable from geothematic sources and statistical databases of regional or national archives available online such as those provided by the ISTAT [93].

Based on the scores of SV, in agreement with [89], we attributed an international or national relevance to geosites with values equal to or greater than 300 and equal to or greater than 200, respectively. To geosites with SV less than 200 (geodiversity sites *sensu* Brilha 2016 [86], and Prosser et al. 2010 [94] in Albani et al. 2020 [89]), we attributed a regional relevance.

3.2. Design and Illustration of the Geoitinerary

Once assessed their SV and PTU and, therefore, having validated the 16 geosites, we designed the geoitinerary by identifying the stops allowing on-site and/or panoramic views at one or more geosites, and tracing the route also through field surveys.

To illustrate the geoitinerary, a map was drawn in GIS environment by overlaying a simplified geological map with other informative layers created on purpose containing the route, the stops and the geosites, respectively. The geological map contains all the basic geological information to facilitate the understanding of the main geological features illustrated by the geosites by a broad audience. The base of this map is a hillshade model derived from a 40 m resolution Digital Elevation Model (DEM).

To visualize their "spatial-temporal position" within the Matese geological framework, the stops were located both on a stratigraphic column and on two cross-sections included in the geoitinerary map. The geological cross-sections are simplified to ideally represent the stratigraphic and tectonic setting and to emphasize particular geological aspects encountered along the itinerary.

Furthermore, a synthetic view of the stops, reporting the names of related geosites, the main geological themes they illustrate and the time intervals they cover, was also prepared. Finally, to make it possible to enjoy the itinerary independently and without a guide, we prepared descriptive cards for each stop. These descriptive cards have been enriched with specific illustrative material consisting in photos, geological sketches, 3D schemes and more, depending on the case, to facilitate the disclosure of geosites and their understanding by people without a geological background.

4. Results

4.1. The Matese National Park Geosite Database

Our Matese National Park Geosite Database contains 59 geosites (Figure 1). The 34 geosites located in the Molise sector are all included in the regional inventory and 16 of them are also present in the ISPRA Italian Geosites Inventory. The other 25 geosites are located in the Campanian sector. Twelve of these geosites are included both in the Geosite Map of Campania and in the ISPRA Geosites Inventory, 11 geosites are included only in the Geosite Map of Campania and other 2 only in the ISPRA Inventory.

These geosites cover an overall time interval from the Mesozoic to Quaternary. Their primary scientific interests (Figure 5) are Geomorphology (more than half of the geosites are geomorphosites), Stratigraphy, Paleontology and Structural Geology, but also Hydrogeology and Geomining are represented. In addition, these geosites can be linked to one or more of the following geological themes: Paleogeography, Tectonics, Hydrogeology, Karst, Long-term landscape evolution, Paleoclimate, Active morphodynamics and Geohistory. Regarding the geosites related to the Geohistory theme, we mean geosites that contribute to/have a meaning for the history of geology [95].

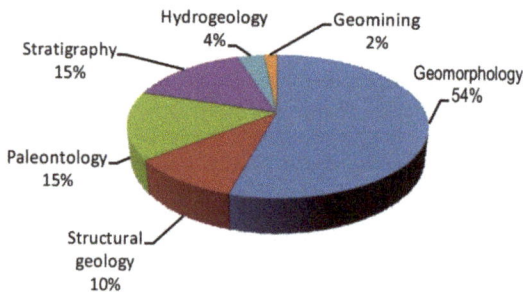

Figure 5. Primary scientific interests and relative percentages of the Matese National Park geosites.

4.2. The Geosites Selected for the Geoitinerary

We have selected 16 geosites for the geoitinerary (Figure 6; Table 1). Some of these geosites have already been included in other itineraries [50–53] and/or the subject of specific excursions dedicated to geology specialists [96].

Consistent with the presence of numerous geomorphological geosites (54%) within the Matese National Park Geosite Database, the majority of the 16 selected geosites, precisely ten of them, are geomorphosites. There are, however, also three paleontological geosites, two stratigraphical ones and a geomining geosite. Four of these geosites refer to the Cretaceous, one to the Cretaceous and the Miocene, and eleven to the Pleistocene and Holocene. All the major geological themes of the Matese National Park Geosite Database listed above are embraced by these geosites. In fact, many of them cover more than one of the topics listed above, with the Regia Piana Bauxite Mines covering even five themes. Based on the recurring themes, two main geosite groups can be distinguished: a group made of 5 geosites that are tightly linked to the theme Paleogeography, and a group of 11 geosites that are mainly an expression of the Long-term landscape evolution.

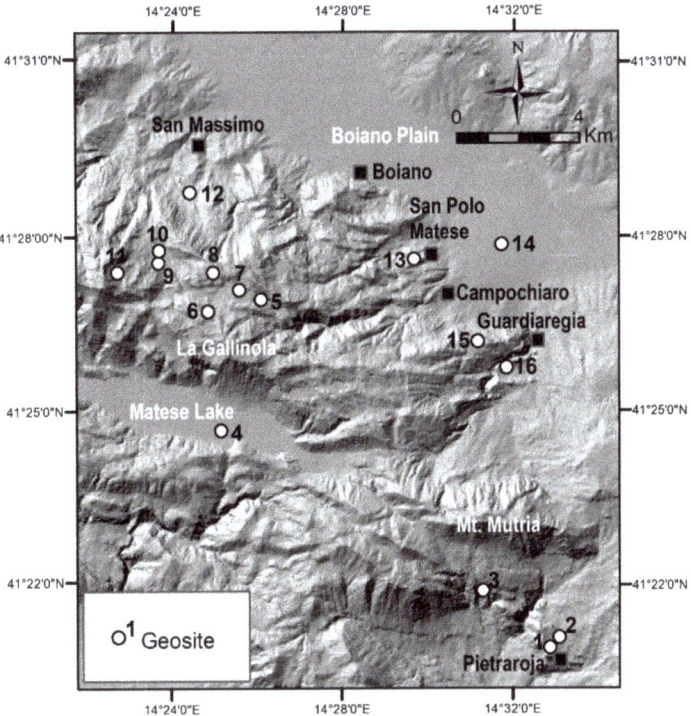

Figure 6. Location of the 16 selected geosites (the location of the area is shown in Figure 2). 1 = Fossiliferous limestones of Pietraroja (Le Cavere); 2 = Cava Canale; 3 = Regia Piana Bauxite Mines; 4 = Matese Lake polje; 5 = La Costa Alta fossiliferous limestones; 6 = La Gallinola fault slope; 7 = Campo Puzzo polje; 8 = Serra Le Tre Finestre karst surface; 9 = Campitello Matese polje; 10 = Campitello Matese moraine deposits; 11 = Mt. Miletto glacialcirques; 12 = Lacustrine deposits of San Massimo; 13 = San Polo Matese rudist limestones; 14 = Campochiaro alluvial fan; 15 = Costa della Defenza fault slope; 16 = Quirino gorge.

Among the five geosites included in the first group (Figure 7), four sites are also an expression of the theme Geohistory. In particular, the *San Polo Matese rudist limestones* geosite (13 in Figure 6 and Table 1, Figure 7e) allows for observing Upper Cretaceous limestones rich in rudist fossils, partly in position of growth, that indicate an open-marginal shelf environment ([97] and reference therein). This geosite is located in the northern central sector of the Matese Mountains and known in the geological-paleontological literature since 1901 [98]. The other three geosites (1, 2 and 3 in Figure 6 and Table 1) are instead located in the southeastern sector of the Matese Park area. Here, one of the first Mesozoic type sequences of the Southern Apennines has been described [99] and three important successions (the Cusano, Longano and Pietraroja formations) referable to the Southern Apennines Miocene transgression have been defined by Selli in 1957 [100]. Among them, the *Fossiliferous limestones of Pietraroja* (*Le Cavere*) site (1 in Figure 6 and Table 1; Figure 7a,b) certainly stands out, known since the end of the eighteenth century [55] for its paleontological richness. This geosite, given its very high scientific interest, is nowadays protected by a fence and managed by the Ente Geopaleontologico di Pietraroja whose aims are to protect and enhance it. This Lower Cretaceous limestone outcrop contains exceptionally well-preserved fossil fishes (Figure 7b), amphibians and reptiles, but became most famous for the discovery of *Scipionix sammiticus* [101], a juvenile theropod dinosaur with an exceptional soft tissue preservation. This very didactic geosite is an excellent testimony of the Lower Cretaceous environments that established in this sector of the Southern Apennines, and indicative for thetropical-subtropical shallow water carbonate domains that bordered emergent isolated lands in the Cretaceous Tethys ([96] and reference therein). The other two geosites are equally noteworthy. The *Cava Canale* geosite (2 in Figure 6 and Table 1) offers a superb three-dimensional exposition of the Miocene transgression in the Southern Apennines. In particular, it allows good observation of the Lower Cretaceous limestones overlain by the sediments of the Cusano Formation [100] (Figure 7c), Burdigalian-Langhian in age [102,103], referred to an open carbonate platform temperate neritic environment [62]. Finally, the *Regia Piana Bauxite Mines* geosite (3 in Figure 6 and Table 1) allows the exceptional observation of bauxite deposits (Figure 7d) in correspondence of a Cretaceous stratigraphic gap that characterizes in a unique way the Southern Apennines Mesozoic carbonate platform successions. These continental deposits formed above a karst surface between the Lower and Upper Cretaceous [97,99] under tropical to subtropical climate conditions, to testify clearly repeated long-lasting phases of emersion that affected the Lower Cretaceous carbonate platform in the Southern Apennines [104]. This site is also of geomining interest as it preserves several traces of the mining activities (such as mining tunnels, Figure 7d) that were carried out during the periods 1919–1925 and 1939–1965 [105].

Table 1. Main features of selected geosites. Number (see location in Figure 6) and name of geosite. Primary (1) and secondary (2) scientific interests, geological themes, ages of rock formations (AR) and ages of genetic processes (AGP). Paleontology = Pa; Stratigraphy = St; Sedimentology = Se; Geomorphology = Gm; Geomining = Gmi; Structural Geology = SG; Hydrogeology = H; Pedology = Pe; Palogeography = Pgeo; Geohistory = Ghis; Paleoclimate = Pcli; Tectonics = T; Karst = K; Long-term landscape evolution = LsEv; Active morphodynamics = Amd; Lower Cretaceous = LCret; Upper Cretaceous = UCret; Lower-Middle Miocene = L-MMio; Lower Pleistocene = LP; Middle Pleistocene = MP; Upper Pleistocene = UP; Pleistocene-Holocene = P-H; Holocene = H.

Geosite	Scientific Interests	Geological Themes	AR	AGP
1 ^* Fossiliferous limestones of Pietraroja	Pa(1), St(2), Se(2)	Pgeo, Ghis, Pcli	LCret	-
2 ^* Cava Canale	St(1), Pa(2), Gm(2)	Pgeo, Ghis	LCret L-MMio	-
3 ^* Regia Piana Bauxite Mines	Gmi(1), Gm(2), St(2), SG(2), Pa(2), Se(2)	Pgeo, Ghis, Pcli, T, K	LCret UCret	-
4 ^* Matese Lake polje	Gm(1), SG(2), H(2)	K, H, LsEv, T	-	MP-H
5 °* La Costa Alta fossiliferous limestones	Pa(1), St(2)	Pgeo	UCret	-
6 ° La Gallinola fault slope	Gm(1), SG(2)	LaEv, Amd	-	MP-H
7 °* Campo Puzzo polje	Gm(1), Pe(2)	K, T, LsEv	-	MP-H
8 ° Serra Le Tre Finestre karst surface	Gm(1), Pa(2), Pe(2)	K, LsEv, T	-	LP-H
9 °* Campitello Matese polje	Gm(1)	K, H, LsEv, T	-	MP-H
10 °* Campitello Matese moraine deposits	Gm(1), St(2)	Pcli, LsEv	UP	-
11 °* Mt. Miletto glacial cirques	Gm(1)	Pcli, LsEv, Amd	-	MP-UP
12 ° Lacustrine deposits of San Massimo	St(1), Gm(2), SG(2)	LsEv, Pgeo, Pcli	MP	-
13 °* San Polo Matese rudist limestones	Pa(1), St(2)	Pgeo, Ghis	UCret	-
14 ° Costa della Defenza fault slope	Gm(1), St(2)	LsEv, Pcli, T	-	MP-H
15 °* Campochiaro alluvial fan	Gm(1), SG(2)	LsEv, T	-	UP-H
16 °* Quirino Gorge	Gm(1), SG(2), H(2)	LsEv, T	-	MP

* Geosite included in the ISPRA Geosite Inventory; ^ Geosite included in the Geosite Map of Campania; ° Geosite included in the Molise Geosite Inventory.

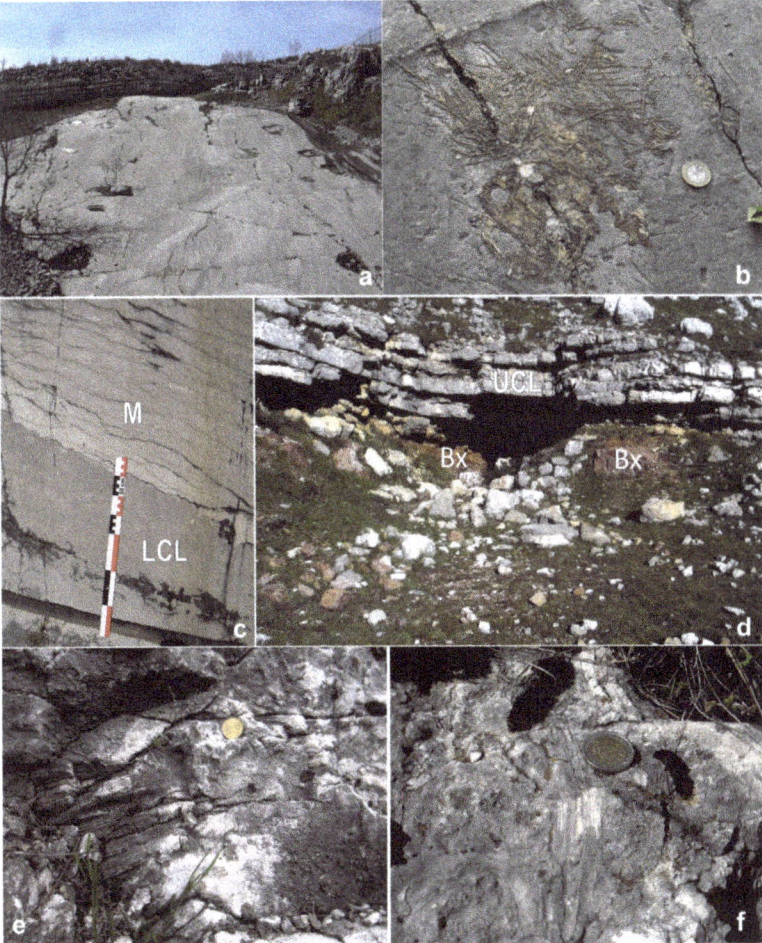

Figure 7. (a) Panoramic view of the Pietraroja fossiliferous limestones geosite (*Le Cavere*); (b) One of the fish fossils visible at Pietraroja site; (c) Lower Cretaceous limestones (LCL) overlain by the Miocene Cusano Formation (M) visible at Cava Canale; (d) Bauxite mining tunnel in the Regia Piana area excavated along the contact between the bauxite deposits (Bx) and the overlyingUpper Cretaceous limestones (ULC); (e) Detail of the *Hippurites colliciatus* (Woodward) Reef at San Polo Matese; (f) Detail of the La Costa Alta limestones with rudists.

Within the second group, made of 11 geosites linked to the theme Long-term landscape evolution, four sites are also expression of the topics Karst and Tectonics. The first one is represented by the *Serra Le Tre Finestre karst surface* geosite (8 in Figure 6 and Table 1; Figure 8a), a surface with a typical karst hummocky morphology generated by the widespread presence of active and inactive dolines as well as open or closed karst depressions, which appear frequently aligned according to NW-SE oriented tectonic lineaments [68,70]. The other three geosites are represented by structurally controlled poljes (4, 7 and 9 in Figure 6 and Table 1), generated during Quaternary extensional tectonics and related block-faulting [70]. Among them, the *Matese Lake polje* (4 in Figure 6 and Table 1; Figure 8b,c) and the *Campitello Matese polje* geosites (9 in Figure 6 and Table 1; Figure 8d) are also good expressions of the Hydrogeology theme. The Matese Lake polje, which is the largest polje of the Matese area, is mainly controlled by WNW-ESE normal faults [70].

This very didactic geosite allows easy observation of numerous geological and geomorphological features. Among the latter, noteworthy are the fault slopes bordering the Matese Lake to the north and the related hanging valleys of fluviokarst origin [70] (Figure 8c), at least two generations of alluvial fans dated to the Late Pleistocene-Holocene [106] and two ponors found along its southern edge (the Scennerato and Brecce ponors). The Matese Lake area is the largest endorheic area (45 km^2) of the Massif. It contributes significantly to the recharge of the Matese aquifer and is believed to supply the Maretto and Torano springs located along its southern slopes [107] (Figure 2). In particular, the connection between the Scennerato ponor and the Torano spring has been demonstrated [108,109].

The *Campitello Matese polje* geosite (Figure 8d) is an active polje, mainly controlled by NW-SE tectonic lineaments. This geosite well illustrates how closely its morphology and hydrology are related with extensional tectonics and karst drainage [68]. In particular, on its floor, which is covered prevailingly by alluvial and moraine deposits of Late Pleistocene-Holocene age [106,110], a couple of ponors ensuring the endorheic drainage are easily observable.

Other three geosites of the second group host also important evidence linked to the Paleoclimate theme. The first two illustrate relict glacial landforms. One is the *Mt. Miletto glacial cirques* geosite (11 in Figure 6 and Table 1), which consists of two armchair-shaped glacial cirques, the so-called Circo Maggiore (major cirque) and the S. Nicola cirque (Figure 8e). The Circo Maggiore, which is the larger and better preserved cirque, placed a few hundreds of meters higher up with respect to the S. Nicola cirque, has been referred to the Last Glacial Maximum [70]. The other one is the *Campitello Matese moraine deposits* geosite (10 in Figure 6 and Table 1), referred to the Late Glacial Maximum [69,111], a very suitable site to illustrate the typical features of glacial deposits (Figure 8f). The third site is the *Lacustrine deposits of San Massimo* geosite (12 in Figure 6 and Table 1), consisting of a relatively small outcrop located along Serra San Giorgio. This outcrop exposes terraced fluvial-lacustrine deposits [112] (Figure 8g), testifying an ancient lake, Middle Pleistocene in age [64,113], along with some moraine deposits (Figure 8h) pre-Last Glacial in age [70], overlying the fluvial-lacustrine succession.

Returning to the *Mt. Miletto glacial cirques* geosite, this also embraces the topic Active morphodynamics. The steep headwalls of the cirques, in fact, are partially covered by active scree slopes produced by cryoclastic processes and associated termoclastic phenomena, which typically reflect the climate conditions that currently characterize the Apennines high mountain areas.

Finally, the last four geosites of this group are also linked to the theme of tectonics. Two of them are represented by structural landforms, the *La Gallinola fault slope* (6 in Figure 6 and Table 1), and the *Coste della Defenza fault slope* (15 in Figure 6 and Table 1) geosites, the first being located in the inner high mountain area, the second along the northern front of the Matese massif. The *La Gallinola fault slope* geosite (Figure 9a) is the result of Pleistocene tectonic uplift and block-faulting, and shows typical geomorphological features such as its rectilinear profile resulting from long-term processes of slope replacement under periglacial conditions. Its footslope and lower backslope are covered by partly still active scree slopes and debris cones, testifying the active morphodynamics affecting the slope related to ongoing intense cryoclastic degradation. The *Coste della Defenza fault slope* (Figure 9b) is evidence of the tectonic uplift that has affected the northern Matese flanks bordering the Boiano basin that, conversely, was affected by tectonic subsidence. The genesis of this E-W trending slope is related to a Pleistocene normal fault [64,114], which also shows evidence of historical to recent activity and is believed to be one of the most hazardous seismogenic structures of Europe [64].

The other two, the *Quirino gorge* (16 in Figure 6 and Table 1) and the *Campochiaro alluvial fan* (14 in Figure 6 and Table 1) geosites, are instead fluvial landforms controlled by tectonics. The *Quirino gorge* geosite (Figure 9c) is a didactic example of a superimposed, deep and narrow gorge [68] that was incised by the Quirino Stream during the Middle Pleistocene uplift of the Matese massif. Finally, the *Campochiaro alluvial fan* geosite (Figure 9d) shows the typical features of Apennine alluvial fan deposits laid down by major streams near the mountain fronts under the influence of tectonic uplift and climate [68] during Late Pleistocene cold periods [64,115,116].

Figure 8. (**a**) Detail of the Serra Le Tre Finestre surface showing several solution dolines; (**b**) Panoramic view of the Matese Lake basin seen from La Gallinola; (**c**) View of the fault slopes of Mt. Crocetta and La Gallinola, bordering the Matese Lake polje to the north and of the edges of some ancient suspended valleys of fluviokarst origin; (**d**) The Campitello Matese polje typically flooded during spring; (**e**) The Circo Maggiore (major cirque) of Mt. Miletto and the S. Nicola cirque; (**f**) The Campitello Matese moraine deposits; (**g**) Detail of the laminated fluvio-lacustrine deposits of San Massimo, made of silts and clays with interbedded reworked volcaniclastic material and rounded pebbles; (**h**) Detail of the Middle Pleistocene glacial deposits overlying the S. Massimo lacustrine deposits, made of angular to subangular heterometric, calcareous clasts in sandy-silty matrix.

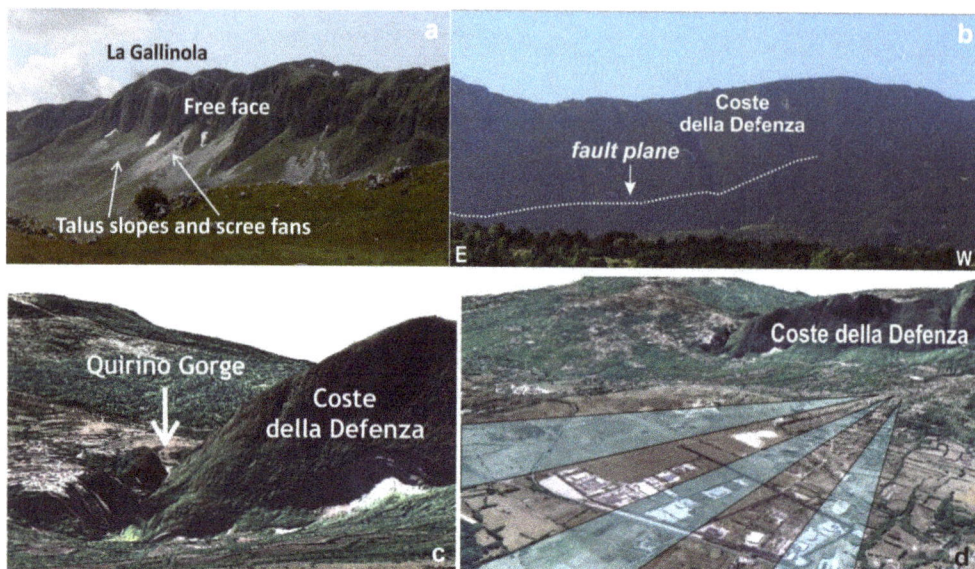

Figure 9. (**a**) The La Gallinola fault slope; (**b**) The Costa della Defenza fault slope; (**c**) The Quirino Stream gorge on 3D scene; (**d**) The Campochiaro alluvial fan on 3D scene.

4.3. SV and PTU of the Selected Geosites

The quantitative assessment of the Scientific Value (SV) carried out for the selected geosites has provided for them values ranging between 150 and 325 (Table 2).

Table 2. Weights, scores and values attributed to the criteria used to assess the Scientific Value of the selected geosites. Rp = Representativeness; KL = Key locality; SK = Scientific knowledge; I = Integrity; GD = Geological diversity; RR = Rarity; UL = Use limitation. The total SV value and the relevance of each geosite are also shown.

Geosite	Criterion	Rp	KL	SK	I	GD	RR	UL	Total Value	Relevance
	Weight	30%	20%	5%	15%	5%	15%	10%		
Fossiliferous limestones of Pietraroja (Le Cavere)	score	4	2	4	4	1	4	2	-	-
	value	120	40	20	60	5	60	20	325	International
Cava Canale	score	4	1	4	4	2	1	4	-	-
	value	120	20	20	60	10	15	40	285	National
Regia Piana Bauxite Mines	score	4	1	4	2	1	4	4	-	-
	value	120	20	20	30	5	60	40	295	National
Matese Lake polje	score	4	1	4	4	4	0	4	-	-
	value	120	20	20	60	20	0	40	280	National
La Costa Alta fossiliferous limestones	score	1	0	4	4	1	0	4	-	-
	value	30	0	20	60	5	0	40	155	Regional
La Gallinola fault slope	score	1	0	4	4	4	0	4	-	-
	value	30	0	20	60	20	0	40	170	Regional

Table 2. Cont.

Geosite	Criterion Weight	Scientific Value (SV)							Total Value	Relevance
		Rp 30%	KL 20%	SK 5%	I 15%	GD 5%	RR 15%	UL 10%		
Campo Puzzo polje	score	1	0	2	4	2	0	4	-	-
	value	30	0	10	60	10	0	40	150	Regional
Serra Le Tre Finestre karst surface	score	1	0	2	4	1	1	4	-	-
	value	30	0	10	60	5	15	40	160	Regional
Campitello Matese polje	score	1	0	4	4	4	0	4	-	-
	value	30	0	20	60	20	0	40	170	Regional
Campitello Matese moraine deposits	score	1	0	4	2	1	2	4	-	-
	value	30	0	20	30	5	30	40	155	Regional
Mt. Miletto glacial cirques	score	2	0	4	4	1	1	2	-	-
	value	60	0	20	60	5	15	20	180	Regional
Lacustrine deposits of San Massimo	score	1	1	4	1	2	2	4	-	-
	value	30	20	20	15	10	30	40	165	Regional
San Polo Matese rudist limestones	score	1	2	4	2	0	2	2	-	-
	value	30	40	20	30	0	30	20	170	Regional
Costa della Defenza fault slope	score	1	1	4	4	2	0	4	-	-
	value	30	20	20	60	10	0	40	180	Regional
Campochiaro alluvial fan	score	1	1	4	4	1	1	2	-	-
	value	30	20	20	60	5	15	20	170	Regional
Quirino Gorge	score	1	0	4	4	2	0	4	-	-
	value	30	0	20	60	10	0	40	160	Regional

In particular, a maximum SV of 325 was attributed to the *Fossiliferous limestones of Pietraroja* (*Le Cavere*) geosite, confirming its prominent position and highlighting its international relevance (Table 2). For the other three geosites, we achieved instead a national relevance. Among these, the *Regia Piana Bauxite Mines* geosite has the highest SV (295), due to its maximum score for the Rarity criterion that compensates the lower scores obtained for the Integrity and Geological Diversity criteria. The remaining twelve geosites obtained SV values less than 200, and therefore are considered of regional relevance. Among them, the highest scientific values were achieved by the *Mt. Miletto glacial cirques* and the *Costa della Defenza fault slope* geosites (SV = 180, Table 2), while the *Campo Puzzo polje* geosite obtained the lowest SV (150). The high scientific value of the *Mt. Miletto glacial cirques* geosite is mainly determined by the high scores of the Integrity and Scientific knowledge criteria combined with the average score of the Representativeness criterion. The high SV of *Costa della Defenza fault slope* geosite, in addition to the high scores of the Integrity and Scientific knowledge criteria, is linked also to the high score of the Use limitations criterion. Instead, the null scores of the Key locality and Rarity criteria, the average scores of the Scientific knowledge and Geological diversity criteria as well as the low score of Representativeness criterion, essentially determined the low SV of the *CampoPuzzo polje* geosite.

The PTU values calculated for the selected geosites range between a maximum of 350 and a minimum of 195 (Table 3). In particular, the best PTU values have been reached by the geosites *Fossiliferous limestones of Pietraroja* (*Le Cavere*), *Matese Lake polje* and *Campitello Matese polje* (respectively 350, 300 and 295 in Table 3). The lowest PTU value of 195 was attributed instead to the *Lacustrine deposits of San Massimo* geosite. With exception to the latter, all other geosites reached PTU values equal to or greater than 230, highlighting a relatively good to very high tourist potential.

Table 3. Weights, scores and values attributed to the criteria used to assess the Potential Tourist Use, and total PTU values of selected geosites. V = Vulnerability; A = Accessibility; UL = Use limitations; Sa = Safety; L = Logistics; Pd = Population density; As = Association with other values; Sc = Scenery; Un = Uniqueness; O = Observation conditions; IP = Interpretative Potential; E = Economic level; RA = Proximity of recreational areas.

Geosite	Criterion	V 10%	A 10%	UL 5%	Sa 10%	L 5%	Pd 5%	As 5%	Sc 15%	Un 10%	O 5%	IP 10%	E 5%	RA 5%	Total Value
Fossiliferous limestones of Pietraroja (Le Cavere)	score	4	4	3	3	3	1	4	4	4	4	4	1	4	-
	value	40	40	15	30	15	5	20	60	40	20	40	5	20	350
Cava Canale	score	3	3	4	2	3	1	4	2	2	4	3	1	4	-
	value	30	30	20	20	15	5	20	30	20	20	30	5	20	265
Regia Piana Bauxite Mines	score	3	3	4	2	3	1	3	2	3	4	3	1	4	-
	value	30	30	20	20	15	5	15	30	30	20	30	5	20	270
Matese Lake polje	score	3	4	4	2	3	1	4	3	2	4	4	1	4	-
	value	30	40	20	20	15	5	20	45	20	20	40	5	20	300
La Costa Alta fossiliferous limestones	score	3	3	4	2	4	2	4	0	2	4	4	1	4	-
	value	30	30	20	20	20	10	20	0	20	20	40	5	20	255
La Gallinola fault slope	score	4	3	4	2	4	2	4	0	1	4	3	1	4	-
	value	40	30	20	20	20	10	20	0	10	20	30	5	20	245
Campo Puzzo polje	score	3	3	4	2	4	2	4	0	1	4	4	1	4	-
	value	30	30	20	20	20	10	20	0	10	20	40	5	20	245
Serra Le Tre Finestre karst surface	score	4	3	4	2	4	2	4	0	1	4	4	1	4	-
	value	40	30	20	20	20	10	20	0	10	20	40	5	20	255
Campitello Matese polje	score	3	4	4	2	4	1	4	3	1	4	4	1	4	-
	value	30	40	20	20	20	5	20	45	10	20	40	5	20	295
Campitello Matese moraine deposits	score	2	3	4	2	4	1	4	0	3	3	4	1	4	-
	value	20	30	20	20	20	5	20	0	30	15	40	5	20	245
Mt. Miletto glacial cirques	score	3	0	4	2	4	1	4	2	3	4	4	1	4	-
	value	30	0	20	20	20	5	20	30	30	20	40	5	20	260
Lacustrine deposits of San Massimo	score	1	3	3	2	4	1	3	0	2	2	3	1	3	-
	value	10	30	15	20	20	5	15	0	20	10	30	5	15	195
San Polo Matese rudist limestones	score	2	4	4	2	4	1	4	1	2	4	4	1	3	-
	value	20	40	20	20	20	5	20	15	20	20	40	5	15	260
Costa della Defenza fault slope	score	3	3	4	2	4	1	4	0	1	4	3	1	4	-
	value	30	30	20	20	20	5	20	0	10	20	30	5	20	230
Campochiaro alluvial fan	score	3	3	4	2	4	1	4	0	1	3	4	1	4	-
	value	30	30	20	20	20	5	20	0	10	15	40	5	20	235
Quirino Gorge	score	3	3	4	2	4	1	4	2	1	4	3	1	4	-
	value	30	30	20	20	20	5	20	30	10	20	30	5	20	260

For the geosites of international and national relevance, a positive correspondence between high SV and PTU values is observed. In fact, the *Fossiliferous limestones of Pietraroja* geosite reaches the highest values both for SV and PTU (Tables 2 and 3), confirming its importance within the Matese National Park area. Furthermore the three geosites of national relevance (*Cava Canale, Regia Piana bauxite Mines* and *Matese Lake polje*) are characterized by PTU values among the highest ones.

Naturally, higher scientific values not necessarily correspond to higher potential touristic use values and vice versa (Figure 10). For example, the *Lacustrine deposits of San Massimo* geosite has obtained the lowest PTU value (195), but a SV of 165 close to the median value (170), which corresponds to the SV obtained by the *Campitello Matese polje* geosite (Table 2) that instead has obtained the third highest PTU value (295). Regarding the *Campitello Matese polje* geosite, its high values in the Accessibility, Logistics and Interpretative Potential criteria (Table 3), combined with the medium value related to the Scenery criterion, contribute to make this site stand out for its total PTU value when compared to the *Cava Canale* and *Regia Piana bauxite Mines* geosites, albeit of national relevance. Conversely, the relatively low PTU value of the *Lacustrine deposits of San Massimo* geosite essentially depends on the fact that this site has obtained the absolute lowest values for the Vulnerability and Observation conditions criteria, some of the lowest values obtained for the Use limitations, Association with other values and Proximity of recreational areas criteria, as well as the value zero related to the Scenery criterion.

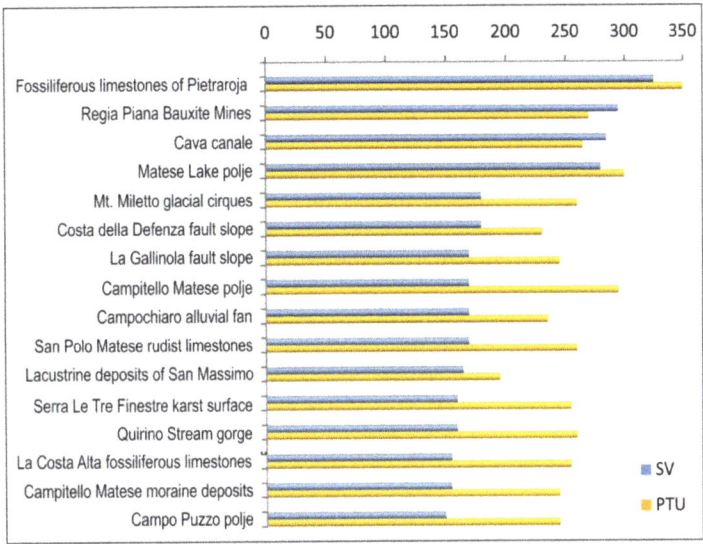

Figure 10. Comparative view of SV and PTU values assessed for the selected geosites.

4.4. The Geoitinerary

By overlaying the specially prepared geological map with the thematic layers containing information about the stops, the itinerary route and the geosites, we obtained the geoitinerary map shown in Figure 11. A stratigraphic column and two simplified schematic geological sections that help to visualize the distribution of stops in space and time accompany this map. A synthetic view of the stops and two examples of descriptive cards realized to illustrate single stops are shown in Figures 12–14, respectively.

Broadly, the geoitinerary crosses the Matese massif from SE to NW, with a deviation for the Lake Matese stops and the last stretch of the path that runs from NW to SE along the southern border of the Boiano Plain. The itinerary consists of 13 stops, allowing the observation of the sixteen geosites onsite and through panoramic views. It is 82 km long and has an overall difference in height of 1100 m (a maximum altitude of 1680 m is reached at stop 8, the Serra Le Tre Finestre surface, a minimum altitude of 570 m in correspondence of stop 12, the Campochiaro alluvial fan and Costa della Defenza slope). The geoitinerary develops mainly along the main road network allowing the visitors to move with the car from stop to stop. It also includes, however, two detours on foot, along CAI (Club Alpino Italiano) trails, to reach respectively stop 3 and stop 8 (Bauxite deposits of Regia Piana and the Serra Le tre Finestre surface, Figure 12).

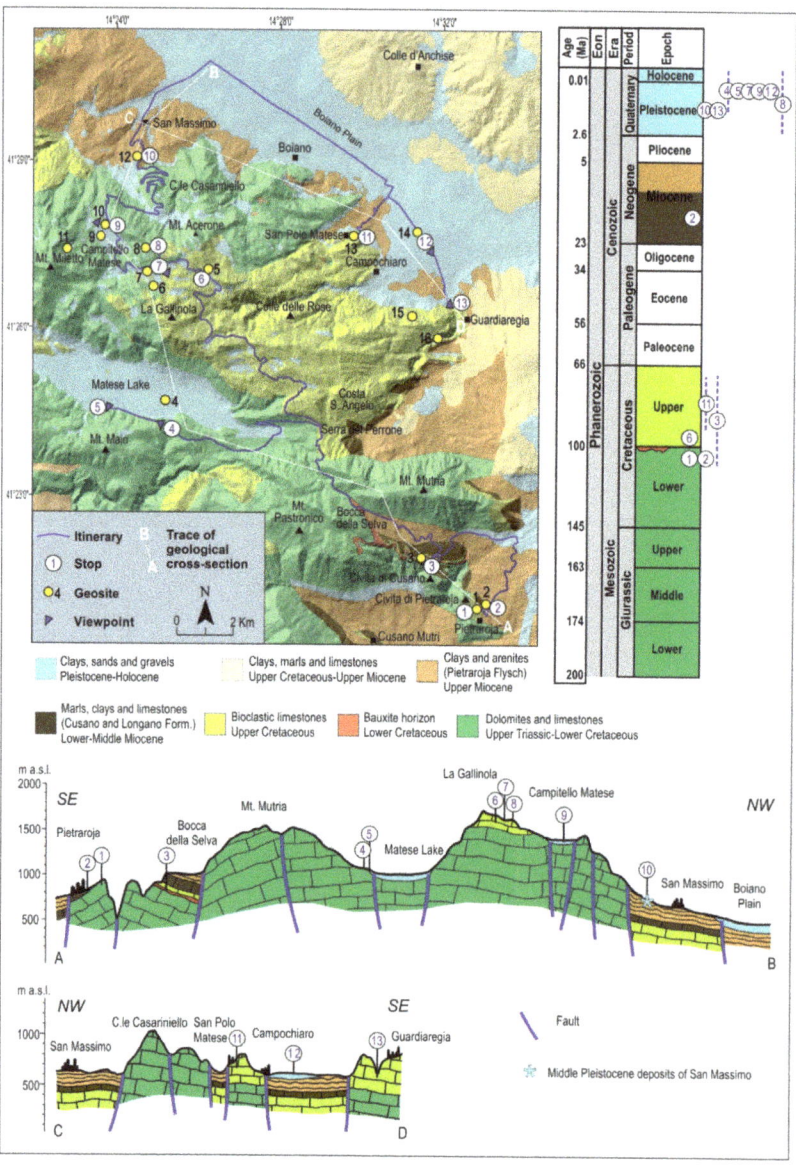

Figure 11. The geoitinerary map with the two simplified schematic geological sections and the stratigraphic column.

It takes at least two days to complete the entire route.

Much of the itinerary crosses high mountain areas. Therefore, due to the more difficult observation conditions of geosites and the difficult accessibility due to even partial closure of some roads (such as the SP106 road connecting Serra del Perrone with Campitello Matese) during autumn and winter, related to adverse weather conditions (i.e., the presence of snow and ice), the recommended visiting season is spring/summer.

Stop N	Name	Geosites	Main themes	Time interval
1	The Pietratoja fossiliferous limestones: a Lower Cretaceous tropical shallow-water environment	1. Fossiliferous limestones of Pietraroja (Le Cavere)	Paleogeography Geohistory Paleoclimate	Lower Cretaceous
2	The Cava Canale site: the Miocene transgression in the Southern Apennine chain	2 Cava Canale	Paleogeography Geohistory	Lower Cretaceous Lower-Middle Miocene
3	The Bauxite deposits of Regia Piana: a tropical continental phase in the Cretaceous marine succession	3 Regia Piana Bauxite Mines	Paleogeography Geohistory Paleoclimate Tectonics Karst	Lower Cretaceous Upper Cretaceous
4-5	The Matese Lake: the Pleistocene extensional tectonics and the karst modeling	4 Matese Lake polje	Tectocnics Karst Hydrogeology Long term landscape evolution	Middle Pleistocene-Holocene
6	La Costa Alta: an Upper Cretaceous shelf lagoon environment	5 La Costa Alta fossiliferous limestones	Paleogeography	Upper Cretaceous
7	Campo Puzzo: a panoramic view on active morphodynamics and typical landscape features of a high-mountain carbonate area	6 Campo Puzzo Polje 7 La Gallinola fault slope	Long term landscape evolution Karst Tectonics Active morphodynamics	Middle Pleistocene-Holocene
8	The Serra Le Tre Finestre surface: a remnant of ancient gentle and low relief erosional landscapes	8 Serra Le Tre Finestre karst surface	Karst Long term landscape evolution Tectonics	Lower Pleistocene-Holocene
9	The Campitello Matese area: the karst modeling and the Pleistocene glaciations	9 Campitello Matese polje 10 Campitello Matese moraine deposits 11 Mt. Miletto glacial cirques	Long term landscape evolution Karst Hydrogeology Tectonics Paleoclimate Active morphodynamics	Middle Pleistocene-Holocene
10	The deposits of San Massimo: remnants of a 600.000 years old lake	12 Lacustrine deposits of San Massimo	Long term landscape evolution Paleogeography Paleoclimate	Middle Pleistocene
11	The Rudists of San Polo Matese: a Upper Cretaceous fossil reef	13 San Polo Matese Rudist limestones	Paleogeography Geohistory	Upper Cretaceous
12	The Campochiaro fan and Costa della Defenza slope: the Quaternary tectonics and climate variations	14 Campochiaro alluvial fan 15 Costa della Defenza fault slope	Long term landscape evolution Paleoclimate Tectonics	Middle Pleistocene-Holocene
13	The Quirino Gorge: an evidence of Quaternary tectonics and stream superimposition	16 Quirino Gorge	Long term landscape evolution Tectonics	Middle Pleistocene

Figure 12. Summary of the stops of the itinerary, with the geosites that allow visiting, the main themes and the time interval embraced.

The two examples of descriptive cards (Figures 13 and 14) concern stops 2 and 9 (Figure 11), located respectively in the southeastern and the central northern sectors of the Matese area. The card on stop 2 is dedicated to the Miocene transgression in the Matese area and the related Cretaceous and Miocene paleogeography (Cava Canale site). The card on stop 9 deals with the Quaternary landscape evolution in the Campitello Matese area. These cards provide some essential information, through text, photos and figures, to illustrate to visitors the major themes on which the stops focus.

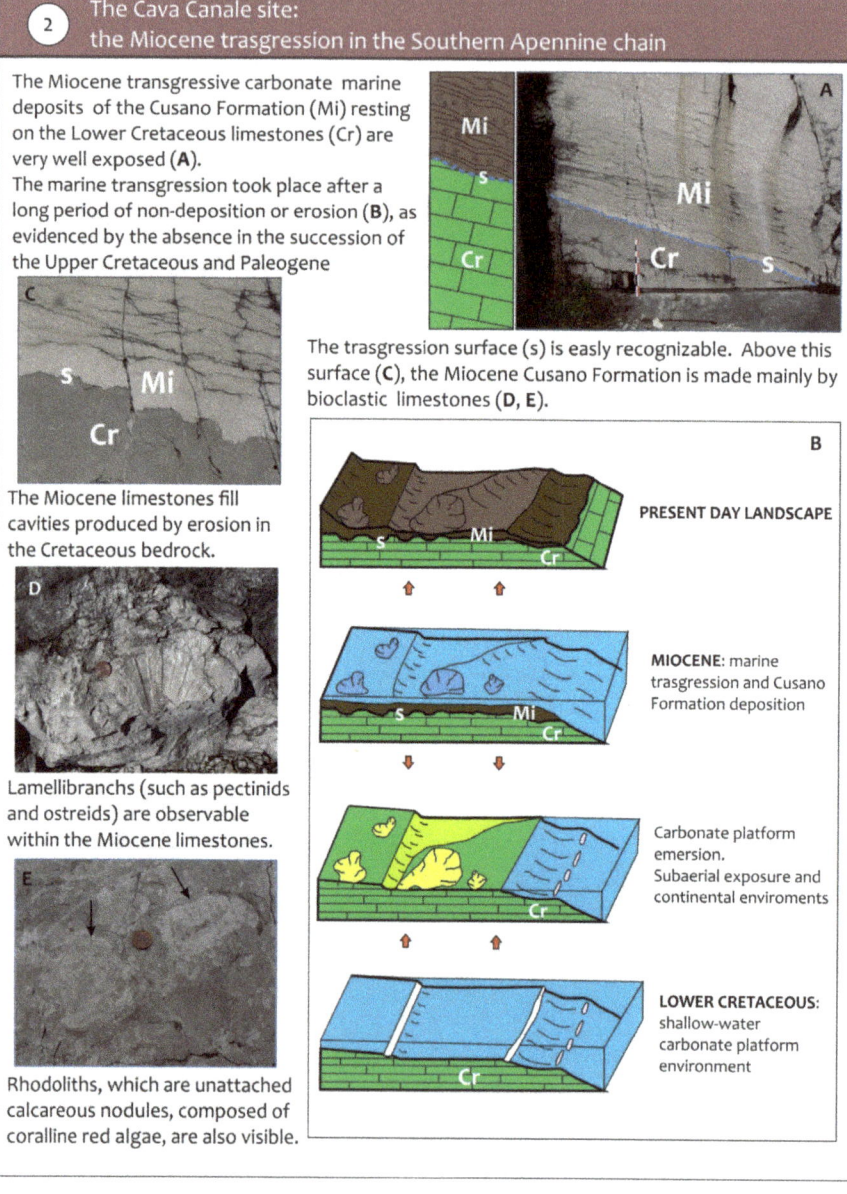

Figure 13. The descriptive card prepared for the visit of the Cava Canale site—Stop 2.

⑨ The Campitello Matese area: the karst modeling and the Pleistocene glaciations

A rich assemblage of glacial, karst, fluvial and slope landforms gives evidence of the Late Quaternary evolution of this area (**A**). The Campitello Plain is a good example of polje: a large, flat-floored karst depression (**B**). It is an active polje, mainly controlled by NW-SE tectonic lineaments. On its floor, some ponors, ensuring the endorheic drainage, and the uppermost portion of its filling are easily observable (**A**).

Phases of climate deterioration during the Pleistocene are testified by several relict glacial landforms. In particular, the «Circo Maggiore» (**C**), the major glacial cirque of Mt. Miletto, dated to the Last Glacial Maximum, and several outcrops of Late Pleistocene moraine deposits (**D, E**) are observable.

The Circo Maggiore shows the typical features of glacial cirques (**E**): the amphitheatre-like shape and the steep headwalls, partially covered by active scree slopes.

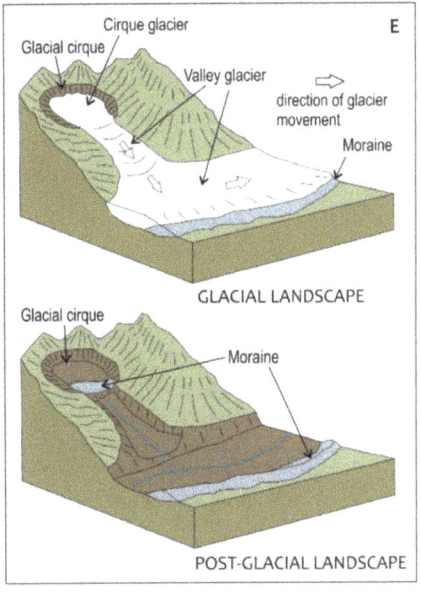

The moraine deposits typically consist of sub-angular, highly heterometric, calcareous clasts in very abundant sandy-silty matrix.

Figure 14. The descriptive card prepared for the visit of the Campitello Matese area—Stop 9.

The card of the Cava Canale stop (Figure 13) provides information on the major geological features observable in the disused quarry, which are also easily recognizable by a non-expert audience. It contains four pictures that show some of the main observations that visitors can make along the quarry walls, and a picture containing a sequence of very simplified 3D evolutive schemes. In particular, picture A, accompanied by an interpretative simplified stratigraphic column, shows the Cretaceous and Miocene limestones separated by the transgression surface. Picture C shows in detail the transgression surface, highlighting its erosive nature, while pictures D and E show the fossils that are easily recognizable in this site. Finally, the simplified 3D evolutive schemes in picture B allow visualizing the evolution of the carbonate platform in the area, with particular reference to what can be observed on the site and the paleoenvironments in which the now outcropping rocks were originally formed.

The card of the Campitello Matese stop (Figure 14) provides information on the major landforms observable, and on the processes responsible for their formation and involved in the Quaternary landscape evolution of this area. It contains five pictures, three of which show the landforms that are easily recognizable in this area even by a non-specialist audience, and representative of the three geosites present in this area: the Campitello Matese polje (B), the Mt. Miletto glacial cirques (C) and the Campitello Matese moraine deposits (D). In particular, regarding the glacial cirques, we have chosen to focus the visitor's attention on the "Circo Maggiore" (major cirque), which is the best preserved one and easier to observe. A 3D view including the different landforms (A) gives an overview of the Campitello Matese landscape. Furthermore, two simplified 3D evolutive schemes (E) are included, showing the passage from a glaciated to a non-glaciated landscape, highlighting how the now-relict glacial landforms originally formed.

5. Discussion and Conclusions

The present study deals with the enhancement of the geoheritage of the Matese Mountains, one of the most suggestive and integral mountain areas of southern Italy. This mountain area largely shares the hardships and limitations that widespread characterize other mountain regions and, in general, inner areas, such as land abandonment [3], population decrease [72], marginality, limits in mobility and inaccessibility [1,6].

Bearing in mind the special characteristics and physiographic features of the Matese Mountains, along with the presence of the Matese National Park currently taking off, the proposed geoitinerary, apart the obvious geotourism purpose, can help to unite separate physiographic units and overcome existing limits connected to difficult access, scarcity of infrastructures, administrative boundaries, etc.

Many of the studies on geological itineraries (see the introduction section for references) highlight the importance of a correct selection of geosites, to choose the most suitable ones for geotourism purposes.

For the geosites included in the Molise geosite census (12 of the selected geosites), the quantitative assessment of the SV provided results on overall consistent with those obtained through the quantitative assessment method used in the Molise regional census. The latter allowed to calculate the so-called intrinsic geosite value [68] resulting from the weighted sum of Representativeness, Rarity and Scenic-aesthetic values. This consistency emphasizes the adequacy of the procedure used in this work to assess the SV of the selected geosites.

In addition, the values obtained for the SV and PTU highlight that the selected geosites not only stand out for their scientific values but are mostly characterized by high to very high PTU values as well. These qualities, together with the selected stops (Figure 11), which include both on-site visits and beautiful panoramic views, allowing the illustration of the geosites in an appealing manner by also highlighting their scenic-aesthetic qualities, represent a remarkable strength of the proposed geoitinerary. The latter also manages to tell in a simply but sufficiently exhaustive manner the main steps of the geological history and long-term landscape evolution of the Matese area, highlighting its rich geoheritage and high geodiversity.

The geoitinerary crosses the municipal territories of Pietraroja, Cusano Mutri, Piedimonte Matese, Castello del Matese, San Gregorio Matese, Bojano, Campochiaro, San Polo Matese, Guardiaregia and San Massimo. The first six of these municipalities are classified as "tourist municipalities not belonging to a specific category" [117], San Massimo is defined a "municipality with a mountain vocation", San Polo Matese, Campochiaro and Guardiaregia instead as belonging to the "non-tourism municipality" category.

Tourism is concentrated mainly in the Campitello Matese district, located in the municipality of San Massimo, which is the home of a winter and ski tourism of predominantly commuting-type [72]. Other places of tourist attraction along the itinerary are Bocca della Selva (Cusano Mutri municipality), with prevailing winter tourism (essentially cross-country skiing), the Matese Lake area with its recreational outdoor activities especially during spring and summer, the WWF Oasis Guardiaregia-Campochiaro Regional Reserve, mainly visited by ecotourists and schools, and the Le Cavere geosite and Paleo-Lab Museum of Pietraroja, which are important attraction poles for geology enthusiasts.

The proposed geoitinerary can represent an opportunity to foster geotourism in the Matese area, both by attracting new tourists and by intercepting geology enthusiasts, who intend to visit or already have visited Pietraroja, bringing them to discover other features of the rich geoheritage of the Matese massif. It also can contribute to promoting a sustainable tourism during the spring and summer seasons in the Campitello Matese area.

In addition, the itinerary can act as a connection between the territories of the Campania and Molise regions, which until now have been managed in an independent and uncoordinated way, substantially in the frameworks of the Matese Regional Park and the Italian Strategy for Inner Areas, respectively. In particular, the geoitinerary can support the development of sustainable geotourism and related associate activities within the nascent Matese National Park, helping to create a tourist offering capable of attracting visitors driven by an interest in geology and in other natural or cultural resources.

Future developments of our study will be aimed at linking the geoitinerary with the visit of other sites of natural/cultural interest, and at creating a network with other trails such as the geological trails already designed for the Molise side [51], so as to encourage tourists to stay for several days and favor overnight stays. Finally, initiatives aimed at exploiting the proposed geoitinerary (for instance, the realization of illustrative panels, brochures, and illustrative materials suitable for mobile technologies), in collaboration with public administrations, private tourism managers and/or local stakeholders, are strongly hoped for and will be a further focus of our future activities.

Author Contributions: Conceptualization F.F. and C.M.R.; methodology, F.F., C.M.R., V.A. and M.C.; formal analysis, F.F. and C.M.R.; investigation, F.F., V.A. and M.C.; writing—original draft preparation, F.F. and C.M.R.; writing—review and editing, F.F. and C.M.R.; supervision, C.M.R. All authors have read and agreed to the published version of the manuscript.

Funding: This research received no external funding.

Institutional Review Board Statement: Not applicable.

Informed Consent Statement: Not applicable.

Data Availability Statement: The data presented in this study are available in the article itself, in the references cited and from the authors, upon request.

Acknowledgments: The authors wish to thank the anonymous reviewers whose suggestions contributed to improving the manuscript.

Conflicts of Interest: The authors declare no conflict of interest.

References

1. Battaglia, M.; Annesi, N.; Pierantoni, I.; Sargolini, M. Future perspectives of sustainable development: An innovative planning approach to inner areas. Experience of an Italian alpine region. *Futures* **2019**, *114*, 102468. [CrossRef]
2. Vendemmia, B.; Pucci, P.; Beria, P. An institutional periphery in discussion. Rethinking the inner areas in Italy. *Appl. Geogr.* **2021**, *135*, 102537. [CrossRef]
3. Varga, D. Are agrarian areas in Mediterranean mountain regions becoming extinct? A methodological approach to their conservation. *Forests* **2020**, *11*, 1116. [CrossRef]
4. Regato, P.; Salman, R. *Mediterranean Mountains in a Changing World: Guidelines for Developing Action Plans*; IUCN Centre for Mediterranean Cooperation: Malaga, Spain, 2008; p. 88.
5. Basile, G.; Cavallo, A. Rural identity, authenticity, and sustainability in Italian inner areas. *Sustainability* **2020**, *12*, 1272. [CrossRef]
6. Nepal, S.K.; Chipeniuk, R. Mountain tourism: Toward a conceptualframework. *Tour. Geogr.* **2005**, *7*, 313–333. [CrossRef]
7. Duglio, S.; Bonadonna, A.; Letey, M.; Peira, G.; Zavattaro, L.; Lombardi, G. Tourism development in inner mountain areas—The local stakeholders' point of view through a mixed method approach. *Sustainability* **2019**, *11*, 5997. [CrossRef]

8. Nistoreanu, P.; Dorobanțu, M.R.; Țuclea, C.E. The trilateral relationship ecotourism—Sustainable tourism—Slow travel among nature in the line with authentic tourism lovers. *J. Tour. Stud. Res. Tour.* **2011**, *11*, 35–38.
9. Dowling, R.K. Global Geotourism—An emerging form of sustainable tourism. *Czech J. Tour.* **2013**, *2*, 59–79. [CrossRef]
10. Hose, T.A.; Vasiljević, D.A. Defining the nature and purpose of modern geotourism with particular reference to the United Kingdom and South-East Europe. *Geoheritage* **2012**, *4*, 25–43. [CrossRef]
11. Ólafsdóttir, R.; Tverijonaite, E. Geotourism: A systematic literature review. *Geosciences* **2018**, *8*, 234. [CrossRef]
12. Dowling, R.K.; Newsome, D. Geotourism: Definition, characteristics and international perspectives. In *Handbook of Geotourism*; Dowling, R., Newsome, D., Eds.; Edward Elgar Publishing: Cheltenham, UK, 2018; pp. 1–22.
13. Widawski, K.; Jary, Z.; Oleśniewicz, P.; Owczarek, P.; Markiewicz-Patkowska, J.; Zaręba, A. Attractiveness of protected areas for geotourism purposes from the perspective of visitors: The example of Babiogórski National Park (Poland). *Open Geosci.* **2018**, *10*, 358–366. [CrossRef]
14. Tomić, N.; Marković, S.B.; Antić, A.; Tešić, D. Exploring the potential for geotourism development in the Danube Region Serbia. *Int. J. Geoherit. Parks* **2020**, *8*, 123–139. [CrossRef]
15. Ateş, H.; Ateş, Y. Geotourism and rural tourism synergy for sustainable development—Marçik Valley Case—Tunceli, Turkey. *Geoheritage* **2019**, *11*, 207–215. [CrossRef]
16. Hamoud, A.; El Hadi, H.; Tahiri, A.; Chakiri, S.; Mehdioui, S.; Baghdad, B.; El Maidani, A.; Bejjaji, Z.; Aoufad, M. Mauritanian geological resources: A lever for sustainableregional development via geotourism. *Int. J. Geoherit. Parks* **2021**, *9*, 415–429. [CrossRef]
17. Farsani, N.; Coelho, C.; Costa, C. Geotourism and geoparks as novel strategies for socio-economic development in rural areas. *Int. J. Tour. Res.* **2011**, *13*, 68–81. [CrossRef]
18. Farsani, N.; Coelho, C.; Costa, C. Rural geotourism: A new tourism product. *Acta Geoturistica* **2013**, *4*, 1–10.
19. Härtling, J.W.; Meier, I. Economic effects of geotourism in geopark TERRA. vita, Northern Germany. *George Wright Forum* **2010**, *27*, 29–39.
20. Lee, Y.; Jayakumar, R. Economic impact of UNESCO global geoparks on local communities in Asia: Comparative analysis of three UNESCO global geoparks in Asia. *Int. J. Geoherit. Parks* **2021**, *9*, 189–198. [CrossRef]
21. Telbisz, T.; Gruber, P.; Mari, L.; Kőszegi, M.; Bottlik, Z.; Standovár, T. Geological heritage, geotourism and local development in Aggtelek National Park (NE Hungary). *Geoheritage* **2020**, *12*, 5. [CrossRef]
22. Dóniz-Páez, J.; Hernández, P.A.; Pérez, N.M.; Hernández, W.; Márquez, A. TFgeotourism: A Project to quantify, highlight, and promote the volcanic geoheritage and geotourism in Tenerife (Canary Islands, Spain). In *Updates in Volcanology. Transdisciplinary Nature of Volcano Science*; Németh, K., Ed.; IntechOpen Book Series; IntechOpen: London, UK, 2021. [CrossRef]
23. Drinia, H.; Tsipra, T.; Panagiaris, G.; Patsoules, M.; Papantoniou, C.; Magganas, A. Geological Heritage of Syros Island, Cyclades Complex, Greece: An Assessment and Geotourism Perspectives. *Geosciences* **2021**, *11*, 138. [CrossRef]
24. Zafeiropoulos, G.; Drinia, H.; Antonarakou, A.; Zouros, N. From geoheritage to geoeducation, geoethics and geotourism: A critical evaluation of the Greek Region. *Geosciences* **2021**, *11*, 381. [CrossRef]
25. Sadry, B.N. *The Geotourism Industry in the 21st Century: The Origin, Principles, and Futuristic Approach*, 1st ed.; Apple Academic Press: Burlington, ON, Canada, 2020; p. 596.
26. Dowling, R.K.; Newsome, D. Setting an agenda for geotourism. In *Geotourism. The Tourism of Geology and Landscape*; Newsome, D., Dowling, R.K., Eds.; Goodfellow Publishers: Oxford, UK, 2010; pp. 1–12.
27. Dowling, R.K.; Newsome, D. *Global Geotourism Perspectives*; Goodfellow Publishers: Oxford, UK, 2010.
28. Dowling, R.K.; Newsome, D. Geotourism Destinations—Visitor Impacts and Site Management Considerations. *Czech J. Tour.* **2017**, *6*, 111–129. [CrossRef]
29. Gordon, J.E. Geoheritage, geotourism and the cultural landscape: Enhancing the visitor experience and promoting geoconservation. *Geosciences* **2018**, *8*, 136. [CrossRef]
30. Magagna, A.; Ferrero, E.; Giardino, M.; Lozar, F.; Perotti, L. A Selection of Geological Tours for Promoting the Italian Geological Heritage in the Secondary Schools. *Geoheritage* **2013**, *5*, 265–273. [CrossRef]
31. Marvinney, R.G.; Anderson, W.A.; Barron, H.F.; Hernández, R. The International Appalachian Trail: The ancient Appalachians as ambassador of the geosciences to modern societies. In Proceedings of the CAG-MAC/Joint Annual Meeting, Geological Association of Canada & Mineralogical Association of Canada, Fredericton, NB, Canada, 21–23 May 2014.
32. Pica, A.; Fredi, P.; Del Monte, M. The ernici mountains Geohritage (Central Apennines, Italy): Assessment of the Geosites for Geotourism development. *Geo J. Tour. Geosites* **2014**, *2*, 193–206.
33. Bertok, C.; d'Atri, A.; Martire, L.; Barale, L.; Piana, F.; Vigna, B. A trip through deep time in the rock succession of the Marguareis Area (Ligurian Alps, South Western Piemonte). *Geoheritage* **2015**, *7*, 5–12. [CrossRef]
34. Balestro, G.; Cassulo, R.; Festa, A.; Fioraso, G.; Giardino, M.; Nicolò, G.; Perotti, L. 3D geological visualizations of geoheritage information in the Monviso Massif (Western Alps). *Rend. Online Soc. Geol. Ital.* **2016**, *39*, 81–84. [CrossRef]
35. Bentivenga, M.; Palladino, G.; Prosser, G.; Guglielmi, P.; Geremia, F.; Laviano, A. A geological itinerary through the Southern Apennine thrust belt (Basilicata—southern Italy). *Geoheritage* **2017**, *9*, 1–17. [CrossRef]
36. Martínez-Graña, A.M.; Serrano, L.; González-Delgado, J.A.; Dabrio, C.J.; Legoinha, P. Sustainable geotourism using digital technologies along a rural georoutein Monsagro (Salamanca, Spain). *Int. J. Digit. Earth* **2017**, *10*, 121–138. [CrossRef]

37. Migoń, P.; Duszyński, F.; Jancewicz, K.; Różycka, M. From plateau to plain—Using space-for-time substitution in geoheritage interpretation, Elbsandsteingebirge, Germany. *Geoheritage* **2019**, *11*, 839–853. [CrossRef]
38. Bucci, F.; Tavarnelli, E.; Novellino, R.; Palladino, G.; Guglielmi, P.; Laurita, S.; Prosser, G.; Bentivenga, M. The history of the Southern Apennines of Italy preserved in the geosites along a geological itinerary in the High Agri Valley. *Geoheritage* **2019**, *11*, 1489–1508. [CrossRef]
39. Filocamo, F.; Di Paola, G.; Mastrobuono, L.; Rosskopf, C.M. MoGeo, a mobile application to promote geotourism in Molise region (Southern Italy). *Resources* **2020**, *9*, 31. [CrossRef]
40. Freire-Lista, D.M.; Becerra Becerra, J.E.; Simões de Abre, M. The historical quarry of pena (Vila Real, north of Portugal): Associated cultural heritage and reuse as a geotourism resource. *Resour. Policy* **2022**, *75*, 102528. [CrossRef]
41. Miccadei, E.; Piacentini, T.; Esposito, G. Geomorphosites and Geotourism in the Parks of the Abruzzo Region (Central Italy). *Geoheritage* **2011**, *3*, 233–251. [CrossRef]
42. Santangelo, N.; Romano, P.; De Santo, A.V. Geo-itineraries in the Cilento Vallo di Diano Geopark: A Tool for Tourism Development in Southern Italy. *Geoheritage* **2014**, *7*, 319–335. [CrossRef]
43. Aoulad-Sidi-Mhend, A.; Maaté, A.; Amri, I.; Hlila, R.; Chakiri, S.; Maaté, S.; Martín-Martín, M. The geological heritage of the Talassemtane National Park and the Ghomara coast Natural Area (NW of Morocco). *Geoheritage* **2019**, *11*, 1005–1025. [CrossRef]
44. Pazari, F.; Dollma, M. Geotourism potential of ZALL GJOÇAJ national park and the area nearby. *Int. J. Geoherit. Parks* **2019**, *7*, 103–110. [CrossRef]
45. Petrosino, P.; Iavarone, R.; Alberico, I. Enhancing social resilience through fruition of geological heritage in the Vesuvio National Park. *Geoheritage* **2019**, *11*, 2005–2024. [CrossRef]
46. Beltrán-Yanes, E.; Dóniz-Páez, J.; Esquivel-Sigut, I. Chinyero Volcanic Landscape Trail (Canary Islands, Spain): A geotourism proposal to identify natural and cultural heritage in volcanic areas. *Geosciences* **2020**, *10*, 453. [CrossRef]
47. Santangelo, N.; Amato, V.; Ascione, A.; Ermolli, E.R.; Valente, E. Geotourism as a tool for learning: A geoitinerary in the Cilento, Vallo di Diano and Alburni Geopark (Southern Italy). *Resources* **2020**, *9*, 67. [CrossRef]
48. Marino Alfonso, J.L.; Poblete Piedrabuena, M.Á.; Beato Bergua, S.; Herrera Arenas, D. Geotourism itineraries and augmented reality in the geomorphosites of the Arribes del Duero Natural Park (Zamora Sector, Spain). *Geoheritage* **2021**, *13*, 16. [CrossRef]
49. Rais, J.; Barakat, A.; Louz, E.; Ait Barka, A. Geological heritage in the M'Goun geopark: A proposal of geo-itineraries around the Bine El Ouidane dam (Central High Atlas, Morocco). *Int. J. Geoherit. Parks* **2021**, *9*, 242–263. [CrossRef]
50. Rosskopf, C.M.; Cesarano, M.; Filocamo, F.; Aucelli, P.P.C.; Brancaccio, L.; Pappone, G. Itinerario 5 Molise. In *Guide Geologiche Regionali. Campania e Molise*; Calcaterra, D., D'Argenio, B., Pappone, G., Petrosino, P., Eds.; Litografia Alcione: Lavis, Italy, 2016; pp. 117–138.
51. Filocamo, F.; Rosskopf, C.M. The geological heritage for the promotion and enhancement of a territory. A proposal of geological itineraries in the Matese area (Molise, Southern Italy). *Rend. Online Soc. Geol. Ital.* **2019**, *49*, 142–148. [CrossRef]
52. Ruggiero, E.; Amore, O.; Anzalone, E.; Barbera, C.; Cavallo, S.; Conte, M.; Fiano, V.; Massa, B.; Raia, P.; Sgrosso, I.; et al. I Geositi del Parco Regionale del Matese: Itinerario da Pesco Rosito a Cerreto Sannita. *Geologiadell'Ambiente* **2003**, *2003*, 181–192.
53. D'Argenio, B.; Ferranti, L.; Carannante, G.; Simone, L. Itinerario 6 Monti del Matese. In *Guide Geologiche Regionali. Campania e Molise*; Calcaterra, D., D'Argenio, B., Pappone, G., Petrosino, P., Eds.; Litografia Alcione: Lavis, Italy, 2016; pp. 139–151.
54. Bartiromo, A. Plant remains from the Lower Cretaceous Fossil-Lagerstatte of Pietraroja, Benevento, southern Italy. *Cretac. Res.* **2013**, *46*, 65–79. [CrossRef]
55. Breislak, S. *Topografia Fisica della Campania*; Nella Stamperia di Antonio Brazzini: Firenze, Italy, 1798.
56. Mostardini, F.; Merlini, S. Appennino centro-meridionale. Sezioni geologiche e porposta di modello strutturale. *Mem. Soc. Geol. Ital.* **1986**, *35*, 177–202.
57. Patacca, E.; Scandone, P.; Bellatalla, M.; Perilli, N.; Santini, U. La zona di giunzione tra l'arco appenninico settentrionale e l'arco appenninico meridionale nell'Abruzzo e nel Molise. In *Studi Preliminari All'acquisizione Dati del Profilo CROP 11 Civitavecchia-Vasto*; Tozzi, M., Cavinato, G.P., Parotto, M., Eds.; Special Issue 1991-2; Studi Geologici Camerti: Camerino, Italy, 1992; pp. 417–441.
58. Hippolyte, J.C.; Angelier, J.; Barrier, E. Compressional and extensional tectonics in an arc system: Example from the Southern Apennines. *J. Struct. Geol.* **1995**, *17*, 1725–1740. [CrossRef]
59. Robustini, P.; Corrado, S.; Di Bucci, D.; Calabrò, R.A.; Tornaghi, M. Comparison between contractional deformation styles in the Matese Mountains: Implications for shortening rates in the Apennines. *Boll. Soc. Geol. Ital.* **2003**, *122*, 295–306.
60. Vitale, S.; Ciarcia, S. Tectono-stratigraphic setting of the Campania region (Southern Italy). *J. Maps* **2018**, *14*, 9–21. [CrossRef]
61. Carannante, G.; Signore, M.; Vigorito, M. Vertebrate-rich Plattenkalk of Pietraroia (Lower Cretaceous, Southern Apennines, Italy): A new model. *Facies* **2006**, *52*, 555–577. [CrossRef]
62. Bassi, D.; Carannante, G.; Checconi, A.; Simone, L.; Vigorito, M. Sedimentological and palaeoecological integrated analysis of a Miocene channelized carbonate margin, Matese Mountains, Southern Apennines, Italy. *Sediment. Geol.* **2010**, *230*, 105–122. [CrossRef]
63. Boncio, P.; Dichiarante, A.M.; Auciello, E.; Saroli, M.; Stoppa, F. Normal faulting along the western side of the Matese Mountains: Implications for active tectonics in the Central Apennines (Italy). *J. Struct. Geol.* **2016**, *82*, 16–36. [CrossRef]
64. Galli, P.; Giaccio, B.; Messina, P.; Peronace, E.; Amato, V.; Naso, G.; Nomade, S.; Pereira, A.; Piscitelli, S.; Bellanova, J.; et al. Middle to Late Pleistocene activity of the northern Matese fault system (southern Apennines, Italy). *Tectonophysics* **2017**, *699*, 61–81. [CrossRef]

65. Valente, E.; Buscher, J.T.; Jourdan, F.; Petrosino, P.; Reddy, S.M.; Tavani, S.; Corradetti, A.; Ascione, A. Constraining mountain front tectonic activity in extensional setting from geomorphology and Quaternary stratigraphy: A case study from the Matese ridge, southern Apennines. *Quat. Sci. Rev.* **2019**, *219*, 47–67. [CrossRef]
66. Esposito, A.; Galvani, A.; Sepe, V.; Atzori, S.; Brandi, G.; Cubellis, E.; De Martino, P.; Dolce, M.; Massucci, A.; Obrizzo, F.; et al. Concurrent deformation processes in the Matese massif area (Central-Southern Apennines, Italy). *Tectonophysics* **2020**, *774*, 228234. [CrossRef]
67. Amato, V.; Aucelli, P.P.C.; Cesarano, M.; Rosskopf, C.M.; Cifelli, F.; Mattei, M. A 900 m-deep borehole from Boiano intermontane basin (Southern Apennines, Italy): Age constraints and palaeoenvironmental features of the Quaternary infilling. *Geol. J.* **2020**, *56*, 2148–2166. [CrossRef]
68. Filocamo, F.; Rosskopf, C.M.; Amato, V. A contribution to the understanding of the Apennine landscapes: The potential role of Molise Geosites. *Geoheritage* **2019**, *11*, 1667–1688. [CrossRef]
69. Giraudi, C. Datazione diretta e correlazione di depositi glaciali con l'uso di tephra e loess: Il caso del Matese (Campania-Molise). *Quaternario* **1999**, *12*, 11–16.
70. Aucelli, P.P.C.; Cesarano, M.; Di Paola, G.; Filocamo, F.; Rosskopf, C.M. Geomorphological map of the central sector of the Matese Mountains (Southern Italy): An example of complex landscape evolution in a Mediterranean mountain environment. *J. Maps* **2013**, *9*, 604–616. [CrossRef]
71. ISPRA 2028. Analisi delle Valenze Ambientali Dell'area di Interesse per L'istituzione del Parco Nazionale del Matese. Available online: https://www.naturacampania.it/Presentazione%20Matese-ISPRA_dic_18.pdf (accessed on 24 November 2021).
72. Forleo, M.B.; Giannelli, A.; Giaccio, V.; Palmieri, N.; Mastronardi, L. Geosites and parks for the sustainable development of inner areas: The Matese Mountain (Italy). *Geo J. Tour. Geosites* **2017**, *20*, 231–242.
73. Rosskopf, C.M.; Di Paola, G.; Filocamo, F. *Carta di Sintesi dei Geositi Molisani 2014*; Regione Molise: Campobasso, Italy, 2014.
74. I Geositi del Molise. Available online: http://www3.regione.molise.it/flex/cm/pages/ServeBLOB.php/L/IT/IDPagina/382 (accessed on 8 January 2022).
75. ISPRA. The Italian Geosites Inventory. Available online: http://sgi.isprambiente.it/GeositiWeb/ (accessed on 23 December 2021).
76. Carta dei Geositi. Available online: http://www.difesa.suolo.regione.campania.it/content/category/6/46/71/ (accessed on 23 December 2021).
77. Piano Territoriale Regionale. Geopotale Regione Campania. Available online: https://sit2.regione.campania.it/content/piano-territoriale-regionale (accessed on 23 December 2021).
78. Taddei, A.; Cotugno, R.; Fraissinet, M.; Massa, B.; Ruggiero, E. The 18 Geosites of Matese Regional Park (Campania, Southern Italy) proposed for inclusion in the APAT Italian Geosites database. In Proceedings of the 32nd International Geological Congress, Florence, Italy, 20–28 August 2004; Abstracts Part 1, pp. 240–241.
79. Ruggiero, E.; Taddei, A. Parco Naturale Regionale del Matese. In *Patrimonio Geologico e Geodiversità. Esperienze ed Attività dal Servizio Geologico d'Italia all'APAT*; D'Andrea, M., Lisi, A., Mezzetti, T., Eds.; Rapporti 51/2005; APAT: Rome, Italy, 2006; pp. 177–186.
80. Coratza, P.; Giusti, C. Methodological proposal for the assessment of the scientific quality of geomorphosites. *Quaternario* **2005**, *18*, 307–313.
81. Reynard, E.; Fontana, G.; Kozlik, L.; Scapozza, C. A method for assessing scientific and additional values of geomorphosites. *Geogr. Helv.* **2007**, *62*, 148–158. [CrossRef]
82. Pereira, P.; Pereira, D.I. Methodological guidelines for geomorphosite assessment. *Geomorphol. Relief Process. Environ.* **2010**, *2*, 215–222. [CrossRef]
83. Bruschi, V.M.; Cendrero, A.; Albertos, J.A.C. A statistical approach to the validation and optimisation of geoheritage assessment procedures. *Geoheritage* **2011**, *3*, 131–149. [CrossRef]
84. Coratza, P.; Bruschi, V.; Piacentini, D.; Saliba, D.; Soldati, M. Recognition and assessment of Geomorphosites in Malta at the Il-Majjistral Natural and History Park. *Geoheritage* **2011**, *3*, 175–185. [CrossRef]
85. Fassoulas, C.; Mouriki, D.; Dimitriou-Nikolakis, P.; Iliopoulos, G. Quantitative assessment of geotopes as an effective tool for geoheritage management. *Geoheritage* **2012**, *4*, 177–193. [CrossRef]
86. Brilha, J. Inventory and quantitative assessment of geosites and geodiversity sites: A review. *Geoheritage* **2016**, *8*, 119–134. [CrossRef]
87. Brilha, J. Geoheritage: Inventories and evaluation. In *Geoheritage: Assessment, Protection and Management*; Reynard, E., Brilha, J., Eds.; Elsevier: Amsterdam, The Netherlands, 2018; pp. 69–86.
88. Cappadonia, C.; Coratza, P.; Agnesi, V.; Soldati, M. Malta and Sicily joined by geoheritage enhancement and geotourism within the framework of land management and development. *Geosciences* **2018**, *8*, 253. [CrossRef]
89. Albani, R.A.; Mansur, K.L.; Carvalho, I.d.S.; Santos, W.F.S.D. Quantitative evaluation of the geosites and geodiversity sites of João Dourado Municipality (Bahia—Brazil). *Geoheritage* **2020**, *12*, 46. [CrossRef]
90. ISPRA. Scheda per L'inentario dei Geositi Italiani. Available online: http://sgi.isprambiente.it/geositiweb/public/scheda_geositi.pdf (accessed on 23 December 2021).
91. Mucivuna, V.C.; Garcia, M.G.M.; Reynard, E. Comparing quantitative methods on the evaluation of scientific value in geosites: Analysis from the Itatiaia National Park, Brazil. *Geomorphology* **2022**, *396*, 107988. [CrossRef]

92. Mucivuna, V.C.; Garcia, M.G.M. Geomorphosites assessment methods: Comparative analysis and typology. *Geoheritage* **2019**, *11*, 1799–1815. [CrossRef]
93. ISTAT. Istituto Nazionale di Statistica. Available online: https://www.istat.it/ (accessed on 15 November 2021).
94. Prosser, C.D.; Burek, C.V.; Evans, D.H.; Gordon, J.E.; Kirkbride, V.B.; Rennie, A.F.; Walmsley, C.A. Conserving geodiversity sites in a changing climate: Management challenges and responses. *Geoheritage* **2010**, *2*, 123–136. [CrossRef]
95. Zorina, S.O.; Silantiev, V.V. Geosites, Classification of. In *Encyclopedia of Mineral and Energy Policy*; Tiess, G., Majumder, T., Cameron, P., Eds.; Springer: Berlin/Heidelberg, Germany, 2014. [CrossRef]
96. Rook, L.; Pandolfi, L. Paleodays 2019. *La Società Paleontologica Italiana a Benevento e Pietraroja. Parte 2: Guida all'e Scursione della XIX Riunione Annuale SPI (Società Paleontologica Italiana)*; Ente GeoPaleontologico di Pietraroja (Benevento): Pietraroja, Italy, 2019; p. 24.
97. Carannante, G.; Puglies, A.; Ruberti, D.; Simone, L.; Vigliotti, M.; Vigorito, M. Evoluzione cretacica di un settore della piattaforma apula da dati di sottosuolo e di afforiamento (Appennino campano-molisano). *Ital. J. Geosci.* **2009**, *128*, 3–31.
98. Parona, C.F. Le Rudiste e le Camacee di S. Polo Matese raccolte da Francesco Bassani. *Mem. Regia Accad. Sci. Torino* **1901**, *50*, 197–214.
99. Catenacci, E.; De Castro, P.; Sgrosso, I. Complessi guida del Mesozoico calcareo-dolomitico nella zona orientale del Massiccio del Matese. *Mem. Soc. Geol. Ital.* **1962**, *4*, 837–856.
100. Selli, R. Sulla trasgressione del Miocene nell'Italia meridionale. *Giorn. Geol.* **1957**, *26*, 1–54.
101. Dal Sasso, C.; Signore, M. Exceptional soft-tissue preservation in a theropod dinosaur from Italy. *Nature* **1998**, *392*, 383–387. [CrossRef]
102. Barbera, C.; Simone, L.; Carannante, G. Depositi circalittorali di piattaforma aperta nel Miocene Campano. Analisi sedimentologica e paleoecologica. *Boll. Soc. Geol. Ital.* **1978**, *97*, 821–834.
103. Carannante, G.; Simone, L. Rhodolith facies in the central-southern Apennines Mountains, Italy. In *Models for Carbonate Stratigraphy from Miocene Reef Complexes of Mediterranean Regions*; Franseen, E.K., Esteban, M., Rouchy, J.M., Eds.; SEPM Concepts in Sedimentology and Paleontology; SEPM: Tulsa, OK, USA, 1966; Volume 5, pp. 261–275.
104. D'Argenio, B.; Mindszenty, A. Bauxites and related paleokarst: Tectonic and climatic event markers at regional unconformities. *Eclogae Geol. Helv.* **1995**, *88*, 453–499.
105. Del Prete, S.; Mele, R.; Allocca, F.; Bocchino, B. Le miniere di bauxite di Cusano Mutri. (Monti del Matese -Campania). *Opera Ipogea* **2002**, *1*, 3–34.
106. Pappone, G.; Aucelli, P.P.C.; Cesarano, M.; Putignano, M.L.; Ruberti, D. *Note Illustrative del Foglio 405 Campobasso della Carta Geologica d'Italia alla Scala 1: 50.000*; ISPRA: Roma, Italy, 2013.
107. Fiorillo, F.; Pagnozzi, M. Recharge processes of Matese karst massif (southern Italy). *Environ. Earth Sci.* **2015**, *74*, 7557–7570. [CrossRef]
108. Ruggero, P. Risultati di alcune indagini sul regime idrologico del Massiccio del Matese. *Ann. Lav. Pubblici* **1926**, *64*, 381–401.
109. Civita, M. Valutazione analitica delle riserve in acque sotterranee alimentanti alcune tra le principali sorgenti del massiccio del Matese (Italia meridionale). *Mem. Soc. Nat. Napoli* **1969**, *78*, 133–163.
110. Giraudi, C.; Zanchetta, G.; Sulpizio, R. A Late-Pleistocene phase of Saharan dust deposition in the high Apennine mountains (Italy). *Alp. Mediterr. Quat.* **2013**, *26*, 110–122.
111. Palmentola, G.; Acquafredda, P. Gli effetti dei ghiacciai quaternari sulla montagna del Matese, al confine molisano-campano. *Geogr. Fis. Din. Quat.* **1983**, *6*, 117–130.
112. Brancaccio, L.; Cinque, A.; Orsi, G.; Pece, R.; Rolandi, G.; Sgrosso, I. Lembi residui di sedimenti lacustri pleistocenici sospesi sul versante settentrionale del Matese, presso S. Massimo. *Boll. Soc. Nat. Napoli* **1979**, *88*, 275–286.
113. Di Bucci, D.; Naso, G.; Corrado, S.; Villa, I. Growth, interaction and sismogenic potential of coupled active normal faults (Isernia Basin, Central-Southern Italy). *Terra Nova* **2005**, *17*, 44–55. [CrossRef]
114. Ferrarini, F.; Boncio, P.; de Nardis, R.; Pappone, G.; Cesarano, M.; Aucelli, P.; Lavecchia, G. Segmentation pattern and structural complexities in seismogenic extensionalsettings: The North Matese Fault System (Central Italy). *J. Struct. Geol.* **2017**, *95*, 93–112. [CrossRef]
115. Russo, F.; Terribile, F. Osservazioni geomorfologiche, stratigrafiche e pedologiche sul Quaternario del Bacino di Bojano (Campobasso). *Quaternario* **1995**, *8*, 239–254.
116. Guerrieri, L.; Scarascia Mugnozza, G.; Vittori, E. Analisi stratigrafica e geomorfologica della conoide tardo-quaternaria di Campochiaro ed implicazioni per la conca di Boiano in Molise. *Quaternario* **1999**, *12*, 237–247.
117. ISTAT—Istituto Nazionale di Statistica. Classificazione dei Comuni in Base alla Densità Turistica. Available online: https://www.istat.it/it/archivio/247191 (accessed on 5 December 2021).

Article

Preservation of the Geoheritage and Mining Heritage of Serifos Island, Greece: Geotourism Perspectives in a Potential New Global Unesco Geopark

Nikolaos Vlachopoulos * and Panagiotis Voudouris *

Faculty of Geology and Geoenvironment, National and Kapodistrian University of Athens, 15784 Athens, Greece
* Correspondence: nvlahopoulos@gmail.com (N.V.); voudouris@geol.uoa.gr (P.V.)

Abstract: Serifos island is characterized by rich geodiversity, industrial and cultural heritage. The present paper focuses on the geological and mining heritage of Serifos, with the aim of integrating the island in the international environment of Geoparks, in the near future. In this geopark, Serifos can highlight the rich geological heritage of the island combined with the rich industrial heritage as expressed by mining activities since prehistoric times and the mining facilities of iron and copper mines. During the present study, six geotrails have been developed to link these cultural and ecological sites with the geological heritage. Along the routes, the geodiversity is explained, including its relationship with the surrounding biodiversity, and the historical and cultural aspects of the region. In the proposed geocultural routes (geotrails), the dialectic relationship between Humans and Nature is determined by historical conditions and by the record of the process that transforms space into a landscape. The geological-mining heritage of Serifos will attract people from all over the world with different kind of interests and will make it known to alternative tourists. The results of this paper are intended to constitute a valuable tool for enhancing and raising awareness of the geological heritage of the island of Serifos.

Keywords: geosite; geotourism; mineralogical heritage; geo-conservation; mineralogical museum; iron mines; skarn-related mineralization; geotrails

1. Introduction

Geoheritage comprises those elements of the Earth's geodiversity (rocks, minerals, fossils, landforms, sediments, water and soils) that are considered to have significant scientific, educational, cultural or aesthetic value [1,2]. Traditionally, the valorization and the use of geological valuable areas as touristic resources has been linked to areas characterized by the beauty of the landscape, the spectacular rock formations or relevant features (mountains, glacier formations, rivers, canyons, caves, etc.) interesting for people loving geology or, at least, nature [1–4]. As stated by Carcavilla et al. [4] "Geological-geomorphological heritage is the collection of geotopes, deposits, forms, and processes that comprise the geological history of each region, and the concept of preserving geological-geomorphological heritage is a cultural concept". Geoheritage aims to highlight the diversity of our planet to illustrate the importance of the biotic and abiotic factors, which document the historical evolution of the Earth [5]. The value of geological heritage is further underlined in a report from UNESCO [6], according to which geological heritage is characterized as the whole of the most interesting geological sites (geotopes, geoparks, and geological natural monuments) that deserve to be preserved for scientific, didactic, historical, aesthetic, and cultural reasons.

In addition to geoheritage, the industrial and mining heritage means all the sources of the industrial past that contribute to the knowledge of the history of the productive activities of a country or a population. Considering a monument or an object of industrial use as an information carrier is important and necessary since it incorporates all the influences of culture and the environment.

Geotourism is connecting the geological heritage with the natural environment and the cultural monuments [7–10]. Geoheritage is the driving force of the geotourism itineraries and cultural heritage is also added to increase the value of the visited regions. The management of industrial and geological heritage has demonstrated a number of special applications, such as geoparks, technology parks, eco-museums or theme parks, cultural centers, recreation sites and industrial museums in order to emerge as tourist destinations.

At present, there are 169 UNESCO Global Geoparks in 44 countries, six among them in Greece [11] (Figure 1). Geoparks are wider areas that contain significant sites of geological monuments and geotopes as well as sites of ecological, archaeological, historical or cultural interest and are tools for environmental education [6,12,13]. The concept of geopark was attributed to areas with special geological appearances, which can contribute to sustainable local development. Thus, the establishment of a geopark not only opens up new opportunities and creates enthusiasm for geoconservation, but the park also becomes a new tourist attraction.

Serifos island, located in the South Aegean Sea, represents a multiple-mineralized district, including porphyry, skarn, carbonate-replacement and vein-type ores [14–18]. The Serifos mineralization is related with the emplacement of I-type granodiorite, considered to be synchronous with Miocene extensional detachment faulting, and intruded gneisses, amphibolites, schists and marbles causing an extensive contact metamorphic aureole [14,15,19–23].

Serifos is well known to collectors for its spectacular skarn-related mineralization with gem quality green quartz crystals with actinolite inclusions (prase or prasem) [24–30]. The Serifos minerals are among the most impressive collectable objects exhibited in museums and private collections worldwide. These minerals at Serifos can be founds either on the surface, but also underground in mining sites and include species of exceptional beauty and scientific value [28].

The Serifos island is also known for its numerous iron mines opened in: (i) numerous magnetite-rich skarns; (ii) hematite/limonite ± barite ore bodies hosted in marbles that were exploited until 1963 [15,31]. Already Ducoux et al. [15] emphasized the role of low-angle detachment fault systems to ore mineralization at the various mining sites, and described the geological and mineralogical characteristics of Serifos. From an archeological point of view, Serifos island is famous for the exploitation of copper ore during ancient times [32,33].

The aim of this article is to summarize the current state of knowledge on the geology, petrology, mineralization, and the primary mineralogy and mining characteristics of Serifos island, and to expand on previous work by highlighting its unique geological, petrological, mineralogical, mining, and educational features. We suggest various geosites and industrial sites on the island, present new geotrails that could be used for geotourism, cultural, education and research activities, and evaluate all data considering Serifos as a best candidate for integration in a potential new Unesco Global Geopark.

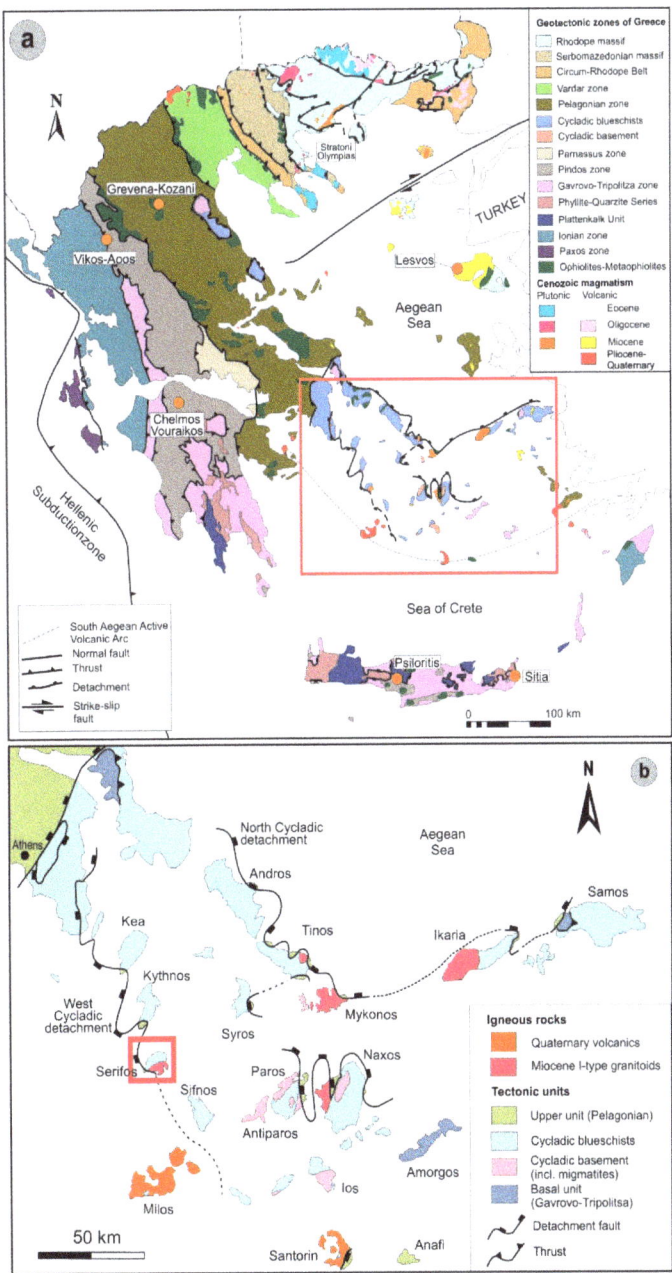

Figure 1. Geological map of (**a**) Greece and (**b**) Attic-Cycladic Crystalline Belt, showing the major tectonic structures, as for example, the North Cycladic Detachment System (NCDS), the West Cycladic Detachment System (WCDS). The location of Lesvos, Psiloritis, Chelmos-Vouraikos, Vikos-Aoos, Sitia and Grevena-Kozani UNESCO Global Geoparks are shown, while the Serifos area is marked with a red square in Figure 1b and described in detail in Figure 2. Other mining sites with exemplary geological and mineralogical characteristics include Stratoni and Olympias at Chalkidiki, and Milos, Syros, and Naxos islands. Modified from Voudouris et al. [34] and references therein.

2. Geological Setting

Serifos island belongs to the Attic-Cycladic crystalline belt, which represents a polymetamorphic terrane of the Hellenides, in the back-arc region of the Hellenic subduction zone [35,36] (Figure 1). Since Oligocene, slab roll-back resulted in the southward migration of magmatic activity from north to south, in orogenic collapse, and the post-orogenic exhumation of the crust as metamorphic core complexes, associated with voluminous magmatism [36]. The Attic-Cycladic belt comprises four major units, the Cycladic basement, the so-called Basal unit, the Cycladic blueschist unit, and the Upper tectonic unit (e.g., [36–38], Figure 1b). The Cycladic basement includes Pre-Carboniferous schists and Carboniferous gneisses tectonically overlain by the Cycladic blueschist unit consisting of a volcano-sedimentary sequence of metasediments, marbles, calc-silicate schists, and meta-igneous rocks of Permo-Carboniferous to latest Cretaceous ages [35,36]. The Upper tectonic unit consists of various unmetamorphosed Upper Permian to Jurassic volcaniclastics, ophiolites, and carbonates, greenschist-facies rocks of Cretaceous to Tertiary age, and Late Cretaceous amphibolite-facies rocks and granitoids (e.g., [39,40]).

All tectonic units in the Cyclades are separated by crustal-scale detachment faults [21,35,41] (Figure 1b). Three stages of metamorphism during the Tertiary characterize the Cycladic area (e.g., [42,43]): blueschist-to-eclogite facies metamorphism during the Eocene (~52–40 Ma), followed by Eocene-Oligocene (ca. 40–30 Ma) upper greenschist-facies metamorphism, and then, by Oligo-Miocene (25–17 Ma) greenschist to upper amphibolite-facies metamorphism (locally with crustal anatexis at about 17 Ma). The Oligo-Miocene event was accompanied by exhumation of Attic-Cycladic metamorphic core complexes along several low-angle extensional detachment faults, such as the Northern and the Western Cycladic detachment systems, the latter also exposed at Serifos island [21,41] (Figure 1b). Intrusion of various granitoids throughout the Attic-Cycladic belt between 15 and 6 Ma was accommodated by movement of detachments [44,45].

Serifos Island is interpreted as a metamorphic core complex exhumed below the Western Cycladic detachment system [19,20]. The northern part of the island is dominated from base to top by (Figure 2): (1) metamorphic rocks of the Cycladic Continental Basements Unit (CCB) composed of gneiss and mylonitic schists, covered with calcite and dolomite mylonitic marbles; (2) the Cycladic Blueschist Unit (CBU) composed of amphibolites at the base overlain by greenschists with marble intercalations; (3) the Upper Cycladic Unit (UCU), composed of marbles, ankeritized cataclastic shales and schists exposed in the southwestern part, the Kyklopas area, and the northern part, at Platy Gialos [15,21]. These three units are separated by two detachments which both show top-to-the-SSW kinematics, which belong to the WCDS [15]: the Megalo Livadi detachment (MLD) separating the CCB from CBU, is well exposed at Megalo Livadi and Koundouros, and the Kavos Kiklopas detachment (KKD) separating the CBU from UCU which is exposed at Kavos Kiklopas and Platy Gialos [15].

On Serifos, blueschist facies metamorphism dated between 38 to 35 Ma based on $^{40}Ar/^{39}Ar$ in phengites [46], while all metamorphic rocks have been affected by retrograde LP/HT greenschist facies metamorphism during the Late Oligocene-Early Miocene [15,47,48]. During the Late Miocene the metamorphic pile is intruded by a I-type hornblende and biotite-bearing granodiorite with an age between 11.6 to 9.5 Ma, based on zircon U–Pb dating [22,44,47–50]. The pluton has two distinct facies [15,47,48]: an inner facies which is fine-grained equigranular unfoliated and a border coarse-grained facies with large biotite flakes and mafic enclaves. The pluton intrudes the MLD and the granodiorite roof is deformed by extensional shear zones in the south (Vagia Bay) and east (Agios Sostis) of Serifos [15].

The Serifos granodiorite was emplaced along the West Cycladic detachment fault in the CCB and the CBU units and caused contact metamorphism with the development of high-temperature Ca-Fe endo- and exoskarns and iron ores (Figure 2). Both the Megalo Livadi and Kavos Kyklopas detachment faults acted as pathways by the magmatic-hydrothermal fluids resulting to the development of medium-temperature Ca-Fe-Mg exoskarn zones in the southwestern part of the island (Figure 2) [15]. Skarns and iron ores demonstrate ductile and brittle structures as a consequence of the activity of the detachment fault [15]. Sulfide ore was formed during the retrograde stage and consists mainly of pyrite, chalcopyrite, galena, sphalerite and is extensively oxidized [14–16,18]. Calcite, barite, garnet, pyroxene, epidote, fluorite, talc, chlorite, adularia are the main gangue minerals [14–16,18].

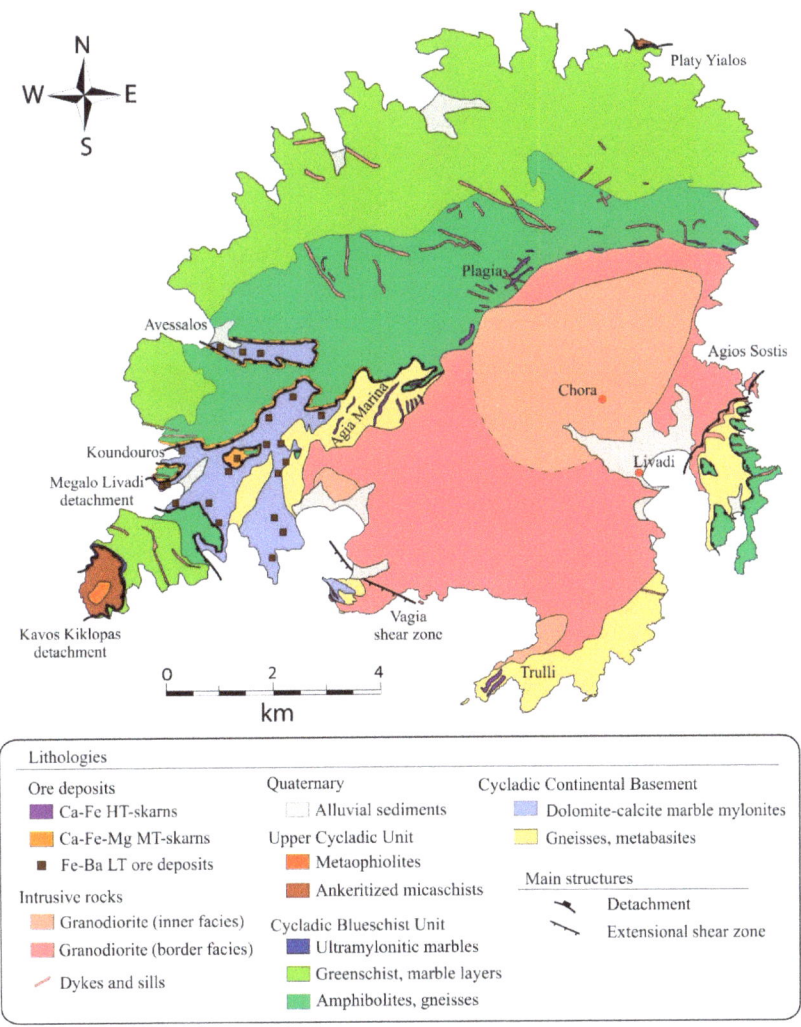

Figure 2. Geological map (modified after Ducoux et al. [15]). HT/MT/LT for high, medium and low temperature, respectively.

3. Results

The data used in this study have derived from several field trips in Serifos island. During this fieldwork, all identified geosites were assessed and ranked. The geosites (and mining sites) considered in this paper are of great scientific interest, as they provide information about the conditions and timing of metamorphism, the subsequent exhumation paths of the metamorphic core complexes, and the magmatic and hydrothermal processes leading to magma emplacement, and subsequent ore deposition and exploitation, as well as of crystallization of rare minerals. The geosites (and industrial sites) have been classified based on their physical and scientific characteristics into various categories such as: (a) Geomorphosites, such as landforms resulting from differential erosion and weathering (tafoni), tectonics, drainage network, sea level changes and depositional processes of the geological formations and karstic geotopes, such as caves; (b) Geological, which are also distinguished in petrological or mineralogical geotopes; (c) Tectonics, including detachments, tectonic covers and contacts, faults and folds; (d) Hydrogeological, concerning the springs; (e) Mining heritage. Localities for the various geosites and mining sites at Serifos island are presented in Figure 3 and described in detail in the following paragraphs.

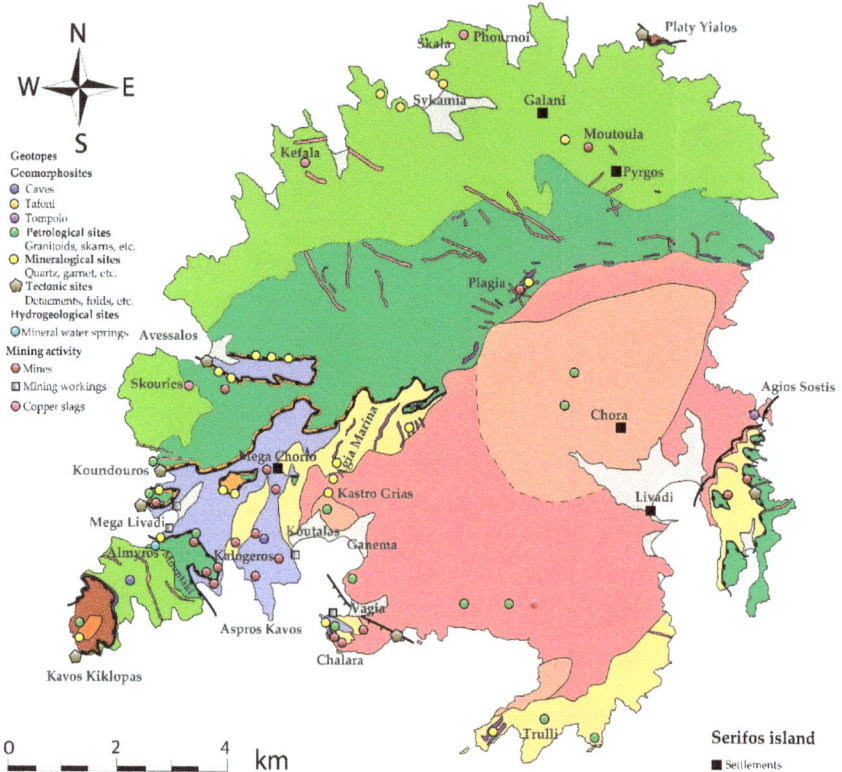

Figure 3. Localities of various geosites and mining sites for the suggested Serifos geopark. Geological map modified after Ducoux et al. [15]).

3.1. Geological-Tectonic Geosites at Serifos

On Serifos Island, geological, tectonic, magmatic, and skarnization processes can be best demonstrated at hundreds of sites, thus making the island a natural geological observatory. Lithological and tectonic contacts can be observed around Megalo Livadi and Kavos Kiklopas, where both the Megalo Livadi and the Kavos Kiklopas detachments are exposed (Figure 4a–d). At these sites, marbles from the CCB, amphibolites from the CBU and schist and serpentinites from the UCU are well exposed. Mylonitic gneisses from the CCB can be observed in the Tsilipaki area and at Trulli (Figure 4e,f).

The I-type granodiorite pluton with its border facies is well exposed around Chora and Vagia (Figure 4g,i), and dacite dikes crosscut marbles of the CCB at Megalo Livadi (Figure 4h). Another igneous body, located in the southern part of the island, was mentioned by Voudouris et al. [51] as an undeformed leucogranite composed of K-feldspar, quartz, minor plagioclase and biotite and hosts molybdenite-pyrite mineralization (Figure 4f).

Figure 4. Field photos showing typical lithologies and tectonic features at Serifos. (**a**) The Megalo Livadi detachment (MLD) separating marbles of the Cycladic Continental Basement (CCB, footwall) from amphibolite and gneisses of the Cycladic Blueschist Unit (CBU, hanging wall). Note the location of Fe-mines beneath and along the detachment fault; (**b**) the MLD separating marbles of the CCB from hedenbergitic skarn hosted in brecciated amphibolite of the CBU at Koundouros area; (**c**,**d**) the Kavos Kiklopas detachment (KKD) separating mylonitic marbles of the CBU (footwall) from metaophiolites and schists of the Upper Cycladic Unit (UCU); (**e**) panoramic view of the mylonitic orthogneiss of the CCB at Trulli area (view from W to E); (**f**) undeformed S-type (?) granitoid (leucogranite) at Trulli area intruding the mylonitic orthogneiss; (**g**) I-type granodiorite (border facies) at Vagia area; (**h**) dacitic dikes crosscutting Fe-oxides stained marbles of the CCB at Megalo Livadi area; (**i**,**j**) I-type granodiorite including high-T endoskarn garnet-pyroxene-magnetite bodies (dark areas) intruding white and grey marbles of the CCB at Chalara area; (**k**) medium-T hedenbergite skarn crosscutting epidotized amphibolite at Avessalos area; (**l**) modern galleries within oxidized ore hosted in schists and marbles of the CBU at Galani area (Moutoula, Pyrgos).

High-temperature endo- and exoskarns, medium-temperature distal exoskarns with prograde and retrograde assemblages including iron oxides and sulfides occur among others at Chalara (Figure 4i,j), Agia Marina, Megalo Livadi, Avessalos (Figure 4k) and Moutoulas (Figure 4l). Additional sites and photos with petrological and mineralogical characteristics are demonstrated in the following paragraphs.

3.2. Mineralogical and Petrological Geotopes at Serifos

Serifos is not only famous because of the mining activity in the past, it also shows unique mineralogical and petrological features: its very rare and worldwide known skarn minerals (e.g., garnets, quartz and its green variety called prase, ilvaite; Figures 5–9), attracted scientists and mineral collectors from all over the world. This has led to a dramatic reduction in the abundance of the mineral occurrences of the island, making their preservation necessary. The Serifos skarn is unique in the world containing the best varieties of prase (e.g., the green quartz variety) and ilvaite [24–30].

Figure 5. Field and hand specimen photos showing typical petrological and mineralogical features of geotopes at Agia Marina (a to j), and Koutalas (k to o) areas. (**a**) Panoramic view of Agia Marina area from N to S with mylonitc orthogneiss outcrop; (**b**) mineralized geode within hedenbergitic skarn surrounding orthogneiss; (**c**) quartz crystals from the previous geode; (**d,e**) garnetite (andraditic skarn) hosted in orthogneiss at the contact to I-type granodiorite (view from N to S and E to W respectively); (**f–h**) garnetite with idiomorphic andradite and late quartz crystals; (**i,j**) geode within hedenbergite filled with quartz crystals, occasionally amethystine at their basis; (**k**) I-type granodiorite (inner facies) at the Kastro Grias area (view from S to N); (**l**) hydrothermal breccia composed of sericite-altered granodiorite fragments within a matrix of Fe-oxides and barite at the Kastro Grias locality; (**m**) Fe-Pb mines within marbles of the CCB; (**n,o**) barite crystals with galena and Fe-oxides at the Koutalas mines.

Figure 6. Field and hand specimen photos showing typical petrological and mineralogical features of geotopes at Megalo Livadi area. (**a–c**) Panoramic view (from S to N) showing marbles of the CCB and hedenbergite skarn separated by the MLD fault. Note location of mine within marbles just beneath the detachment fault; (**d,e**) hand specimens with calcite rhombohedron coated by quartz associated with hedenbergite; (**f**) hand specimen with quartz crystals coated with Fe-oxides; (**g**) hand specimen with barite crystals covered with calcite; (**h**) panoramic view (from N to S) showing location of ilvaite in the hedenbergitic skarn on the road towards Megalo Livadi; (**i**) hedenbergite followed by ilvaite and then by quartz being part of breccia cement of the cockade megabreccia in the previous locality; (**j**) amethyst and prase as late open-space filling in geodes of the hedenbergite bearing megabreccia.

Quartz is a very common mineral in the skarns of Serifos. Combinations of amethyst and prase forming scepter growths at Serifos are worldwide unique specimens. Garnet is a major constituent of skarn and represented by several varieties at Serifos. The Serifos andradites are famous due to their zonal growth with colors ranging from deep brown to orange. Ilvaite, up to 50 cm long in association with hedenbergite at Serifos, represent the best occurrence of this mineral worldwide. Calcite crystals up to 35 cm are intergrown with prase at Serifos. The carbonate-replacement deposits at Serifos contain large crystals of fluorite (up to 5 cm) and barite (up to 50 cm). Native bismuth in grains up to 5 cm occurs at Moutoulas, near Galani in the northern part of Serifos island. A list of all known collective minerals in Serifos, including primary, supergene as well as slag minerals are presented in Table 1.

Figure 7. Field photos showing typical petrological and mineralogical features of geotopes along (**a**) the Megalo Livadi-Koundouros geotrail; (**b**) magnetite replacing marbles of the CCB; (**c**) panoramic view (from SW to NE) of the Koundouros geosite, showing mineralized marbles of the CCB, and hedenbergitic skarn on both sides; (**d,e**) hedenbergitic skarn developed in several growth zones around fragments of amphibolite; (**f**) prase and hematite (iron roses) in cavities of hedenbergite; (**g**) the Megalo Livadi detachment (red dotted line) separating Fe-oxide mineralized marbles of the CCB from amphibolite-hosted hedenbergitic skarn of the CBU. Note the location of mine entrances just beneath the detachment fault; (**h,i**) ilvaite crystals developed in geodes within hedenbergitic skarn at the Koundouros geosite.

In addition, under the microscope, the minerals arsenopyrite, sphalerite, cobaltite, glaucodot, gersdorffite, marcasite, native Au and native Bi and sulfosalts of the bismuthinite-aikinite series were identified in southwestern Serifos mineralization by Korisidis et al. [18] and Bi- and Te-rich tetrahedrite-tennantite solid solutions, greenockite tellurides, tetradymite, hessite, and melonite are additionally mentioned by Fitros et al. [16]. Beyerite, bismutite, and bismite replaced native bismuth at Moutoulas, and together with covellite, cerussite, anglesite, chalcocite, goethite, are products of the supergene stage.

Figure 8. Field and hand specimen photos showing typical petrological and mineralogical features of geotopes at north (**a–f**) and south (**g–m**) Avessalos area. (**a**) Panoramic view from W to E with location of dark green prase crystals and red-colored andradite within hedenbergite skarn. Marbles of the CCB can also be seen; (**b**) panoramic view from E to W with outcrops of mineralized geodes within hedenbergite skarn; (**c**,**d**) hedenbergite with geodes filled by red andradite; (**e**) hand specimen with red andradite on hedenbergite both covered by green quartz (prase) (photo courtesy of B. Ottens); (**f**) hand specimen of dark green variety of quartz with hematite roses at its base; (**g**) panoramic view from W to E with location of light green prase crystals and amethyst scepters within hedenbergite skarn at south Avessalos; (**h–j**) goesite with huge geodes within hedenberite filled with prase, at the previous locality; (**k**) hand specimen of prase crystals with orange-colored upper parts due to iron oxide inclusion; (**l**) hand specimen with amethyst scepter on prase; (**m**) hand specimen with intergrowth between prase and platy calcite.

Figure 9. Field photos showing typical petrological, mineralogical and mining features of the geotope at Chalara area. (**a**) Panoramic view (from S to N) showing marbles of the CCB intruded by I-type granodiorite. Note location of garnet-pyroxene-magnetite endoskarn within the granodiorite; (**b**) granodiorite veins intruding marbles of the CCB; (**c**) garnet-bearing endoskarn of granodiorite sill intruding marbles of the CCB; (**d**) garnet-pyroxene exoskarn within layers of CCB marbles; (**e**) pyrite-magnetite mineralization associated with garnet-pyroxene endoskarn within granodiorite; (**f**) pyrite-magnetite mineralization within exoskarn.

The SW part of Serifos island, including the Agia Marina-Koutalas, Megalo Livadi-Koundouro and Avessalos subareas, displays all the criteria necessary to be characterized as a mineralogical and petrological geotope [25–27,52,53].

The Agia Marina area is dominated by mylonitic gneisses of the CCB intruded by granodiorite veins and sills transformed to granitic endoskarn bodies (Figure 5a–l; see also Ducoux et al. [15]). The geotope includes splendid occurrences of andraditic garnets (crystals up to 5 cm in size) formed during the prograde stage in association with quartz crystals crystallized during the retrograde skarn stage (Figure 5c–h). Quartz (often amethystine) is associated with hematite in quartz veins crosscutting garnetite or hedenbergitic skarn (Figure 5i,j). Towards the Kastro Grias locality, a brecciated granodiorite with iron oxide and barite in the breccia matrix is exposed (Figure 5k,l). At the adjacent Koutalas area, the Fe-Pb mines within marbles of the CCB contains large barite crystals (up to 30 cm), in associated with galena (Figure 5m–o).

The Megalo Livadi-Koundouro area is dominated by several Fe mines hosted within medium temperature hedenbergitic skarns and CCB marbles, following the trace of the Megalo Livadi detachment (Figure 6a–c). Magnetite was formed during the retrograde stage and followed by pyrite-arsenopyrite-chalcopyrite-Bi-Au mineralization [18]. Retrograde minerals also include calcite, quartz and barite (Figure 6d–g). At about 1 km NE of Megalo Livadi, a cockade megabreccia is composed of epidotized amphibolite fragments, rimmed by several bands of prograde hedenbergite in association with ilvaite, followed by quartz (also amethystine) and late calcite (Figure 6h–j). The Koundouros locality is characterized by hedenbergitic skarn including the best ilvaite crystals worldwide. Geodes within the skarn are filled by idiomorphic crystals of ilvaite, hematite (iron roses), quartz and calcite. The ilvaite crystals are associated with hedenbergite, forming radial aggregates reaching sizes of up to 50 cm (Figure 7).

Table 1. List of collective minerals from Serifos island (data from Ottens and Voudouris [28] and this study).

Mineral	Chemical Formula
Primary minerals	
Actinolite	$Ca_2(Mg_{4.5-2.5}Fe^{2+}_{0.5-2.5})_5Si_8O_{22}(OH)_2$
Epidote	$Ca_2Fe^{3+}Al_2(Si_2O_7)(SiO_4)O(OH)$
Andradite	$Ca_3Fe_2^{3+}(SiO_4)_3$
Hedenbergite	$CaFe^{2+}Si_2O_6$
Ilvaite	$CaFe^{3+}Fe_2^{2+}O(Si_2O_7)(OH)$
Quartz	SiO_2
Chalcopyrite	$Cu^{1+}Fe^{3+}S_2$
Galena	PbS
Pyrite	FeS_2
Hematite	Fe_2O_3
Magnetite	$Fe^{2+}Fe_2^{3+}O_4$
Aragonite	$CaCO_3$
Barite	$BaSO_4$
Calcite	$CaCO_3$
Fluorite	CaF_2
Supergene minerals	
Cuprite	Cu_2O
Malachite	$Cu_2CO_3(OH)_2$
Brochantite	$Cu_4SO_4(OH)_6$
Azurite	$Cu_3(CO_3)_2(OH)_2$
Chrysocolla	$(Cu_{2-x}Al_x)H_{2-x}Si_2O_5(OH)_4 \cdot nH_2O$
Limonite	Fe,O,OH,H_2O
Slag minerals	
Atacamite	$Cu_2Cl(OH)_3$
Malachite	$Cu_2CO_3(OH)_2$
Nantokite	$CuCl$
Paratacamite	$Cu_3^{2+}(Cu,Zn)(OH)_6Cl_2$
Rouaite	$Cu_2NO_3(OH)_3$
Spangolite	$Cu_6AlSO_4(OH)_{12}Cl \cdot 3H_2O$
Brochantite	$Cu_4SO_4(OH)_6$
Buttgenbachite	$Cu_{36}(NO_3)_2Cl_6(OH)_{64} \cdot nH_2O$
Chalcanthite	$CuSO_4 \cdot 5H_2O$
Connellite	$Cu_{36}(SO_4)(OH)_{62}Cl_8 \cdot 6H_2$
Chrysocolla	$(Cu_{2-x}Al_x)H_{2-x}Si_2O_5(OH)_4 \cdot nH_2O$
Cuprite	Cu_2O
Delafosite	$Cu^{1+}Fe^{3+}O_2$
Clinoatacamite	$Cu_2Cl(OH)_3$
Langite	$Cu_4SO_4(OH)_6 \cdot 2H_2O$
Linarite	$PbCuSO_4(OH)_2$
Fayalite	$Fe_2^{2+}SiO_4$
Goethite	$FeO(OH)$
Gypsum	$CaSO_4 \cdot 2H_2O$

The Avessalos area is the best site in the world in respect to the mineral green quartz (prase). Similarly to Megalo Livadi-Koundouro, at Avessalos, the mineralization occurs in geodes developed within a cockade megabreccia just below the Megalo Livadi detachment, formed around epidotized amphibolite, subsequently covered by hedenbergite and garnet during prograde skarnization and by quartz during the retrograde stage. The northern Avessalos geosite was discovered 20 years ago and represents the best locality of the mineral prase, the green variety of quartz [24]. The crystal forms, intergrowths and sizes (up to 40 cm) of green quartz specimens from this locality are unique. Similar crystals have never been observed elsewhere in the world. The very deep green-colored prase crystals are accompanied by iron roses. The northern Avessalos area is characterized by a granatitic and hedenbergitic exoskarn and by the development of huge geodes filled by prograde and retrograde skarn minerals (Figure 8a–f). Zoned andraditic garnets occur in this location, but the spectacular specimens of green quartz and amethyst are those which attracted the interest of mineral collectors in that geosite. In the double-colored crystals of prase-amethyst, the transition between these two crystals is abrupt within the same crystal, where prase occurs at the base and amethyst at the top of the crystal. The amethysts are transparent and of gemstone quality [30,54–56].

The second geosite is located at the southern part of the Avessalos area (Figure 8g,h). This geosite includes large geodes, containing unique quartz crystals, not only in respect to their quality but also for their crystal forms, which reflect very special growth conditions (Figure 8i–l).

Rare combinations of prase-amethyst scepter crystals contain phase alternations, including transitions from prase towards amethyst, and finally, to prase even within a single composite crystal [30,54–56]. Scepters include both normal and reverse forms. Calcite-prase intergrowths, abundant within the southern Avessalos geodes, were found for the first time in Serifos island: calcite crystals, either as rhombohedron, or in platy forms alternate with the prase suggesting contemporaneous deposition probably during boiling processes (Figure 8m). The Avessalos area underwent extensive mineral exploitation by local and foreign dealers, often destroying scientific information on the geological evolution and valuable elements of the geocultural heritage.

The Chalara locality best demonstrates outcrops of high-temperature endo- and exoskarn and of typical prograde and retrograde skarn sequences as described in the literature (e.g., [57,58]). Skarn formation is related to intrusion of granodiorite (Figure 9a–d), whereas at places granodiorite sills intruding marbles of the CCB are totally transformed to garnetites (Figure 9c,d). Pyroxene accompanies garnet during prograde skarn formation followed by magnetite and then by pyrite in the retrograde stage (Figure 9d–f).

3.3. Geomorphological–Hydrogeological Geosites at Serifos

The geotopes considered in this paper as geomorphosites include a variety of geomorphological landscapes. The lithology of Serifos consists of various lithological formations with different resistance in erosion and weathering, which in combination with the action of the hydrographic network and tectonics, has created the geomorphosites (tafoni) located in the NW part of Serifos' coast (Figures 3 and 10). The increased presence of halite indicates a salt-induced weathering (humidity and seawater spray) that plays an important role in their development.

Additionally, karstic geotopes, such as caves are areas with unique characteristics and archaeological findings. The Koutalas cave is decorated with stalactites and columns, while the next chamber is covered almost along its entire length by a small lake.

In the coastal zone, the most characteristic geomorphosites are due to deposition processes with notable examples of coastal dunes, tombolo (Figure 10), as well as lagoon in the Tsilipaki area.

The hydrogeological geotopes include rivers, a number of springs (created by the discontinuities of the granite), fountains and wells and the presence of hot springs such as the Almyros springs near the old loading facility of Megalo Livadi (Figures 3 and 10). During the mining period, the Mining Company had built stone baths for therapeutic purposes.

Figure 10. Field photos showing typical geomorphological features of geotopes at Serifos. (**a**) Tombolo at Agios Sostis; (**b–d**) landforms resulting from differential erosion and weathering (tafoni) at Skala, Moutoula and Sykamia, respectively; (**e,f**) the hot springs of Almyros area.

3.4. Mining Heritage of Serifos

Serifos was aptly named "the iron island" among the Cycladic islands, due to the flourishing of the iron industry. Iron ore mining activities have been witnessed in Serifos since antiquity. Minoans and Mycenaeans developed mining activities here, which continued during Roman times and Venetian domination from the 14th to 16th century. Prehistoric clay kilns have been identified in Avessalos in the Phournoi area and on the Kefala peninsula, a fact that testifies the extraction and processing of ore in the early stages [32,33]. Copper minerals are more frequent in the southern part of the island in association with the magnetite-hematite deposits [18]. Unique in Greece are the kilns that melted iron and copper in various places on the island [32,33]. By far the largest known copper slag heap in the Aegean area is that of Skouries at Avessalos with estimates of about 100,000 tons of slags present. Kefala and Phournoi with several hundred tons and a few tens of tons of slags, respectively, are minor compared to Skouries [32,33]. In addition to copper and iron ores, galena-rich ores are known in the northeastern part of Serifos at Moutoula (Figure 4l) and Pyrgos close to Galani. The Moutoula deposit was exploited in the 19th century for galena, while a possibly earlier undated gallery has also been noted [31].

Modern mining of the district started from 1861 and the iron ore mines finally ceased operations in 1965 [59,60]. From 1861, and systematically, from 1869, began the extraction of iron ores by the Hellenic Mining Company, which remained there until 1875. After 1880, the French company Serifos-Spiliazeza proceeded to intensive exploitation of the deposits of Serifos [59,60]. Andreas Sygros and Giovanni Baptista Serpieri were involved in its operation. In 1886, the German miner Emilios Groman took over the management of the company Serifos-Spiliazeza and essentially all the mines of Serifos [59–63]. The Gromans effectively controlled the island, while building extensive infrastructure for the extraction, transport and loading of ore on ships. From 1869 to 1940, a total export of 6.59×10^6 tons of iron ore (e.g., hematite, limonite and magnetite) from several mines mainly in the southwestern part of the island was made (Figure 11). When Grohmann undertook the operation of the mines, the headquarters of the company were transferred to Megalo Livadi, where a two-story neoclassical building with architectural elements of the "Ziller" style were created, ruins of which still stand at the end of the beach today (Figure 11d). Megalo Livadi was the main iron ore export harbor of Serifos, equipped with all the necessary sorting and shipment facilities (Figure 11a,b).

Figure 11. Field and underground photos showing evidence of mining activity at Serifos. (**a**,**b**) Loading bridges at Megalo Livadi; (**c**) entrance of gallery at Megalo Livadi; (**d**) the headquarters of the iron ore mines at Megalo Livadi; (**e**) loading bridge at Koutalas; (**f**) mining wagon in a Koutalas mine.

The Ministry of Culture declared the following as historical monuments: the Headquarters of the mines, the loading bridge in Megalo Livadi, the ore loading bridge in Koutalas, the workers' residencies, as well as any kind of equipment that remains to provoke memories of the flourishing of the island in another era (Figure 11). The activities of the mines in their three-thousand-year history have left behind monuments and residential complexes, which are an important part of the pre-industrial history of Greece [63]. They are an important testimony of the industrial activity that had been developed on the island and are of historical, architectural, and sociological interest.

Serifos is in itself an Open Air Museum of mining activity since its hinterland is engraved with mining galleries, its ports have been turned into loading stations of its minerals, and its social and economic history is timelessly tied to the mines [63]. All the facilities created from 1869 to 1964, including the loading ladder, the hydromechanical enrichment complex, the engine room with the equipment of the workers' houses of the 19th century and a newer house complex from 1950 are preserved in the mining center of Koutalas. The ruins of a residential complex and a loading ladder are preserved along the coast in Chalara. West of Koutalas, in Aspros Kavos, dozens of galleries reach sea level. Dozens of galleries and transport routes are maintained throughout the area. In the bay of Avessalos, there are ruins of a loading ladder and mining facilities, and in the place Aerata, traces of ancient piles of slag and carved basins in the rocks for the cleaning of the ore can be found. The installations date from the 4th century BC. In the area of Mountaki, a large gallery was built, 1400 m long, which connected it with Kalogeros. The gallery served the transport of the ore from Kalogiros to the loading ladder in Mountaki.

4. Discussion

4.1. The Possibilities of Geotourism Development in Serifos—Geotourism Perspectives

Geoparks are wider areas that contain significant sites of geological monuments and geotopes as well as sites of ecological, archaeological, historical or cultural interest and are tools for environmental education [6,11–13,64]. Geoparks are of particular scientific interest since the purpose of their existence is to explore the relationships between geological, natural and cultural heritage. Serifos has been included in the Atlas of the Aegean geological monuments of the Ministry of the Aegean since 2002 with the Koutalas Cave, the Mineral occurrences and Iron Mines [65].

The inclusion of Serifos as a future UNESCO Global Geopark will highlight the concept of geotopes, the scientific–educational–touristic interest they raise, and the values they advocate (geo-conservation, geoprotection) contributing to the development of geotourism.

Green values of the cultural and natural heritage of the Serifos island, and therefore, the interest about it may be briefly highlighted as follows:

Geological and mineralogical value: Interesting geology with noticeable scientific and educational values. The area presents an impressive variety of minerals. In Serifos, they also found some rare and highly developed crystals for their kind of minerals, such as green quartz, amethyst, ilvaite, hedenbergite, garnet, calcite and barite.

Mining and metallurgical value: By far the largest known copper slag heap in the Aegean area is that of Avessalos on Serifos, with estimates of about 100,000 tons of slag present.

Environmental value: Interesting and important types of ecosystems and habitats as well as numerous and/or important flora and fauna species are also found. Areas of Serifos Island have joined the Natura 2000 network, such as GR4220009 South Serifos and GR4220029 coastal zone and the islands Serifopoula, Piperi and Vous.

Industrial value: The mines of Serifos are an example of an industrial monument and present a uniqueness, since the mineral areas of the island were mainly exploited in three different periods (prehistoric period, 14th century and from 1869 to 1963).

Historical value: Serifos is in itself an Open Air Museum of mining activity since its hinterland is engraved with mining galleries, its ports have been turned into loading stations of its minerals, and its social and economic history is timelessly tied to the mines.

Social value: With the influx of workers, Serifos became the center of the early Greek trade union movement contributing greatly to the shaping of the Greek trade union culture [61,62].

Archaeological and cultural value: In historical times, the presence of circular towers, such as Aspropyrgos in the bay of Koutalas, and other buildings may be associated with mining and metallurgical activity on the island. Prehistoric clay kilns have been identified at Moutoula, on the northern slope of Vigla hill, on Avessalos in the Phournoi area and on the Kefala peninsula, which testify to the extraction and processing of ore in the early stages.

Aesthetic value: The two-story neoclassical building, with architectural elements of the "Ziller" style in Megalo Livadi. Mineral species of extraordinary aesthetic and scientific value suitable for exhibition in museums and collections.

Nowadays, geotourism has proved to be a new and much promising trend for the whole district. The geotouristic development of mineralogical and petrological geotopes at Serifos ensures the preservation of the geological heritage of Serifos Island and also offers the opportunity for sustainable development.

The Serifos geotopes belong to the Greek mineralogical and geological heritage and can be considered as mineralogical treasures, some of them unique throughout world, as listed below (see also Figures 2–11):

- Serifos geomorphosites, such as lagoons, sand dunes, tompolo, landforms resulting from differential erosion and weathering (tafoni) (Figure 10).
- The Serifos granodiorite considered to be synchronous with Miocene extensional detachment faulting, intruded gneisses, schists and marbles causing an extensive contact metamorphic aureole.

- Tectonics geotopes, including Megalo Livadi and Kavos Kiklopas detachments.
- The Koutalas cave, between the bays of Megalo Livadi and Koutalas with stalactites and stalagmites, at Stavrakopoulos, discovered in 1893 during excavations. In one of the rooms formed by the cave, utensils and pottery were found, which prove that in ancient times, it was a place of worship. The first room of the cave is decorated with stalactites and columns, while the next chamber is covered almost along its entire length by a small lake. In the chamber that follows is an altar framed by various utensils, traces of fire and bones, covered with stalagmite material. A protection zone of 500 m has been defined around the cave (Government Gazette 29/B/26/1/63).
- The cave of the Cyclops. It is located near the monastery of Evangelistria and Kavos Kiklopas detachments.
- The hot springs at the bay of Almyros, near the sea (Figure 10). There, tabular translucent barite and rhombohedral calcite have been found in exceptional crystals.
- Agia Marina-Koutalas. The Agia Marina area is characterized by splendid occurrences of andraditic garnets in association with quartz crystals. The Koutalas area is characterized by the operations of mining activity, such as rail systems, ore transport, wagons and the workers' residencies.
- Avessalos. The Avessalos area is the best site in the world in respect to the mineral green quartz (prase). The crystal forms, intergrowths and sizes (up to 40 cm) of green quartz specimens from this locality are spectacular.
- Chalara with iconic mining infrastructure and also best development of proximal high-T skarn.
- Miners pathway. The hiking trail leading from "Giftika" area to Ano Chora was the road that the miners used to take in order to reach the western areas of the island and get to work. It was built in 1858, it is still well-preserved and it constitutes one of the most beautiful paths of the island.
- Prehistoric clay kilns have been identified on Avessalos in the Phournoi area and on the Kefala peninsula, which testify to the extraction and processing of ore in the early stages. The presence of circular towers, such as Aspropyrgos in the bay of Koutalas, and other buildings may be associated with mining and metallurgical activity on the island.
- Moutoula sulfide ore deposits (galena).

4.2. The Proposed Network of Geocultural Routes of Serifos (Geotrails of Serifos)

During the present study, six geotrails—georoutes of exceptional mineralogical and petrological interest—have been developed during field and laboratory work using GPS and geographical information systems (GIS) to link these geotopes with the cultural and ecological sites. Figure 12 and Table 2 demonstrate the proposed network of georoutes (total length: 30 km) and the geotouristic comparative advantage of the island. Along the routes, the geodiversity is interpreted, including its relationship with the surrounding historical and industrial activity of the region. These geotrails—georoutes include the Agia Marina-Koutalas, Megalo Livadi-Koundouros, Pyrgos-Galani and the Avessalos area—form part of the proposed Serifos geopark.

The geotrails are involving stopping places which help us find out more about the variety of geology in the area and see a different side to the scenery. This includes site interpretation panels, training courses and other educational and interpretative activities that do not demand large financial and/or organizational investments, but could significantly benefit visitors' activities in the park [66,67]. Good interpretation and educational activities could attract more visitors, especially non-experts or casual geotourists [67,68], who do not possess great knowledge of geology and other similar topics. Interpretation is a vital part of how people experience the places they visit, as it explains the natural and cultural heritage and brings them to life. Interpretative methods could be divided in two categories [68], depending on the location of their implementation:

1. In situ interpretive form implemented at the geosites which provides direct and visual aspect (information signs, panels, geotrails, guided tours, etc.). These provide information about the attractions and the geoheritage and their significance [69]. The information signs will be placed in central points of the settlements and at the entry points of an important junction of paths, where they will inform the visitor of all the geotopes and the routes. Specifically, these signs will contain informative texts and photos and a thematic map that will inform the visitor about its location in the geopark and will lead him to the geotopes through a map that will be designed and will depict the existing roads and paths, the main place names, the settlements, and any useful information for the visitor [69,70].
2. Ex situ interpretive form used in related facilities (visitor centers, museums, etc.), such as popular lectures, interactive and video presentations, museum artefacts, laboratories, etc. [70].

Table 2. Geotrails of Serifos.

Route Number/Km	Route Description	Route Title	Geoheritage and Cultural Points of Interest	Petrological Types/Minerals
1/5, 8	Pyrgos–Galani (galena mines)–Sykamia–Skala–Phournoi	Discovering the first mines, the geomorphosites and the copper slags	Paved path: Traditional settlement Galani (named after the ancient galena mines), Tafoni (cellular honeycomb geomorphological formations of weathering), copper slags (Phournoi).	Amphibolites with gneiss intercalations, greenschists, alluvial sediments/galena, sphalerite, pyrite, etc.
2/4, 8	Megalo Chorio–Skouries–Avessalos	On the copper road	Avessalos area is the best site in the world in respect to the mineral green quartz. Skouries: By far the largest known copper slag heap in the Aegean with estimates for 100,000 tons of slag present.	Amphibolites, mylonitic orthogneiss, dolomitic and calcitic marbles with boudinaged quartzites and schists/ilvaite, hedenbergite, green quartz, etc.
3/4, 1	Agia Marina–Kastro Grias–Koutalas–Megalo Chorio	Protecting the mines	Skarn-related mineralization: The area is characterized by splendid occurrences of garnets in association with quartz crystals. Koutalas area is characterized by the operations of mining activity, such as rail systems, ore transport, wagons and the workers residencies. Koutalas cave between the bays of Megalo Livadi and Koutalas with stalactites and stalagmites at Stavrakopoulos.	Granodiorite, amphibolites with gneiss intercalations, mylonitic orthogneiss, dolomitic and calcitic marbles with boudinaged quartzites and schists/andraditic garnets, quartz, barite crystals with galena and Fe-oxides, etc.
4/3, 5	Kastro Grias–Ganema–Chalara–Vagia Bay	Outdoor mining museum	Area with iconic mining infrastructure, and also best development of proximal high-T skarn. An outdoor mining museum, with underground mining galleries, iron rails, semi-destroyed bridges, wagons and a loading bridge.	Granodiorite, mylonitic orthogneiss, calcitic and impure marbles, alluvial sediments/andraditic garnets, quartz, epidote, etc.
5/9, 4	Megalo Chorio–Kalogeros–Kavos Kiklopas–Megalo Livadi–Koundouros	On the iron road	Ore mineralization along various extensional low-angle detachment fault systems. (Command) Village with miners: An outdoor mining museum, with underground mining galleries, iron rails, semi-destroyed bridges, wagons and a loading bridge, hot water springs. Koundouros area is characterized by hedenbergitic skarn including the best ilvaite crystals worldwide. Megalo Livadi and Kavos Kiklopas detachments.	Amphibolites with gneiss intercalations, mylonitic orthogneiss, dolomitic and calcitic marbles with boudinaged quartzites and schists, greenschists/ilvaite, green quartz, hematite, magnetite, barite, etc.
6/2, 4	Chora–Gyftika	The road of the miners	Stone-paved path used to be the only way for miners to get to the mines.	Granodiorite, mylonitic orthogneiss

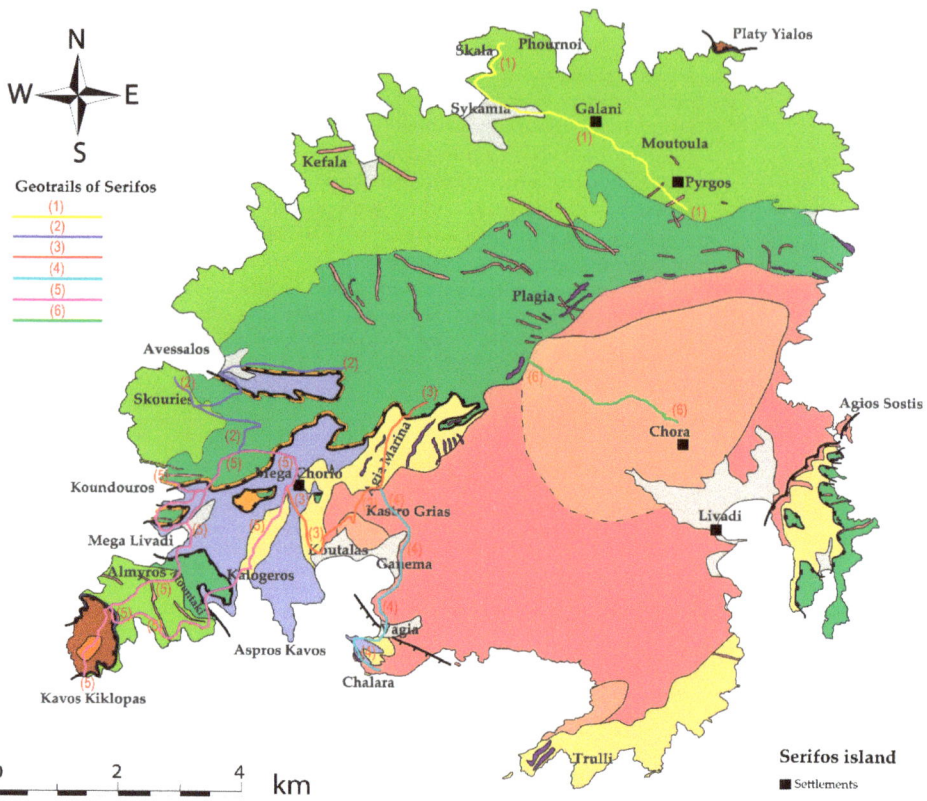

Figure 12. The routes of geotrails suggested for the Serifos geopark. Geological map modified after Ducoux et al. [15].

In this case study of the Serifos geopark, both of these methods could be used. Visitors should be able to see and hear many scientific facts not only about mining, but also about history and the time period in which they worked. Other educational activities could include different programs for visitors, such as the junior mineralogist program in which children take part in simulated excavations. The main idea behind these projects would be to enable learning through practice. The most effective tools for interpretation and visitor involvement could include guided tours to the mining sites, workshops (e.g., simulated mining, mineral identification, etc.), or multimedia performances. Besides interpretation and education, the park should also support the local people in producing and selling their local products and souvenirs [67,71].

Interpretation therefore needs to meet the requirements of a broad spectrum of audiences from the specific site-related geological and educational information for the dedicated geotourist and those actively seeking to learn about the geology of an area.

The suggested routes are the first geo-walking paths of the area, in which an Outdoor Geological Museum can operate, in the context of the proclamation of the wider part of the island as a geopark. Many of the entrances to the old network of underground mines remain open on the hillsides around. The terrain on the routes consists mainly of a variety of quiet roads and paths across fields. These paths were further examined in order to highlight the geological sites and to characterize both the individual sites as geotopes and the wider area as a geopark. The georoutes, where possible, followed the existing network of the island. The old Hiking Network of Serifos Island has a logo that connects the history and nature of the island, as in ancient times, the frog was the symbol of the island. In addition, the title "Iron and stone paths" reflects the geology of the island, but also its mythology, during which Perseus with the head of Medusa petrified the whole island [63].

4.3. Serifos: A Potential New Unesco Global Geopark

There are six UNESCO Global Geoparks in Greece: Lesvos island, Psiloritis and Sitia in Crete, Chelmos-Vouraikos at Peloponnese, and Vikos-Aoos in Epirus and Grevena-Kozani (e.g., [11,72–75]) (Figure 1). These geoparks combine extraordinary geological, paleontological, and natural features. The proposed Serifos geopark is an area enclosing features of special geological significance, rarity or beauty. However, the potential Serifos geopark combines spectacular mineralogical, petrological, geological, mining, cultural characteristics and can be described as a polythematic geopark.

Besides the above mentioned six geoparks, similar geological projects in Greece include: the Syros island, Cyclades complex, as a prime locality for the study of processes active in deep levels of orogeny and is world famous for its exceptionally well-preserved glaucophane schist-to eclogite-facies lithologies [76]; the Meteora and Mount Olympos in central Greece with legendary geological history and mythology [77,78]; the Kefalonia, Ithaki Lefkas and Meganisi islands, in the active geotectonic region of the Ionian Sea, Western Greece and in the convergent zone between the African and Eurasian plates [79]; the Naxos island in the Cyclades, characterized by intense mining and quarrying activities, since the antiquity and by distinctive geological and mineral wealth [80].

The suggested Serifos geopark is also comparable to the Lavrion geopark, which in addition to mineralogical, petrological, geological, mining and cultural features also displays worldwide unique archaeological characteristics [34].

Worldwide, a research on mineralogical, petrological and mining heritage, comparable to that of Serifos has been demonstrated: in post-mining areas of central Mexico [81], Poland (Gold Mine in Zloty Stok), Spain (La Tortilla Mine in Linares), UK (King Edward Mine, in Cornwall) [82], Italy (the Traversella Mining Site at Piemonte, Italy) [83]; in petrological and mineralogical geosites such as granite landforms of Seoraksan Mountains, Republic of Korea [84]; the Monviso Massif and the Cottian Alps in Piemonte, Italy, with some of the best preserved ophiolites in the Alps and associated Cu–Fe mineralizations, the first primary source of jade in the Alps, the world-famous minerals such as coesite and giant pyrope, as well as type localities for new minerals [85]; the eastern limb of the Bushveld Igneous Complex, South Africa, with layered ultramafic-mafic rocks, metamorphic aureole outcrops, orebodies of chromitites, PGE reefs, etc. [86].

The Serifos geopark is designated with a focus on three main components: the protection and the conservation of Serifos' heritage, the tourism-related infrastructural development and the socio-economically sustainable development of the local community using a sustainable territorial development strategy. Heritage sites within a geopark can be related not only to geology but also to archaeology, ecology, history and culture. All these sites in the Serifos geopark constitute thematic parks and will be linked in a network with routes, trails and sections that should be protected and managed. The mineralogical "treasures" of Serifos Island, featuring in a worldwide unique hedenbergite-ilvaite-garnet-prase-bearing skarn, and the unique geological structure of the island, due to specific tectonic processes, makes it a special site with great archaeological, as well as aesthetical, cultural and scientific value. Some specific actions should be developed such as a management plan of the Serifos geopark and a clear presentation of the geotopes of Serifos and their quantitative assessment. The aim of a quantitative assessment is to decrease the subjectivity associated with any evaluation procedure. The result of this numerical assessment is a sorted list of sites, which is a powerful tool for the establishment of management priorities. Inventorying and quantitative assessment, statutory protection, conservation, promotion and interpretation and monitoring of sites are some specific methods used to promote geoconservation.

The suggested six geotrails–geocultural routes are of exceptional mineralogical and petrological interest, and have been developed in order to discover the natural, geological and cultural treasure of the island. Along the routes, the geodiversity is explained, including its relationship with the surrounding biodiversity, and the historical and cultural aspects of the region. The southwestern part of Serifos forms a wide network of habitats for the flora and the rich fauna of the region because of the combination of its geomorphological and hydrogeological features. Serifos and particularly its southwestern part is also included in the European network of protected areas NATURA 2000, which is the main national means for the purpose of Directive 92/43/EEC of the European Council "for the conservation of natural habitats and wild fauna and flora" which are significant at the European level [87]. The region, as the whole of the Cyclades islands, is a passage of sea turtles and migratory birds and there are seasonal hunting restrictions. Almost half of the island is covered by typical low Aegean phrygana. There are also sand dunes with characteristic vegetation. The hydrographic net of the site includes rivers with N–S direction and a total length of about 14 km. It also includes a number of springs, fountains and wells.

The combination of its rich morphological relief and the ancient and historical monuments—in the western part of the region, there are abandoned mines, which have been characterized as historical monuments—gives great value to the region, which requires special protection.

The suggested geotrails can be important management and interpretative tools for geotourism development. The selection of the content is not limited to natural and historical information but incorporates the geoenvironment with the cultural character of the landscape, as the design goal of this touristic product is the perception of timelessness of space. While designing a matrix of geotrails–geocultural routes the old path networks are revealed, activating at the same time the local community. In these routes, tourists through experiential activities turn into travelers, exploring the features of a place, creating experiences and, finally, emotional binding.

A necessary condition for the integration of the proposed network of georoutes with the aim of highlighting the geodiversity and the mining history of Serifos is the perfect organization of the marking and its annual maintenance by the municipality.

The municipality of Serifos will install new signs (information signs and direction signs). In addition, it will be instrumental the procurement and placement of interpretive panels for geoheritage and display of the relief of the island using 3D printing technique. Signage is an integral part of the construction and operation of a Hiking Trail Network. In order to hike the geotrails, a walking map is recommended that contains detailed maps and short descriptions of the waypoints on the route.

The inclusion of Serifos in the network of Global Unesco Geoparks is a basic goal of the municipality. In Megalo Livadi, there is a small museum of minerals and rocks of Serifos, while some samples of minerals are also exhibited among the other exhibits of the folklore museum of Chora. As a place for the creation of an open-air museum, Megalo Livadi is proposed, in which all the facilities that were built between 1869–1875 by the Mining Company are preserved. The facilities of the neoclassical building (command post), the workers' residences, towers, explosives depot, rails, wagons, galleries, stairs will be restored and will be highlighted with modern museum teaching materials and will be the central core and the open part of the Museum [63]. Megalo Livadi is located in the center of an extensive area of special historical interest.

The works that will be required for the creation of the museum include restorations of buildings, landscaping, fixing and maintenance of metal structures (e.g., loading ladder, tipping pylons, etc.), reconstruction of railway networks, fixing and highlighting selected galleries, decoration uniting the basic cores of the museum, etc.

Nowadays, the historic building of the neoclassical building is being reconstructed and upgraded to a cultural/conference center and a museum of mineral wealth and mining history. A series of exhibits will be exhibited in the historic building, including collections of rocks and minerals, information and educational materials will be displayed and educational seminars and workshops will be organized. Exhibition sections will be developed at the headquarters regarding the geological and mining image of Serifos, its mineral wealth, the history of quarrying operations from antiquity to modern times, the productive process of exploitation of mineral wealth, social and economic context, as well as the workforce. With its operation, Serifos will be promoted and will be promoted worldwide through conferences, seminars and other special events, in order to attract visitors with special interests but also to inform the local community about its history and the identity of its place.

The touristic infrastructure of Serifos is well developed. It includes, besides the famous beaches, a lot of hotels and private accommodations, traditional restaurants, and water sport possibilities, as well as other leisure and sportive activities (diving).

Serifos island is a unique site for education and research in several disciplines, such as geological investigation, prospecting, mineralogy, geochemistry, mining, metallurgical processes, economic, political, and social sciences. Serifos provides an almost inexhaustible field of activity for scientists and the public, and also offers ideal opportunities for educational and geotourism. We conclude that Serifos Island meets all the scientific, educational, cultural, and touristic criteria that makes the area highly suitable for its development as a UNESCO Global Geopark.

5. Conclusions

Serifos is a geographical unity which represents the combined work of nature and humans depicting the evolution of the local society under the influence of the physical constraints of this island area. It is characterized by a worldwide unique geological, petrological and mineralogical diversity, combined with a very rich cultural heritage, biodiversity and folk tradition. The Serifos geotopes belong to the world mineralogical and geological heritage and should be protected from further commercial exploitation. The geotouristic development of the geological, mineralogical and petrological geotopes at Serifos, combined with the foundation of a mineralogy-petrology museum, ensures the preservation of the geological heritage of Serifos Island and also offers the opportunity for sustainable development.

The present paper focuses on exemplary mineralogical, geological and mining features of Serifos, and presents six geotrails which have been developed to link the geological heritage with cultural and ecological sites. The suggested Serifos geopark focuses on the promotion of the geological and mining heritage of the island, with the aim of integrating Serifos in the network of Global Unesco Geoparks, in the near future.

Author Contributions: Conceptualization, N.V. and P.V.; methodology, N.V.; investigation, N.V.; resources, N.V. and P.V.; writing—original draft preparation, N.V.; writing—review and editing, N.V. and P.V.; supervision, P.V. All authors have read and agreed to the published version of the manuscript.

Funding: This research received no external funding.

Institutional Review Board Statement: Not applicable.

Informed Consent Statement: Not applicable.

Data Availability Statement: The data presented in this study are available on request from the corresponding author.

Acknowledgments: The authors gratefully thank Maxime Ducoux for kindly providing his geological Serifos map. We also gratefully thank the journal editor and the two reviewers for their thorough consideration of this paper.

Conflicts of Interest: The authors declare no conflict of interest.

References

1. Crofts, R. Putting Geoheritage Conservation on All Agendas. *Geoheritage* **2018**, *10*, 231–238. [CrossRef]
2. Crofts, R.; Tormey, D.; Gordon, J. Introducing New Guidelines on Geoheritage Conservation in Protected and Conserved Areas. *Geoheritage* **2021**, *13*, 33. [CrossRef]
3. Santangelo, N.; Valente, E. Geoheritage and Geotourism. *Resources* **2020**, *9*, 80. [CrossRef]

4. Carcavilla, L.; Durán Valsero, J.; García-Cortés, Á.; López-Martínez, J. Geological Heritage and Geoconservation in Spain: Past, Present, and Future. *Geoheritage* **2009**, *1*, 75–91. [CrossRef]
5. Gordon, J.E. Geoheritage, Geotourism and the Cultural Landscape: Enhancing the Visitor Experience and Promoting Geoconservation. *Geosciences* **2018**, *8*, 136. [CrossRef]
6. UNESCO Geoparks Program—A New Initiative to Promote a Global Network of Geoparks Safeguarding and Developing Selected Areas Having Significant Geological Features; Document 156 EX/11 Rev., Executive Board, 156th session; UNESCO: Paris, France, 1999; p. 4.
7. Farsani, N.D.; Coelho, C.; Costa, C. Geotourism and Geoparks as novel strategies for socio-economic development in rural areas. *Int. J. Tour. Res.* **2011**, *13*, 68–81. [CrossRef]
8. Dowling, R.; Newsome, D. (Eds.) *Geotourism*; Elsevier/Heineman: Oxford, UK, 2006.
9. Dowling, R.K. Global geotourism—An emerging form of sustainable tourism. *Czech J. Tour.* **2013**, *2*, 59–79. [CrossRef]
10. Hose, T.A. 3G's for Modern. Geotourism. *Geoheritage* **2012**, *4*, 7–24. [CrossRef]
11. UNESCO. Available online: https://en.unesco.org/global-geoparks (accessed on 22 June 2021).
12. UNESCO. *Global Geoparks Network*; Division of Ecological and Earth Sciences: Paris, France, 2006.
13. Mc Keever, P.; Zouros, N. Geoparks: Celebrating earth heritage, sustaining local communities. *Episodes* **2005**, *28*, 274–278. [CrossRef]
14. Salemink, J. Skarn and Ore Formation at Serifos, Greece as a Consequence of Granodiorite Intrusion. Ph.D. Thesis, University of Utrecht, Utrecht, The Netherlands, 1985.
15. Ducoux, M.; Branquet, Y.; Jolivet, L.; Arbaret, L.; Grasemann, B.; Rabillard, A.; Gumiaux, C.; Drufin, S. Synkinematic skarns and fluid drainage along detachments: The West Cycladic Detachment System on Serifos Island (Cyclades, Greece) and its related mineralization. *Tectonophysics* **2017**, *695*, 1–26. [CrossRef]
16. Fitros, M.; Tombros, S.; Williams-Jones, A.E.; Tsikouras, B.; Koutsopoulou, E.; Hatzipanagiotou, K. Physicochemical controls on bismuth mineralization: An example from Moutoulas, Serifos island, Cyclades, Greece. *Amer. Miner.* **2017**, *102*, 1622–1631. [CrossRef]
17. Voudouris, P.; Mavrogonatos, C.; Spry, P.G.; Baker, T.; Melfos, V.; Klemd, R.; Haase, K.; Repstock, A.; Djiba, A.; Bismayer, U.; et al. Porphyry and epithermal deposits in Greece: An overview, new discoveries, and mineralogical constraints on their genesis. *Ore Geol. Rev.* **2019**, *107*, 654–691. [CrossRef]
18. Korosidis, J.; Voudouris, P.; Kouzmanov, K. Distal Fe skarn deposits of Serifos Island: New mineralogical and geochemical constrains on the retrograde assemblage and associated ore mineralization. In *The Critical Role of Minerals in the Carbon-Neutral Future, Proceedings of the 16th SGA Biennial Meeting, Rotorua, New Zealand, 28–31 March 2022*; Society Geology Applied: Rotorua, New Zealand, 2022; in press.
19. Grasemann, B.; Zamolyi, A.; Petrakakis, K.; Rambousek, C.; Igelseder, C. Ein neuer metamorphic core complex in den West-Kykladen (Serifos, Greichenland). *Erlanger Geol. Abh.* **2002**, *3*, 36–37.
20. Grasemann, B.; Petrakakis, K. Evolution of the Serifos metamorphic core complex. *J. Virtual Explor.* **2007**, *27*, 1–18. [CrossRef]
21. Grasemann, B.; Schneider, D.A.; Stockli, D.F.; Iglseder, C. Miocene bivergent crustal extension in the Aegean: Evidence from the western Cyclades (Greece). *Lithosphere* **2012**, *4*, 23–39. [CrossRef]
22. Iglseder, C.; Grasemann, B.; Schneider, D.A.; Petrakakis, K.; Miller, C.; Klötzli, U.S.; Thöni, M.; Zámolyi, A.; Rambousek, C. I and S-type plutonism on Serifos (W-Cyclades, Greece). *Tectonophysics* **2009**, *473*, 69–83. [CrossRef]
23. Rabillard, A.; Arbaret, L.; Jolivet, L.; Le Breton, N.; Gumiaux, C.; Augier, R.; Grasemann, B. Interactions between plutonism and detachments during metamorphic core complex formation, Serifos Island (Cyclades, Greece). *Tectonics* **2015**, *34*, 1080–1106. [CrossRef]
24. Gauthier, G.; Albandakis, N. Minerals from the Serifos skarn, Greece. *Miner. Rec.* **1991**, *22*, 303–308.
25. Voudouris, P.; Katerinopoulos, A.; Christofalou, F.; Kassimi, G. Serifos island, Aegean Sea/Greece: A worldwide unique mineralogical and petrological geotope. *ProGeo. News* **2007**, *1*, 7–8.
26. Voudouris, P.; Voulgaris, N.; Christophalou, F.; Kassimi, P. Development of Mineralogical-Petrological Geotopes on the Serifos Island, using Geographic Information Systems (GIS). In Proceedings of the 11th Conference of the Greek Geological Society, Special Session of the Geological and Geomorphological Heritage Conservation Committee, Athens, Greece, 24–26 May 2007; pp. 49–52.
27. Voudouris, P.; Katerinopoulos, A.; Magganas, A. Mineralogical geotopes in Greece: Preservation and promotion of museum specimens of minerals and gemstones. Sofia Initiative "Mineral diversity preservation". In Proceedings of the IX International Symposium Mineral Diversity Research and Reservation, Sofia, Bulgaria, 16–18 October 2017; pp. 149–159.
28. Ottens, B.; Voudouris, P. *Griechenland: Mineralien-Fundorte-Lagerstätten*; Christian Weise Verlag: Munich, Germany, 2018; p. 480. ISBN 978-3-921656-86-0.
29. Klemme, S.; Berndt, J.; Mavrogonatos, C.; Flemetakis, S.; Baziotis, I.; Voudouris, P.; Xydous, S. On the Color and Genesis of Prase (Green Quartz) and Amethyst from the Island of Serifos, Cyclades, Greece. *Minerals* **2018**, *8*, 487. [CrossRef]
30. Voudouris, P.; Mavrogonatos, C.; Graham, I.; Giuliani, G.; Tarantola, A.; Melfos, V.; Karampelas, S.; Katerinopoulos, A.; Magganas, A. Gemstones of Greece: Geology and Crystallizing Environments. *Minerals* **2019**, *9*, 461. [CrossRef]
31. Marinos, G. Geology and metallogeny of Serifos island. *Geol. Geophys. Res.* **1951**, *1*, 95–127. (In Greek)

32. Georgakopoulou, M.; Bassiakos, Y.; Philaniotou, O. Seriphos surfaces: A study of copper slag heaps and copper sources in the context of Early Bronze Age Aegean metal production. *Archaeometry* **2011**, *53*, 123–145. [CrossRef]
33. Philaniotou, O.; Bassiakos, Y.; Georgakopoulou, M. Early Bronze Age copper production on Seriphos (Cyclades, Greece). In *Metallurgy: Understanding How, Learning Why. Studies in Honour of James D. Muhly*; Betancourt, P.P., Ferrence, S.C., Eds.; Prehistory Monographs 29; INSTAP Academic Press: Philadelphia, PA, USA, 2011; pp. 157–164.
34. Voudouris, P.; Melfos, M.; Mavrogonatos, C.; Photiades, A.; Moraiti, E.; Rieck, B.; Kolitsch, U.; Tarantola, A.; Scheffer, C.; Morin, D.; et al. The Lavrion mines: A unique site of geological and mineralogical heritage. *Minerals* **2021**, *11*, 76. [CrossRef]
35. Ring, U.; Glodny, J.; Will, T.; Thomson, S. The Hellenic subduction system: High-pressure metamorphism, exhumation, normal faulting, and large-scale extension. *Ann. Rev. Earth Planet. Sci.* **2010**, *38*, 45–76. [CrossRef]
36. Jolivet, L.; Faccenna, C.; Huet, B.; Labrousse, L.; Le Pourhiet, L.; Lacombe, O.; Lecomte, E.; Burov, E.; Denèle, Y.; Brun, J.-P.; et al. Aegean tectonics: Strain localization, slab tearing and trench retreat. *Tectonophysics* **2013**, *597*, 1–33. [CrossRef]
37. Wind, S.C.; Schneider, D.A.; Hannington, M.D.; McFarlane, C.R.M. Regional similarities in lead isotopes and trace elements in galena of the Cyclades Mineral District, Greece with implications for the underlying basement. *Lithos* **2020**, *366*, 105559. [CrossRef]
38. Seman, S.; Stockli, D.F.; Soukis, K. The provenance and internal structure of the Cycladic Blueschist Unit revealed by detrital zircon geochronology, Western Cyclades, Greece. *Tectonics* **2017**, *36*, 1407–1429. [CrossRef]
39. Reinecke, T.; Altherr, R.; Hartung, B.; Hatzipanagiotou, K.; Kreuzer, H.; Harre, W.; Klein, H.; Keller, J.; Geenen, E.; Boeger, H. Remnants of a Late Cretaceous high temperature belt on the island of Anafi (Cyclades, Greece). *N. Jahrbuch Miner. Abhandl.* **1982**, *145*, 157–182.
40. Stouraiti, C.; Pantziris, I.; Vasilatos, C.; Kanellopoulos, C.; Mitropoulos, P.; Pomonis, P.; Moritz, R.; Chiaradia, M. Ophiolitic Remnants from the Upper and Intermediate Structural Unit of the Attic-Cycladic Crystalline Belt (Aegean, Greece): Fingerprinting Geochemical Affinities of Magmatic Precursors. *Geosciences* **2017**, *7*, 14. [CrossRef]
41. Jolivet, L.; Brun, J.P. Cenozoic geodynamic evolution of the Aegean region. *Int. J. Earth Sci.* **2010**, *99*, 109–138. [CrossRef]
42. Coleman, M.J.; Schneider, D.A.; Grasemann, B.; Soukis, K.; Lozios, S.; Hollinetz, M.S. Lateral termination of a Cycladic-style detachment system (Hymittos, Greece). *Tectonics* **2020**, *39*, e2020TC006128. [CrossRef]
43. Kruckenberg, S.C.; Vanderhaeghe, O.; Ferré, E.C.; Teyssier, C.; Whitney, D.L. Flow of partially molten crust and the internal dynamics of a migmatite dome, Naxos, Greece. *Tectonics* **2011**, *30*, TC3001. [CrossRef]
44. Altherr, R.; Kreuzer, H.; Wendt, I.; Lenz, H.; Wagner, G.A.; Keller, J.; Harre, W.; Höndorf, A. A late Oligocene/early Miocene high temperature belt in the Attic-Cycladic crystalline complex (SE Pelagonian, Greece). *Geol. Jahrb.* **1982**, *23*, 97–164.
45. Menant, A.; Jolivet, L.; Vrielynck, B. Kinematic reconstructions and magmatic evolution illuminating crustal and mantle dynamics of the eastern Mediterranean region since the late Cretaceous. *Tectonophysics* **2016**, *675*, 103–140. [CrossRef]
46. Schneider, D.A.; Senkowski, C.; Vogel, H.; Grasemann, B.; Iglseder, C.; Schmitt, A.K. Eocene tectonometamorphism on Serifos (western Cyclades) deduced from zircon depth-profiling geochronology and mica thermochronology. *Lithos* **2011**, *125*, 151–172. [CrossRef]
47. Salemink, J. On the Geology and Petrology of Serifos island (Cyclades, Greece). *Ann. Geol. Pays Hell.* **1980**, *30*, 342–365.
48. Stouraiti, C. Geochemistry and Petrogenesis of the Serifos Granite, in Relation to Other Aegean Granitoids, Greece. Ph.D. Thesis, University of Leicester, Leicester, UK, 1995.
49. Stouraiti, C.; Mitropoulos, P.; Tarney, J.; Barreiro, B.; McGrath, A.M.; Baltatzis, E. Geochemistry and petrogenesis of late Miocene granitoids, Cyclades, southern Aegean: Nature of source components. *Lithos* **2010**, *114*, 337–352. [CrossRef]
50. Stouraiti, C.; Baziotis, I.; Asimow, P.D.; Downes, H. Geochemistry of the Serifos calc-alkaline granodiorite pluton, Greece: Constraining the crust and mantle contributions to I-type granitoids. *Int. J. Earth Sci.* **2018**, *107*, 1657–1688. [CrossRef]
51. Voudouris, P.; Melfos, V.; Moritz, R.; Spry, P.G.; Ortelli, M.; Kartal, T. Molybdenite Occurrences in Greece: Mineralogy, Geochemistry and Rhenium Content. In *Scientific Annals of the School of Geology AUTH, Proceedings of the XIX Congress of the Carpathian-Balkan Geological Association, Thessaloniki, Greece, 23–26 September 2010*; Charis Ltd.: Thessaloniki, Greece, 2010; pp. 369–378.
52. Vlachopoulos, N.; Voudouris, P. Geological and mining history of Serifos island, Greece: Current state and perspectives for protection of mineralogical and petrological geotopes. In Proceedings of the 15th Congress of the Geological Society of Greece, Athens, Greece, 22–24 May 2019.
53. Voudouris, P.; Photiades, A.; Tarantola, A.; Scheffer, C.; Vanderhaeghe, O.; Morin, D.; Vlachopoulos, N. The Lavrion and Serifos mining centers: Two worldwide unique mineralogical and geological monuments and perspectives for their protection. In *The Value Framework for the Protection and Management of Sites and Monuments Extracted during the Antiquity: Current Uses and Future Synergies, Proceedings of the Greek ICOMOS Conference, Athens-Lavrion, Greece, 29–30 November 2019*; ResearchGate: Berlin, Germany, 2019. [CrossRef]
54. Voudouris, P.; Katerinopoulos, A. New occurences of mineral megacrysts in tertiary magmatic-hydrothermal and epithermal environments in Greece. *Doc. Nat.* **2004**, *151*, 1–21.
55. Maneta, V.; Voudouris, P. Quartz megacrysts in Greece: Mineralogy and environment of formation. *Bull. Geol. Soc. Greece* **2010**, *43*, 685–696. [CrossRef]
56. Voudouris, P.; Maneta, V. *Quartz in Greece*; CreateSpace Publ.: Seattle, WA, USA; Amazon.com, Inc.: Seattle, WA, USA, 2017. (In Greek)

57. Meinert, L.D. Variability of skarn deposits—Guides to exploration. In *Revolution in the Earth Sciences*; Boardman, S.J., Ed.; Kendall-Hunt Publishing: Dubuque, IA, USA, 1983; pp. 301–316.
58. Meinert, L.; Dipple, G.; Nicolescu, S. World skarn deposits. In *Economic Geology 100th Anniversary Volume*; Hedenquist, J.W., Thompson, J.F.H., Goldfarb, R.J., Richards, J.P., Eds.; Society of Economic Geologists: Littleton, CO, USA, 2005; pp. 299–336.
59. Mavrokordatou, D.; Balodimou, M.; Belavilas, N.; Papastefanaki, L.; Frangiskos, A.Z. Historical Mines in the Aegean, Laboratory of Urban Environment, School of Architecture, NTUA, Athens 2000–2006. Available online: https://www.elke.ntua.gr/en/research_project/historical-aegean-mines/ (accessed on 22 June 2021). (In Greek)
60. Belavilas, N.; Papastefanaki, L. *Historical Mines in the Aegean*, 1st ed.; Melissa: Athens, Greece, 2009. Available online: http://courses.arch.ntua.gr/111992.html (accessed on 22 June 2021). (In Greek)
61. Skonis, N. *The Bloody Strike of the Miners of Serifos–21 August 1916*, 1st ed.; Federation of Miners of Greece: Athens, Greece, 1990. (In Greek)
62. Speras, K. *The Strike of Serifos*, 3rd ed.; Bibliopelagos: Athens, Greece, 2001. (In Greek)
63. Katsilieri, M.; Louvi, A.; Mavrokordatou, D.; Belavilas, N.; Economou, M.; Trova, V.; Frangiskos, A.Z. *Open Air Museum of Mining Activities and Mineral Wealth of Serifos*, 1st ed.; Piraeus Group Cultural Foundation: Athens, Greece, 1998. (In Greek)
64. Zouros, N. The European Geoparks Network. *Episodes* **2004**, *27*, 165–171. [CrossRef]
65. Velitzelos, E.; Mountrakis, D.; Zouros, N.; Soulakellis, N. *Atlas of the Geological Monuments of the Aegean*; Ministry of the Aegean: Athens, Greece, 2003; p. 352. (In Greek)
66. Bruno, B.C.; Wallace, A. Interpretive Panels for Geoheritage Sites: Guidelines for Design and Evaluation. *Geoheritage* **2019**, *11*, 1315–1323. [CrossRef]
67. Hose, T.A. European 'geotourism'—Geological interpretation and geoconservation promotion for tourists. In *Geological Heritage: Its Conservation and Management*; Barretino, D., Wimbledon, W.P., Gallego, E., Eds.; Instituto Tecnologico Geominero de Espana: Madrid, Spain, 2000; pp. 127–146.
68. Hose, T.A. Geotourism and interpretation. In *Geotourism*; Dowling, R.K., Newsome, D., Eds.; Elsevier: Oxford, UK, 2005; pp. 221–241.
69. Hose, T.A.; Vasiljevic, D.A. Defining the nature and purpose of modern geotourism with particular reference to the United Kingdom and south-east Europe. *Geoheritage* **2012**, *4*, 25–43. [CrossRef]
70. Tomić, N.; Marković, S.B.; Korać, M.; Mrđić, N.; Hose, T.A.; Vasiljević, D.A.; Jovičić, M.; Gavrilov, M.B. Exposing mammoths: From loess research discovery to public palaeontological park. *Quat. Int.* **2015**, *372*, 142–150. [CrossRef]
71. Antić, A.; Tomić, N.; Đorđević, T.; Marković, S.B. Promoting palaeontological heritage of mammoths in Serbia through a cross-country thematic route. *Geoheritage* **2021**, *13*, 7. [CrossRef]
72. Zouros, N. Lesvos petrified forest geopark, Greece: Geoconservation, geotourism, and local development. In *The George Wright Forum*; George Wright Society: Hancock, MI, USA, 2010; pp. 19–28.
73. Fassoulas, C.; Zouros, N. Evaluating the influence of Greek geoparks to the local communities. *Bull. Geol. Soc. Greece* **2010**, *43*, 896–906. [CrossRef]
74. Fassoulas, C.; Staridas, S.; Perakis, V.; Mavrokosta, C. Revealing the geoheritage of Eastern Crete, through the development of Sitia Geopark, Crete, Greece. *Bull. Geol. Soc. Greece* **2013**, *47*, 1004–1016. [CrossRef]
75. Zafeiropoulos, G.; Drinia, H.; Antonarakou, A.; Zouros, N. From Geoheritage to Geoeducation, Geoethics and Geotourism: A Critical Evaluation of the Greek Region. *Geosciences* **2021**, *11*, 381. [CrossRef]
76. Drinia, H.; Tsipra, T.; Panagiaris, G.; Patsoules, M.; Papantoniou, C.; Magganas, A. Geological Heritage of Syros Island, Cyclades Complex, Greece: An Assessment and Geotourism Perspectives. *Geosciences* **2021**, *11*, 138. [CrossRef]
77. Rassios, A.E.; Ghikas, D.; Dilek, Y.; Vamvaka, A.; Batsi, A.; Koutsovitis, P. Meteora: A Billion Years of Geological History in Greece to Create a World Heritage Site. *Geoheritage* **2020**, *12*, 83. [CrossRef]
78. Rassios, A.E.; Krikeli, A.; Dilek, Y.; Ghikas, C.; Batsi, A.; Koutsovitis, P.; Hua, J. The Geoheritage of Mount Olympus: Ancient Mythology and Modern Geology. *Geoheritage* **2022**, *14*, 15. [CrossRef]
79. Spyrou, E.; Triantaphyllou, M.V.; Tsourou, T.; Vassilakis, E.; Asimakopoulos, C.; Konsolaki, A.; Markakis, D.; Marketou-Galari, D.; Skentos, A. Assessment of Geological Heritage Sites and Their Significance for Geotouristic Exploitation: The Case of Lefkas, Meganisi, Kefalonia and Ithaki Islands, Ionian Sea, Greece. *Geosciences* **2022**, *12*, 55. [CrossRef]
80. Periferakis, A. The emery of Naxos: A multidisciplinary study of the effects of mining at a local and national context. *J. NX-Multidiscip. Peer Rev. J.* **2021**, *7*, 93–115.
81. García-Sánchez, L.; Canet, C.; Cruz-Pérez, M.Á.; Morelos-Rodríguez, L.; Salgado-Martínez, E.; Corona-Chávez, P. A comparison between local sustainable development strategies based on the geoheritage of two post-mining areas of Central Mexico. *Int. J. Geoheritage Parks* **2021**, *9*, 391–404. [CrossRef]
82. Kaźmierczak, U.; Strzałkowski, P.; Lorenc, M.W.; Szumska, E.; Sánchez, A.A.P.; Baker, K.A. Post-mining Remnants and Revitalization. *Geoheritage* **2019**, *11*, 2025–2044. [CrossRef]
83. Costa, E.; Dino, G.A.; Benna, P.; Rossetti, P. The Traversella Mining Site as Piemonte Geosite. *Geoheritage* **2019**, *11*, 55–70. [CrossRef]
84. Migoń, P.; Kasprzak, M.; Woo, K.S. Granite Landform Diversity and Dynamics Underpin Geoheritage Values of Seoraksan Mountains, Republic of Korea. *Geoheritage* **2019**, *11*, 751–764. [CrossRef]

85. Rolfo, F.; Benna, P.; Cadoppi, P.; Castelli, D.; Favero-Longo, S.E.; Giardino, M.; Balestro, G.; Belluso, E.; Borghi, A.; Cámara, F.; et al. The Monviso Massif and the Cottian Alps as Symbols of the Alpine Chain and Geological Heritage in Piemonte, Italy. *Geoheritage* **2015**, *7*, 65–84. [CrossRef]
86. Scoon, R.N.; Viljoen, M.J. Geoheritage of the Eastern Limb of the Bushveld Igneous Complex, South Africa: A Uniquely Exposed Layered Igneous Intrusion. *Geoheritage* **2019**, *11*, 1723–1748. [CrossRef]
87. Natura 2000. Available online: https://natura2000.eea.europa.eu/Natura2000/SDF.aspx?site=GR4220009 (accessed on 22 June 2021).

Article

Salinas and "Saltscape" as a Geological Heritage with a Strong Potential for Tourism and Geoeducation

Katia Hueso-Kortekaas [1,2] and Emilio Iranzo-García [3,*]

1. Mechanical Engineering Department, ICAI, Comillas Pontifical University, 28015 Madrid, Spain; khueso@comillas.edu
2. Institute of Saltscapes and Salt Heritage, IPAISAL, 28450 Madrid, Spain
3. Cátedra de Participación Ciudadana y Paisajes Valencianos, Universitat de València, 46010 Valencia, Spain
* Correspondence: emilio.iranzo-garcia@uv.es

Abstract: Salinas and saltscapes are relevant geoheritage sites with important implications on socioeconomic activities beyond the production of salt, particularly tourism and education. As cultural landscapes, they also have implications related to the identity of their communities. This work presents the study of the patrimonialization processes of four sites in Europe (Añana in Spain, Guérande in France, Læsø in Denmark, and Sečovlje in Slovenia). Lessons obtained from these processes may contribute to the recovery and valuation of similar saltscapes and other forms of geoheritage. The study is based on interviews with relevant stakeholders, a survey of the related grey and scientific literature, and a simplified SWOT analysis. Despite their differences in historical background and current management, all four sites share features that have contributed to the success of their patrimonialization processes, such as having a dedicated entity for this purpose or being protected in some way. They also share common threats that need to be addressed, such as the banalization of the heritage discourse. Other saltscapes and geoheritage sites in general may benefit from these common features, which should serve as an inspiration and not as a template. In the end, shifting from a little-known productive, (proto-)industrial activity toward a sustainable, multifunctional landscape in which geoeducation and tourism are paramount contributes to a more resilient and educated society.

Keywords: salinas; saltscapes; geoheritage; cultural landscapes; geotourism; geoeducation; local development

1. Introduction

Salinas are a cultural landscape type. Within the context of this paper, salinas can be defined as productive landscapes in which salt that is present in nature and forms a saline ecosystem is harvested by humans by means of different techniques (solar evaporation, seething, etc.). Cultural landscapes, on the other hand, are a blend of nature and culture, tangible and intangible heritage, and geological, biological, and cultural diversity, embodying a framework of complex relationships. The recognition of landscape as a cultural heritage, as the perceived and interpreted manifestation of a territorial reality loaded with values, reveals its protagonism in different spheres: environmental, cultural, identitarian, economic, and educational. The preamble of the European Landscape Convention (2000) indicates that the landscape plays an important role of general interest in the cultural, ecological, environmental, and social fields, and that it constitutes a favorable resource for economic activity. Its protection, management, and planning can stimulate the creation of employment. It also indicates that the landscape has a fundamental role in shaping local cultures and that it is an essential component of the European natural and cultural heritage, contributing to human well-being and to the consolidation of European identity [1]. Therefore, landscapes can and should be considered as a key element for the integral development of the individual and the community [2,3].

Saltscapes or salt landscapes are the result of the interaction between geological-geographical processes and socio-economic dynamics that, because of their multiple values, turn the territory into a "cultural geoheritage" that must be conserved, rationally managed, and taught [4]. These areas, full of geodiversity, interconnect territories, their people, and their culture. Salinas, a term that in this context refers to the salt architecture and productive activity that gives rise to saltscapes, are the result of an ancestral economic activity, based on geological resources, which in their development have generated unique ecosystems. However, not all salinas have the same characteristics, present the same model of productive activity, or have the same state of conservation. While some are in full production and are well preserved, others are in decline, endangering ecosystems, cultural heritage, and, ultimately, the landscape. In both cases it is necessary to devise valorization strategies, either to maintain the ecosystem and reinforce its social recognition, or to provide the salinas with new uses, such as tourism or education, that prevent their degradation and disappearance.

Tourism has established itself as an important economic and social activity in almost all scenarios, with emphasis on the segments related to nature and culture [5]. It is a well-known fact that tourism and cultural heritage do not always get along well, but they are mutually interdependent. How to solve this paradox? How can tourism contribute to the conservation of heritage, and to the education and livelihoods of residents? Heritage can be "activated" for the purpose of tourism [6] and heritage tourism, in turn, can contribute to create identities, not only among tourists but also within the local community. This is no trivial issue: in places where the identity of a community was linked to a certain economic activity (industry, mining, agriculture, etc.) that has subsequently disappeared, there is a threat of "dissolution of the society" (unemployment, emigration, ageing, etc.) and the local community's identity may need to be reactivated. Heritage tourism is one of the possible tools to do so [7].

However, tourism in heritage sites or cultural landscapes is an activity that carries controversy among conservationists because they see it as a threat to the values they defend, but also as a source of necessary income to preserve cultural heritage [8–10]. This tension needs special attention when taking into consideration sustainability [9]. It is also relevant to see to what extent the profits of heritage tourism reach the local community and the resources they depend on [11]. From the point of view of tourism development, heritage has some virtues: it can be overtly promoted by public administrations; it is—in principle—free of charge and owned by society in general; it can be visited almost any time of the year; and it offers an air of respectability to the travel experience. Despite these tensions, as said, authorities and the private sector tend to see heritage tourism as a tool to enhance the economic activity in their area of reference, and even as an educational tool. Current tourists originating from a post-industrial society, often seek meaning in their activity. This type of visitor is better informed and prefer destinations with ecological, ethical, and social values. Some of today's tourists want to differentiate themselves from mass tourists, although authors such as Urry and Larsen or Dujmovic and Vitasovic argue that postmodern tourism also has its shadows, such as experiential travel with excessive cultural simulations. Nevertheless, tourists in general have become more demanding and there has been an increase in the creation of new services based on emotions and cultural and landscape experiences [12,13]. In this sense, it is important to identify, analyze and enjoy the benefits of educational potential of this multidimensional phenomenon, in the formation of sensitive citizens aware of the role of the cultural landscapes as something inherent to the well-being and individual and collective identity. With cultural tourism the public can understand the environmental processes, appreciate the different aspects intervening in the history of the local community visited, and demand products and services that require the conservation and management of heritage sites [14,15].

It has been shown that cultural tourism brings potential benefits to other sectors of activity [10]. Direct profits may go to companies presenting heritage to the public (e.g., communications, engineering, and design) and indirect benefits to those that take advantage of the presence of public enjoying leisure activities (e.g., hospitality, fashion, and design), and these will, at the end of the line, revert to the conservation of the heritage and landscapes they depend on. Among the benefits of the link between heritage tourism and local development, two groups may be distinguished: a priori benefits, among which are the creation of employment, increased profits, and improved training of stakeholders. Ultimately, benefits include an improvement in quality of life, a better-quality cultural tourism, enhanced social inclusion, and stronger local development [10]. One important benefit of heritage tourism is the preservation of the landscape that host(-ed) the heritage assets in question. Whatever the activity (agriculture, mining, industry, etc.), the public will be able to gain a better insight if the landscape can be "read". In the case of operating salinas, the heritage will be alive. Hence, from a paradigm of "What can heritage do for tourism?" we have moved on to "What can tourism to for heritage?" [16].

Cultural tourism may favor sites with sensitivity toward their heritage and that actively contribute to its conservation [17]. Visitors also want to actively participate in the experience, engaging in "creative tourism", that is, becoming producers of their own consumer goods and services on site [18]. The global economic crisis and later the pandemic have spurred an upsurge of so-called slow travelers, characterized by travels within the region, especially to rural areas, the search for wellness and health, and short breaks and day trips [19–23]. An important motivation found in slow travelers is the quest for meaningful and participatory experiences rather than for witnessing people and places passively [18,24]. Cultural tourism is no longer just an activity that presents artifacts to tourists, but one that takes into account the relationship between the visitor and the heritage presented, especially when including intangible heritage, which in turn helps visitors get "emotionally involved" in a "heritage experience" [18,25,26]

Salinas epitomize the complexity of cultural landscapes in which human, cultural, and natural features are intimately linked and are mutually dependent for achieving and maintaining sustainability. In addition, salinas are or can be living landscapes without the need to change them into surrogate or fossilized "heritagescapes". This strength of salinas has also been its weakness. The built heritage in salinas is pragmatic, modest, and, because of the materials used, easily degraded, hence not attracting the attention of architects and other heritage specialists. Unfortunately, changes in society and history in the past decades or even centuries have motivated the disappearance of numerous salt-making sites worldwide. A loss of around 90% has been registered [27]. By all measures, this is a serious threat to the remaining sites. Those that have not been transformed into industrial saltworks (e.g., roughly half of the remaining coastal sites in Spain) are under threat of abandonment. Hueso-Kortekaas presents an extensive review of the fate of Spanish and European salinas [27,28].

This research as a threefold goal: first, to characterize the patrimonialization process of four well-managed saltscapes across Europe. This process describes the gradual transformation of a resource-based productive activity (in this case, salt), usually in decline, into a multifunctional landscape with a heritage-based economic activity. Typically, the first stage patrimonialization takes place after the abandonment or irreversible decline of the production of salt and is then activated by interested stakeholders, usually NGOs or public administrations. After activation follows a stage of professionalization, in which the heritage asset or landscape is taken care of by individuals or organizations that have expertise in the topic, although still somewhat patchy. This stage merges onto the consolidation of the process in which a specific, dedicated entity is created and a budget is allotted for the conservation, promotion and use of the site. This is the optimal stage from the point of view of patrimonialization, as it provides stability and a strategic long-term vision [27].

The sites have been selected upon the basis of their heritage values (all of them enjoy a certain degree of protection and have a relevant historical background) and their dedicated heritage management, which consolidates their success in the transition from saltscapes in decline to complex geoheritage sites with a multifunctional character (production of salt, tourism, wellness, and education). Second, common features are sought by means of looking at the lessons learnt in the patrimonialization process, and a simplified spatial SWOT analysis is performed. Third, the work aims at extrapolating the results to other saltscapes and similar forms of geoheritage, which may contribute to better-targeted efforts and more efficient results in the protection and valuation of other saltscapes and geoheritage in general. This paper is organized as follows. Section 2 introduces the data collection techniques and SWOT analysis methods used in the study, before Section 3 exposes the results. Finally, in Section 4, the results are discussed and some ideas for future work explained.

2. Materials and Methods

2.1. Selection of Study Sites

Four study sites were chosen, namely Valle Salado de Añana in Spain, Marais salants de Guérande in France, Sečovlje soline in Slovenia, and the Læsø saltworks in Denmark (Table 1). The selection of all cases responds to one criterion: that is, whether they are or have been in the process of patrimonialization. This means that the selected sites do not only produce salt, but also have other areas of economic activity focused on the public: tourism, health services, or educational activities. They harbor significant heritage values, which are acknowledged by different instruments of legal protection of natural and/or cultural assets, and count with a relevant historical background at regional and national level. These are sites also known for their successful management of the artisanal salt-making activity in a balanced combination with the protection of natural and cultural values, as well as the provision of a livelihood for the local community. The cases selected also have in common that the patrimonialization process is found in an advanced stage and can serve as an example or paradigm for others. In addition, these sites produce salt by traditional methods, reinforcing the idea of heritage conservation and transmission. The four areas are thriving examples of heritage recovery at a regional level and are well known on an international scope. Having said this, the sites differ considerably from each other in their past and recent history and how the process is driven. There are more cases in Europe and elsewhere, but their diversity of patrimonialization processes showcases different possible pathways to success.

Table 1. Description of case study sites according to different criteria. Own elaboration.

Site	Geophysical Features				Productive Features			
	Location	Landscape	Hydrogeol. Origin	Production Method	Energy Source	Scale	State of Facilities
Añana	Inland salina	Mountain	Diapir	Trad. Solar evaporation	Sun & wind	Artisanal	Active
Guérande	Coastal salina	Marsh	Sea	Trad. Solar evaporation	Sun & wind	Artisanal	Active
Læsø	Coastal salina	Marsh	Groundwater (marine intrusion)	Seething	Biomass	Artisanal	Active
Sečovlje	Coastal salina	Coast	Sea	Trad. solar evaporation	Sun & wind	Artisanal	Active

2.2. Bibliographic Survey

An important part of the work relied on the consultation of written literature. The sources covered both scientific as non-scientific literature. The search has been, therefore, eclectic by nature (from systematic key-word use in Google Scholar and Google News to the websites of the companies or organizations in charge of the sites, including cross references from any written document or oral referral). Google Scholar was preferred above other scientific portals because it also provides references to grey literature, which is very relevant in this context. The latter include unpublished reports, plans, and projects or internal documents and have usually been published by non-profit organizations and authorities.

2.3. Interviews with Stakeholders

Understanding local development around a saltscape (or any other form of landscape-based heritage) requires the appropriate identification of stakeholders. Stakeholders are any person or organization that feels affected by any event related to this form of heritage or landscape.

The main challenge of this part of the research has been finding the right stakeholders in each of the study sites, given the variety of roles and profiles [29,30]. They are organized or not in formal, known structures, or may be informal opinion leaders without apparent filiation. Some of them may have the right information, but not the capacity or willingness to participate in the research [11]. Stakeholders who have been relevant in the recent past but are now disengaged from the site have also been considered, especially those who have inspired or triggered management practices that have been used for some time or still are. The main method to find stakeholders in the field was the snowball, by which first-level contacts provided new contacts that were deemed relevant in the context of this work.

Interviews have been performed with 10–12 key stakeholders per site in relation to past, present, and future plans and projects in the study sites involved. Their roles included owners and managers of the salinas, local public administrations, tourism authorities and businesses, (nature) guides, academics and scholars, spa and wellness managers, and, of course, salt makers.

The interviews were semi-structured. The reasons to choose this format were threefold [31]. First, the situations tackled were very different between and even within sites, which required flexibility in the design and development of the interview.

Second, the information needed was rather complex and the responses were expected to differ significantly, both in tone and in content, depending on the stakeholder involved. Third, the need to create a relaxed atmosphere, especially in group settings or with biased stakeholders, recommended this user-friendly format.

2.4. Field Visits and Observations

The field visits were intended to observe first-hand the state of the salt-making site and the surrounding landscape, to visit the businesses and other facilities associated to salt, and to perform the interviews with local stakeholders. Direct observation allowed us to improve the understanding of the site, the decisions of its owners or managers, and the relations among stakeholders. The field visits provided an opportunity to register the most relevant features and significant events related to or resulting from the management of the salt-making site and also a holistic, integrated view of its hinterland.

2.5. Spatial SWOT Analysis

To understand the current situation of artisanal salinas and how they face the future, it is important to have a deep knowledge of them and of their hinterland. Spatial SWOT analysis is a method frequently used to elaborate on strategic territorial diagnoses and make decisions. The SWOT is a classic analysis tool for strategic management enterprise proposed by Kenneth Andrews (1971) [32] that has been transferred to territorial planning and cultural management. Spatial SWOT analysis plays a dual role. It fills in the gap of scientific knowledge and local data and is an approach that synthesizes the information collected through different sources [33–35].

Its name comes from the four ideas that it focuses on: strengths, weaknesses, opportunities, and threats. The items "strengths" and "weaknesses" refer to the current intrinsic aspects of the sites, whereas the items "opportunities" and "threats" are related to their sociocultural and business environment and typically hint at situations that will arise in the future. SWOT is an appropriate diagnostic and evaluation tool to obtain an initial idea of the state of the saltscapes and their possible future evolution. The benefit of this method is its simplicity. Moreover, it is user friendly and does not require computer systems or software [36,37].

In this work, the SWOT matrix analyzes the internal strengths and weaknesses, as well as external threats and opportunities, to guide the future expected strategies. The objective of this matrix is to determine all applicable strategies. First, by using the internal strengths we attempt to exploit the external opportunities and maximizes them. Second, using the present opportunities in the internal environment we attempt to improve weaknesses. Third, using the strengths counteracts the effects of current threats and, fourth, the internal weaknesses are minimized and threats caused by the external environment are prevented.

3. Results

3.1. The Values of Saltscapes

Common salt (NaCl) is an essential constituent for living beings. It is found in nature in the form of rock salt in layers at different depths of the Earth's surface; or dissolved in surface water and groundwater (oceans, lakes, springs, or saline aquifers). Sea waters contain an average of 35 g of salt in solution per liter, which varies between warmer seas such as the Mediterranean and the Red Sea, which have a salinity of 37 and more than 40 g per liter, respectively, and colder seas such as the Baltic, which barely reach 10 g of salt per liter of water. In the case of salt sources, the circulation of water between salt deposits dissolves and transports the salt in solution over long distances in saline aquifers until it reaches the surface, sometimes reaching salinities of over 200 g of salt per liter.

Halophile vegetation absorbs NaCl in solution and incorporates it into the food chain. Wildlife and humans consume vegetables and other animals, incorporating salt into their diet. However, that being insufficient, they collect brackish water or lick rock salt, because salt is essential for their survival. It has specific functions in each of the metabolic cycles and in cellular nutrition. Our body has no reserves of sodium chloride, so it must constantly regulate the amount of salt present through the kidneys and urine [38].

But in addition to the importance of sodium chloride at an organic level, we cannot ignore all the culture that has been generated around its use: food, therapeutics, salting, preservation of skins, industry, religious rites, etc. This is the reason why salt, together with cereals and wine, was considered the basic trilogy of the Mediterranean economy from antiquity until practically the 20th century. Since Neolithic times, humans have collected salt for a variety of purposes. Salt production, trade, and use are at the basis of the so-called saltscapes: environmentally unique sites with a very powerful historical, cultural, and symbolic significance. Therefore, the historical value of salt has left its imprint on the territory and on culture: the salinas themselves, but also trade relations, taxes, by-products, etc. The salinas have allowed a social and ecological coexistence by reconciling the exploitation and sustainable use of the territory. They have contributed to human supply and environmental heterogeneity, and increased biological diversity, playing a key role at the ecological, anthropological, landscape, socio-economic, and cultural levels [39]. Saline water generates ecosystems with special features. They are wetlands in which salinity acts as a limiting factor. However, the organisms that occupy these fragile and unique ecosystems serve as food for many species of birds that nest and rest in them. When humans develop an economic activity to produce salt, environmental and landscape conditions are modified. The surface area occupied by brackish water increases and a semi-industrial architecture is created (ponds, pools, threshing floors, canals, roads, warehouses, etc.) with ecological, cultural, and heritage implications. Saltpans become habitats for organisms that thrive in a range of extreme salinity, temperature, pH, nutrient concentration, oxygen availability, and solar radiation. The production process involves increasing the concentration of salt in the water until it becomes brine. As salinity increases, some organisms are replaced by others: halophilic micro-organisms, including bacteria, archaea, and fungi, which are important in the biogeochemical functioning of salinas [40,41]. Halophilic algae incorporate energy into the system, feeding crustaceans such as *Artemia* sp., which in turn feed the birds [42], making the salt pans important sites for various species of flora and fauna.

However, it should not be forgotten that these unique habitats have their origin in a productive anthropic activity, which began with the collection of salt precipitated in the hollows of coastal rocks or in the beds of salty rivers, and which has evolved into the construction of semi-industrial facilities. The geographical features of a territory are determinant in salt production [43,44]. Therefore, the production process involves knowledge of the environmental processes and the construction of an infrastructure for the management of salt water. The salinas are made up of a set of shallow artificial ponds at different topographical levels, to which seawater or water from saline springs is transported by means of a network of canals. The salt is obtained in these ponds after a first concentration stage and a second crystallization stage.

The physical process of the evaporation of brackish water requires an external energy source that increases the temperature of the water to the point where the dissolved salts begin to precipitate, forming salt crystals. The latitude and location provide favorable climatic characteristics (high annual sunshine and low rainfall during the harvesting season; winds to accelerate evaporation) so that salt production depends exclusively on the use of direct solar energy [45,46]. In sites where climatic conditions are not suitable for solar evaporation, the alternative has been to induce brine evaporation by a wood combustion process. On the other hand, the topographical characteristics are relevant in its location, since it is a process that requires important extensions of land on which to build the evaporation ponds.

Obtaining salt does not depend only on the production infrastructure. The experience and know-how of the salt makers has also been relevant when it comes to achieving greater production and higher levels of purity in the salt (NaCl). Sodium chloride is not the only salt present in the water. The construction of a succession of ponds of different depths and at different elevations makes it possible, on the one hand, to saturate the water, converting it into a brine, and on the other to precipitate—in the first set of ponds—other salts that are less soluble than sodium chloride, such as $CaCO_3$ or $CaSO_4$, which are present in the water. Finally, the practically saturated brine (25.7 °Baumé) is redistributed to the crystallizers. These are square or rectangular ponds, no more than 15 centimeters deep and with a flat bottom, also known as saltpans, where NaCl precipitates at 28–29 °Baumé.

In both coastal and inland salinas, the production process does not end with the crystallization of sodium chloride. Once the NaCl has precipitated, the brine contains other more soluble salts, so the water can be reused to obtain sodium and magnesium sulphates, which are used in the glass and soap industries or in cosmetics, respectively. Finally, the salt itself must be accumulated, dried, and protected from possible inclement weather. Thus, a building linked to the salt installation is needed: the warehouse, where the tools of the salt workers were also kept.

In some inland areas, the abrupt relief makes it almost impossible to have flat surfaces to build the crystallizers. To make up for this lack of horizontality, a system of artificial terraces is set up on the slopes, based on stone and wood structures. The upper part of the terraces is used as a saltpan, while the space between the embankment and the structure is used to store the extracted salt. A salina is therefore a natural, scenic, and cultural heritage that demonstrates the human use of geological resources and shows the intimate relationship between humans and nature that goes beyond the sheer extraction of a raw material.

According to Quesada and Malpica, the production of salt in the salinas resembles an agricultural process because, first, the land must be conditioned and transformed as if it were cultivated fields. As in agriculture, it is necessary to build structures (terraces, banks, and pipelines). Second, there is an assimilation of hydraulism, as found in irrigated agriculture. Water management is typically based on four pillars: catchment, transport, storage, and distribution. Third, it does not require a complex infrastructure, nor investment in manpower or their training because expertise is acquired with practice. Fourth, an idea of "harvesting" is generated from the use and management of natural resources [45–47]. Having made this analogy, it is important to highlight not only the ecosystem services offered by salinas, but also the economic and cultural ones. They are productive cultural landscapes with multiple values, so they must be managed in terms of conservation of geological and geomorphic processes, conservation of biological processes, production of goods, and production of tourist and educational services [48,49].

3.2. Landscape and Geoheritage Characterization of Study Cases

Each case study has, as expected, unique features and narratives. In this section, a brief description, and a summary of the recent history of each site is provided, in relation to its patrimonialization process. Table 2 summarizes the main features of the patrimonialization process of each site. However, patterns can be found with respect to these processes that allow the identification of strengths and weaknesses, as well as lessons taught, for saltscapes and salt heritage in general. These are offered at the end of this section.

Table 2. Main features of the four study sites. Own elaboration.

Site	Patrimonialization Process	Approach	Cause of Change	Owner	Manager	Main Source of Funding
Añana	Institutional	Top down	Abandonment	Private	Public/Private	Public
Guérande	Social	Bottom up	Threat of land use change	Private	Coope-rative	Largely private
Læsø	Institutional	Bottom up	Historical reconstruction	Private	SME	Private (initially public)
Sečovlje	Corporate	Top down	Abandonment/Political change	Public	Large corporate	Largely private

3.2.1. Valle Salado de Añana (Basque Country, Spain)

The cultural landscape of Valle Salado de Añana (42°48′ N, 2°59′ W, 531 m a.s.l) is located in the southwest of the Basque Autonomous Region. The salinas lie in a deep Y-shaped valley formed by the River Muera (brine, in Spanish) and occupy a surface of ca 10 hectares (Figure 1). Brine is obtained from wells that tap groundwater that has been in contact with the diapir just below. The brine is then distributed via aerial wooden carved channels to the crystallizers located in wooden terraces built on the slopes. The origin of the saline aquifer lies in the presence of a diapiric structure derived from the intrusion and tectonic uplift of evaporite rocks dating from the Triassic period—facies Keuper (salts, gypsums, and clays) through denser materials [50]. When surface runoff waters infiltrate and circulate through these materials, they are loaded with salts and become brackish or brine. These brackish waters arise to the surface in the form of saline springs. From there, the brine, which flows with a salinity level of 210–240 gr/L, is conducted by gravity through channels toward the crystallization pans, which are arranged in terraces on the slopes of the valley. Once the brine is distributed over the saltpans, the sun evaporates the water and salt crystallizes, just as in other solar evaporation salinas. The geological origin of the salt and the geomorphology of the valley explain the uniqueness of this landscape from an environmental perspective [50,51].

Salt making in the area has been documented from the year 822 C.E., but recent archaeological research has found evidence of salt-making activity as early as 7000 B.C.E. [52]. The saltworks flourished in medieval times, under the control of the salt workers-cum-owners organization known as the Community of Heirs. As occurred with many other salinas in Spain, in 1564, under King Philip II, they became state-owned during a prolonged period. In 1869, the Community of Heirs recovered the power over the management of the salinas and the traditional salt-making methods they had been using in the past. Due to competition from other salinas, productivity was being increased by irrationally enlarging the surface of crystallizers, building them on dangerously steep slopes, above the level of the sources or using new materials that proved useless, such as concrete. The latter caused major damage to the wooden structures and pollution from debris in the valley. Despite these modernization efforts, salt was still being harvested by hand [53].

Figure 1. Maps of the study area and location of the Valle Salado de Añana (Spain). The perimeter of the salina is delimited in red. Source: Own elaboration.

In 1960, the valley had about 5000 crystallizers in operation, which went down to 150 in 2000. Production decreased from 4000 tons to hardly 3. In the years 1999–2000, the salinas were practically inactive [54–57]. The Valle Salado was declared a BIC (Good of Cultural Interest, in its Spanish acronym) in 1984 with the category of Monument. The Diputación Foral de Álava (provincial administration) initiated a series of actions to recover the valley. In the years 1998 and 1999 the Comunidad de Herederos de las Reales Salinas de Añana (Community of Heirs) became a private company, the Sociedad de Salineros Gatzagak, S.L., which gathered all the owners of the crystallizers. With a contemporary legal structure, the ownership became unified, and third parties had one single representative to address, thereby facilitating the recovery of the valley. In 2009, the Fundación Valle Salado (Valle Salado Trust) was founded, its trustees being the provincial government, the Basque regional government, the municipality of Salinas de Añana, and the Company Gatzagak S.L.

A 20-year Master Plan with a budget of EUR 20 million was devised, setting the physical limits of the monument in order to better determine its functional and landscape recovery and to organize the management and activities of the salinas and its environment to enhance its use and enjoyment by all [58–60]. With the turn of the century, the first measures to create public access to the salina were taken (Figure 2). The public was invited to visit the works, under the motto "Open for repairs". The visitor program of Añana is an ever-growing activity, with ca 100,000 annual visitors in the last decade. The tourism offers range from regular guided tours to specialized tours for schools or special interest groups. Visitors can also book brine foot or hand baths and soon a flotarium will be available.

Figure 2. View of some of the restored saltpans and channels in the Salado valley of Añana. This section can be visited on guided tours only. Own elaboration.

The informants generally agree that the project has brought significant benefits to the village, aside from visibility and local pride. The initial stages of the patrimonialization process were difficult, as stakeholders needed to find a common ground to agree upon and the top-down approach did not contribute to motivate the local community.

3.2.2. Marais Salants de Guérande (Bretagne, France)

The Guérande salt marshes (47°17′ N, 2°27′ W, 0 m a.s.l.) are located in the southern half of Brittany (France) between the mouths of the rivers Loire and Vilaine, facing the Atlantic Ocean. They form a very large wetland zone in western Loire-Atlantique and occupy a surface of 2000 ha (Figure 3) [61]. The current relief is the result of the razing of the Hercynian mountains, the fracturing in inclined blocks, of the post-Hercynian razing surface during the Cenozoic, and of differential coastal erosion during the last marine transgression. The geographical area has a lithological variety typical of geological history (Brioverian schists, gneiss, granite, migmatites, and Quaternary sediments) [62]. The disposition of the inclined blocks of Le Croisic-Batz and Guérande, as well as that of the isthmuses of Pen-Bron and La Baule, have created the conditions for the formation of a maritime marsh, whose clogging by sand and mud intensifies toward the base of the slope of Guérande (where it is more than 20 m thick) [61–63].

Salt has been harvested on the peninsula since the Iron Age. The first saltworks to use the storage capacity of the lagoon goes back to the 3rd century, shortly after the Roman conquest. The first salt marshes as are known today were shaped by the monks from Landévennec Abbey, who, in 945, carved them out by studying the tides, wind, and sun. The salinas brought prosperity to Guérande for many centuries and opened the first trading routes in Europe. Today, at least five saltworks from the Carolingian period are still in operation. After a period in the mid-20th century when the salinas were threatened with urban sprawl, certain sectors of civil society sensitive to the cultural and natural values of the site managed to reverse this threat and recover the salt marshes as they had always been. A key issue in the empowering of salt makers was the strong union formed by the different stakeholders that fought the development plans in the early 1970s, including Breton nationalists and environmentalists. This formed the seed of a strong social and political awareness in the area, that has now grown to become a solid, well-organized supporting tissue in the region [30,64].

Figure 3. Maps of the study area and location of the Marais salants de Guérande (France). The perimeter of the salina is delimited in red. Source: Own elaboration.

The decade of 1975 to 1985 was characterized by the reconstruction of the salt-making activity and the recovery of the marshes (Figure 4). The main challenge was to find replacements for the ageing salt makers, as few young people wanted to take this profession.

Thanks to the arrival of interested apprentices from other regions, the activity gradually regained momentum. Because of this interest, in 1979, a training center for young salt makers was opened. The tradition of the salt worker's profession was thus recovered, and the preservation of these skills have allowed the Guérande marshes to survive through to modern times. Today, about 16,000 tons of coarse salt and 700 tons of *fleur de sel* are produced each year.

The public interest in hand-harvested (as opposed to industrial) salt was gaining strength and so did the tourism pressure in the area. Salt makers still saw tourism as the main threat to their livelihood but slowly started to see visitors as partners rather than enemies in their quest to defend their profession and their landscape. The relationship between salt makers and visitors gradually improved over time and the creation of the visitor center Terre de Sel further contributed to regulate and ease the previous tensions between them [64]. It organizes thematic guided visits, also catering to school groups. Two other museums exist around the world of salt making in the neighboring locations of Batz-sur-Mer (Musée des Marais Salants) and Saillé (Maison des Paludiers), the three of them receiving more than 130,000 visitors annually. Today, the area of La Baule–Presqu'île de Guérande has a well-developed tourism industry. Guérande receives 1.2 million visitors per year. It is calculated that one-fifth of the revenues generated in the area are related to tourism and more than 8100 people are employed in this sector [63,65].

Figure 4. Winter view of the salt pans of one productive unit in the marshes of Guérande. These pans will be cleaned before the next salt making season. Own elaboration.

The informants generally present positive feelings about the patrimonialization process, which was initially rough. However, there remain some differences in the focus, as some *paludiers* believe the essence of the site is being lost because of the commodification of the salt itself and the concept of *fleur de sel* in particular. In fact, some salt masters do not wish to belong to the cooperative for this reason (G.P., pers. comm.). On the other hand, nature conservationists perceive a pressure to recover more surface for salt production and feel it will cause a detrimental effect on habitats and birds (D.M., pers. comm.). In such a large site, striking the right balance seems difficult.

3.2.3. Læsø Saltworks (North Jutland, Denmark)

The Læsø saltworks (57°15′ N, 11°2′ W, 0 m a.s.l.) are located in the southeast of the island of Læsø in northernmost Denmark (Figure 5). As Jørgensen explains, the area is the marine foreland composed of four minor low islands, a belt of coastal meadows and salt marshes and wide areas of sand in shallow water [66]. The terrain is practically flat (low altitude and minimal slope) and the existing vegetation depends on the seawater level. From the inner part toward the coast, the meadows gradually transition into salt marshes. The geology present in the area is composed of a thin cover of sands and silts intertwined by layers of sands and coarse gravels. Beneath these materials is interglacial marine stiff clay [66,67]. The saline water to produce brine and salt is captured from the aquifer formed by post-glacial marine sediments, which has salinity percentages above 17%.

Salt is being produced by seething, using wood as fuel. The brine is pumped from the salty water table of Rønnerne, in the nearby sandbanks of the southern edge of the island. This brine is twice, or three times as concentrated as seawater and is collected in wells to be further concentrated. The brine is then boiled or seethed to obtain a product of high-quality, equal to the famous salt from Lüneburg in Germany.

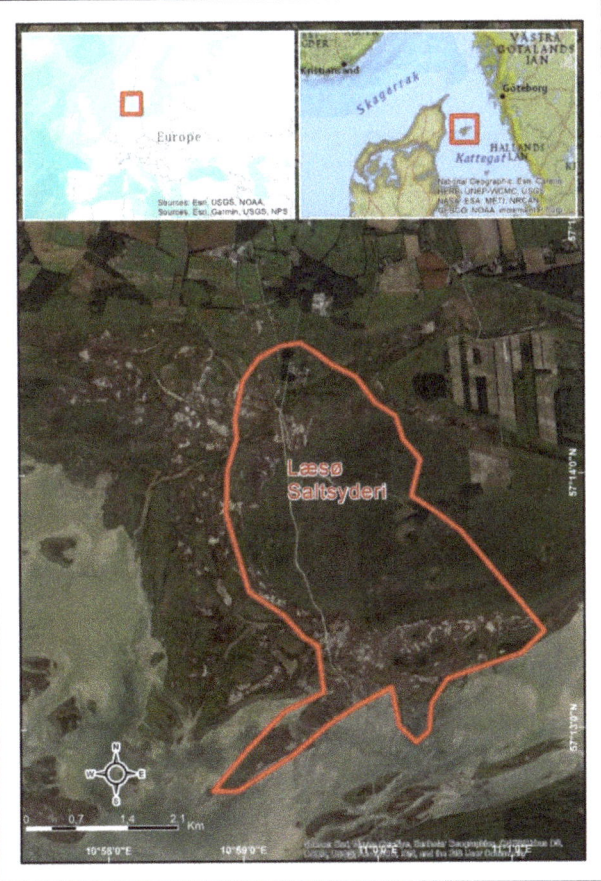

Figure 5. Maps of the study area and location of the Læsø saltworks (Denmark). The perimeter of the salina is delimited in red. Source: Own elaboration.

In the Middle Ages the Læsø saltworks were the most important workplace of the island and were considered the first industry of the time. Salt production stopped in 1652 because seething salt in the huts required large amounts of biomass. By then, the salt industry had used up all the fuel wood on the island and the island was transformed into a windswept desert. The ruins of the old huts where the salt was boiled are still standing as low, square embankments. There are an estimated 1000 of them on the island.

Archaeological research in the mid-20th century revealed how salt making was done a few centuries ago. From the results of the excavations in 1990 it was decided that the Municipality of Læsø would rebuild a salt seething hut, originally with an educational purpose. The goal of the project was partly to put Læsø on the map by telling its unique story about salt, and partly to contribute to the archaeological knowledge on seething by restarting the salt-making activity on the island according to 16th century methods [68–71]. Initially supported by the public authorities, it quickly grew into a thriving productive activity with a keen interest from visitors.

It was soon decided that the visits should be free of charge. In exchange, the salt was sold at a high price, but that was justified as a support to cover the costs of the project. In the first years, the usual high costs derived from salaries were cut because of the participation of the school workshop of the island, whose employees were hired to work in the huts.

Læsø Saltworks uses between 1000–1500 cubic meters of firewood every year, less than 10% of the current harvest of forest products, well below the limit of sustainability. Today the salt making activity is highly organized and successful.

The saltworks are making an important contribution to the economy of Læsø. As a tourist attraction, the saltworks receive more than 60,000 visitors per year, half of the total amount of visitors to the island. The saltworks produce ca. 70 tons of salt per year, selling both locally and all over Scandinavia. The salt is highly valued by customers and visitors and has become a culinary reference in high-end restaurants in the region. The new saltworks were never conceived as a museum, even though its main revenues come from tourism (Figure 6). Visits are still free of charge, because the site is considered a living place of production, in full operation and visitors come for the experience [72].

Figure 6. Restored salt-making hut in the saltworks of Læsø, according to the findings of the archaeological excavations conducted in 1990. Own elaborate.

Perhaps the single most relevant new business associated with salt is the thalassotherapy center, Læsø Kur, which opened in 2008 in a deconsecrated church. In just over 1 year, the center offered numerous therapy services (sauna, steam bath, cool water pool, mother lay baths, jacuzzi, and massage), leisure, and beauty treatments. The center has an agreement with the Danish health system to offer packages for patients and, of course, anyone interested can purchase their own wellness or therapy packages.

All informants agree upon the benefits of the patrimonialization process and do not manifest critical views of it. It seems to have had a net positive impact on the economy of the island, also beyond the salt-making and tourism activities themselves (B.B., P.C., P.S. *pers. comm.*). Some initiatives did not survive (e.g., a salt-themed restaurant), but generally speaking there is a broad consensus about the success of the process.

3.2.4. Sečovlje Soline (Istria, Slovenia)

The Sečovlje salt pans (45°29′ N, 13°36′ E, 0 m a.s.l.) are located in the southwest part of Slovenia, (Gulf of Trieste, northern Adriatic), next to the border with the Republic of Croatia, on the Istria Peninsula. The Slovenian coast, albeit short, is highly varied. There are cliffs, shingle beaches, and coastal plains (lagoons and wetlands). This last type of coast is the result of the accumulation of large quantities of fine sediments, deposited by the

Soca, Rizana, Badasevica, and Dragonja rivers, facing a shallow sea with a shelving sea bottom. The Slovenian coast evolved in the Holocene. Vahtar explains that the valleys have been transformed into bays, with alluvial sediment deposition still ongoing, while ridges changed into peninsulas developing cliffs at the coastline [73]. The rock materials are Eocene flysch, while in the plains, fine-grained alluvial sediments predominate. The Sečovlje salina is a coastal marsh wetland developed on a sedimentary plain of the Dragonja River [73,74]. It consists of two parts. Its northern section, where salt is still being actively produced and harvested, is known as Lera. The southern section, called Fontanigge, is separated by the Grande–Drnica channel (Figure 7).

Figure 7. Maps of the study area and location of the Secovlje soline (Slovenia). The perimeter of the salina is delimited in red. Source: Own elaboration.

The Sečovlje salinas are today the largest coastal marsh wetlands (650 ha) in the country, and at the same time the most important Slovenian locality from the ornithological point of view. Today, 272 bird species have been found in the salinas, with some 90 breeders among them. Based on these facts, the Government of the Republic of Slovenia in the year 2001 declared the Sečovlje Salina Natural Park and the adjacent Museum of Salt-making as a cultural monument of national importance. In 1993, the salinas became the first Slovene wetland, inscribed on the list of internationally important marshes under the auspices of the Ramsar Convention. The salina represents different ecosystems, from marine to brackish, fresh water and land ecosystems.

The traditional manual harvesting of salt in these salinas, over 700 years old, is a representative feature of the cultural heritage of Mediterranean Slovenia. Until the beginning of the 20th century, the saltworks were owned by wealthy families, churches, monasteries, and charitable institutions. The salt worker was merely the tenant of the salt field and the producer of the salt. The golden age of salt making in Sečovlje lasted from the 15th century to the end of the 18th century, under the control of the Venetian Republic [75]. In 2000, the Sečovlje Salinas Nature Park was designated the first protected area in Slovenia when the concession for its management had been given to a business company (SOLINE Pridelava soli d.o.o.), which is owned by the national biggest phone company (Mobitel d.d.). The company is responsible for the management of the state-designated Nature Park and use of its natural resources. The company also is responsible for the protection of nature in the state-owned property of the Sečovlje Salina Nature Park. In return, the Republic of Slovenia provides funding for the management of the protected area [76]. The park receives 30,000 to 45,000 visitors per year, mainly during the summer [77]. Their salt is well known in the Eastern Mediterranean and the site constitutes an example of good management practices and smooth transition from a communist to a capitalist economic system.

In general, the informants agree that the saltworks have a positive influence on the local economy, although there seems to be certain lack of coordination between stakeholders. There are pending issues, such as the access to the salt museum in the border between Croatia and Slovenia, or the management of visitors to the protected area within the productive area of Lera (*F.B. and D.C., pers. comm.*).

At the turn of the millenium, 593 ha of salt-making surface were recovered in Fontanigge (only for the provision of brine) and in Lera (for the whole salt-making process, which accounted for 25 salt-making units). In 2002, 18 men were employed who produced 100 tons of salt. A decade later, more than 94 hired workers produce up to 5000 tons of salt and 30 tons of *fleur de sel* per year (Figure 8). In 2013, the company SOLINE Pridelava d.o.o. decided to invest in a thalassotherapy center to take advantage of the two subproducts of salt making with healing properties, namely the mud (also known as *fango* or peloid) and the mother lay or *acqua madre*. The complex, named Lepa Vida, was built within the natural protected area and has a total surface of 4000 square meters.

Figure 8. Salt worker harvesting salt in Lera, according to methods used in the salinas of the Adriatic basin in the 15th century onwards. Own elaboration.

3.3. Common Features of the Patrimonialization Processes of the Four Study Sites

The four study sites can be considered consolidated examples of patrimonialization and sound use of geoheritage. Despite their differences in ownership, management, heritage assets, and funding, they share some common features that may be contributing to their success. All four sites have some form of protection status, some more diverse, including both natural and cultural values (Añana, Guérande), than others that focus on the natural aspects (Sečovlje, Læsø). From the point of view of management, the four sites count with a specific, dedicated entity: the Valle Salado Trust in Añana, the Cooperative de Salines de Guérande in Guérande, the company Læsø Salt in Læsø, and the company Soline Pridelava in the case of Sečovlje. These entities oversee the management, financing and long-term strategies for the protection of geoheritage, education, and tourism in their respective locations. They are all but financially self-sufficient, with gradually decreasing government support in the cases of Añana and Sečovlje and none in Guérande or Læsø. In all cases, albeit with slight differences in priorities, they form multifunctional landscapes with a focus on four aspects: artisanal salt making, tourism, education, and wellness products and services. At this point in their geoheritage development and use, two risks need to be taken care of.

On the one hand, these entities are dependent on income and need to devise mechanisms to secure financing. To this end, the initial priority to protect and educate about their heritage may shift to a more commercial one. As the protagonist role of the stakeholders involved in the early stages of patrimonialization fades, the new managers may lose this initial perspective. Examples of the banalization of this geoheritage are the organization of mass events such as mountain races (e.g., in Añana, which to be fair was held only once) or the export of artisanal salt to places where this heritage is unknown (especially in Guérande and Læsø), thereby counteracting the discourse of local identity and heritage. This effect is under further stress because of competition from industrial saltworks across the Mediterranean region, which now use a narrative and aesthetics that imitate artisanal salt-making. Consolidated artisanal salt-making sites need to constantly shift their discourse to distinguish themselves from such imitators.

3.4. Saltscapes' Spatial SWOT Analysis

Table 3 summarizes the main strengths, weaknesses, opportunities, and threats of studied saltscapes. From the point of view of the structural strengths, as compared with other productive activities, salt is a well-known commodity that may trigger the interest of a broad sector of the public. These landscapes provide a harmonious and serene combination of natural and artificial elements, in which water is protagonist. The apparent simplicity of the structure of the wetland allows easy reading and understanding, especially if there are recent remains of the activity. Since these sites have been traditionally isolated, they have not often been visited prior to their use as tourism and educational assets. The members of the public who seek new, rewarding sensorial experiences can find them in (former) salt-making sites.

Saltscapes are usually located in rural areas, which are experiencing an ever-growing appreciation of their culture, as a token of authenticity and a return to one's roots. The local community, on the other hand, shows a pride in their traditions, practices, and products, with a stronger sense of belonging to the area. Salt has the advantage of being a universal, everyday item. It is therefore relatively easy to raise the interest of the public in it.

There are numerous protection and planning instruments available that can be applied to the management of these sites. Given the multifunctionality of saltscapes, the variety of instruments is one of the largest possible. All this will also increase the chances to obtain public investments, although the global economic crisis is hitting hard, especially in the socio-cultural and environmental sectors. Perhaps the strongest and most specific strength of saltscapes is that they can be recovered for the original purpose they were made for. This is not common in former industrial or mining sites and happens only occasionally in certain rural activities such as bakeries, lime kilns, or charcoal-from-biomass.

Table 3. SWOT analysis of salt heritage and saltscapes.

Strengths	Weaknesses
Structural	*Structural*
Salt as a universally known commodity	Isolated geographical location
Scenic beauty of saltscapes	Decay and fragility of salt heritage
Visible remains of the salt making activity	Salt making as a physical challenging activity
Appreciation of rural life	Seasonality
Managerial	Costly recovery and maintenance
Planning and protection instruments	*Managerial*
Attractive for public investments	Lack of support for artisanal salt as a product
Sense of belonging	Conflicts in the uses of saltscapes
Can be recovered to original function	Vandalism
Opportunities	**Threats**
Managerial	*Managerial*
Development of sustainable tourism around saltscapes	Low population density
Potential innovative uses (also R&D)	Global economic crisis/pandemic
Synergies with other heritage assets and activities	Plans and projects beyond the scope of control
Empowerment and motivation of the local community around heritage	Political shortsightedness
	Dependence on public funds
Flexible financing mechanisms	Lack of entrepreneurial culture
	Climate change

The main structural weaknesses of saltscapes have to do with location and technology. Salt can only be made where certain geological, climatic, and topographical conditions exist. Salt making itself is a strenuous activity that requires a young, fit, and motivated workforce. Bearing in mind that salt can only be harvested a couple of months per year, salt workers need to combine this job with other activities. The maintenance of salinas is also demanding, especially those that had some importance in the past, which host more complex infrastructures, a larger productive surface, plus housing, offices, and several warehouses. These sites are especially vulnerable to climate change, not only in terms of flooding (if at the coast), but also because of their intimate dependence on functional natural processes [78]. Delays in planning and performing recovery activities rapidly increase the costs, and maintenance after that is very costly in terms of manpower.

From the point of view of management, salt making is not a priority activity for authorities and institutions that may provide (financial) support, such as rural development agencies or chambers of commerce. The little institutional willingness to invest in these facilities is usually aggravated when the sites are in private hands. Planning and implementing recovery projects in these sites is costly and the global economic crisis has significantly decreased the funding opportunities for the upkeep and rehabilitation of rural heritage, both from public as from private bodies.

Nevertheless, the increasing sensitivity toward sustainable tourism initiatives may benefit tourism around saltscapes, as the sites lend themselves well to a slow, conscious, experience-based form of tourism [79]. There are also numerous potential uses of a saltscape that are compatible with the conservation of their natural, cultural, and human values. They may even lend themselves to new economic activities around research, innovation, and development and commercialization of by-products. There is also a growing flexibility in funding practices that do not require strong investments. As well, authorities seem to be more open to transfer tasks to private organizations, e.g., with land stewardship agreements, volunteer work, etc., making the management of the sites more diverse. This enhanced flexibility is also perceived in society in general, with a growing diversity of products and services, where the traditional dichotomies client-customer, resident-visitor, or student-teacher, to name a few, are becoming blurred. Also, synergies with other heritage assets or like-minded initiatives in the area can be found. The diversification of activities increases the opportunities of participation for the local community, thereby empowering them around their heritage. All this may create unimaginable synergies and collaborations that may provide new opportunities for heritage to (re-) emerge. Hence, despite the gloomy economic scenario, it may become an opportunity rather than a threat.

4. Discussion and Conclusions

Salinas and saltscapes are geoheritage sites whose existence cannot be understood without the human use of geological resources. They are places where the economic exploitation of salt has contributed to increase their environmental and cultural values and to their legal recognition under different forms of protection. This sum of circumstances, geological processes, biodiversity, habitats, cultural heritage, landscape, etc., has led to an increase in the significance of these areas, which have become more than just a wetland or a salt production area. There is a greater identification by the local society, which demands preservation policies from public authorities. A process of patrimonialization of the site takes place with consequences on the environmental, cultural, political, and social dimensions. However, there are different circumstances that make these sites fragile enclaves, which require attention, sensitivity, and creativity in their treatment so that they can continue to fulfil their multifunctional role. There are no homogeneous measures for their management because their diversity (geographical, functional, and social) means that specific actions are required in each case. However, some lessons can be learned from patrimonialized saltscapes, which could serve as a reference for other geoheritage sites. In these cases, the importance lies in being inspired by the processes behind their patrimonialization and socio-economic valuation, rather than by the specific products and outcomes of each site.

One of the approaches being proposed is the social and economic revaluation of territorial resources and landscapes. There are various public policies that rely on cultural and natural assets as instruments for the differentiation and recognition of regions and places to be valued, as criteria on which to base the distribution of facilities, as elements for the promotion of tourism and other services, as sources of employment, and as places for learning and the creation of collective identities.

Among the factors that contribute to the patrimonialization process of geoheritage sites are the range of cultural ecosystem services they offer. These services are associated with the ecological, cultural, and symbolic valuation of the sites, in addition to the production necessary for their proper functioning. In this way, many key sites are protected and recognized for their aesthetic contributions, the beauty they inspire, the spirituality they trigger, the cultural identity they establish, the knowledge they represent, and the health, education, recreation, and tourism services they provide for human well-being. However, there is a need for greater recognition of the importance of geology and geomorphology focused on conservation, education, and sustainable development. Indeed, although geosites and their landscapes synthesize a whole set of structures and socio-environmental processes, their educational use has not received much attention in research [80], nor in outdoor teaching activities [81]. However, the interpretation of the landscape in situ, together with the use of historical events and material and immaterial cultural heritage, can facilitate the understanding of places and the learning of the environmental, socio-economic, and cultural processes that have shaped them. The importance of geo-environmental education for the promotion and preservation of geological heritage and geo-ethical values should be emphasized [4].

The value of geoheritage sites should be promoted among civil society and taken advantage of by teachers and planners—hence the need to promote them publicly, develop methods for their valuation, and define their qualities and character to establish their vocation. This is the way to achieve effective management under an appropriate legal framework [82,83] and a dedicated entity could provide stability and focus. During the last 2 decades, international networks and organizations have worked to promote binding protection of geosites (such as the Geosites project, promoted by the IUGS, or the European Geoparks Network). The states and regions have also opted for their legal recognition and regulation [84]. The protected status should not be seen as a limiting element but should serve as a catalyst for preservation and local development. Salt-making sites that have developed legal planning and management mechanisms, sometimes with public-private investment, eco-labels, marketing strategies, or land stewardship mechanisms, are a magnificent example. In addition to guaranteeing the preservation of the values that are at their origin, they have made possible the creation of new services and products. Geotourism stands out, which involves visits by tourists interested in consuming knowledge and products linked to the salinas, which translates into new investments, more infrastructure, businesses, and job creation that contribute to the diversification of the local economy.

Author Contributions: Conceptualization, K.H.-K. and E.I.-G.; methodology, K.H.-K. and E.I.-G.; software, E.I.-G.; validation, K.H.-K. and E.I.-G.; formal analysis, K.H.-K. and E.I.-G.; investigation, K.H.-K. and E.I.-G.; data curation, K.H.-K.; writing—original draft preparation, K.H.-K. and E.I.-G.; writing—review and editing, K.H.-K. and E.I.-G.; visualization, K.H.-K. and E.I.-G.; supervision, K.H.-K. and E.I.-G. All authors have read and agreed to the published version of the manuscript.

Funding: This research received no external funding.

Conflicts of Interest: The authors declare no conflict of interest.

References

1. Déjeant-Pons, M. The European Landscape Convention. *Landsc. Res.* **2006**, *31*, 363–384. [CrossRef]
2. Scazzosi, L. Reading and assessing the landscape as cultural and historical heritage. *Landsc. Res.* **2004**, *29*, 335–355. [CrossRef]
3. Kyvelou, S.S.; Gourgiotis, A. Landscape as Connecting Link of Nature and Culture: Spatial Planning Policy Implications in Greece. *Urban Sci.* **2019**, *3*, 81. [CrossRef]
4. Zafeiropoulos, G.; Drinia, H.; Antonarakou, A.; Zouros, N. Geoheritage to Geoeducation, Geoethics and Geotourism: A Critical Evaluation of the Greek Region. *Geoscience* **2021**, *11*, 381. [CrossRef]
5. Fonsêca, F.O.; dos Santos, J.C.; Vieira, L.V.L.; Ferreira, F.A. Pedagogical Tourism in National Parks: Relations Between Brazil and Portugal. In *Advances in Tourism, Technology and Systems: Selected Papers from ICOTTS20 (Smart Innovation, Systems and Technologies)*, 1st ed.; Abreu, A., Liberato, D., González, E.A., Garcia Ojeda, J.C., Eds.; Springer: Singapore, 2021; Volume 209, pp. 560–571. [CrossRef]
6. Prats, L. *Antropología y Patrimonio*, 3rd ed.; Ariel: Barcelona, Spain, 1997; p. 176.
7. Ruiz, E.; Hernández, M. Identity and community—Reflections on the development of mining heritage tourism in Southern Spain. *Tour. Manag.* **2007**, *28*, 677–687. [CrossRef]
8. Nuryanti, W. Heritage and postmodern tourism. *Ann. Tour. Res.* **1996**, *23*, 249–260. [CrossRef]
9. Garrod, B.; Fyall, A. Managing heritage tourism. *Ann. Tour. Res.* **2000**, *27*, 682–708. [CrossRef]
10. Greffe, X. Es el patrimonio un incentivo para el desarrollo? *Rev. PH* **2003**, *42*, 43–50. [CrossRef]
11. Aas, C.; Ladkin, A.; Fletcher, J. Stakeholder collaboration and heritage management. *Ann. Tour. Res.* **2005**, *32*, 28–48. [CrossRef]
12. Urry, J.; Larsen, J. *The Tourist Gaze 3.0.*; Sage: Thousand Oaks, CA, USA, 2011.
13. Dujmović, M.; Vitasović, A. Postmodern society and tourism. *J. Tour. Hosp. Manag.* **2015**, *3*, 192–203. [CrossRef]
14. Lussetyowati, T. Preservation and conservation through cultural heritage tourism: Case study: Musi riverside Palembang. *Procedia Soc. Behav. Sci.* **2015**, *184*, 401–406. [CrossRef]
15. Chong, K.Y.; Balasingam, A.S. Tourism sustainability: Economic benefits and strategies for preservation and conservation of heritage sitesin Southeast Asia. *Tour. Rev.* **2019**, *74*, 268–279. [CrossRef]
16. Silberberg, T. Cultural tourism and business opportunities for museums and heritage sites. *Tour. Manag.* **1995**, *16*, 361–365. [CrossRef]
17. Wearing, S.; Neil, J. *Ecoturismo: Impacto, Tendencias y Posibilidades*; Síntesis: Madrid, Spain, 2000; p. 269.
18. Richards, G.; Wilson, J. Developing creativity in tourist experiences: A solution to the serial reproduction of culture? *Tour. Manag.* **2006**, *27*, 1209–1223. [CrossRef]
19. European Commission. *Using Natural and Cultural Heritage to Develop Sustainable Tourism in Non-Traditional Tourist Destinations*; European Commission: Luxembourg, 2000; p. 136.
20. Mitkova-Todorova, R. *Traditional Salt-Works and Tourism: A Practitioners Guide*; ALAS Technical Letter: Koper, Greece, 2002.
21. Skumov, M. Salinas and tourism: The ALAS experience. In Proceedings of the ALAS All About Salt Final Conference, Mytilini, Greece, 29 November–2 December 2002; Petanidou, T., Dahm, H., Vayanni, L., Eds.; University of the Aegean: Mytilini, Greece, 2002; pp. 53–56.
22. Vodenska, M.; Popova, N.; Mitkova-Todorova, R. Sustainable Regional Development of Salinas and Salt Production based Tourism. ALAS Interregional Study: Pomorie, Bulgaria, 2002.
23. Seyfi, S.; Hall, C.M. COVID-19 pandemic, tourism and degrowth. In *Degrowth and Tourism: New Perspectives on Tourism Entrepreneurship, Destinations and Policy*; Hall, C.M., Lundmark, L., Zhang, J., Eds.; Routledge: London, UK, 2020; pp. 220–238.

24. Bosshart, D.; Frick, K. *The Future of Leisure Travel—Trend Study*; Gottlieb Duttweiler Institute: Zürich, Switzerland, 2006; p. 67.
25. Poria, Y.; Butler, R.; Airey, D. Links between tourists, heritage and reasons for visiting heritage sites. *J. Travel Res.* **2004**, *43*, 19–28. [CrossRef]
26. Alivizatou, M. Museums and intangible heritage: The dynamics of an unconventional relationship. *Pap. Inst. Archaeol.* **2006**, *17*, 47–57. [CrossRef]
27. Hueso, K. Salt in Our Veins. The Patrimonialization Processes of Artisanal Salt and Saltscapes in Europe and Their Contribution to Local Development. Ph.D. Thesis, Univeritat de Barcelona, Barcelona, Spain, 2019.
28. Hueso, K. Gente Salada. Las Salinas de Interior, ¿Un Patrimonio Vivo? IPAISAL: Collado Mediano, Spain, 2015; p. 164.
29. Ander-Egg, E. *Métodos y Técnicas de Investigación Social IV. Técnicas para la Recogida de Datos e Información*; Lumen Humanitas: Buenos Aires, Argentina, 2003; p. 384.
30. Steyaert, P.; Barzman, M.; Billaud, J.P.; Brives, H.; Hubert, B.; Ollivier, G.; Roche, B. The role of knowledge and research in facilitating social learning among stakeholders in natural resources management in the French Atlantic coastal wetlands. *Environ. Sci. Policy* **2007**, *10*, 537–550. [CrossRef]
31. Corbetta, P. *Metodología y Técnicas de Investigación Social*, 1st ed.; McGraw-Hill: Madrid, Spain, 2007; p. 448.
32. Andrews, K.R. *The Concept of Corporate Strategy*; Dow Jones Irwin: Homewood, IL, USA, 1971.
33. Praveena, S.M.; Aris, A.Z. A review of groundwater in islands using SWOT analysis. *World Rev. Sci. Technol. Sustain. Dev.* **2009**, *6*, 186–203. [CrossRef]
34. Diamantopoulou, P.; Voudouris, K. Optimization of water resources management using SWOT analysis: The case of Zakynthos Island, Ionian Sea, Greece. *Environ. Geol.* **2008**, *54*, 197–211. [CrossRef]
35. Comino, E.; Ferretti, V. Indicators-based spatial SWOT analysis: Supporting the strategic planning and management of complex territorial systems. *Ecol. Indic.* **2016**, *60*, 1104–1117. [CrossRef]
36. Beeho, A.J.; Prentice, R.C. Conceptualizing the experiences of heritage tourists: A case study of New Lanark World Heritage Village. *Tour. Manag.* **1997**, *18*, 75–87. [CrossRef]
37. Benzaghta, M.A.; Elwalda, A.; Mousa, M.M.; Erkan, I.; Rahman, M. SWOT analysis applications: An integrative literature review. *J. Glob. Bus. Insights* **2021**, *6*, 54–72. [CrossRef]
38. Monckeberg, B. Salt is indispensable for life, but how much? *Rev. Chil. Nutr.* **2012**, *39*, 192–195.
39. Martín, J. Las salinas: Socioecosistemas que conectan vidas. In *Paisajes de la Sal en Iberoamérica. Cultura, Territorio y Patrimonio*; Moreno, O., Román, E., Eds.; Instituto Juan de Herrera: Madrid, Spain, 2021; pp. 55–74.
40. Davis, J.S. Biological management of solar saltworks. In Proceedings of the 5th International Symposium Salt, Northern Ohio Geological Society Inc, Cleveland, OH, USA, 19–22 April 1980; Volume 1, pp. 265–268.
41. Mani, K.; Salgaonkar, B.B.; Das, D.; Bragança, J.M. Community solar salt production in Goa, India. *Aquat. Biosyst.* **2012**, *8*, 1–8. [CrossRef]
42. Korovessis, N.A.; Lekkas, T.D. Solar saltworks production process evolution—wetland function. In Proceedings of the Post Conference Symposium SALTWORKS: Preserving Saline Coastal Ecosystems, Samos, Greece, 11–30 September 1999.
43. Hocquet, J.C. Explotation et appropiation des salines de la Méditerranée occidentale (1250 1350 env.). In Proceedings of the XI Congresso di Storia della Corona d'Aragona, Palermo, Italy, 11–15 April 1984; Volume III, pp. 219–248.
44. Weller, O. First salt making in Europe: An overview from Neolithic times. *Doc. Praehist.* **2015**, *XLII*, 185–196. [CrossRef]
45. Quesada, T. Las salinas de interior de Andalucía Oriental: Ensayo de tipología. In Proceedings of the II Coloquio de Historia y Medio Físico. Agricultura y regadío en al-Andalus, Almería, Spain, 9–10 June 1996; pp. 317–333.
46. Iranzo-García, E. *Las Salinas Continentales en la Provincia de Valencia. Aproximación al Estudio de un Elemento Singular del Patrimonio Rural*; Departament de Geografía. Universitat de Valencia: Valencia, Spain, 2005.
47. Quesada, T.; Malpica-Cuello, A. Las salinas de Andalucía oriental en epoca medieval. Planteamientos generales y perspectivas de investigación. *J. Salt Hist.* **1994**, *2*, 144–169.
48. Iranzo-García, E.; Kortekaas, K.H.; López, E.R. Inland Salinas in Spain: Classification, Characterisation, and Reflections on Unique Cultural Landscapes and Geoheritage. *Geoheritage* **2021**, *13*, 1–10. [CrossRef]
49. Georgousis, E.; Savelidi, M.; Savelides, S.; Holokolos, M.-V.; Drinia, H. Teaching Geoheritage Values: Implementation and Thematic Analysis Evaluation of a Synchronous Online Educational Approach. *Heritage* **2021**, *4*, 3523–3542. [CrossRef]
50. Ruiz Urrestarazu, E.; Galdós Urrutia, R. Patrimonio e innovación en el Valle Salado de Añana, País Vasco. *Ciud. Territ. Estud. Territ.* 2015 *47*, 73–88.
51. Corella, J.P.; Stefanova, V.; El Anjoumi, A.; Rico, E.; Giralt, S.; Moreno, A.; Valero-Garcés, B.L. A 2500-year multi-proxy reconstruction of climate change and human activities in northern Spain: The Lake Arreo record. *Palaeogeograph. Palaeoclimatol. Palaeoecol.* **2012**, *386*, 555–568. [CrossRef]
52. Plata, A. *La Comunidad de Propietarios del Valle Salado de Añana*, 2nd ed.; Publicaciones de la Diputación Foral de Álava/Arabako Foru Aldundiaren Argitalpenak: Vitoria, Spain, 2019; p. 360.
53. Torres, J.M. La recogida de la sal en Salinas de Añana. *Narria: Estud. Artes Costumbr. Pop.* **1991**, *53*, 23–29.
54. Lasagabaster, J.I. El Valle Salado de Salinas de Añana o donde la historia se hace paisaje. In Proceedings of the I Biennal de la Restauració Monumental, L'Hospitalet de Llobregat, Barcelona, Spain, 23–26 November 2000.

55. Landa, M.; Lasagabaster, J.I. La recuperación integral del valle salado de Salinas de Añana: Gestión y método. In *Las Salinas y la Sal de Interior en la Historia: Economía, Medio Ambiente y Sociedad*; Morère, N., Ed.; Universidad Rey Juan Carlos: Madrid, Spain, 2007; pp. 1021–1042.
56. Plata, A.; Landa, M.; Lasagabaster, J.I. Salinas de Añana, Álava. In *Los Paisajes Ibéricos de la Sal 1: Las Salinas de Interior*; Carrasco, J.F., Hueso, K., Eds.; Asociación de Amigos de las Salinas de Interior: Guadalajara, Spain, 2008; pp. 45–57.
57. Plata, A. La recuperación y el estudio de una fábrica de sal: Las salinas de Añana-Alava. In *La Explotación Histórica de la Sal: Investigación y Puesta en Valor*; SEHA, Ed.; Sociedad Española de Historia de la Arqueología: Ciempozuelos, Spain, 2009; pp. 15–36.
58. Landa, M.; Ochandiano, A. El valle salado de Salinas de Añana, recuperación integral. *Akobe Restaur. Conserv. Bienes Cult.* **2002**, *3*, 43–46.
59. Mallarach, J.M. La conservació dels paisatges singulars i rellevants del territori històric d'Àlaba: Una estratègia comprensiva i integradora. *Espais Rev. Dept. Política Territ. Obres Públ.* **2005**, *50*, 150–155.
60. López de Eguilaz, R. Paisaje cultural del Valle Salado de Añana (Alava). Candidato a patrimonio mundial de la Unesco en 2014. In *Paisajes Culturales, Patrimonio Industrial y Desarrollo Regional*; Álvarez, M.A., Ed.; Colección Los ojos de la Memoria no. 13; INCUNA: Gijón, Spain, 2013; pp. 603–609.
61. Lemonnier, P. Le marais salant de Guérande: Un écosysteme transformé en moyen de production. *Étud. Rural.* **1977**, *66*, 7–22. [CrossRef]
62. Sellier, D. Un moyen de vulgarisation de la géomorphologie: Le triptyque explicatif des géomorphosites (application au pays de Guérande, Loire-Atlantique). *Cahiers Nant.* **2010**, *1*, 119–126.
63. Harduin, R.; Ragot, C.; Trichet, L.; Andreu-Boussut, V.; Chadenas, C. Gérer la fréquentation humaine, protéger le patrimoine naturel: Étude de cas sur les marais salants de Guérande. *Cahiers Nant.* **2016**, *2*, 31–42.
64. Perraud, C. La renaissance du sel marin de l'Atlantique en France (1970–2004). In *I Seminário Internacional Sobre o Sal Português*; Amorim, I., Ed.; Instituto de História Moderna da Universidade do Porto: Porto, Portugal, 2005; pp. 423–430.
65. Chambre d'Agriculture Loire-Atlantique. *Diagnostic Salicole avec la Participation du Comité Professionnel Salicole*; CAP Atlantique: Nantes, France, 2011.
66. Jørgensen, N.O. Origin of shallow saline groundwater on the Island of Læsø, Denmark. *Chem. Geol.* **2002**, *184*, 359–370. [CrossRef]
67. Hansen, J.M.; Aagaard, T.; Stockmarr, J.; Moller, I.; Nielsen, L.; Binderup, M.; Larsen, B. Continuous record of Holocene sea-level changes and coastal development of the Kattegat island Læsø (4900 years BP to present). *Bull. Geol. Soc. Den.* **2016**, *64*, 1–55. [CrossRef]
68. Vellev, J. *Salt Produktion på Læsø, i Danmark og i Europa*, 3rd ed.; Forlaget Hikuin: Højberg, Denmark, 2000; p. 108.
69. Vellev, J. En rejse 1597 til Læsøs salt. In *Rænæssancens Verden. Tænkning, Kulturliv, Dagligliv og Efterliv*, 4th ed.; Høiris, O., Vellev, J., Eds.; Århus Universitetsforlag: Århus, Denmark, 2006; pp. 371–400.
70. Mørtensen, M.D.; Olsen, N.F. *Kulturhistoriske Værdier på Læsø. Pilotprojekt Marin Nationalpark Læsø*; Center for Kulturanalyse, Københavns Universitet: Copenhagen, Denmark, 2005; pp. 1–56.
71. Tanvig, H.W. Læsø-lokalsamfund og natur på det globale marked. *Videnblade Planlægning Af By Og Land* **2007**, *4*, 25–26.
72. Lorentzen, A. Leisure, culture and experience economy in the periphery. Does Northern Jutland benefit from the Experience economy? In *Regional Studies Association International Conference*; University of Newcastle: Tyne, UK, 2011; pp. 1–20.
73. Vahtar, M. *Slovenian Coast (Slovenia). Report of UAB Pilot Sites*; Draft; EUROSION: Rotterdam, The Netherlands, 2002; Volume 3, pp. 1–20.
74. Ogorelec, B.; Mišič, M.; Šercelj, A.; Cimerman, F.; Faganeli, J.; Stegnar, P. Sediment of the salt marsh of Sečovlje. *Geologija* **1981**, *24*, 179–216.
75. Žagar, Z.; Benčič, E.; Bonin, F. *Muzej Solinarrstva/Museo delle Saline/Museum of Salt-Making/Salzgartenmuseum*; Maritime Museum Sergej Mašera: Piran, Slovenia, 2006.
76. Držek, P. A rescue plan for the traditional Salinas: Vision of a long-term development of the Landscape Park of Sečovlje Saltworks, Piran. In *Salt and Salinas as Natural Resources and Alternative Poles for Local Development*; Petanidou, T., Dalm, H., Eds.; University of the Aegean: Mytilene, Greece, 2002; pp. 58–61.
77. Sovinc, A. Sečovlje Salina Nature Park, Slovenia: Latest developments and important cultural activities. In *Culture and Wetlands in the Mediterranean: An Evolving Story*; Papayannis, T., Pritchard, D., Eds.; Med-INA: Athens, Greece, 2011; pp. 227–234.
78. Adam, P. Saltmarshes in a time of change. *Environ. Conserv.* **2002**, *29*, 39–61. [CrossRef]
79. Wu, T.C.E.; Xie, P.F.; Tsai, M.C. Perceptions of attractiveness for salt heritage tourism: A tourist perspective. *Tour. Manag.* **2015**, *51*, 201–209. [CrossRef]
80. Stewart, A. Whose place, whose history? Outdoor environmental education pedagogy as reading the landscape. *J. Adv. Educ. Outdoor Learn.* **2008**, *8*, 79–98. [CrossRef]
81. Fägerstam, E. High school teachers' experience of the educational potential of outdoor teaching and learning. *J. Adv. Educ. Outdoor Learn.* **2014**, *14*, 56–81. [CrossRef]
82. Panizza, M. Geomorphosites: Concepts, methods and examples of geomorphological survey. *Chinna Sci. Bull.* **2001**, *46*, 4–5. [CrossRef]
83. Henriques, M.H.; dos Reis, R.P.; Brilha, J.; Mota, T. Geoconservation as an emerging geoscience. *Geoheritage* **2011**, *3*, 117–128. [CrossRef]
84. Palacio, J.L. Geositios, geomorfositios y geoparques: Importancia, situación actual y perspectivas en México. *Investig. Geogr. Bolet. Inst. Geogr.* **2013**, *82*, 24–37. [CrossRef]

Article

How Greek Students Perceive Concepts Related to Geoenvironment: A Semiotics Content Analysis

Efthymios Georgousis [1,*], Maria Savelidi [2], Socrates Savelides [3], Spyros Mosios [1], Maximos-Vasileios Holokolos [4] and Hara Drinia [1,*]

1. Department of Geology and Geoenvironment, National and Kapodistrian University of Athens, 15784 Athens, Greece; spymosi@geol.uoa.gr
2. Faculty of Economics and Business, University of Ghent, 9000 Ghent, Belgium; maria.savelidi@ugent.be
3. Hellenic Ministry of Education and Religious Affairs, Directorate of Secondary Education of Magnesia, 38333 Volos, Greece; ssavelidis@uth.gr
4. Department of Culture, Creative Media and Industries, University of Thessaly, 38221 Volos, Greece; mcholokolos@uth.gr
* Correspondence: egeorgousis@geol.uoa.gr (E.G.); cntrinia@geol.uoa.gr (H.D.); Tel.: +30-210-727-4394 (H.D.)

Abstract: In order to design a geoeducation program in the context of the possibilities given to the Experimental Schools of Greece of Lower Secondary Education, teachers identified the need for diagnostically assess students' understanding of basic concepts of the geoenvironment and particularly the concepts of geodiversity, geoheritage, geoethics and geotourism. In addition, there was a need to apply the educational technique of creating cognitive conflicts in order to promote the scientific perceptions of these concepts. Thus, research questions were identified which led the research to assess the current latent state of students' perceptions regarding the thematic areas of the concepts and to identify concepts whose perceptions can be used in the educational process in order to achieve effective cognitive conflicts in order to promote scientific perceptions of them. The students briefly answered a four-question questionnaire, wherein each question examined their perceptions regarding the four concepts of geoenvironment: geodiversity, geoheritage, geoethics and geotourism. All 45 students of the geoeducation program that took part in the survey were aged between 12 and 15 years old. The qualitative research strategy approach was selected and specifically the hybrid technique of semiotics content analysis in combination with thematic analysis. This technique was selected due to the need to identify, code, categorize and count both obvious and latent meanings in the students' written answers; these meanings were related to the four concepts under examination. The results of the research show that the current latent state of students' perceptions regarding the thematic fields of the four concepts of the geoenvironment can be considered as particularly confused since the majority of students did not understand the concepts as they are employed in the international literature. The research also highlighted concepts that can be used by teachers in their efforts to develop students' clear or even scientifically acceptable perceptions for the concepts of geodiversity, geoheritage, geoethics and geotourism in the thematic field of the geoenvironment.

Keywords: geodiversity; geoheritage; geoethics; geotourism; geoeducation; semiotics content analysis; Greece

1. Introduction

Students' education in geosciences is connected to the neglected component of the geoenvironment [1], which includes a variety of concepts. Among all these concepts, the ones that were examined in the present research were students' perceptions of the concepts of geodiversity, geoheritage, geoethics and geotourism.

Geodiversity represents multiple values (intrinsic, cultural, aesthetic, economic, educational [2]) and is perceived as the variability of abiotic nature elements (such as geomorphological, tectonic, soil, hydrological and topographical) and physical processes, both on the surface of the Earth as well as in the sea, together with endogenous and exogenous systems which cover the diversity of places, elements and particles that are generated by either natural or human processes [3] (p. 144).

Geoheritage is an integral part of natural heritage [4] (p. 7) which presupposes the "complete perception of man for nature and the environment" [5,6] and must be preserved for the benefit of future generations [7]. The concept of geoheritage includes the valuable and important geological and geomorphological elements of the landscape [8] with significant scientific, educational, cultural, aesthetic and/or tourist value. These elements of natural geodiversity are of great value to man [9], to cultural development and sense of place. These are characterized by great importance for education and research due to their special geological characteristics, the types of rocks or minerals they contain, unusual fossils or other geological elements. Additionally, they include places that have played a role in cultural or historical events or are aesthetically appealing landscapes [10,11].

Geoethics, like geological heritage, is a relatively new topic in geosciences, so the relationships it comprises are not yet fully understood [12]. Geoethics, in addition to the awareness of geoscientists, refers to the re-examination of the relationship between humans and the Earth system [13], therefore encouraging and promoting ethical values [14] in order to raise public awareness concerning the problems related to geoenvironment [15]. As defined by the International Association for Promoting Geoethics (IAPG) "geoethics provides a reference and guidelines for behavior in addressing concrete problems of human life by trying to find socio-economic solutions compatible with respect for the environment and the protection of nature and land" [16].

Finally, geotourism reveals the economic value of geological heritage [17] (p. 147). Among the various approaches, because there is no generally accepted definition [6], geotourism focuses particularly on the geological and geomorphological aspects of the landscape [18] and refers to a new more holistic type of sustainable tourism [19] arising from two very different disciplines, namely geology and of tourism [6]. According to Newsome and Dowling, "Geotourism is a form of natural area tourism that specifically focuses on geology and landscape" [20] (p. 4); [21]. From a different standpoint, the concept of geotourism is defined as "geographical" tourism [22] and refers to tourism which contributes to the preservation or promotion of the special geographical character of a place [23] (p. 1) in terms of culture, heritage, environment, aesthetics and the well-being of its inhabitants [24]. Therefore, its goal is to "extend the principles of ecotourism" [25] (p. 21) and contribute to regional economic development [25] (p. 24).

The reason for this research is the hardly optimistic education of students in geosciences [26] in compulsory education in Greece, although the need for their education in knowledge offered by geosciences has been repeatedly pointed out. Geosciences knowledge is useful for everyday life [27] and helps to understand the natural environment and the interaction between people and the environment so that students can eventually develop a sense of responsibility for their environment and a moral code for its protection and preservation [28].

In contrast to the prevailing situation in the Greek educational system, over the last two decades, society's interest in the geoenvironment has been constantly increasing internationally. Concepts such as "geosites" [29] (p. 25); [30,31], "geodiversity" [32–35] and "geoheritage" [36] have become more widely known [7] (p. 20), while the establishment of "geoparks" and the development of "geotourism" contribute to the economic and cultural development of visited areas [6,37–39]. At the core of interest is formal or informal "geoeducation" as part of sustainable development education and the promotion of geosciences [40], because in geosciences, it is imperative to infuse the ethical way of behavior in teaching from the very first module of Earth science in primary school [41]. However, both the concept of geological heritage and the concepts associated with it are absent from the school curricula of the geology–geography course of the Greek educational system, so one way of introducing the concept and its meaning in schools is environmental education [42] (p.112); [28]—and its evolution into education for the environment and sustainability.

In the Greek educational system, the development of geological thinking is provided in primary education through a few teaching hours in the context of the course of geography [43] as well as in lower secondary education through a few hours in the course of geology–geography. Education in the field of geosciences and the geoenvironment is considered incomplete [26] (p. 74) and does not help students understand the history of the Earth and explain natural processes [44]. The lack of geological knowledge can be covered by the development of environmental education programs [27,43,45]/education for the environment and sustainability, which, however, are implemented on a voluntary basis by both teachers and students. In Greek schools, teachers of all specialties have the ability to conduct environmental education programs, but largely ignore the importance of geological formations [46], geodiversity and geoheritage, which is why the environmental groups of Greek schools that choose to develop an environmental program with a geoenvironmental theme are very limited [46,47]. Nevertheless, what contributes to the need for students' education on geosciences are the positive examples of designing and implementing environmental educational programs and educational activities in geologically protected areas which are organized for primary and lower secondary education students [38,39,48].

A recent study, one which investigated the understanding of geocultural heritage and the relative values of lower secondary school students (gymnasium) and university students, found that in contrast to the aesthetic value perceived by the participants and the importance they attach to cultural, anthropocentric, utilitarian and economic values, participants understand geological value to a moderate degree, and ecological, ecocentric and intrinsic values to a fairly low degree [47]. This is why, in the Greek school reality, the design and implementation of geoenvironmental education programs for lower secondary school (gymnasium) students is promising in the aims of empowering their geocultural values [49,50].

According to the above, it is possible to argue that the education of students in geosciences is incomplete in the curricula of the cognitive field of geology–geography of lower secondary school (gymnasium), whilst the concept of the geoenvironment as well as witnessing geological phenomena and their processes are absent from the topics of the educational programs of environmental education, thus preventing students' understanding values of the geological heritage. However, students' perceptions of basic concepts of the geoenvironment have not been explored.

The present research explores students' perceptions of the concepts of geodiversity, geoheritage, geoethics and geotourism in order to design a geoeducation program within the framework of the possibilities given to an experimental school. It is noted that this type of educational unit supports the experimentation and pilot implementation of educational innovations mainly through the creation and operation of creativity and innovation groups that concern various cognitive fields [51].

The geoeducation program was designed with the aim of broadening students' ethical concerns concerning the recognition of geodiversity's intrinsic value [9], which essentially means that people do not have the right to reduce geodiversity [52] and that students are expected to realize their personal values' framework which may signal their transition to a higher stage of ethical thinking [53]. The expected learning outcomes of the geoeducation program for the students are:

To recognize the geoenvironment as a witness to geological phenomena and its relation to socio-economic reality.

To understand geodiversity and the fact that the elements constituting geoheritage are of value to society [9], so that they understand both the geological heritage and the cultural values of geodiversity associated with mythological, historical, archaeological, spiritual and religious aspects [2,33].

To discover the geological peculiarities of an area as a geotourism product and the potential of geotourism as the basis of promoting the development of sustainable tourism [54].

To develop a moral code and a sense of responsibility for the protection and conservation of the environment [28].

As a result of exploration and empowerment in geocultural values, students will evaluate monuments of geological and cultural heritage and propose their own geotours through digital narratives on Internet maps.

The geoeducational program was undertaken by two teachers and 45 students of the experimental school who expressed the desire to participate as members of the respective Creativity and Innovation Group.

In order to design this geoeducational program, teachers realized the need for a diagnostic assessment of students regarding their perception of the basic concepts of the geoenvironment and specifically the concepts of geodiversity, geoheritage, geoethics and geotourism [55,56]. Teachers, aiming for the effective reconstruction of students' perceptions about these concepts, designed the use of the educational technique of creating cognitive conflicts [57]. This reconstruction aimed to change students' latent perceptions of these concepts, with perceptions accepted by so-called school science [58]. This signaled the need, through this diagnostic assessment, to identify those students' latent perceptions, which should be reconstructed through the educational process.

2. Materials and Method

Teachers, based on fragmented attempts to communicate with students, suspect that the cognitive structure [59] of their knowledge and perceptions of geoenvironment probably do not correspond to the corresponding scientific knowledge and perceptions, which leads them to seek a conceptual change in their students. This conceptual change leads them to the acquisition of knowledge and the formation of perceptions in the real world on the subject of the geoenvironment [60].

Therefore, they agree that an effective way for students to construct their new own models or conceptions which are in line with modern scientific conceptions of the geoenvironment is to rely on elements of knowledge and perceptions that students already have prior conceptions of [61]. Thus, using "bridging analogies" will allow students to perceive the issue of geoenvironment in the "real world" in a new way—in relation to the system of geoenvironment—based on students' conceptual understanding [61] (p. 485).

In other words, teachers want to cause a cognitive conflict in their students so that its resolution, through the implementation processes of the geoeducation program, causes the expected result of building a new conceptual construction on a scientific basis for each of them. It is noted here that the concept of cognitive conflict is understood as "the starting point to promote any change in the conceptual network ... [leading] ... the individual to be aware of the differences between their own beliefs, concepts or theories and the new information" [62] (p. 374).

This is how the question/concern firstly arises: Do the cognitive structures of students' knowledge and perceptions of the geoenvironment correspond to the respective scientific knowledge and perceptions of science? Additionally, if these do not correspond, what are those elements (beliefs, concepts or theories) of their cognitive structure which we should use with the technique of cognitive conflict to instigate the desired conceptual change in our students?

In order to successfully implement this geoeducation training program through the promotion of student learning and the corresponding diagnostic feedback [63] (p. 1), teachers recognize the need to collect detailed information on students' latent ability to acquire knowledge in the thematic area of the geoenvironment, i.e., in the area which includes the topic and the educational object of the program [64] (p. 1). In practice, the teachers were interested in assessing the current (latent) learning situation as students' knowledge and perception in the thematic subareas of the geoenvironment: geodiversity, geoheritage, geoethics and geotourism in the most objective way possible.

This problem identified the research questions of this survey:

Q1: What is the current latent state of students' perceptions regarding the thematic areas of the concepts of the geoenvironment, namely geodiversity, geoheritage, geoethics and geotourism?

Q2: Which concepts can be used in the educational process to achieve cognitive conflicts in order to promote scientific understanding of the concepts of geodiversity, geoheritage, geoethics and geotourism to students?

Thus, the sensitizing concepts of the survey were identified. It is noted that the similarity between the obvious or latent meanings of the students' statements with the meaning of the sensitizing concepts constitutes a measure of concordance between the teachers regarding the analysis of these statements.

Naturally, sensitizing concepts include the concepts of geodiversity, geoheritage, geoethics and geotourism:

"Geodiversity" is defined as the abiotic equivalent of biodiversity and describes the variety of geological, geomorphological, pedological and hydrological features and processes [33,34].

"Geoheritage" refers to those elements of the planet's geodiversity that are assessed as worthy of conservation [34].

The concept of "geoethics" refers to research and reflection on the values that underpin appropriate behaviors and practices, wherever human activities interact with the Earth system [65] (p. 4) [66].

"Geotourism" is a sustainable form of tourism [5] focusing on the geological and landscape component [67].

In order to answer the research question of the present survey, the qualitative approach of semiotics content analysis was followed [68] (p. 25); [64], always within the framework of the educational process of cognitive diagnostic assessments (CDAs) and with the corresponding adaptation to the objectives of this geoeducation program (Figure 1).

As mentioned above, the objectives of this educational program in addition to knowledge are oriented towards the development of perceptions and the empowerment of students in the values of geoheritage [69] (p. 32) in relation to the issue of geoenvironmental sustainability [70], especially regarding the concepts of geodiversity, geoheritage, geoethics and geotourism. This orientation in the development of perceptions and values in students showed teachers that such a diagnostic assessment technique should be followed which can deepen students' perceptions and knowledge and provide the necessary information to teachers to shape and properly implement the program in order to promote the empowerment of values in students. The latter instructed teachers to follow a similar assessment technique and process to that of Georgousis, Savelidi, Savelides, Holokolos and Drinia (2021) [50,71].

Thus, the semiotics content analysis was followed due to the interest of teachers in the search for deeper meaning in students' answers [72] (p. 716). Teachers were particularly interested in highlighting *obvious* or *latent* meanings as *conceptual patterns* and their correlation with the meaning of the characteristic key concepts which are designed to have an effect upon the actual syllabus and the educational objectives of the course [73] (p. 102). The key concepts were described as the sensitizing concepts of the present survey [72] (p. 716). This analysis used elements of thematic analysis in order to better manage and comprehend the data and to facilitate the exportation and reporting of results [74]. These elements are thematic categorization based on data and coding based on the text elements of the teachers' field of interest and the relevant reporting of the survey [75].

Four categories were identified which correspond to the four educational objectives of the syllabus of the course. These are the main sensitizing concepts of this survey, namely geodiversity, geoheritage, geoethics and geotourism.

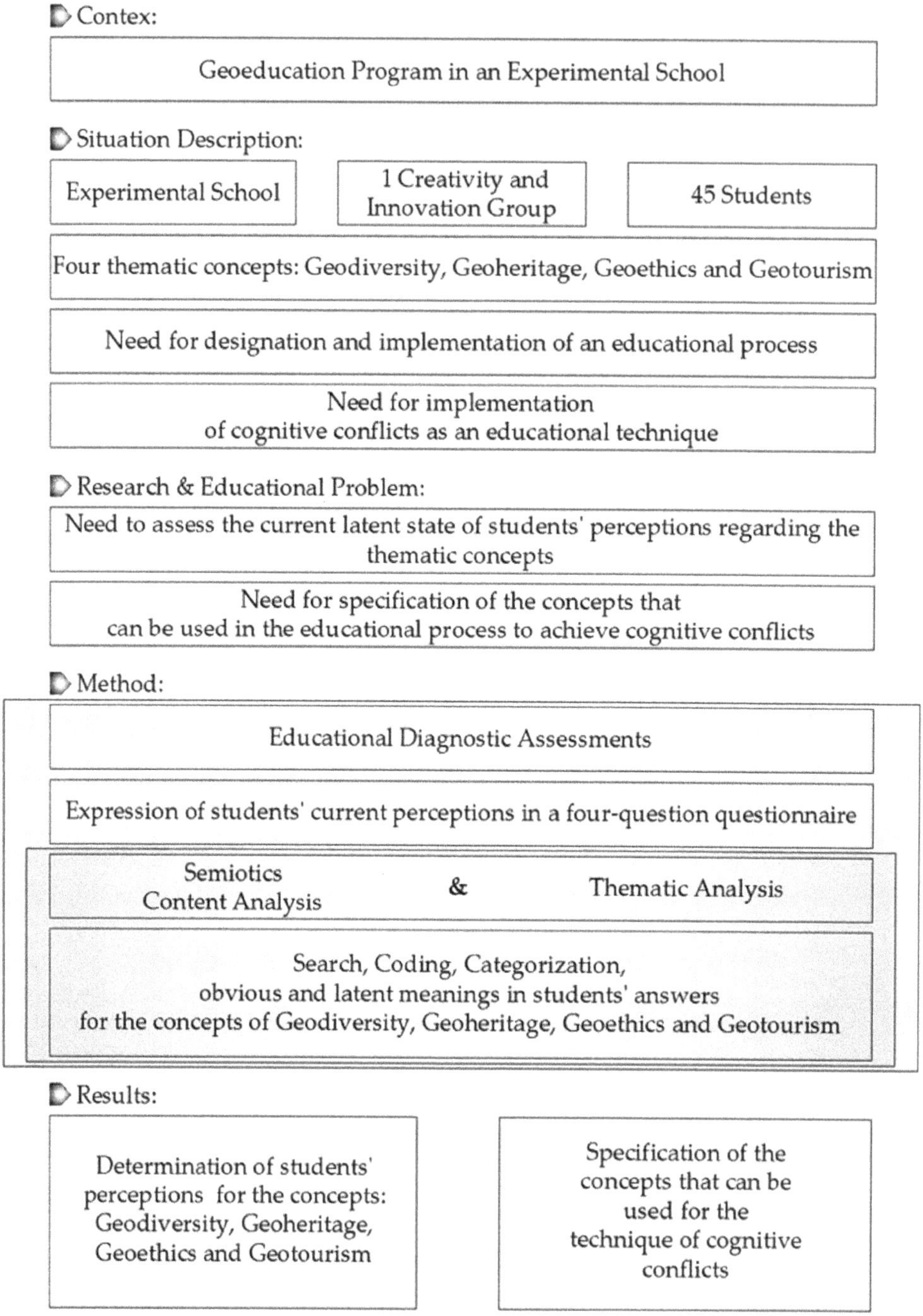

▷ Context:

| Geoeducation Program in an Experimental School |

▷ Situation Description:

| Experimental School | 1 Creativity and Innovation Group | 45 Students |

| Four thematic concepts: Geodiversity, Geoheritage, Geoethics and Geotourism |

| Need for designation and implementation of an educational process |

| Need for implementation of cognitive conflicts as an educational technique |

▷ Research & Educational Problem:

| Need to assess the current latent state of students' perceptions regarding the thematic concepts |

| Need for specification of the concepts that can be used in the educational process to achieve cognitive conflicts |

▷ Method:

| Educational Diagnostic Assessments |

| Expression of students' current perceptions in a four-question questionnaire |

| Semiotics Content Analysis & Thematic Analysis |

| Search, Coding, Categorization, obvious and latent meanings in students' answers for the concepts of Geodiversity, Geoheritage, Geoethics and Geotourism |

▷ Results:

| Determination of students' perceptions for the concepts: Geodiversity, Geoheritage, Geoethics and Geotourism | Specification of the concepts that can be used for the technique of cognitive conflicts |

Figure 1. Context and methodology for answering the research questions.

The teachers estimated that the knowledge and perceptions of all participating students should be explored as we are interested in "mining" data that will help in the education and pedagogical empowerment of all students with no exceptions. In addition, a thorough qualitative examination of these data (as opposed to a quantitative one) may have given teachers the opportunity to draw interesting "internally generalized" conclusions, even if the source of the data was the knowledge and perceptions of a small minority of students [76] (p. 6). For these reasons, it was decided that the diagnostic assessments of students should be addressed to all students. Thus, an inventory survey was carried out [77] (p. 020023-2) with a sample of all 45 students of the geoeducation program of the Experimental School of Volos, who were aged between 12 and 15 years old.

A questionnaire of four questions was typed which can be answered openly. The questionnaire was addressed to students who are encouraged to use a small number of words to express their understanding of what the concepts of geodiversity, geoheritage, geoethics and geotourism mean to them. Specifically, the urge to express their perception of what these concepts mean to them refers to what "comes to mind" when they read these words. The questionnaire was in the format of a digital form and it was filled in optionally.

Students' answers were examined on the basis of the linguistic unit of the sentence [78] (p. 50). The sentence of each answer was approached semiotically, i.e., it was semantically analyzed in order to highlight obvious concepts and reveal latent concepts, which are found in it and are related to the identified sensitizing concepts [72] (p. 559, 716); [73] (p. 102). In their answers seeking to express their perception of the general meaning of one of the four concepts, the students instilled points of meaning which reflected this perception. Thus, the search for these instilled points was sought which referred to the students' perception of the meaning of these four concepts, the examination of which is of interest to the teachers. These points are defined as conceptual patterns and are grouped—and coded—according to their relevance to each other in the codes of analysis [72] (p. 599).

An "a priori" code [79] was identified in each category which was expected to include those conceptual patterns (students' answers) that are consistent with the concept (general meaning) of the category as described—based on the selected bibliography—in the corresponding sensitizing concept. The four (4) a priori codes, in their name, were characterized by the word consciousness. The a priori codes are geodiversity consciousness, geoheritage consciousness, geoethics consciousness and geotourism consciousness.

The following is a thorough semiotic examination of the concepts in the students' answers simultaneously and collaboratively conducted by the two teachers of the course.

Related conceptual patterns were grouped and integrated "in processus" codes (corresponding to the term "a priori"). Conceptual patterns which were not consistent with the concept (general meaning) of a category were included in a code named *Uncategorized*.

The examination of the linguistic unit (sentences) of the answers was carried out with the following systematization:

- Words were sought to convey the meaning identified in the description of the corresponding sensitizing concept in an obvious or latent way [72] (p. 559, 716); [73] (p. 102). The explanation of the meaning was also examined with regard to the meaning of the definition which was mentioned in the corresponding sensitizing concept and always according to the teachers' assessment.
- It was examined whether the words in correlation with the corresponding context of the text unit give the meaning of the respective sensitizing concept in an obvious or latent way [80] (p. 204).

The genre of the sentence was examined in order to determine the semantic "intensity" of the morpheme [78] (p. 10, 13). The genre was examined, as a school textual genre, according to the categorization: the genre of describing, the genre of instructing, the genre of explaining, the genre of arguing and the genre of narrating [81] (p. 13). On an intensity scale, according to the teachers' assessment, linguistic unit with an arguing character is considered to be of high semantic intensity, while a linguistic unit with a describing character is considered as low [82].

What followed was a quantitative drawing of the results and conclusions. The results were reported in quantitative form using tables and reports, and documented using Sankey diagrams and verbally according to linguistic scales [83].

The claim that the survey can be characterized as descriptive, interpretive and theoretical in its environment validity could be supported. This is inferred as the data were consistently recorded by the participants themselves under the supervision of the teachers. Additionally, the sensitizing concepts, as the foundation of the survey, are based on modern theoretical perceptions of academic and research bodies and the aim of this survey was intertwined with daily practice [84] (pp. 284-285); [72] in the club of an experimental school in Greece. It can also be argued that the research was characterized by credibility, since the research design, consistent data collection and documentation of the results extraction process led to findings which, as considered by the authors, can be trusted by the reader [85] (p. 1057).

The survey was conducted in compliance with the basic ethical principles of research [86]. The questionnaires were anonymous and were submitted in digital form, thus ensuring the required respect, confidentiality, data protection and students' personality protection. Of course, the whole process was aimed at the students' benefit (it was aimed at improving the educational and pedagogical processes with them as recipients). The latter was perceived by them and it emerged with their informed consent as there was no student who did not submit the questionnaire, despite it being optional for them to complete it. Finally, the whole process was not motivated by any self-centered interest since—as mentioned above—it aimed to optimize the educational work for the benefit of the students only.

The method was implemented as follows.

The classification of the conceptual patterns of the linguistic units into codes and corresponding categories, as well as the extraction of results, was performed via relevant software computer assisted qualitative data analysis (CAQDAS), namely Atlas.ti 9 [87].

Thematic analysis techniques were applied and upon their basis, one (1) theme was identified, namely "geoenvironmental concepts and perceptions" and four (4) categories were identified: geodiversity, geoheritage, geoethics and geotourism. In addition, four (4) "a priori" codes were also identified with the names geodiversity consciousness, geoheritage consciousness, geoethics consciousness and geotourism consciousness.

The followed process sought to be understood by the reader through the following examples:

- Linguistic units are examined in the students' answers to some of the questions, for example, that referring to geodiversity. The examination was performed according to the above in order to identify conceptual patterns with a meaning related to that of geodiversity, as mentioned in the description of the respective sensitizing concept (of geodiversity). Related concepts were classified in codes, which were either created for this purpose ("a priori" codes) or when the need to create a code was present ("in processus" codes). It was noted that, during the coding through QACDAS software, the linguistic unit receives a characteristic numeric code (e.g., 1: 6 p 1 in 01_Geodiversity), which in the present study is used for the exact reference to the linguistic unit only.
- For example, the linguistic unit: 1: 6 p 1 in 01_Geodiversity was mentioned, to which the student answers: "It is the many species of flora and fauna". Here, we observe that the student's expressed perception of the concept of geodiversity is a conceptual pattern, which according to the teachers, was neither identified as obvious nor as latent (with the concept of geodiversity). It is a descriptive text, whose obvious meaning basically derives from the words "many species", "flora" and "fauna", according to the teachers, and that can be identified with the concept of biodiversity. Thus, another code was added to the thematic analysis map, in the category geodiversity, namely that of biodiversity. Of course, the conceptual pattern of the linguistic unit 1:6 p 1 in 01_Geodiversity was coded in code biodiversity, of the category geodiversity.

- Another example refers to the latent conceptual pattern of a linguistic unit. Specifically, in linguistic unit: 2:17 p 1 in 02_Geoheritage, the student, expressing his perception of Geoheritage, answers: "It is what people have inherited from the land, forests and plains". Here, first of all, it seems that the meaning of the conceptual pattern does not refer to the meaning of the concept of geoheritage, as it was defined in the corresponding sensitizing concept. On the contrary, the teachers assessed that the latent meaning of the linguistic unit was consistent with that of natural heritage. This was deduced both from the identification of words that refer to heritage concerning "land, forests and plains" and more generally from the examination of their context, things which—according to the teachers—refer to the concept of natural heritage. Thus, linguistic unit 2:17 p 1 in 02_Geoheritage was classified in code natural heritage of the category geoheritage.
- An example of a linguistic unit, coded as uncategorized, is 4:18 p 1 in 04_Geoetourism. The student, expressing his understanding of the concept meaning of geotourism, stated: "It refers to young people working as farmers". Here, the teachers considered that the meaning of the answer was not in line with the topic of geoenvironment, nor even with any of the examined concepts and for this reason, it was coded as uncategorized.

Finally, it was noted that the linguistic scale [81] was used which characterizes the number of students who answered about the meaning of a concept as many (percentage $\geq 25\%$), several (percentage $> 25\%$ and $\geq 10\%$), some (percentage $> 10\%$ and $\geq 5\%$), few (percentage $> 5\%$ and $\geq 2\%$) and minimal (percentage $> 2\%$).

3. Results and Discussion

In the application of the procedures of the aforementioned methodology, the course teachers received 45 answered digital questionnaires from an equal number of students. Thus, 180 answers were received (45 questionnaires × 4 answers per questionnaire) which means that the linguistic units were also identical at the same number ($n = 180$). The 180 students' answers were redistributed into four documents according to the concept which the question explored and which each answer dealt with. Thus, the answers were redistributed in documents: 01_Geodiversity.pdf, 02_Geoheritage.pdf, 03_Geoethics.pdf and 04_Geotourism.pdf. Naturally, their redistribution yielded 45 responses in each document, distributed by negotiated concept, which is also evident in the names of the documents. These four (4) documents were input in the CAQDAS software, Atlas.ti [87], for the purpose of categorization, coding and further drawing of conclusions. A part of document 02_Geoheritage.pdf and some of the students' relevant answers are shown in Figure 2.

In this way, the semiotics content analysis of the linguistic units of the students' answers highlighted 170 conceptual patterns which were estimated to be related to the theme. These patterns were included in the 4 a priori and another 15 "in processus" codes (codes that arose during the analysis process). Conceptual patterns which were judged as unrelated to the theme (23 conceptual patterns) were included in a (twentieth) code named uncategorized.

Figure 3 depicts the thematic map of analysis (theme, categories, codes and link routing).

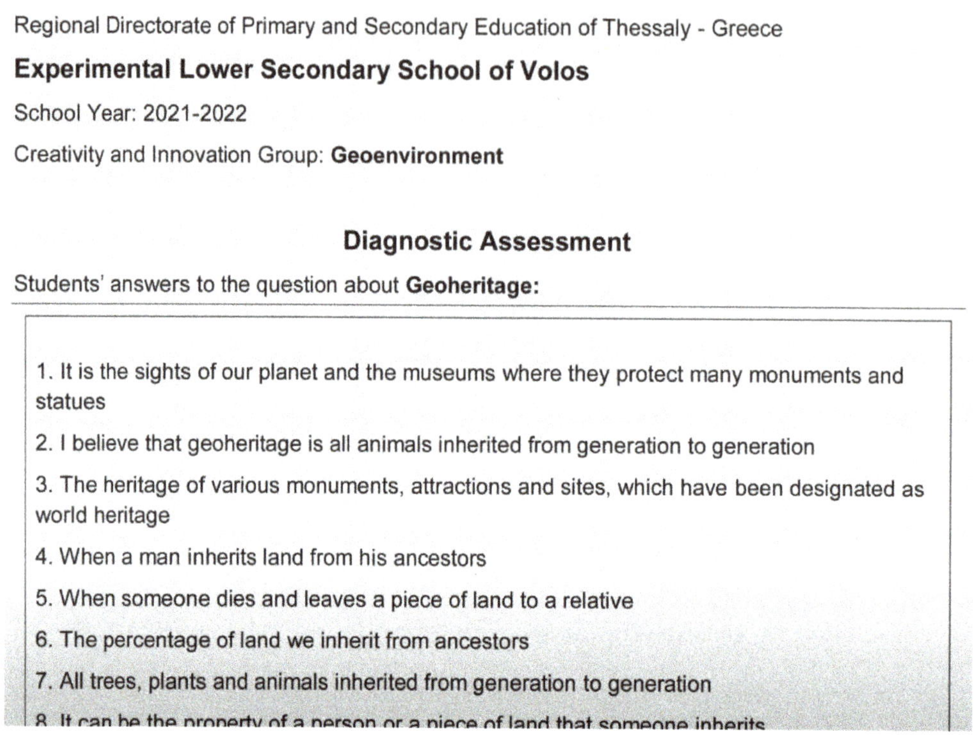

Regional Directorate of Primary and Secondary Education of Thessaly - Greece

Experimental Lower Secondary School of Volos

School Year: 2021-2022

Creativity and Innovation Group: **Geoenvironment**

Diagnostic Assessment

Students' answers to the question about **Geoheritage**:

1. It is the sights of our planet and the museums where they protect many monuments and statues
2. I believe that geoheritage is all animals inherited from generation to generation
3. The heritage of various monuments, attractions and sites, which have been designated as world heritage
4. When a man inherits land from his ancestors
5. When someone dies and leaves a piece of land to a relative
6. The percentage of land we inherit from ancestors
7. All trees, plants and animals inherited from generation to generation
8. It can be the property of a person or a piece of land that someone inherits

Figure 2. Part of the document with students' answers (on the concept of geoheritage).

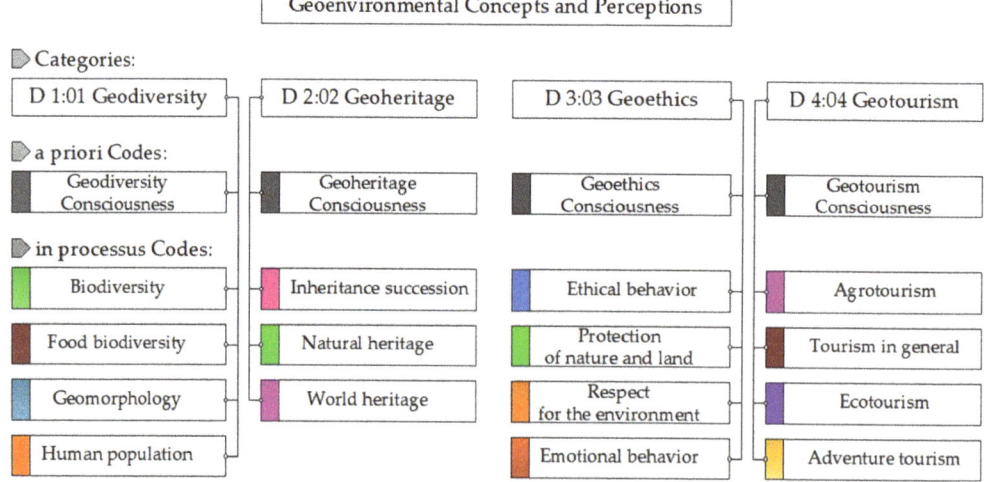

Figure 3. Thematic map: theme, categories, codes and coding color index.

In Table 1, the number of conceptual patterns was presented, which were identified per document, category and code. What was also presented was the marking color index of the codes in the CAQDAS software [87].

Table 1. Semiotics content analysis/thematic analysis: number of conceptual patterns per document/category and code. Coding color index.

Code s/n	Category	Code	01_G [1]	02_G [1]	03_G [1]	04_G [1]	Totals
1	Geodiversity	•—Geodiversity consciousness	0				0
2		•—Biodiversity	19				19
3		•—Geomorphology	12				12
4		•—Food biodiversity	12				12
5		•—Human population	3				3
6	Geoheritage	•—Geoheritage consciousness		1			1
7		•—World heritage		3			3
8		•—Inheritance succession		21			21
9		•—Natural heritage		15			15
10	Geoethics	•—Geoethics consciousness			4		4
11		•—Respect for the environment			10		10
12		•—Protection of nature and land			11		11
13		•—Ethical behavior			13		13
14		•—Emotional behavior			2		2
15	Geotourism	•—Geotourism consciousness				4	4
16		•—Ecotourism				7	7
17		•—Agrotourism				16	16
18		•—Adventure tourism				6	6
19		•—Tourism in general				11	11
20	Uncategorized	•—Uncategorized	5	7	9	2	23
		Totals (minus uncategorized quantities):	46	40	40	44	170
		Totals:	51	47	49	46	193

[1] 01_G: 01_Geodiversity.pdf (Gr = 45), 02_G: 02_Geoheritage.pdf (Gr = 45), 03_G: 03_Geoethics (Gr = 45), 04_G: 04_Geoetourism.pdf (Gr = 45).

The results of the examination of all the students' answers per category (which exactly corresponds to the corresponding document) are presented below, per category. Here, we must note that the absolute number of total frequencies of the codes is higher than the number of students' answers (45 × 4), since in any one answer, more than one different conceptual patterns may have been identified, which were naturally coded in more than one corresponding codes.

3.1. Category Geodiversity

Based on the examination of the respective linguistic units of their conceptual patterns answers, it was found that the students did not understand the concept of geodiversity, as was found in the international literature and was described in the corresponding sensitizing concept (Table 2).

It seems that many students (37.26%) confused it with the concept of biodiversity, that is "that part of nature which includes the variety and richness of all the plant and animal species at different scales" [88]. For example, the linguistic unit 1:25 p 2 in 01_Geodiversity was mentioned: "Geodiversity is the variety of products of the earth, such as fruit and vegetables and especially the organic products produced by the land", which was evaluated as a descriptive text with latent meaning, which resembles the concept of biodiversity and not geodiversity.

Many students (23.53%) perceived it as food biodiversity, that is, as "the diversity of plants, animals and other organisms used for food" [89] and also many (23.53%) perceived it as geomorphology, that is, they referred to features found on Earth, such as mountains, hills, plains and rivers. [90].

Table 2. Frequencies/distribution of codes (students' perceptions) in the category geodiversity.

01_Geodiversity (n = 45)			Frequencies	
	Code	Absolute	Relative (in Category)	Relative (Within All Codings)
1	•—Biodiversity	19	37.26%	9.84%
2	•—Food biodiversity	12	23.53%	6.22%
3	•—Geomorphology	12	23.53%	6.22%
4	•—Uncategorized	5	9.80%	2.59%
5	•—Human population	3	5.88%	1.55%
6	•—Geodiversity consciousness	0	0.00%	0.00%
	Totals:	51	100.00%	26.42%

For example, linguistic unit 1:21 p 1 in 01_Geodiversity "It is the variety in various plants and animals used for food" seems to suggest that the student's perception that the concept of geodiversity referred to food biodiversity. Additionally, linguistic unit 1:20 p 1 in 01_Geodiversity "The morphology of the soil" rather refers to the meaning of geomorphology, perhaps because in Greek, the word "soil" is also called "Earth" ("Gi" from the ancient Greek word "Tαία", "Gea").

Some students (5.88%) understood the meaning of geodiversity as *Human population*. For example, in linguistic unit 1:31 p 2 in 01_Geodiversity, the student answers: "Geodiversity refers to the population of the earth", which refers to the concept of human population, where it was codified. This perception seems to derive from the fact that in the Greek language, the word "Earth" is called—this one too—"Tή" ("Gi" from the ancient Greek word "Tαία", "Gea") and associatively, the student understood it as the prefix of the word "(Geo)poikilotita", which is the word (Geo)diversity in Greek.

It appears remarkable that no student (0.00%) understood the concept of geodiversity (code: geodiversity consciousness), as it was found in the international literature, even approximately. It is also noteworthy that several students (9.80%) gave answers which did not even refer to the topic of geoenvironment and which were coded in code uncategorized.

Finally, it seems that the concepts that represented the cognitive structure of the students' knowledge and perceptions about the concept of geodiversity were basically biodiversity, food biodiversity and geomorphology, and secondarily, the concept of human population (Figure 4).

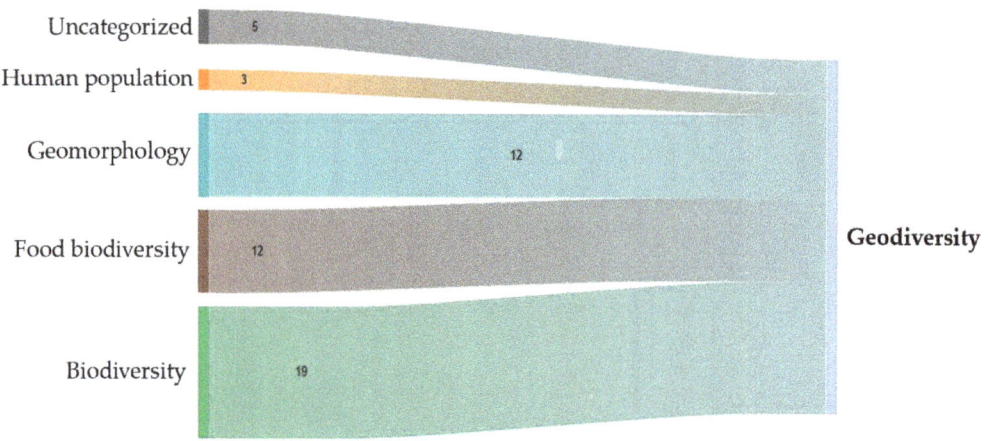

Figure 4. Graph of students' perceptions (codes) on the concept of geodiversity (Sankey diagram).

3.2. Category Geoheritage

Here, the students' answers are examined in relation with the concept of geoheritage. Here, as well, the students appeared to fail to understand the concept of geoheritage, as it is found in the literature, but on the contrary, they confused it with other concepts. Examining the answers' conceptual patterns, as linguistic units, displayed perceptions which were coded by the teachers in five different codes of category geoheritage (Table 3).

Table 3. Frequencies/distribution of codes (students' perceptions) in the category geoheritage.

	02_Geoheritage (*n* = 45)		Frequencies	
	Code	Absolute	Relative (In Category)	Relative (Within All Codings)
1	•—Inheritance succession	21	44.68%	10.88%
2	•—Natural heritage	15	31.92%	7.77%
3	•—Uncategorized	7	14.89%	3.63%
4	•—World heritage	3	6.38%	1.55%
5	•—Geoheritage consciousness	1	2.13%	0.52%
	Totals:	47	100.00%	24.35%

It was found that many students (44.68%) understood the concept of geoheritage as *Inheritance succession*. For example, linguistic unit 2:19 p 1 in 02_Geoheritage was mentioned. Here, the student expressed his perception of geoheritage by answering: "It is the heritage of estates located in towns or villages". In this descriptive text of the answer, it was estimated that there is a latent meaning, which refers to inheritance succession, that is, the transfer of assets, rights and obligations by reason of death [91], as well as in many other answers by his classmates (more than 20). Thus, the conceptual pattern of linguistic unit 2:19 p 1 in 02_Geoheritage led to its codification in the code inheritance succession of the category geoheritage.

Additionally, many students (31.92%) seemed to understand the concept of geoheritage as natural heritage [92]. This was also found in linguistic unit 2:29 p 2 in 02_Geoheritage, as the student replied: "Geoheritage is everything beautiful that exists on earth, rivers, waterfalls, lakes, animals and fishes". Here, it was estimated that the student's perception of geoheritage was more in line with the meaning of the concept natural heritage. Some students (6.38%) perceived the concept of geoheritage as world heritage, i.e., places on Earth that are of outstanding universal value to humanity [93]. What is particularly interesting is that several students expressed completely "irrelevant" perceptions to the meaning of theme and category. For instance, in linguistic *unit 2:10 p 1 in 02_Geoheritage*, the student's answer "I believe that it is the wealth that animals and plants offer us. Some animals offer meat and others milk, while plants offer us beauty and serenity" was considered "irrelevant" to the meaning of the concept geoheritage and was coded as uncategorized.

A characteristic feature of the differentiation of students' perceptions from the accepted bibliographic perception of geoheritage was the assessment that only 1 in 45 students understood—albeit approximately—its meaning. The laconic but comprehensive answer of linguistic unit 2:11 p 1 in 02_Geoheritage was mentioned, for which it was estimated that its latent meaning was in line with the meaning of geoheritage: "It is the geographical heritage".

Hence, it seems that the concepts that can be used for the conceptual change of students regarding the concept of geoheritage are inheritance succession and natural heritage, while, secondarily, world heritage can be used as well, which is visually presented in Figure 5.

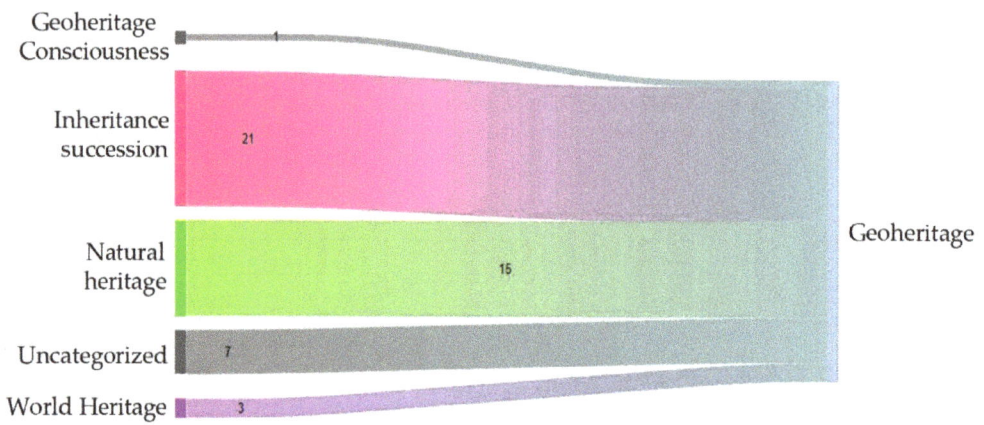

Figure 5. Graph of students' perceptions (codes) on the concept of geoheritage (Sankey diagram).

3.3. Category Geoethics

Unlike the previous two categories, the results in the category geoethics present a different image. Specifically, based on the examination of the respective linguistic units of the students' answers' conceptual patterns, it seems that on the one hand, their perceptions bears greater resemblance—more than the two previous concepts—to the concept of geoethics as it is expressed in the literature (Table 4). On the other hand, there were a significant number of "irrelevant" answers, which seems to indicate two different groups of students in relation to their perception of the subject of geoethics (Table 4).

Table 4. Frequencies/distribution of codes (students' perceptions) in category geoethics.

03_Geoethics ($n = 45$)			Frequencies	
	Code	Absolute	Relative (In Category)	Relative (Within All Codings)
1	•—Ethical behavior	13	26.53%	6.74%
2	•—Protection of nature and land	11	22.45%	5.70%
3	•—Respect for the environment	10	20.41%	5.18%
4	•—Uncategorized	9	18.37%	4.66%
5	•—Geoethics consciousness	4	8.16%	2.07%
6	•—Emotional behavior	2	4.08%	1.04%
	Totals:	49	100.00%	25.39%

A large number of students (26.53%) was identified as perceiving geoethics as ethical behavior, which shows that their perception can be easily formed in order to provoke the planned conceptual change for them with regard to the concept of geoethics. The conceptual patterns of their answers seem to show an increased sense of morality towards the Earth and the environment. For example, in linguistic unit 3:10 p 1 in 03_Geoethics, the student stated that "[Geoethics is the] . . . Ethical behavior for the variety that exists on earth".

In a similar (latent) spirit of ethics thinking and sustainability promotion, in linguistic unit 3:27 p 2 in 03_Geoethics, another student stated that geoethics " . . . means that people protect the earth so that future generations will have what we have today". These linguistic units were included in the code ethical behavior.

Additionally, several students (22.45%) understood the meaning of geoethics as having the meaning of the protection of nature and land. For example, the student's answer to linguistic unit 3:27 p 2 in 03_Geoethics was "It means that people protect the earth so that future generations will have what we have today". Here too, there is the concept of the protection of nature and land and of sustainability promotion. Thus, linguistic unit 3:27 p 2 in 03_Geoethics was included (codified) in the code protection of nature and land of category geoethics.

Students' perceptions of the meaning of the concept of geoethics is consistent with that of the concept of respect for the environment which appears at a similar number (20.41%). To illustrate this, the answer 3: 5 p 1 in 03_Geoethics is noted: "It means that we must have respect for the environment, that is, not to throw away garbage and not to pollute the environment".

Few students (4.08%) responded emotionally (they used the word *love*). It should be noted that "love" constitutes one of the principal emotions [94] (p. 366). These linguistic units were included in the code of emotional behavior.

As mentioned above, a characteristic element of coding in the geoethics category was the increased number of perceptions which are in line with the sensitizing concept to which it corresponds and is examined. It was found, therefore, that four students (8.16%) perceived the concept of geoethics as expressed in the literature. For example, a student (Linguistic *Unit 3:25 p 2 in 03_Geoethics*) stated "It is the morals that people must have towards the earth, towards the environment". This unit is one of 4, which were coded in code geoethics consciousness.

Another characteristic element of the geoethics category is the large number of uncategorized linguistic units (18.37%). Indeed, several linguistic units were assessed whose conceptual patterns are rather "irrelevant" to the spirit of the geoethics concept. For example, linguistic unit 3:22 p 2 in 03_Geoethics "Respect for foreign land ownership" was mentioned. The conceptual pattern of this linguistic unit was deemed unrelated to the spirit of the theme and was coded as uncategorized.

Thus, it can be argued that the concepts that can serve to provoke a cognitive conflict, which will lead to the development of values in the field of geoethics, are essentially ethical behavior, protection of nature and land and respect for the environment and, secondarily, emotional behavior (Figure 6).

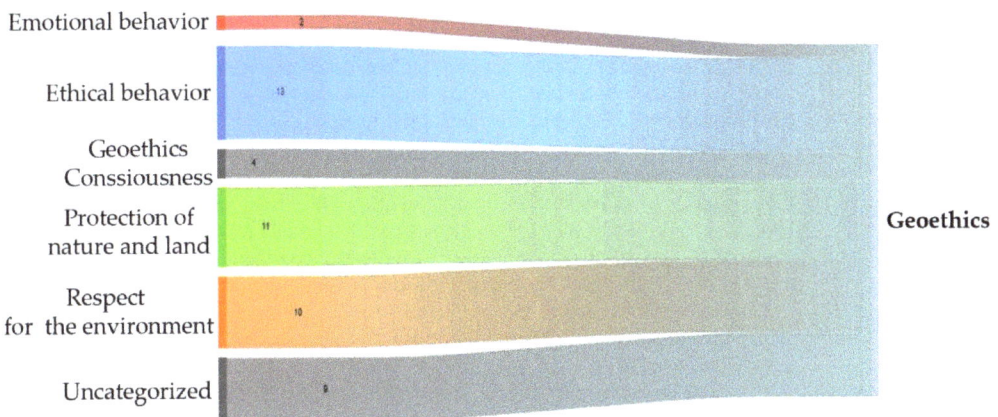

Figure 6. Graph of students' perceptions (codes) on the concept of geoethics (Sankey diagram).

3.4. Category Geotourism

In the category of geotourism, it seems that students' perceptions resemble this concept (of geotourism) in a similar way to that of the concept of geoethics. This means that they refer to concepts related to the word tourism and its various manifestations and sometimes their perceptions typically "approach" the spirit of the concept of geotourism (as it appears in the literature). In general, it should be noted that Greek students are familiar with many concepts of tourism since they live in a country which can be characterized as a tourist destination and in which tourism can be defined as an agent of development [95]. The conceptual patterns' examination of the students' answers led the teachers to the codification of the linguistic units of these conceptual patterns in the category geotourism. The occurrence frequencies of these conceptual patterns per code are shown in Table 5.

Table 5. Frequencies/distribution of codes (students' perceptions) in the category geotourism.

04_ Geotourism (n = 45)		Frequencies		
	Code	Absolute	Relative (In Category)	Relative (Within All Codings)
1	•—Agrotourism	16	34.78%	8.29%
2	•—Tourism in general	11	23.91%	5.70%
3	•—Ecotourism	7	15.22%	3.63%
4	•—Adventure tourism	6	13.04%	3.11%
5	•—Geotourism consciousness	4	8.70%	2.07%
6	•—Uncategorized	2	4.35%	1.04%
	Totals:	46	100.00%	23.83%

Hence, it seems that many students (34.78%) perceived the meaning of geotourism as agrotourism, meaning tourism wherein the main motivation is recreation in rural areas [96]. This is probably due to them hearing about this type of tourism which seems to be developing in their area [97] (p. 1472), but it may also be due to the fact that the first part of the compound word "agrotourism" ("Agro-") derives from the Greek word "αγρός" ("Agros"), which means "field" and "field" in Greece, and is often called "Γή" (Earth) ("Gi" from the ancient Greek word "Γαία", "Gea"), whence the international prefix "Geo-". Thus, this may mean that they confused the prefix "Agro-" with the prefix "Geo-" in the words agrotourism and geotourism. An example of categorization of such a conceptual pattern is the linguistic unit (answer) 4: 8 p 1 in 04_Geoetourism "When a tourist rents a small field for some short time".

Several students (23.91%) perceived geotourism as a general form of tourism. For example, in linguistic unit 4: 5 p 1 in 04_Geotourism, the student expressed his perception by answering: "It is tourism in a beautiful landscape visited by many tourists". Based on the assessment of the latent meaning of this linguistic unit, it was codified in code Tourism in general.

Lower percentages of students seem to perceive the concept as Ecotourism (15.22%), that is, as a form of "tourism that attempts to minimize its impact upon the environment" [98] or Adventure tourism (13.04%), which entails travelling to an unusual, exotic, remote or wild destination [99] (p. 28). An illustration of linguistic unit coding in code ecotourism is 4:27 p 1 in 04_Geotourism, "It is tourism in nature that requires proper behavior towards the environment" and an illustration of codification in the code adventure tourism is 4:34 p 2 in 04_Geoetourism, "It refers to the tourism of an unusual place" (as assessed by its latent meaning).

What appears to be characteristic of the examination of the Geotourism concept is that four (4) students (8.70%) perceived it even approximately (according to the teachers). An example of such an answer is the answer 4:26 p 1 in 04_Geoetourism, "Tourism related to visits to wonderful landscapes". The concept pattern of this linguistic unit based on its latent meaning was codified by the teachers as geotourism consciousness.

In conclusion, it seems that the concepts that can be used in the construction of "bridging analogies" [61] (p. 485) of a cognitive conflict [62] (p. 374) for the development of perceptions on the subject of geotourism are agrotourism, tourism in general, and secondarily, those of ecotourism and adventure tourism (Figure 7).

Figure 7. Graph of students' perceptions (codes) on the concept of geotourism (Sankey diagram).

3.5. Answers to the Research Questions

Regarding the first Research Question (Q1), summarizing the above results, we found that the current latent state of students' perceptions in the thematic fields of geoenvironment—namely the concepts of geodiversity, geoheritage, geoethics and geotourism—can be considered as very confused. This is reminiscent of the research results of Vasconcelos, Torres, Vasconcelos and Moutinho (2016), in a related subject, which found a "low level of knowledge on the three pillars of Sustainable Development" in 187 Portuguese citizens which was attributed to the lack of relevant education of these citizens [100] (p. 514). Similar conclusions were reached by Almeida and Vasconcelos (2015) in a study of 36 higher education students pursuing a Master's of Geology [101] (p. 904). Additionally, Georgousis, Savelidi, Savelides, Holokolos and Drinia (2021) came to similar conclusions about the lack of education in related subjects in approximately 600 Greek pupils and students [47]. We also noted the conclusion of Comănescu and Nedelea (2020) that there is a need to intensify public awareness and education at all levels regarding geoenvironmental factors [102].

More specifically, the vast majority of students did not seem to have a clear understanding—especially regarding a perception that is consistent with the literature—of any of the concepts of the theme "geoenvironmental concepts and perceptions". They attribute meaning to the concept of geodiversity which was considered to be in line with the concepts of biodiversity, food biodiversity, geomorphology and human population. Similarly, concerning the concept of geoheritage, students attribute meaning which can be considered to be in line with the concepts of inheritance succession, natural heritage and world heritage. In essence, the same is true with the concept of geoethics to which they attribute meaning that seems to correspond to the concepts of ethical behavior, protection of nature and land, respect for the environment and emotional behavior. However, it should be noted, especially with regard to the students' perceptions of this concept, that perceptions were observed which either very closely resembled the bibliographic perceptions of geoethics or only "approached" them.

The same was observed with the concept of geotourism. Here too, there are students' perceptions which for the most part "revolve" around the international bibliographic perceptions of geotourism or closely resemble them. However, it seems that the majority of students perceived the concept of geotourism as agrotourism, tourism in general, and secondarily, as ecotourism and adventure tourism.

Regarding the second Research Question (Q2) about which concepts' perceptions can be used in the educational process, in order to achieve cognitive conflicts aiming to promote scientific perceptions about the concepts of geodiversity, geoheritage, geoethics and geotourism among students, it seems that teachers, in order to create cognitive conflicts in the process of their educational program on aspects of geoenvironment, have the opportunity to utilize concepts which students perceive more or less as the concepts of geodiversity, geoheritage, geoethics and geotourism. The present research has shown that some concepts can be utilized, and by implementing the appropriate educational strategy, these can provide students with opportunities to clarify their ideas and then to challenge, develop or replace them by approaching acceptable scientific perceptions of these concepts in the field of geoenvironment [61] (p. 485). These concepts can be:

Biodiversity, food biodiversity and geomorphology, and secondarily, the concept of the human population, in order to develop perceptions of the concept of geodiversity.

Inheritance succession and natural heritage, and secondarily, world heritage, in order to develop scientific insights into the concept of geoheritage.

Ethical behavior, protection of nature and land, respect for the environment, and secondarily, emotional behavior, in order to promote scientifically substantiated perceptions of the concept of geoethics.

Agrotourism and tourism in general, and secondarily, those of ecotourism and adventure tourism, in order to promote perceptions of the concept of geotourism.

4. Conclusions

In order to design a geoeducation educational program in the context of the creation and operation of an experimental junior high school club (Creativity and Innovation Group), the need to gather detailed information on the current latent state of knowledge and perceptions of students was recognized on basic geoenvironment concepts. Essentially, the need for diagnostic assessments of the 45 students of the program was identified regarding their perception of the basic concepts of the geoenvironment and specifically the concepts of geodiversity, geoheritage, geoethics and geotourism. The main goal of the educational process was the replacement of students' latent perceptions of these concepts with modern scientific perceptions of these concepts. To achieve this goal, we chose to apply the educational technique of creating cognitive conflicts (based on this technique, the teacher seeks the conflict of students' latent—and usually incorrect—knowledge and perceptions of some concepts with modern scientific knowledge and perceptions of the same concepts). Thus, research questions were identified which led the research to assess the current latent state of students' perceptions regarding the thematic areas of the concepts and to identify concepts whose perceptions can be used in the educational process with a view to achieve effective cognitive conflicts aiming to promote scientific insights into them.

The results of the research showed that the majority of the participating students confused perceptions about the concepts of geodiversity, geoheritage, geoethics and geotourism in the thematic field of the geoenvironment. This confusion was attributed to the lack of knowledge and therefore to the lack of students' education on the subject which has been found in previous research as well and not only in Greek students.

Concepts have also emerged which can be used by teachers in their efforts to develop students' clear or even scientifically acceptable perceptions of the concepts of geodiversity, geoheritage, geoethics and geotourism in the field of the geoenvironment. Students' current latent perceptions, as expressed by the concepts that emerged, can be put to good use by teachers to develop "targeted" educational techniques. These can be based on the creation of cognitive conflicts in students, resulting in the reconstruction of perceptions and the development of scientific concepts in them. Thus, a problem which is attributed to educational deficiencies can be transformed into an educational opportunity.

The authors of the article consider that the research method followed herein yielded useful and reliable results for the design and implementation of the educational process in this geoeducation program. Combined with its simplicity and ease of implementation, this leads the authors to recommend method and techniques, as used in this research, as a process for educators and designers of environmental education syllabi and curricula, and especially of geoenvironmental education.

Finally, the authors support the opinion that if any educational policy aims to promote education for sustainability and especially holistic environmental education, it should also promote geoenvironmental education since students do not understand the aspects of the geoenvironment and this is a rather major shortcoming of the desired modern holistic environmental education. Our proposal is therefore the creation and operation of educational programs or entire curricula which, through holistic approaches and interdisciplinary connections, can play an important role in the enhancement of environmental education with geoenvironmental education, with the objective of enhancing understandings of the geoenvironment as a system of multiple components that directly affect the existence and development of human societies.

Author Contributions: Conceptualization, E.G., S.S. and H.D.; methodology, E.G., M.S. and S.S.; software, M.S.; validation, M.S.; formal analysis, M.S.; investigation, E.G. and S.S.; resources, E.G., M.S., S.S., S.M. and M.-V.H.; data curation, E.G. and S.S.; writing—original draft preparation, E.G., S.S., S.M. and M.-V.H.; writing—review and editing, E.G., S.S., S.M. and M.-V.H.; project administration, H.D. and S.S.; supervision, H.D. All authors have read and agreed to the published version of the manuscript.

Funding: This research received no external funding.

Institutional Review Board Statement: Not applicable.

Informed Consent Statement: Not applicable.

Data Availability Statement: The data presented in this study are available on request from the corresponding author.

Acknowledgments: The authors gratefully thank the journal academic editor and the three reviewers for their thorough consideration of this paper.

Conflicts of Interest: The authors declare no conflict of interest.

References

1. Kaláb, Z. The contribution of geophysics to geoenvironmental studies. In Proceedings of the Geoinformatics 2021, Kyiv, Ukraine, 11–14 May 2021; European Association of Geoscientists & Engineers: Kyiv, Ukraine, 2021; pp. 1–6. [CrossRef]
2. Kubalíková, L. Geomorphosite assessment for geotourism purposes. *Czech J. Tour.* **2013**, *2*, 80–104. [CrossRef]
3. Serrano, E.; Ruiz-Flaño, P. Geodiversity: A theoretical and applied concept. *Geogr. Helv.* **2007**, *62*, 140–147. [CrossRef]
4. Vegas, J.; Díez-Herrero, A. *Best Practice Guidelines for the Use of the Geoheritage in the City of Segovia: A Sustainable Model for Environmental Awareness and Urban Geotourism*; Ayuntamiento de Segovia: Segovia, Spain, 2018.
5. Santangelo, N.; Valente, E. Geoheritage and Geotourism. *Resources* **2020**, *9*, 80. [CrossRef]
6. Zafeiropoulos, G.; Drinia, H.; Antonarakou, A.; Zouros, N. From Geoheritage to Geoeducation, Geoethics and Geotourism: A Critical Evaluation of the Greek Region. *Geosciences* **2021**, *11*, 381. [CrossRef]
7. Carcavilla, L.; Díaz-Martínez, E.; García-Cortés, Á.; Vegas, J. *Geoheritage and Geodiversity*; Instituto Geológico y Minero de España (IGME): Madrid, Spain, 2019.
8. Faccini, F.; Gabellieri, N.; Paliaga, G.; Piana, P.; Angelini, S.; Coratza, P. Geoheritage map of the Portofino Natural Park (Italy). *J. Maps* **2018**, *14*, 87–96. [CrossRef]
9. Sharples, C. *Concepts and Principles of Geoconservation*; Tasmanian Parks and Wildlife Service: Hobart, Australia, 2002. Available online: http://www.dpipwe.tas.gov.au/Documents/geoconservation.pdf (accessed on 6 October 2021).
10. Dixon, G. *A Reconnaissance Inventory of Sites of Geoconservation Significance on Tasmanian Islands. A Report to the Parks & Wildlife Service, Tasmania and Australian Heritage Commission*; Parks and Wildlife Service: Hobart, Australia, 1996.
11. Geological Society of America. Geoheritage. The Geological Society of America, Inc. Available online: https://www.geosociety.org/gsa/positions/position20.aspx (accessed on 8 November 2021).
12. DeMiguel, D.; Brilha, J.; Meléndez, G.; Azanza, B. Geoethics and geoheritage. In *Teaching Geoethics: Resources for Higher Education*; Vasconcelos, C., Schneider, S., Peppoloni, S., Eds.; U. Porto Edições: Porto, Portugal, 2020; pp. 57–72. ISBN 978-989-746-254-2. [CrossRef]
13. Peppoloni, S.; Di Capua, G. Geoethics as global ethics to face grand challenges for humanity. In *Geoethics: Status and Future Perspectives*; Di Capua, G., Bobrowsky, P.T., Kieffer, S.W., Palinkas, C., Eds.; Special Publications; Geological Society: London, UK, 2020; Volume 508, pp. 13–29. [CrossRef]
14. Potthast, T. Toward an Inclusive Geoethics-Commonalities of Ethics in Technology, Science, Business, and Environment. In *Geoethics-Ethical Challenges and Case Studies in Earth Sciences*; Elsevier: Amsterdam, The Netherlands, 2015; pp. 49–56.
15. Antić, A.; Peppoloni, S.; Di Capua, G. Applying the Values of Geoethics for Sustainable Speleotourism Development. *Geoheritage* **2020**, *12*, 73. [CrossRef]

16. IAPG (International Association for Promoting Geoethics). Geoethics Themes. Available online: https://www.geoethics.org/themes (accessed on 9 November 2021).
17. Martini, G. Gological Heritage and Geo-tourism. In *Geological Heritage: Its Conservation and Management*; Barettino, D., Wimbledon, W.A.P., Gallego, E., Eds.; Instituto Technológico Geominero de España: Madrid, Spain, 2000; pp. 147–156. ISBN 84-7840-417-1.
18. Kubalíková, L. Assessing Geotourism Resources on a Local Level: A Case Study from Southern Moravia (Czech Republic). *Resources* 2019, *8*, 150. [CrossRef]
19. Dowling, R.K. Global Geotourism—An Emerging Form of Sustainable Tourism. *Czech J. Tour.* 2013, *2*, 59–79. [CrossRef]
20. Dowling, R.K.; Newsome, D. Geotourism: Definition, characteristics and international perspectives. In *Handbook of Geotourism*; Dowling, R., Newsome, D., Eds.; Edward Elgar Publishing: Cheltenham, UK, 2018. [CrossRef]
21. Olson, K.; Dowling, R. Geotourism and Cultural Heritage. *Geoconserv. Res.* 2018, *1*, 37–41. Available online: https://ro.ecu.edu.au/cgi/viewcontent.cgi?article=6297&context=ecuworkspost2013 (accessed on 4 March 2022).
22. Dowling, R. Geotourism. In *Encyclopedia of Tourism*; Jafari, J., Xiao, H., Eds.; Springer: Cham, Switzerland, 2014. [CrossRef]
23. Stokes, A.M.; Cook, S.D.; Drew, D. *Geotourism: The New Trend in Travel*; Travel Industry America and National Geographic Traveler: Washington, DC, USA, 2003.
24. National Geographic Society. Geotourism. National Geographic Partners, LLC. Available online: https://www.nationalgeographic.com/maps/topic/geotourism (accessed on 10 November 2021).
25. Pralong, J.P. Geotourism: A new Form of Tourism utilising natural Landscapes and based on Imagination and Emotion. *Tour. Rev.* 2006, *61*, 20–25. [CrossRef]
26. Trikolas, K.; Ladas, I. The necessity of teaching earth sciences in secondary education. In Proceedings of the 3rd International GEOschools Conference, Teaching Geosciences in Europe from Primary to Secondary School, Athens, Greece, 28–29 September 2013; pp. 73–76. (In Greek).
27. Fermeli, G.; Meléndez, G.; Calonge, A.; Dermitzakis, M.; Steininger, F.; Koutsouveli, A.; Neto de Carvalho, C.; Rodrigues, J.; D'Arpa, C.; Di Patti, C. GEOschools: Innovative Teaching of Geosciences in Secondary Schools and Raising Awareness on Geoheritage in the Society. In *Avances y Retos en la Conservación del Patrimonio Geológico en España. Actas de la IX Reunión Nacional de la Comisión de Patrimonio Geológico (Sociedad Geológica de España)*; Fernández-Martínez, E., Castaño de Luis, R., Eds.; Universidad de León: León, Spain, 2011; pp. 120–124. ISBN 978-84-9773-578-0. Available online: http://naturtejo.com/ficheiros/conteudos/files/fic2.pdf (accessed on 8 November 2021).
28. Fermeli, G.; Markopoulou-Diakantoni, A. Selecting Pedagogical Geotopes in Urban Environment. *Bull. Geol. Soc. Greece* 2004, *36*, 649–658. Available online: https://ejournals.epublishing.ekt.gr/index.php/geosociety/article/view/16770 (accessed on 8 November 2021).
29. Huggett, R.J. *Fundamentals of Geomorphology*, 4th ed.; Routledge: New York, NY, USA, 2017; ISBN 9781138940659.
30. Zorina, S.O.; Silantiev, V. Geosites, Classification of. In *Encyclopedia of Mineral and Energy Policy*; Tiess, G., Majumder, T., Cameron, P., Eds.; Springer: Berlin/Heidelberg, Germany, 2015. [CrossRef]
31. Bruno, D.E. Geosite, Concept of. In *Encyclopedia of Mineral and Energy Policy*; Tiess, G., Majumder, T., Cameron, P., Eds.; Springer: Berlin/Heidelberg, Germany, 2015. [CrossRef]
32. Gray, M. Geodiversity and geoconservation: What, why and how? In *Geodiversity & Geoconservation*; Santucci, V.L., Ed.; George Wright Forum: Hancock, MI, USA, 2005; Volume 22.
33. Gray, M. *Geodiversity: Valuing and Conserving Abiotic Nature*, 2nd ed.; Willey Blackwell: Chichester, UK, 2013.
34. Gray, M. Geodiversity: The Backbone of Geoheritage and Geoconservation. In *Geoheritage: Assessment, Protection, and Management*; Reynard, E., Brilha, J., Eds.; Elsevier: Amsterdam, The Netherlands, 2018; pp. 13–25. [CrossRef]
35. Panizza, M. The geomorphodiversity of the Dolomites (Italy): A key of geoheritage assessment. *Geoheritage* 2009, *1*, 33–42. [CrossRef]
36. Coratza, P.; Reynard, E.; Zwoliński, Z. Geodiversity and Geoheritage: Crossing Disciplines and Approaches. *Geoheritage* 2018, *10*, 525–526. [CrossRef]
37. Zouros, N.; Martini, G. Introduction to the European Geoparks Network, Proceedings of the 2nd European Geoparks Network Meeting, Lesvos, Greece, 3–7 October 2003; Zouros, N., Martini, G., Frey, M.-L., Eds.; Natural History Museum of the Lesvos Petrified Forest: Lesvos, Greece, 2003; pp. 17–21.
38. Zouros, N.; Valiakos, I. Geoparks management and assessment. *Bull. Geol. Soc. Greece* 2010, *43*, 965–977. [CrossRef]
39. Fassoulas, C.; Zouros, N. Evaluating the influence of Greek Geoparks to the local communities. *Bull. Geol. Soc. Greece* 2010, *43*, 896–906. [CrossRef]
40. Andrășanu, A. Basic Concepts in Geoconservation. In *Mesozoic and Cenozoic Vertebrates and Paleoenvironments–Tributes to the Career of Dan Grigorescu*; Csiki, Z., Ed.; Ars Docendi: Bucharest, Romania, 2006; pp. 37–41. ISBN (10) 973-558-275-9. Available online: https://www.academia.edu/10715520/Basic_concepts_in_Geoconservation (accessed on 10 November 2021).
41. Bobrwsky, P.; Cronin, V.S.; Di Capua, G.; Kieffer, S.W.; Peppoloni, S. The emerging field of geoethics. In *Scientific Integrity and Ethics in the Geosciences*; Wiley: Hoboken, NJ, USA, 2017; pp. 175–212. [CrossRef]
42. Theodossiou-Drandaki, I. No Conservation without Education. In *Geological Heritage: Its Conservation and Management*; Barettino, D., Wimbledon, W.A.P., Gallego, E., Eds.; Instituto Tecnológico Geominero de España: Madrid, Spain, 2000; pp. 111–125. ISBN 84-7840-417-1.

43. Rokka, A.C. Geology in Primary Education: Potential and Perspectives. *Bull. Geol. Soc. Greece* **2018**, *34*, 819–823. (In Greek) [CrossRef]
44. Meléndez, G.; Fermeli, G.; Koutsouveli, A. Analyzing Geology textbooks for secondary school curricula in Greece and Spain: Educational use of geological heritage. *Bull. Geol. Soc. Greece* **2007**, *40*, 1819–1832. [CrossRef]
45. Fermeli, G.; Markopoulou-Diakantoni, A. Geosciences in the Curricula and Students Books in Secondary Education. *Bull. Geol. Soc. Greece* **2004**, *36*, 639–648. [CrossRef]
46. Spartinou, M.; Zerlentis, I. The geological heritage of Cyclades and the Environmental Education. In Proceedings of the 6th Pan-Hellenic Geographical Conference of the Hellenic Geographical Society, Thessaloniki, Greece, 3–6 October 2002; Volume III. (In Greek). Available online: http://geolib.geo.auth.gr/digeo/index.php/pgc/article/view/9413/9164 (accessed on 11 November 2021).
47. Georgousis, E.; Savelides, S.; Mosios, S.; Holokolos, M.-V.; Drinia, H. The Need for Geoethical Awareness: The Importance of Geoenvironmental Education in Geoheritage Understanding in the Case of Meteora Geomorphes, Greece. *Sustainability* **2021**, *13*, 6626. [CrossRef]
48. Drinia, H.; Tsipra, T.; Panagiaris, G.; Patsoules, M.; Papantoniou, C.; Magganas, A. Geological Heritage of Syros Island, Cyclades Complex, Greece: An Assessment and Geotourism Perspectives. *Geosciences* **2021**, *11*, 138. [CrossRef]
49. Savelides, S.; Georgousis, E.; Fasouraki, R.; Papadopoulou, G.; Drinia, H. "Storm Tossed Sea Rocks in Pelion" an environmental synchronous online education program. In Proceedings of the 13th Conference on Informatics in Education (13th CIE2021), Athens, Greece, 9–10 October 2021; Greek Computer Society: Athens, Greece, 2021; pp. 577–593, ISBN 978-960-578-084-5. Available online: http://events.di.ionio.gr/cie/images/documents21/CIE2021_OnLineProceedings/CIE2021_Binder1.pdf (accessed on 12 November 2021).
50. Georgousis, E.; Savelidi, M.; Savelides, S.; Holokolos, M.-V.; Drinia, H. Teaching Geoheritage Values: Implementation and Thematic Analysis Evaluation of a Synchronous Online Educational Approach. *Heritage* **2021**, *4*, 3523–3542. [CrossRef]
51. Eurydice. Organisational Variations and Alternative Structures in Primary Education. 2022. Available online: https://eacea.ec.europa.eu/national-policies/eurydice/content/organisational-variations-and-alternative-structures-primary-education-20_en (accessed on 27 January 2022).
52. Nikitina, N. *Geoethics: Theory, Principles, Problems*, 2nd ed.; Geoinformmark Ltd.: Moscow, Russia, 2016; ISBN 978-5-98877-061-9.
53. Oser, F.K. Moral Perspectives on Teaching. *Rev. Res. Educ.* **1994**, *20*, 57–127. [CrossRef]
54. Dowling, R.K.; Newsome, D. Geotourism's Issues and Challenges. In *Geotourism*; Dowling, R., Newsome, D., Eds.; Elsevier Butterworth-Heinemann: Burlington, MA, USA, 2006; pp. 242–254. ISBN 0750662158.
55. Treagust, D.F. Diagnostic assessment in science as a means to improving teaching, learning and retention. In Proceedings of the Assessment in Science Teaching and Learning Symposium, Sydney, Australia, 28 September 2006; The University of Sydney: Sydney, Australia, 2006.
56. Wu, X.; Zhang, Y.; Wu, R.; Chang, H.H. A comparative study on cognitive diagnostic assessment of mathematical key competencies and learning trajectories. *Curr. Psychol.* **2021**, 1–13. [CrossRef]
57. Aikenhead, G.S.; Jegede, O.J. Cross-cultural science education: A cognitive explanation of a cultural phenomenon. *J. Res. Sci. Teach.* **1999**, *36*, 269–287. [CrossRef]
58. Osborne, J.; Erduran, S.; Simon, S. Enhancing the quality of argumentation in school science. *J. Res. Sci. Teach.* **2004**, *41*, 994–1020. [CrossRef]
59. Gabora, L.; Steel, M. Autocatalytic networks in cognition and the origin of culture. *J. Theor. Biol.* **2017**, *431*, 87–95. [CrossRef] [PubMed]
60. Gabora, L.; Beckage, N.M.; Steel, M. An Autocatalytic Network Model of Conceptual Change. *Top. Cogn. Sci.* **2022**, *14*, 163–188. [CrossRef]
61. Driver, R. Students' conceptions and the learning of science. *Int. J. Sci. Educ.* **1989**, *11*, 481–490. [CrossRef]
62. Limón, M. On the cognitive conflict as an instructional strategy for conceptual change: A critical appraisal. *Learn. Instr.* **2001**, *11*, 357–380. [CrossRef]
63. Zhan, P. Longitudinal Learning Diagnosis: Minireview and Future Research Directions. *Front. Psychol.* **2020**, *11*, 1185. [CrossRef]
64. Tang, F.; Zhan, P. Does Diagnostic Feedback Promote Learning? Evidence from a Longitudinal Cognitive Diagnostic Assessment. *AERA Open* **2021**, *7*, 1–15. [CrossRef]
65. Peppoloni, S.; Di Capua, G. The Meaning of Geoethics. In *Geoethics*; Wyss, M., Peppoloni, S., Eds.; Elsevier: Amsterdam, The Netherlands, 2015; Volume 419, pp. 3–14. ISBN 9780127999357.
66. Bohle, M.; Marone, E. Geoethics, a Branding for Sustainable Practices. *Sustainability* **2021**, *13*, 895. [CrossRef]
67. Newsome, D.; Dowling, R. *Geoheritage and Geotourism*; Elsevier: Amsterdam, The Netherlands, 2018; ISBN 9780128095423.
68. Roberts, M.R.; Gierl, M.J. Developing score reports for cognitive diagnostic assessments. *Educ. Meas. Issues Pract.* **2010**, *29*, 25–38. [CrossRef]
69. Maran, A. Geoconservation in Serbia—State of Play and Future Perspectives. *Eur. Geol.* **2012**, *34*, 1–72. Available online: http://eurogeologists.eu/wp-content/uploads/2015/09/a_Magazine-Dec2012.pdf#page=29 (accessed on 15 October 2021).
70. Yong, R.N.; Mulligan, C.N.; Fukue, M. *Geoenvironmental Sustainability*; CRC Press: Boca Raton, FL, USA, 2007; ISBN 0-8493-2841-1. [CrossRef]

71. Savelidi, M.; Savelides, S.; Georgousis, E.; Papadopoulou, G.; Fasouraki, R.; Drinia, H. Microcontroller Systems in Education for Sustainable Development Service. A Qualitative Thematic Meta-Analysis. *Eur. J. Eng. Technol. Res.* **2022**, *CIE 2758*, 53–60. [CrossRef]
72. Bryman, A. *Social Research Methods*; Oxford University Press: Oxford, UK, 2012.
73. Thyme, K.E.; Wiberg, B.; Lundman, B.; Graneheim, U.H. Qualitative content analysis in art psychotherapy research: Concepts, procedures, and measures to reveal the latent meaning in pictures and the words attached to the pictures. *Arts Psychother.* **2013**, *40*, 101–107. [CrossRef]
74. Issari, P.; Pourkos, M. *Qualitative Research Methods in Psychology and Education*; Hellenic Academic Libraries Link: Athens, Greece, 2015; (In Greek). Available online: https://repository.kallipos.gr/handle/11419/5826 (accessed on 28 November 2021).
75. Braun, V.; Clarke, V. Using thematic analysis in psychology. *Qual. Res. Psychol.* **2006**, *3*, 77–101. [CrossRef]
76. Winter, G. A Comparative Discussion of the Notion of 'Validity' in Qualitative and Quantitative research. *Qual. Rep.* **2000**, *4*, 1–14. [CrossRef]
77. Mishra, R.K.; Mishra, V.K. *Modelling and Analysis of Inventory Model for Items under Asymmetrical Substitutability and Complementarity, Proceedings of the AIP Conference, Jamshedpur, India, 21–22 December 2020*; Sharma, R., Nandkeolyar, R., Eds.; AIP Publishing LLC.: Melville, NY, USA, 2022; Volume 2435, pp. 020023-1–020023-10. [CrossRef]
78. Osgood, C.E.; Sebeok, T.A.; Gardner, J.W.; Carroll, J.B.; Newmark, L.D.; Ervin, S.M.; Saporta, S.; Greenberg, J.; Walker, D.; Jenkins, J.; et al. Psycholinguistics: A survey of theory and research problems. *J. Abnorm. Soc. Psychol.* **1954**, *49*, i-203. [CrossRef]
79. Hegland, M. The apriori algorithm–A tutorial. *Math. Comput. Imaging Sci. Inf. Process.* **2007**, *11*, 209–262. [CrossRef]
80. Skalski, P.D.; Neuendorf, K.A.; Cajigas, J.A. Content Analysis in the Interactive Media Age. In *The Content Analysis Guidebook*, 2nd ed.; Neundorf, K.A., Ed.; Sage: Thousand Oaks, CA, USA, 2017; pp. 201–242.
81. Kekia, A.M. School Written Genres as Social Processes: Theoretical Analysis and Teaching Practices. *Hell. Educ. Soc.* **2017**, *58*. (In Greek)
82. Kim, J.K.; de Marneffe, M.C.; Fosler-Lussier, E. *Adjusting Word Embeddings with Semantic Intensity Orders, Proceedings of the 1st Workshop on Representation Learning for NLP, Berlin, Germany, 11 August 2016*; Blunsom, P., Cho, K., Cohen, S.B., Grefenstette, E., Hermann, K.M., Rimell, L., Weston, J., Yih, S.W.T., Eds.; Association for Computational Linguistics: Berlin, Germany, 2016; pp. 62–69.
83. Poleshchuk, O.M. Creation of linguistic scales for expert evaluation of parameters of complex objects based on semantic scopes. In Proceedings of the 2018 International Russian Automation Conference (RusAutoCon–2018), Sochi, Russia, 9–16 September 2018; Institute of Electrical and Electronics Engineers: Piscataway, NJ, USA, 2018; pp. 1–6. [CrossRef]
84. Maxwell, J. Understanding and validity in qualitative research. *Harv. Educ. Rev.* **1992**, *62*, 279–301. [CrossRef]
85. Symeou, L. Validity and credibility in qualitative research: The example of a research on school-family collaboration. In Proceedings of the 9th Conference of the Cyprus Pedagogical Association, Nicosia, Cyprus, 2–3 June 2006; Phtiaka, H., Gagatsis, A., Elia, I., Modestou, M., Eds.; Cyprus Pedagogical Association: Nicosia, Cyprus, 2006; pp. 1055–1064. (In Greek).
86. Pietilä, A.M.; Nurmi, S.M.; Halkoaho, A.; Kyngäs, H. Qualitative Research: Ethical Considerations. In *The Application of Content Analysis in Nursing Science Research*; Kyngäs, H., Mikkonen, K., Kääriäinen, M., Eds.; Springer: Cham, Switzerland, 2020; pp. 49–69. ISBN 978-3-030-30198-9. [CrossRef]
87. Friese, S. *ATLAS.ti 9 User Manual*; ATLAS.ti Scientific Software Development GmbH: Berlin, Germany, 2021.
88. Bharucha, E. *Textbook of Environmental Studies for Undergraduate Courses*; University Grants Commission, New Delhi and Bharati Vidyapeeth Institute of Environmental Education and Research: Pune, India, 2004.
89. Kennedy, G.; Lee, W.T.K.; Termote, C.; Charrondière, R.; Yen, J.; Tung, A. *Guidelines on Assessing Biodiverse Foods in Dietary Intake Surveys*; Food and Agriculture Organization of the United Nations (FAO); Bioversity International: Rome, Italy, 2017.
90. Shukla, D.P. Geomorphology. In *Hydro-Geomorphology-Models and Trends*; Shukla, D.P., Ed.; IntechOpen: London, UK, 2017; pp. 3–8. [CrossRef]
91. EUR-Lex. Regulation EU No 650/2012 on Jurisdiction, Applicable Law, the Recognition and Enforcement of Decisions and the Acceptance and Enforcement of Authentic Instruments in Matters of Succession, and on the Creation of a European Certificate of Succession. European Parliament and of the Council of 4 July 2012. Available online: https://eur-lex.europa.eu/legal-content/EN/TXT/?uri=celex%3A32012R0650 (accessed on 7 December 2021).
92. UNESCO. Natural Heritage. Institute for Statistics, 2009 UNESCO Framework for Cultural Statistics and UNESCO, Convention Concerning the Protection of the World Cultural and Natural Heritage, 1972. Available online: http://uis.unesco.org/en/glossary-term/natural-heritage (accessed on 9 December 2021).
93. UNESCO. What is World Heritage? Available online: https://whc.unesco.org/en/faq/19 (accessed on 9 December 2021).
94. Ekman, P.; Cordaro, D. What is Meant by Calling Emotions Basic. *Emot. Rev.* **2011**, *3*, 364–370. [CrossRef]
95. Balomenou, C.; Lagos, D.; Maliari, M.; Semasis, S.; Mamalis, S. Tourism Development in North Greece. In *Tourism Management and Sustainable Development*; Springer: Cham, Switzerland, 2021; pp. 5–26. [CrossRef]
96. Baranova, A.; Kegeyan, S. Agrotourism as an element of the development of a green economy in a resort area. *E3S Web Conf.* **2019**, *91*, 08006. [CrossRef]
97. Karampela, S.; Kavroudakis, D.; Kizos, T. Agrotourism networks: Empirical evidence from two case studies in Greece. *Curr. Issues Tour.* **2019**, *22*, 1460–1479. [CrossRef]

98. Wearing, S.; Neil, J. *Ecotourism: Impacts, Potentials and Possibilities*; Butterworth-Heinemann: Woburn, MA, USA, 1999; ISBN 0750641371.
99. Swarbrooke, J.; Beard, C.; Leckie, S.; Pomfret, G. *Adventure Tourism: The New Frontier*; Butterworth-Heinemann: Burlington, MA, USA, 2003; ISBN 0 7506 5186 5.
100. Vasconcelos, C.; Torres, J.; Vasconcelos, L.; Moutinho, S. Sustainable development and its connection to teaching geoethics. *Episodes* **2016**, *39*, 509–517. [CrossRef]
101. Almeida, A.; Vasconcelos, C. Geoethics: Master's Students Knowledge and Perception of Its Importance. *Res. Sci. Educ.* **2014**, *45*, 889–906. [CrossRef]
102. Comănescu, L.; Nedelea, A. Geoheritage and Geodiversity Education in Romania: Formal and Non-Formal Analysis Based on Questionnaires. *Sustainability* **2020**, *12*, 9180. [CrossRef]

MDPI
St. Alban-Anlage 66
4052 Basel
Switzerland
Tel. +41 61 683 77 34
Fax +41 61 302 89 18
www.mdpi.com

Geosciences Editorial Office
E-mail: geosciences@mdpi.com
www.mdpi.com/journal/geosciences

www.ingramcontent.com/pod-product-compliance
Lightning Source LLC
LaVergne TN
LVHW070223100526
838202LV00015B/2082